TRANSIENT ANALYSIS OF ELECTRIC POWER CIRCUITS HANDBOOK

Transient Analysis of Electric Power Circuits Handbook

by

ARIEH L. SHENKMAN

Holon Academic Institute of Technology,
Holon, Israel

A C.I.P. Catalogue record for this book is available from the Library of Congress.

ISBN-13 978-1-4419-3955-5 (PB)
ISBN-13 978-0-387-28799-7 (e-book)

Published by Springer,
P.O. Box 17, 3300 AA Dordrecht, The Netherlands.

www.springeronline.com

Printed on acid-free paper

Printed in the Netherlands.

This book is dedicated to my dearest wife Iris
and wonderful children
Daniel, Elana and Joella

CONTENTS

INDEX 559

FOREWORD

Every now and then, a good book comes along and quite rightfully makes itself a distinguished place among the existing books of the electric power engineering literature. This book by Professor Arieh Shenkman is one of them.

Today, there are many excellent textbooks dealing with topics in power systems. Some of them are considered to be classics. However, many of them do not particularly address, nor concentrate on, topics dealing with transient analysis of electrical power systems.

Many of the fundamental facts concerning the transient behavior of electric circuits were well explored by Steinmetz and other early pioneers of electrical power engineering. Among others, *Electrical Transients in Power Systems* by Allan Greenwood is worth mentioning. Even though basic knowledge of transients may not have advanced in recent years at the same rate as before, there has been a tremendous proliferation in the techniques used to study transients. The application of computers to the study of transient phenomena has increased both the knowledge as well as the accuracy of calculations.

Furthermore, the importance of transients in power systems is receiving more and more attention in recent years as a result of various blackouts, brownouts, and recent collapses of some large power systems in the United States, and other parts of the world. As electric power consumption grows exponentially due to increasing population, modernization, and industrialization of the so-called third world, this topic will be even more important in the future than it is at the present time.

Professor Arieh Shenkman is to be congratulated for undertaking such an important task and writing this book that singularly concentrates on the topics related to the transient analysis of electric power systems. The book successfully fills the long-existing gap in such an important area.

Turan Gonen, Ph.D., Fellow IEEE
Professor and Director
Electric Power Educational Institute
California State University, Sacramento

PREFACE

Most of the textbooks on electrical and electronic engineering only partially cover the topic of transients in simple *RL*, *RC* and *RLC* circuits and the study of this topic is primarily done from an electronic engineer's viewpoint, i.e., with an emphasis on low-current systems, rather than from an electrical engineer's viewpoint, whose interest lies in high-current, high-voltage power systems. In such systems a very clear differentiation between steady-state and transient behavior of circuits is made. Such a division is based on the concept that steady-state behavior is normal and transients arise from the faults. The operation of most electronic circuits (such as oscillators, switch capacitors, rectifiers, resonant circuits etc.) is based on their transient behavior, and therefore the transients here can be referred to as "desirable". The transients in power systems are characterized as completely "undesirable" and should be avoided; and subsequently, when they do occur, in some very critical situations, they may result in the electrical failure of large power systems and outages of big areas. Hence, the Institute of Electrical and Electronic Engineers (IEEE) has recently paid enormous attention to the importance of power engineering education in general, and transient analysis in particular.

It is with the belief that transient analysis of power systems is one of the most important topics in power engineering analysis that the author proudly presents this book, which is wholly dedicated to this topic.

Of course, there are many good books in this field, some of which are listed in the book; however they are written on a specific technical level or on a high theoretical level and are intended for top specialists. On the other hand, introductory courses, as was already mentioned, only give a superficial knowledge of transient analysis. So that there is a gap between introductory courses and the above books.

The present book is designed to fill this gap. It covers the topic of transient analysis from simple to complicated, and being on an intermediate level, this book therefore is a link between introductory courses and more specific technical books. In the book the most important methods of transient analysis, such as the classical method, Laplace and Fourier transforms and state variable analysis are presented; and of course, the emphasis on transients in three-phase systems and transmission lines is made.

The appropriate level and the concentration of all the topics under one cover make this book very special in the field under consideration. The author believes that this book will be very helpful for all those specializing in electrical engineering and power systems. It is recommended as a textbook for specialized undergraduate and graduate curriculum, and can also be used for master and doctoral studies. Engineers in the field may also find this book useful as a handbook and/or resource book that can be kept handy to review specific points. Theoreticians/researchers who are looking for the mathematical background of transients in electric circuits may also find this book helpful in their work.

The presentation of the covered material is geared to readers who are being exposed to (a) the basic concept of electric circuits based on their earlier study of physics and/or introductory courses in circuit analysis, and (b) basic mathematics, including differentiation and integration techniques

This book is composed of eight chapters. The study of transients, as mentioned, is presented from simple to complicated. Chapters 1 and 2 are dedicated to the classical method of transient analysis, which is traditional for many introductory courses. However, these two chapters cover much more material giving the mathematical as well as the physical view of transient behavior of electrical circuits. So-called incorrect initial conditions and two generalized commutation laws, which are important for a better understanding of the transient behavior of transformers and synchronous machines, are also discussed in Chapter 2.

Chapters 3 and 4 give the transform methods of transient analysis, introducing the Laplace as well as the Fourier transforms. What is common between these two methods and the differences are emphasized. The theoretical study of the transform methods is accompanied by many practical examples.

The state variable method is presented in Chapter 5. Although this method is not very commonly used in transient analysis, the author presumes that the topic of the book will not be complete without introducing this essential and interesting method. It should be noted that the state variable method in its matrix notation, which is given here, is very appropriate for transient analysis using computers.

Naturally, an emphasis and a great amount of material are dedicated to transients in three-phase circuits, which can be found in Chapter 6. As power systems are based on employing three-phase generators and transformers, the complete analysis of their behavior under short-circuit faults at both steady-state and first moment operations is given. The overvoltages following switching-off in power systems are also analyzed under the influence of the electric arc, which accompanies such switching.

In Chapter 7 the transient behavior of transmission lines is presented. The transmission line is presented as a network with distributed parameters and subsequently by partial differential equations. The transient analysis of such lines is done in two ways: as a method of traveling waves and by using the

Laplace transform. Different engineering approaches using both methods are discussed.

Finally, in Chapter 8 an overview of the static and dynamic stability of power systems is given. Analyzing system stability is done in traditional ways, i.e., by solving a swing equation and by using an equal area criterion.

Throughout the text, the theoretical discussions are accompanied by many worked-out examples, which will hopefully enable the reader to get a better understanding of the various concepts.

The author hopes that this book will be helpful to all readers studying and specializing in power system engineering, and of value to professors in the educational process and to engineers who are concerned with the design and R&D of power systems.

Last but not least, my sincere appreciation goes to my wife, Iris, who prodigiously supported and aided me throughout the writing of this book. I am also extremely grateful for her assistance in editing and typing in English.

Chapter #1

CLASSICAL APPROACH TO TRANSIENT ANALYSIS

1.1 INTRODUCTION

Transient analysis (or just transients) of electrical circuits is as important as steady-state analysis. When transients occur, the currents and voltages in some parts of the circuit may many times exceed those that exist in normal behavior and may destroy the circuit equipment in its proper operation. We may distinguish the transient behavior of an electrical circuit from its steady-state, in that during the transients all the quantities, such as currents, voltages, power and energy, are changed in time, while in steady-state they remain invariant, i.e. constant (in d.c. operation) or periodical (in a.c. operation) having constant amplitudes and phase angles.

The cause of transients is any kind of changing in circuit parameters and/or in circuit configuration, which usually occur as a result of switching (commutation), short, and/or open circuiting, change in the operation of sources etc. The changes of currents, voltages etc. during the transients are not instantaneous and take some time, even though they are extremely fast with a duration of milliseconds or even microseconds. These very fast changes, however, cannot be instantaneous (or abrupt) since the transient processes are attained by the interchange of energy, which is usually stored in the magnetic field of inductances or/and the electrical field of capacitances. Any change in energy cannot be abrupt otherwise it will result in infinite power (as the power is a derivative of energy, $p = dw/dt$), which is in contrast to physical reality. All transient changes, which are also called transient responses (or just responses), vanish and, after their disappearance, a new steady-state operation is established. In this respect, we may say that the transient describes the circuit behavior between two steady-states: an old one, which was prior to changes, and a new one, which arises after the changes.

A few methods of transient analysis are known: the classical method, The Cauchy-Heaviside (C-H) operational method, the Fourier transformation method and the Laplace transformation method. The C-H operational or symbolic (formal) method is based on replacing a derivative by symbol s ($(d/dt) \leftrightarrow s$) and an integral by $1/s$

$$\left(\int dt \leftrightarrow \frac{1}{s}\right).$$

Although these operations are also used in the Laplace transform method, the C-H operational method is not as systematic and as rigorous as the Laplace transform method, and therefore it has been abandoned in favor of the Laplace method. The two transformation methods, Laplace and Fourier, will be studied in the following chapters. Comparing the classical method and the transformation method it should be noted that the latter requires more knowledge of mathematics and is less related to the physical matter of transient behavior of electric circuits than the former.

This chapter is concerned with the classical method of transient analysis. This method is based on the determination of differential equations and splitting the solution into two components: natural and forced responses. The classical method is fairly complicated mathematically, but is simple in engineering practice. Thus, in our present study we will apply some known methods of steady-state analysis, which will allow us to simplify the classical approach of transient analysis.

1.2 APPEARANCE OF TRANSIENTS IN ELECTRICAL CIRCUITS

In the analysis of an electrical system (as in any physical system), we must distinguish between the stationary operation or steady-state and the dynamical operation or transient-state.

An electrical system is said to be in **steady-state** when the variables describing its behavior (voltages, currents, etc.) are either invariant with time (d.c. circuits) or are periodic functions of time (a.c. circuits). An electrical system is said to be in **transient-state** when the variables are changed non-periodically, i.e., when the system is not in steady-state. The transient-state vanishes with time and a new steady-state regime appears. Hence, we can say that the transient-state, or just transients, is usually the transmission state from one steady-state to another.

The parameters L and C are characterized by their ability to store energy: *magnetic energy* $w_L = \frac{1}{2}\psi i = \frac{1}{2}Li^2$ (since $\psi = Li$), in the magnetic field and *electric energy* $w_C = \frac{1}{2}qv = \frac{1}{2}Cv^2$ (since $q = Cv$), in the electric field of the circuit. The voltage and current sources are the elements through which the energy is supplied to the circuit. Thus, it may be said that an electrical circuit, as a physical system, is characterized by certain energy conditions in its steady-state behavior. Under steady-state conditions the energy stored in the various inductances and capacitances, and supplied by the sources in a d.c. circuit, are constant; whereas in an a.c. circuit the energy is being changed (transferred between the magnetic and electric fields and supplied by sources) *periodically*.

When any sudden change occurs in a circuit, there is usually a redistribution of energy between *L*-s and *C*-s, and a change in the energy status of the sources, which is required by the new conditions. **These energy redistributions cannot take place instantaneously, but during some period of time, which brings about the transient-state.**

The main reason for this statement is that an instantaneous change of energy would require infinite power, which is associated with inductors/capacitors. As previously mentioned, power is a derivative of energy and any abrupt change in energy will result in an infinite power. Since infinite power is not realizable in physical systems, the energy cannot change abruptly, but only within some period of time in which transients occur. Thus, from a physical point of view it may be said that the transient-state exists in physical systems while the energy conditions of one steady-state are being changed to those of another.

Our next conclusion is about the current and voltage. To change *magnetic energy* requires a change of current through inductances. Therefore, currents in inductive circuits, or inductive branches of the circuit, cannot change abruptly. From another point of view, the change of current in an inductor brings about the induced voltage of magnitude $L(di/dt)$. An instantaneous change of current would therefore require an infinite voltage, which is also unrealizable in practice. Since the induced voltage is also given as $d\psi/dt$, where ψ is a magnetic flux, the magnetic flux of a circuit cannot suddenly change.

Similarly, we may conclude that to change the *electric energy* requires a change in voltage across a capacitor, which is given by $v = q/C$, where q is the charge. Therefore, neither the voltage across a capacitor nor its charge can be abruptly changed. In addition, the rate of voltage change is $dv/dt = (1/C)\, dq/dt = i/C$, and the instantaneous change of voltage brings about infinite current, which is also unrealizable in practice. Therefore, we may summarize that **any change in an electrical circuit, which brings about a change in energy distribution, will result in a transient-state.**

In other words, by any switching, interrupting, short-circuiting as well as any rapid changes in the structure of an electric circuit, the transient phenomena will occur. Generally speaking, **every change of state leads to a temporary deviation from one regular, steady-state performance of the circuit to another one.** The redistribution of energy, following the above changes, i.e., the transient-state, theoretically takes infinite time. However, in reality the transient behavior of an electrical circuit continues a relatively very short period of time, after which the voltages and currents almost achieve their new steady-state values.

The change in the energy distribution during the transient behavior of electrical circuits is governed by the principle of energy conservation, i.e., the amount of supplied energy is equal to the amount of stored energy plus the energy dissipation. The rate of energy dissipation affects the time interval of the transients. The higher the energy dissipation, the shorter is the transient-state. Energy dissipation occurs in circuit resistances and its storage takes place in inductances and capacitances. In circuits, which consist of only resistances, and neither inductances nor capacitances, the transient-state will not occur at all and the change from one steady-state to another will take place instantaneously. However, since even resistive circuits contain some inductances and capacitances the transients will practically appear also in such circuits; but these transients are very short and not significant, so that they are usually neglected.

Transients in electrical circuits can be recognized as either desirable or undesirable. In power system networks, the transient phenomena are wholly undesirable as they may bring about an increase in the magnitude of the voltages and currents and in the density of the energy in some or in most parts of modern power systems. All of this might result in equipment distortion, thermal and/or electrodynamics' destruction, system stability interferences and in extreme cases an outage of the whole system.

In contrast to these unwanted transients, there are desirable and controlled transients, which exist in a great variety of electronic equipment in communication, control and computation systems whose normal operation is based on switching processes.

The transient phenomena occur in electric systems either by intentional switching processes consisting of the correct manipulation of the controlling apparatus, or by unintentional processes, which may arise from ground faults, short-circuits, a break of conductors and/or insulators, lightning strokes (particularly in high voltage and long distance systems) and similar inadvertent processes.

As was mentioned previously, there are a few methods of solving transient problems. The most widely known of these appears in all introductory textbooks and is used for solving simpler problems. **It is called the classical method.** Other useful methods are Laplace (see Chap. 3) and Fourier (see Chap. 4) transformation methods. These two methods are more general and are used for solving problems that are more complicated.

1.3 DIFFERENTIAL EQUATIONS DESCRIBING ELECTRICAL CIRCUITS

Circuit analysis, as a physical system, is completely described by *integrodifferential equations* written for voltages and/or currents, which characterize circuit behavior. For linear circuits these equations are called linear differential equations with constant coefficients, i.e. in which every term is of the first degree in the dependent variable or one of its derivatives. Thus, for example, for the circuit of three basic elements: R, L and C connected in series and driven by a voltage source $v(t)$, Fig. 1.1, we may apply Kirchhoff's voltage law

$$v_R + v_L + v_C = v(t),$$

in which

$$v_R = Ri$$

$$v_L = L\frac{di}{dt}$$

$$v_C = \int i\,dt,$$

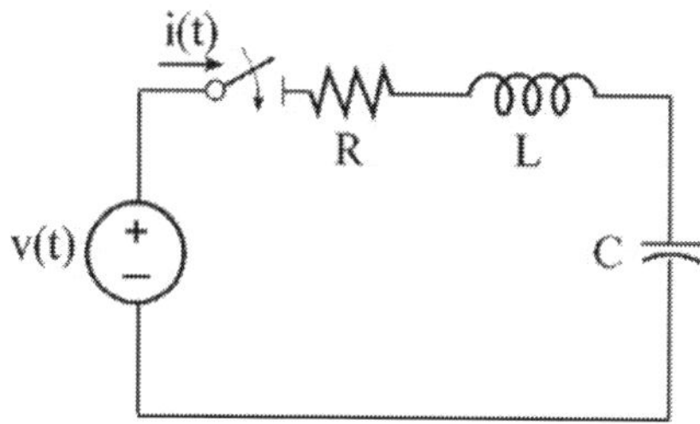

Figure 1.1 Series *RLC* circuit driven by a voltage source.

and then we have

$$L\frac{di}{dt} + Ri + \frac{1}{C}\int i\,dt = v(t). \tag{1.1}$$

After the differentiation of both sides of equation 1.1 with respect to time, the result is a second order differential equation

$$L\frac{d^2t}{dt^2} + R\frac{di}{dt} + \frac{1}{C}i = \frac{dv}{dt}. \tag{1.2}$$

The same results may be obtained by writing two simultaneous first order differential equations for two unknowns, i and v_C:

$$\frac{dv_C}{dt} = \frac{1}{C}i \tag{1.3a}$$

$$Ri + L\frac{di}{dt} + v_C = v(t). \tag{1.3b}$$

After differentiation equation 1.3b and substituting dv_C/dt by equation 1.3a, we obtain the same (as equation 1.2) second order singular equation. The solution of differential equations can be completed only if the initial conditions are specified. It is obvious that in the same circuit under the same commutation, but with different initial conditions, its transient response will be different.

For more complicated circuits, built from a number of loops (nodes), we will have a set of differential equations, which should be written in accordance with Kirchhoff's two laws or with nodal and/or mesh analysis. For example, considering the circuit shown in Fig. 1.2, after switching, we will have a circuit, which consists of two loops and two nodes. By applying Kirchhoff's two laws, we may write three equations with three unknowns, i, i_L and v_C,

$$C\frac{dv_C}{dt} + i_L - i = 0 \tag{1.4a}$$

$$L\frac{di_L}{dt} + R_1 i_L + Ri = 0 \tag{1.4b}$$

$$L\frac{di_L}{dt} + R_1 i_L - v_C = 0 \tag{1.4c}$$

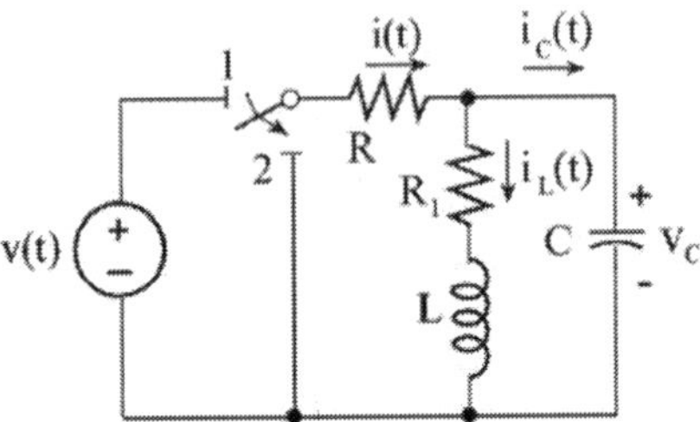

Figure 1.2 A two-loop circuit.

These three equations can then be redundantly transformed into a single second order equation. First, we differentiate the third equation of 1.4c once with respect to time and substitute dv_C/dt by taking it from the first one. After that, we have two equations with two unknowns, i_L and i. Solving these two equations for i_L (i.e. eliminating the current i) results in the second order homogeneous differential equation

$$LCR\frac{d^2i_L}{dt^2} + (L + CRR_1)\frac{di_L}{dt} + (R + R_1)i_L = 0. \tag{1.5}$$

As another example, let us consider the circuit in Fig. 1.3. Applying mesh analysis, we may write three *integro-differential equations* with three unknown mesh currents:

$$\begin{aligned} &L\frac{di_1}{dt} - L\frac{di_2}{dt} + R_1 i_1 = v(t) \\ &L\frac{di_2}{dt} - L\frac{di_1}{dt} + (R_2 + R_3)i_2 - R_3 i_3 = 0 \\ &-R_3 i_2 + R_3 i_3 + \frac{1}{C}\int i_3 dt = 0. \end{aligned} \tag{1.6}$$

In this case it is preferable to solve the problem by treating the whole set of equations 1.6 rather than reducing them to a single one (see further on).

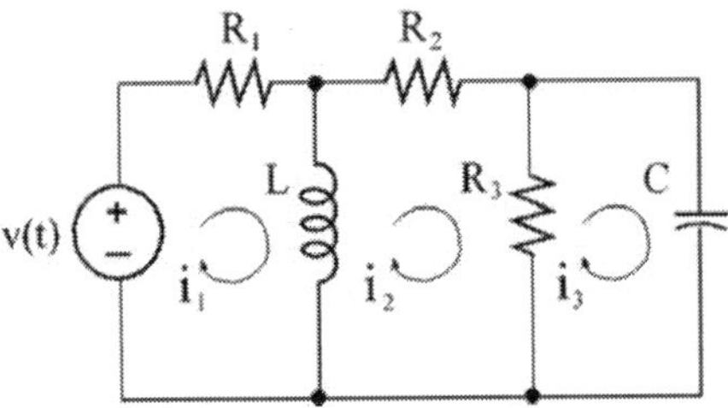

Figure 1.3 A three-loop circuit.

From mathematics, we know that there are a number of ways of solving differential equations. Our goal in this chapter is to analyze the transient behavior of electrical circuits from the physical point of view rather than applying complicated mathematical methods. (This will be discussed in the following chapters.) Such a way of transient analysis is in the formulation of differential equations in accordance with the properties of the circuit elements and in the *direct* solution of the obtained equations, using only the necessary mathematical rules. Such a method is called the **classical method or classical approach** in transient analysis. We believe that the classical method of solving problems enables the student to better understand the transient behavior of electrical circuits.

1.3.1 Exponential solution of a simple differential equation

Let us, therefore, begin our study of transient analysis by considering the simple series *RC* circuit, shown in Fig. 1.4. After switching we will get a source free circuit in which the precharged capacitor *C* will be discharged via the resistance *R*. To find the capacitor voltage we shall write a differential equation, which in accordance with Kirchhoff's voltage law becomes

$$Ri + v_C = 0, \quad \text{or} \quad RC\frac{dv_C}{dt} + v_C = 0. \tag{1.7}$$

A direct method of solving this equation is to write the equation in such a way that the variables are separated on both sides of the equation and then to integrate each of the sides. Multiplying by dt and dividing by v_C, we may arrange the variables to be separated.

$$\frac{dv_C}{v_C} = -\frac{1}{RC}dt. \tag{1.8}$$

The solution may be obtained by integrating each side of equation 1.8 and by adding a constant of integration:

$$\int \frac{dv_C}{v_C} = -\frac{1}{RC}\int dt + K,$$

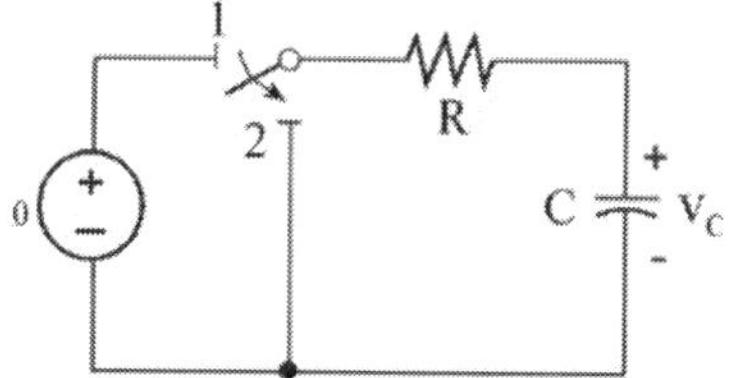

Figure 1.4 A series *RC* circuit.

and the integration yields

$$\ln v_C = -\frac{1}{RC}t + K \tag{1.9}$$

Since the constant can be of any kind, and we may designate $K = \ln D$, we have

$$\ln v_C = -\frac{1}{RC}t + \ln D,$$

then

$$v_C = De^{-\frac{t}{RC}}. \tag{1.10}$$

The constant D cannot be evaluated by substituting equation 1.10 into the original differential equation 1.7, since the identity $0 \equiv 0$ will result for any value of D (indeed: $D(-1/RC)RCe^{-t/RC} + De^{-t/RC} = 0$). The constant of integration must be selected to satisfy the initial condition $v_C(0) = V_0$, which is the initial voltage across the capacitance. Thus, the solution of equation 1.10 at $t = 0$ becomes $v_C(0) = D$, and we may conclude that $D = V_0$. Therefore, with this value of D we will obtain the desired response

$$v_C = V_0 e^{-\frac{t}{RC}}. \tag{1.11}$$

We shall consider the nature of this response by analyzing the curve of the voltage change shown in Fig. 1.5. At zero time, the voltage is the assumed value V_0 and, as time increases, the voltage decreases and approaches zero, following the physical rule that any condenser shall finally be discharged and its final voltage therefore reduces to zero.

Let us now find the time that would be required for the voltage to drop to

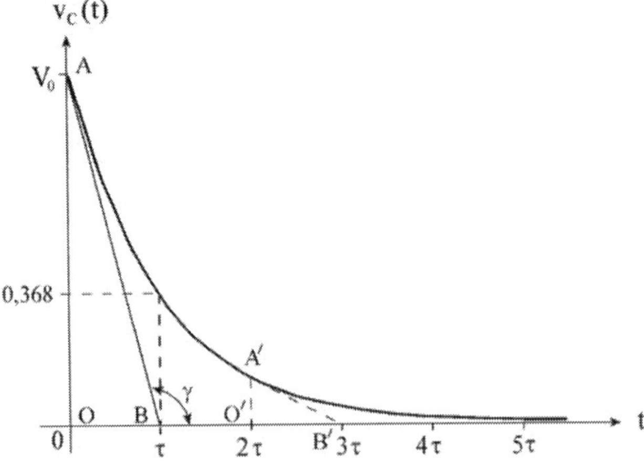

Figure 1.5 The exponential curve of the voltage changing.

zero if it continued to drop linearly at its initial rate. This value of time, usually designated by τ, is called the **time constant**. The value of τ can be found with the derivative of $v_C(t)$ at zero time, which is proportional to the angle γ between the tangent to the voltage curve at $t=0$, and the t-axis, Fig. 1.5, i.e.,

$$\tan\lambda \propto -\frac{V_0}{\tau} = \frac{d}{dt}\left(V_0 e^{-\frac{t}{RC}}\right)_{t=0} = \frac{-V_0}{RC},$$

or

$$\tau = RC$$

and equation 1.11 might be written in the form

$$V_C = V_0 e^{-\frac{t}{\tau}}. \tag{1.12}$$

The units of the time constant are seconds ($[\tau]=[R][C]=\Omega\cdot F$), so that the exponent t/RC is dimensionless, as it is supposed to be. The time constant may be easily found graphically from the response curve, as can be seen from Fig. 1.5: the interception point, B, of the tangent line $\overline{AB}$ with the time axis determine the time constant τ. This line segment $\overline{OB}$ is called under-tangent. It is interesting to note that the under-tangent remains the same no matter at which point the tangent to the curve is drawn (see under-tangent $\overline{O'B'}$).

Another interpretation of the time constant is obtained from the fact that in the time interval of one time constant the voltage drops relatively to its initial value, to the reciprocal of e; indeed, at $t=\tau$ we have $(v_C/V_0)=e^{-1}=0.368$ (36.8%). At the end of the 5τ interval the voltage is less than one percent of its initial value. Thus, it is usual to presume that in the time interval of three to five time constants, the transient response declines to zero or, in other words, we may say that the duration of the transient response is about five time constants. Note again that, precisely speaking, the transient response declines to zero in infinite time, since $e^{-t}\to 0$, when $t\to\infty$.

Before we continue our discussion of a more general analysis of transient circuits, let us check the power and energy relationships during the *period of transient response*. The power being dissipated in the resistor R, or its reciprocal G, is

$$p_R = G v_C^2 = G V_0^2 e^{-2t/RC}, \tag{1.13}$$

and the total dissipated energy (turned into heat) is found by integrating equation 1.13 from zero time to infinite time

$$w_R = \int_0^\infty p_R dt = V_0 G \int_0^\infty e^{-2t/RC} = -V_0^2 G \frac{RC}{2} e^{-2t/RC}\Big|_0^\infty = \frac{1}{2} C V_C^2.$$

This is actually the energy being stored in the capacitor at the beginning of the transient. This result means that all the initial energy, stored in the capacitor, dissipates in the circuit resistances during the transient period.

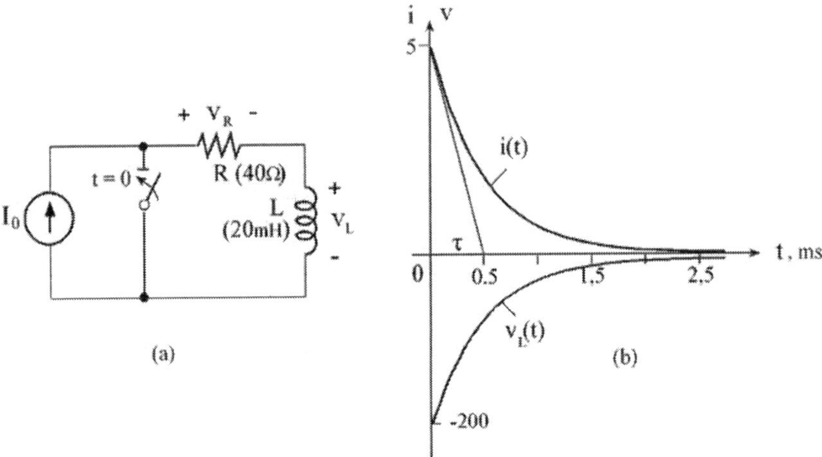

Figure 1.6 A circuit of Example 1.1 (a) and two plots of current and voltage (b).

Example 1.1

Consider a numerical example. The RL circuit in Fig. 1.6(a) is fed by a d.c. current source, $I_0 = 5\,\text{A}$. At instant $t = 0$ the switch is closed and the circuit is short-circuited. Find: 1) the current after switching, by separating the variables and applying the *definite integrals*, 2) the voltage across the inductance.

Solution

1) First, we shall write the differential equation:

$$v_L + v_R = L\frac{di}{dt} + Ri = 0,$$

or after separating the variables

$$\frac{di}{i} = \frac{R}{L}dt.$$

Since the current changes from I_0 at the instant of switching to $i(t)$, at any instant of t, which means that the time changes from $t = 0$ to this instant, we may perform the integration of each side of the above equation between the corresponding limits

$$\int_{I_0}^{i(t)} \frac{di}{i} = \int_0^t -\frac{R}{L}dt.$$

Therefore,

$$\ln i|_{I_0}^{i(t)} = -\frac{R}{L}t|_0^t$$

and

$$\ln i(t) - \ln I_0 = \frac{R}{L}t, \quad \text{or} \quad \ln\frac{i(t)}{I_0} = -\frac{R}{L}t,$$

which results in

$$\frac{i(t)}{I_0} = e^{-\frac{R}{L}t}$$

Thus,

$$i(t) = I_0 e^{-\frac{R}{L}t} = 5e^{-2000t},$$

or

$$i(t) = I_0 e^{-\frac{t}{\tau}} = 5e^{-\frac{t}{0.5\cdot 10^{-3}}}.$$

where

$$R/L = \frac{40}{20\cdot 10^{-3}} = 2000\ \mathrm{s}^{-1},$$

which results in time constant

$$\tau = \frac{L}{R} = 0.5\ \mathrm{ms}.$$

Note that by applying the definite integrals we avoid the step of evaluating the constant of the integration.

2) The voltage across the inductance is

$$v_L = L\frac{di}{dt} = L\frac{d}{dt}(5e^{-2000t}) = 20\cdot 10^{-3}\cdot 5\cdot(-2000)e^{-2000t} = -200e^{-\frac{t}{0.5}},\ \mathrm{V}$$

(time in ms).

Note that the voltage across the resistance is $v_R = Ri = 40\cdot 5e^{-t/0.5} = 200e^{-t/0.5}$, i.e., it is equal in magnitude to the inductance voltage, but opposite in sign, so that the total voltage in the short-circuit is equal to zero. The plots of the current and voltage are shown in Fig. 1.6(b).

1.4 NATURAL AND FORCED RESPONSES

Our next goal is to introduce a general approach to solving differential equations by the classical method. Following the principles of mathematics we will consider the complete solution of any linear differential equation as composed of two parts: the complementary solution (or natural response in our study) and the particular solution (or forced response in our study). To understand these principles, let us consider a first order differential equation, which has already been derived in the previous section. In a more general form it is

$$\frac{dv}{dt} + P(t)v = Q(t). \tag{1.14}$$

Here $Q(t)$ is identified as a forcing function, which is generally a function of time (or constant, if a d.c. source is applied) and $P(t)$, is also generally a function of time, represents the circuit parameters. In our study, however, it will be a constant quantity, since the value of circuit elements does not change during the transients (indeed, the circuit parameters do change during the transients, but we may neglect this change as in many cases it is not significant).

A more general method of solving differential equations, such as equation 1.14, is to multiply both sides by a so-called *integrating factor*, so that each side becomes an exact differential, which afterwards can be integrated directly to obtain the solution. For the equation above (equation 1.14) the integrating factor is $e^{\int P dt}$ or e^{Pt}, since P is constant. We multiply each side of the equation by this integrating factor and by dt and obtain

$$e^{Pt}dv + vPe^{Pt}dt = Qe^{Pt}dt.$$

The left side is now the exact differential of ve^{Pt} (indeed, $d(ve^{Pt}) = e^{Pt}dv + vPe^{Pt}dt$), and thus

$$d(ve^{Pt}) = Qe^{Pt}dt.$$

Integrating each side yields

$$ve^{Pt} = \int Qe^{Pt}dt + A, \tag{1.15}$$

where A is a constant of integration. Finally, the multiplication of both sides of equation 1.15 by e^{-Pt} yields

$$v(t) = e^{-Pt}\int Qe^{Pt}dt + Ae^{-Pt}, \tag{1.16}$$

which is the solution of the above differential equation. As we can see, this complete solution is composed of two parts. The first one, which is dependent on the forcing function Q, is the *forced response* (it is also called the steady-state response or the particular solution or the particular integral). The second one, which does not depend on the forcing function, but only on the circuit parameters P (the types of elements, their values, interconnections, etc) and on the initial conditions A, i.e., on the "nature" of the circuit, is the *natural response*. It is also called the solution of the homogeneous equation, which does not include the source function and has anything but zero on its right side.

Following this rule, we will solve differential equations by finding natural and forced responses separately and combining them for a complete solution. This principle of dividing the solution of the differential equations into two components can also be understood by applying the superposition theorem. Since the differential equations, under study, are linear as well as the electrical circuits, we may assert that superposition is also applicable for the transient-state. Following this principle, we may subdivide, for instance, the current into two components

$$i = i' + i'',$$

and by substituting this into the set of differential equations, say of the form

$$\sum \left(L\frac{di}{dt} + Ri + \frac{1}{C}\int i\,dt \right) = \sum v_s,$$

we obtain the following two sets of equations

$$\sum \left(L\frac{di'}{dt} + Ri' + \frac{1}{C}\int i'\,dt \right) = \sum v_s,$$

$$\sum \left(L\frac{di''}{dt} + Ri'' + \frac{1}{C}\int i''\,dt \right) = 0.$$

It is obvious that by summation (superimposition) of these two equations, the original equation will be achieved. This means that i'' is a natural response since it is the solution of a homogeneous equation with a zero on the right side and develops without any action of any source, and i' is a steady-state current as it develops under the action of the voltage sources v_s (which are presented on the right side of the equations).

The most difficult part in the classical method of solving differential equations is evaluating the particular integral in equation 1.16, especially when the forcing function is not a simple d.c. or exponential source. However, in circuit analysis we can use all the methods: node/mesh analysis, circuit theorems, the phasor method for a.c. circuits (which are all given in introductory courses on steady-state analysis) to find the forced response. In relation to the natural response, the most difficult part is to formulate the characteristic equation (see further on) and to find its roots. Here in circuit analysis we also have special methods for evaluating the characteristic equation simply by inspection of the analyzed circuit, avoiding the formulation of differential equations.

Finally, it is worthwhile to clarify the use of exponential functions as an integrating factor in solving linear differential equations. As we have seen in the previous section, such differential equations in general consist of the second (or higher) derivative, the first derivative and the function itself, each multiplied by a *constant factor*. If the sum of all these derivatives (the function itself might be treated as a derivative of order zero) achieves zero, it becomes a homogeneous equation. A function whose derivatives have the same form as the function itself is an exponential function, so it may satisfy these kinds of equations. Substituting this function into the differential equation, whose right side is zero (a homogeneous differential equation) the exponential factor in each member of the equation might be simply crossed out, so that the remaining equation's coefficients will be only circuit parameters. Such an equation is called a characteristic equation.

1.5 CHARACTERISTIC EQUATION AND ITS DETERMINATION

Let us start by considering the simple circuit of Fig. 1.7(a) in which an *RL* in series is switching on to a d.c. voltage source.

Let the desired response in this circuit be current $i(t)$. We shall first express it as the sum of the natural and forced currents

$$i = i_n + i_f.$$

The form of the natural response, as was shown, must be an exponential function, $i_n = Ae^{st}$(*). Substituting this response into the homogeneous differential equation, which is $L(di/dt) + Ri = 0$, we obtain $Ls\,e^{st} + R\,e^{st} = 0$, or

$$Ls + R = 0. \tag{1.17a}$$

This is a characteristic (or auxiliary) equation, in which the left side expresses the input impedance seen from the source terminals of the analyzed circuit.

$$Z_{in}(s) = Ls + R. \tag{1.17b}$$

We may treat s as the complex frequency $s = \sigma + j\omega$ (for more about complex frequencies see any introductory course to circuit analysis and further on in Chap. 3). Note that by equaling this expression of circuit impedance to zero, we obtain the *characteristic equation*. Solving this equation we have

$$s = -\frac{R}{L} \quad \text{and} \quad \tau = \frac{L}{R}. \tag{1.18}$$

Hence, the natural response is

$$i_n = Ae^{-\frac{R}{L}t}. \tag{1.19}$$

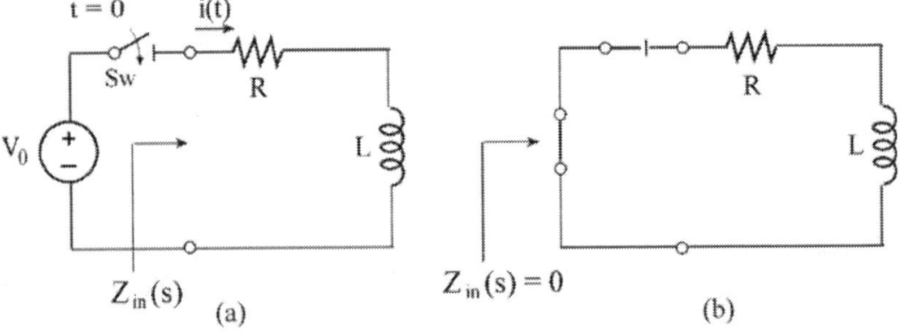

Figure 1.7 An *RL* circuit switching to a d.c. voltage source (a) and after "killing" the source (b).

(*) Here and in the future, we will use the letter s for the circuit parameters' dependent exponent.

Subsequently, the root of the characteristic equation defines the exponent of the natural response. The fact that the input impedance of the circuit should be equaled to zero can be explained from a physical point of view.(*) Since the natural response does not depend on the source, the latter should be "killed". i.e. short-circuited as shown in Fig. 1.7(b). This action results in short-circuiting the entire circuit, i.e. its input impedance.

Consider now a parallel *LR* circuit switching to a d.c. current source in which the desired response is $v_L(t)$, as shown in Fig. 1.8(a). Here, "killing" the current source results in open-circuiting, as shown in Fig. 1.8(b).

This means that the input admittance should be equaled to zero. Thus,

$$\frac{1}{R}+\frac{1}{sL}=0,$$

or

$$sL+R=0,$$

which however gives the same root

$$s=-\frac{R}{L} \quad \text{and} \quad \tau=\frac{L}{R}. \tag{1.20}$$

Next, we will consider a more complicated circuit, shown in Fig. 1.9(a). This circuit, after switching and short-circuiting the remaining voltage source, will be as shown in Fig. 1.9(b). The input impedance of this circuit "measured" at the switch (which is the same as seen from the "killed" source) is

$$Z_{in}(s)=R_1+R_3//R_4//(R_2+sL),$$

or

$$Z_{in}(s)=R_1+\left(\frac{1}{R_3}+\frac{1}{R_4}+\frac{1}{R_2+sL}\right)^{-1}.$$

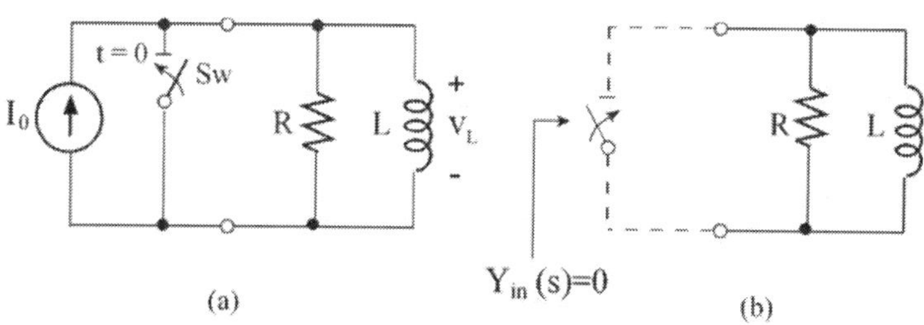

Figure 1.8 A parallel *RL* circuit switching to d.c. current source (a) and after " killing" the source (b).

(*)This fact is proven more correctly mathematically in Laplace transformation theory (see further on).

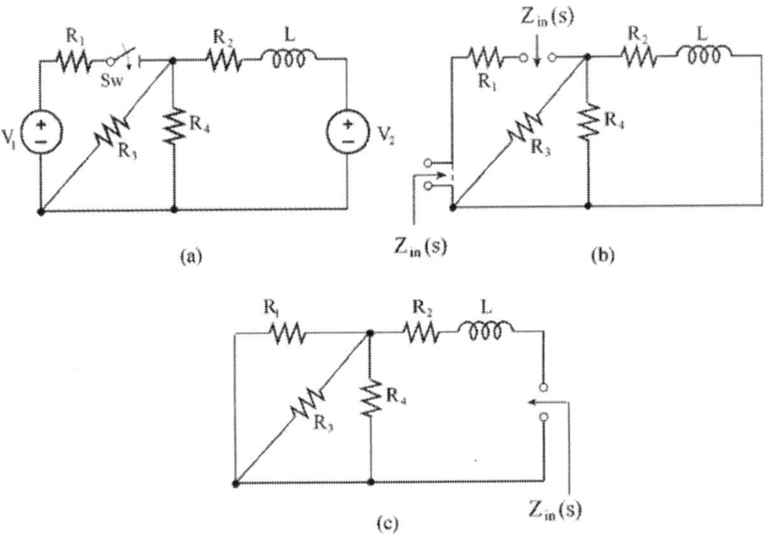

Figure 1.9 A given circuit (a), determining the input impedance as seen from the switch (b) and as seen from the inductance branch (c).

Evaluating this expression and equaling it to zero yields

$$(R_1R_3 + R_1R_4 + R_3R_4)(R_2 + sL) + R_1R_3R_4 = 0,$$

and the root is

$$s = -\frac{R_{eq}}{L}, \quad \text{where} \quad R_{eq} = \frac{R_1R_3R_4 + R_1R_2R_3 + R_1R_2R_4 + R_2R_3R_4}{R_1R_3 + R_1R_4 + R_3R_4}.$$

It is worthwhile to mention that the same results can be obtained if the input impedance is "measured" from the inductance branch, i.e. the energy-storing element, as is shown in Fig. 1.9(c).

The characteristic equation can also be determined by inspection of the differential equation or set of equations. Consider the second-order differential equation like in equation 1.2

$$L\frac{d^2i(t)}{dt} + R\frac{di(t)}{dt} + \frac{1}{C}i(t) = g(t). \tag{1.21}$$

Replacing each derivative by s^n, where n is the order of the derivative (the function by itself is considered as a zero-order derivative), we may obtain the characteristic equation:

$$Ls^2 + Rs^1 + \frac{1}{C}s^0 = 0, \quad \text{or} \quad s^2 + \frac{R}{L}s + \frac{1}{LC} = 0. \tag{1.22}$$

This characteristic equation is of the second order (in accordance with the second order differential equation) and it possesses two roots s_1 and s_2.

If any system is described by a set of integro-differential equations, like in equation 1.6, then we shall first rewrite it in a slightly different form as homogeneous equations

$$\begin{aligned}\left(L\frac{d}{dt}+R\right)i_1-L\frac{d}{dt}i_2+0\cdot i_3&=0\\-L_1\frac{d}{dt}i_1+\left(L_2\frac{d}{dt}+R_2+R_3\right)i_2-R_3i_3&=0\\0\cdot i_1-R_3i_2+\left(\frac{1}{C}\int dt\right)i_3&=0.\end{aligned}\tag{1.23}$$

Replacing the derivatives now by s^n and an integral by s^{-1} (since an integral is a counter version of a derivative) we have

$$\begin{aligned}(Ls+R_1)i_1-sLi_2+0\cdot i_3&=0\\-Lsi_1+(Ls+R_2+R_3)i_2-R_3&=0\\0\cdot i_1-R_3i_2+\left(\frac{1}{sC}+R_3\right)i_3&=0.\end{aligned}\tag{1.24}$$

We obtained a set of algebraic equations with the right side equal to zero. In the matrix form

$$\begin{bmatrix}Ls+R_1 & -sL & 0\\ -sL & Ls+R_2+R_3 & -R_3\\ 0 & -R_3 & \dfrac{1}{Cs}+R_3\end{bmatrix}\begin{bmatrix}i_1\\ i_2\\ i_3\end{bmatrix}=\begin{bmatrix}0\\0\\0\end{bmatrix}\tag{1.24a}$$

With Cramer's rule the solution of this equation can be written as

$$i_{1,n}=\frac{\Delta_1}{\Delta}\quad i_{2,n}=\frac{\Delta_2}{\Delta}\quad i_{3,n}=\frac{\Delta_3}{\Delta},\tag{1.24b}$$

where Δ is the determinant of the system matrix and determinants Δ_1, Δ_2, Δ_3 are obtained from Δ, by replacing the appropriate column (in Δ_1 the first column is replaced, in Δ_2 the second column is replaced, and so forth), by the right side of the equation, i.e. by zeroes. As is known from mathematics such determinants are equal to zero and for the non-zero solution in equation 1.24 the determinant Δ in the denominator must also be zero. Thus, by equaling this determinant to zero, we get the characteristic equation:

$$\begin{vmatrix}sL+R_1 & -sL & 0\\ -sL & sL+R_2+R_3 & -R_3\\ 0 & -R_3 & \dfrac{1}{sC}+R_3\end{vmatrix}=0,$$

or

$$(sL+R_1)(sL+R_2+R_3)\left(\frac{1}{sC}+R_3\right)-R_3^2(sL+R_1)-s^2L^2\left(\frac{1}{sC}+R_3\right)=0$$

Simplifying this equation yields a second-order equation

$$s^2+\left(\frac{R_{1,eq}}{L}+\frac{1}{R_{2,eq}C}\right)s+\frac{1}{LC}\xi=0, \tag{1.25}$$

where

$$R_{1,eq}=\frac{R_1R_2}{R_1+R_2}\quad R_{2,eq}=\frac{R_1+R_2}{R_1/R_3+R_2/R_3+1}\quad \xi=\frac{1+R_2/R_3}{1+R_2/R_1}.$$

We could have achieved the same results by inspecting the circuit in Fig. 1.3 and determining the input impedance (we leave this solution as an exercise for the reader). The characteristic equation 1.25 is of second order, since the circuit (Fig. 1.3) consists of two energy-storing elements (one inductance and one capacitance).

There is a more general rule, which states that the order of a characteristic equation is as high as the number of energy-storing elements. However, we should distinguish between the elements, which cannot be replaced by their equivalent and those which can be eliminated by simplifying the circuit. We therefore shall first combine the inductances and capacitances, which are connected in series and/or in parallel, or can be brought to such connections. For instance, in the circuit in Fig. 1.10(a) we may account for five *L*-s/*C*-s elements. However, after simplification their number is reduced to only two energy-storing elements, as shown in Fig. 1.10(b). Therefore, we may conclude that the given circuit and its characteristic equation are of second order only. Another example is the circuit in Fig. 1.10(c), which contains three inductive elements and two resistances (after switching). By inspection of this circuit, we may simplify it to only one equivalent inductance:

$$L_{eq}=L_1+\frac{L_1L_2}{L_1+L_2}.$$

Therefore, the circuit is of the first order. The equivalent resistance is $R_{eq}=R_1+R_2$.

In such "reduced" circuits, the inductances and capacitances are associated with their currents (through inductances) and voltages (across capacitances), which at $t=0$ define the independent initial conditions (see further on). The number of these initial conditions must comply with the order of the characteristic equation, so that we will be able to determine the integration constant, the number of which is also equal to the order of the characteristic equation.

In more complicated circuits we may find that a few, let us say k inductances are connected in a so-called "inductance" node, as shown in Fig. 1.11(a) and (b). Taking into consideration that, in accordance with KCL, the sum of the

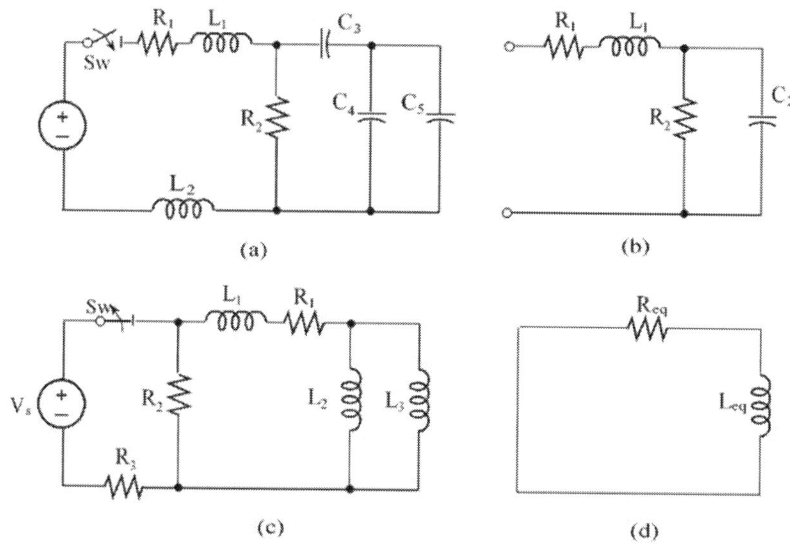

Figure 1.10 A given circuit of five L/C elements (a) and its equivalent of only two L/C elements (b), a circuit of three L elements (c) and its equivalent of only one L element.

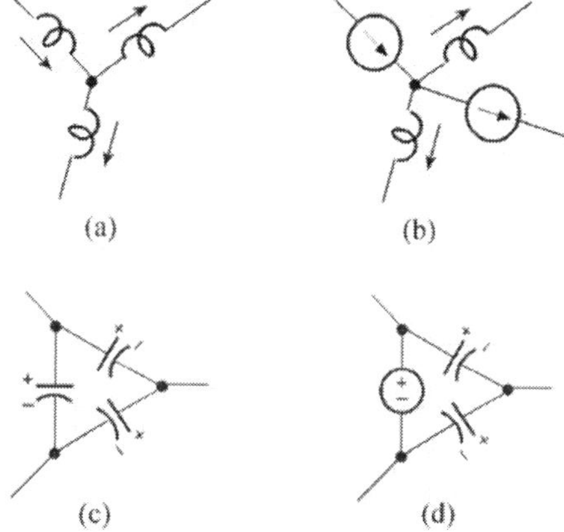

Figure 1.11 An "inductance" node of three inductances (a), an "inductance" node of two inductances and two current sources (b), a "capacitance" loop of three capacitances (c) and a "capacitance" loop of two capacitances and one voltage source.

currents in a node is zero, we may conclude that only $k-1$ inductance currents are independent. This means that the contribution to the order of the characteristic equation, which will be made by the inductances, is one less than the number of inductances. The "*capacitance*" *loop*, Fig. 1.11(c) and (b) is a dual to the "inductance" node, so that the number of independent voltages across the capacitances in the loop will be one less than the number of capacitances. Thus, if the total number of inductances and capacitances is n_L and n_C respectively, and the number of "inductance" nodes and "capacitance" loops is m_L and m_C respectively, then *the order of the characteristic equation* is $n_s = n_L + n_C - m_L - m_C$. Finally, it must be mentioned that the mutual inductance does not influence the order of the characteristic equation.

By analyzing the circuits in their transient behavior and determining their characteristic equations, we should also take into consideration that the natural responses might be different depending on the kind of applied source: voltage or current. Actually, we have to distinguish between two cases:

1) If the voltage source, in its physical representation (i.e. with an inner resistance connected in series) is replaced by an equivalent current source (i.e. with the same resistance connected in parallel), the transient responses will not change. Indeed, as can be seen from Fig. 1.12, the same circuit A is connected in (a) to the voltage source and in (b) to the current source. By "killing" the sources (i.e. short-circuiting the voltage sources and opening the current sources) we are getting the same passive circuits, for which the impedances are the same. This means that the characteristic equations of both circuits will be the same and therefore the natural responses will have the same exponential functions.

2) However, if the *ideal voltage source is replaced by an ideal current source*, Fig. 1.13, the passive circuits in (a) and (b), i.e. after killing the sources, are different, having different input impedances and therefore different natural responses.

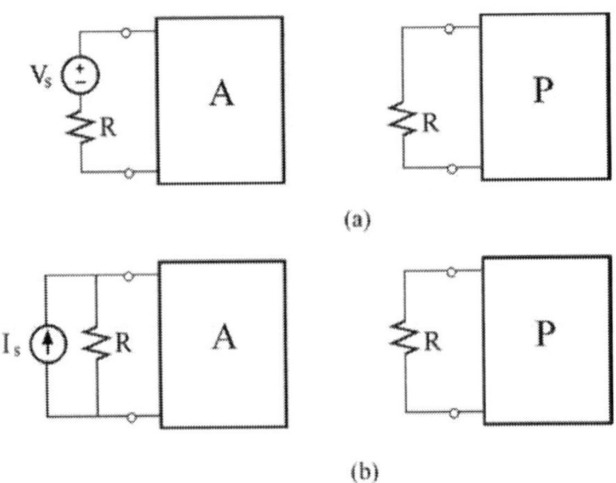

Figure 1.12. A circuit with an applied voltage source (a) and with a current source (b).

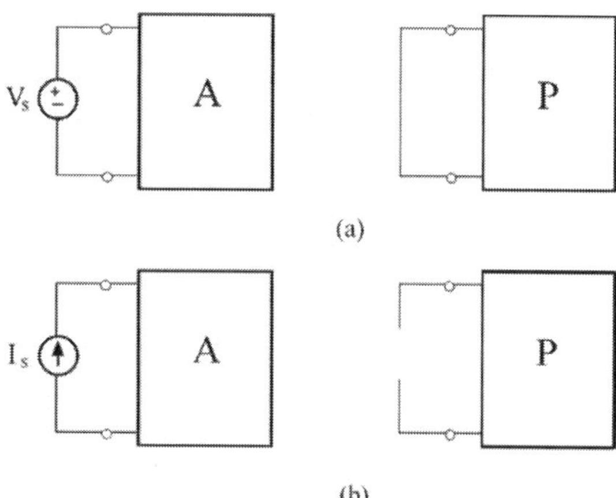

Figure 1.13 Circuit with an applied ideal voltage source (a) and an ideal current source (b).

1.6 ROOTS OF THE CHARACTERISTIC EQUATION AND DIFFERENT KINDS OF TRANSIENT RESPONSES

1.6.1 First-order characteristic equation

If an electrical circuit consists of only one energy-storing element (L or C) and a number of energy dissipation elements (R's), the characteristic equation will be of the first order:

For an RL circuit

$$Ls + R_{eq} = 0 \tag{1.26a}$$

and its root is

$$s = -\frac{R_{eq}}{L} = -\frac{1}{\tau}, \tag{1.26b}$$

where

$$\tau = \frac{L}{R_{eq}}$$

is a time constant.

For an RC circuit

$$\frac{1}{sC} + R_{eq} = 0 \tag{1.27a}$$

and its root is

$$s = -\frac{1}{R_{er}C} = -\frac{1}{\tau}, \tag{1.27b}$$

where $\tau = R_{eq}C$ is a time constant. In both cases the natural solution is

$$f_n(t) = Ae^{st}, \tag{1.28a}$$

or

$$f_n(t) = Ae^{-\frac{t}{\tau}}, \tag{1.28b}$$

which is a decreasing exponential, which approaches zero as the time increases without limit. However, as we have seen earlier (in Fig. 1.5), during the time interval of five times τ the difference between the exponential and zero is less than 1%, so that practically we may state that the duration of the transient response is about 5τ.

1.6.2 Second-order characteristic equation

If an electrical circuit consists of two energy-storing elements, then the characteristic equation will be of the second order. For an electrical circuit, which consists of an inductance, capacitance and several resistances this equation may look like equations 1.22, 1.25 or in a generalized form

$$s^2 + 2\alpha + \omega_d^2 = 0. \tag{1.29}$$

The coefficients in the above equation shall be introduced as follows: α as the exponential *damping coefficient* and ω_d as a *resonant frequency*. For a series RLC circuit $\alpha = R/2L$ and $\omega_d = \omega_0 = 1/\sqrt{LC}$. For a parallel RLC circuit $\alpha = 1/2RC$ and $\omega_d = \omega_0 1/\sqrt{LC}$, which is the same as in a series circuit. For more complicated circuits, as in Fig. 1.3, the above terms may look like $\alpha = \frac{1}{2}(R_{1,eq}/L + 1/R_{2,eq}C)$, which is actually combined from those coefficients for the series and parallel circuits and $\omega_d = \omega_0\xi$, where ξ is a distortion coefficient, which influences the resonant/oscillatory frequency.

The two roots of a second order (quadratic) equation 1.29 are given as

$$s_1 = -\alpha + \sqrt{\alpha^2 - \omega_d^2} \tag{1.30a}$$

$$s_2 = -\alpha - \sqrt{\alpha^2 - \omega_d^2}, \tag{1.30b}$$

and the natural response in this case is

$$f_n(t) = A_1 e^{s_1 t} + A_2 e^{s_2 t}. \tag{1.31}$$

Since each of these two exponentials is a solution of the given differential equation, it can be shown that the sum of the two solutions is also a solution (it can be shown, for example, by substituting equation 1.31 into the considered equation. The proof of it is left for the reader as an exercise.)

As is known from mathematics, the two roots of a quadratic equation can be one of three kinds:

1) negative real different, such as $|s_2| > |s_1|$, if $\alpha > \omega_d$;
2) negative real equal, such as $|s_2| = |s_1| = |s|$, if $\alpha = \omega_d$ and

3) complex conjugate, such as $s_{1,2} = -\alpha \pm j\omega_n$, if $\alpha < \omega_d$ and then $\omega_n = \sqrt{\omega_d^2 - \alpha^2}$ is the frequency of oscillation or **natural frequency** (see further on).

A detailed analysis of the natural response of all three cases will be given in the next chapter. Here, we will restrict ourselves to their short specification.

1) *Overdamping*. In this case, the natural response (equation 1.31) is given as the sum of two decreasing exponential forms, both of which approach zero as $t \to \infty$. However, since $|s_2| > |s_1|$, the term of s_2 has a more rapid rate of decrease so that the transients' time interval is defined by s_1 ($t_{tr} \approx 5(1/|s_1|)$). This response is shown in Fig. 1.14(a).

2) *Critical damping*. In this case, the natural response (equation 1.31) converts into the form

$$f(t) = (A_1 t + A_2)e^{-st}, \tag{1.32}$$

which is shown in Fig. 1.14(b).

3) *Underdamping*. In this case, the natural response becomes oscillatory, which may be imaged as a decaying alternating current (voltage)

$$f(t) = Be^{-\alpha t}\sin(\omega_n t + \beta), \tag{1.33}$$

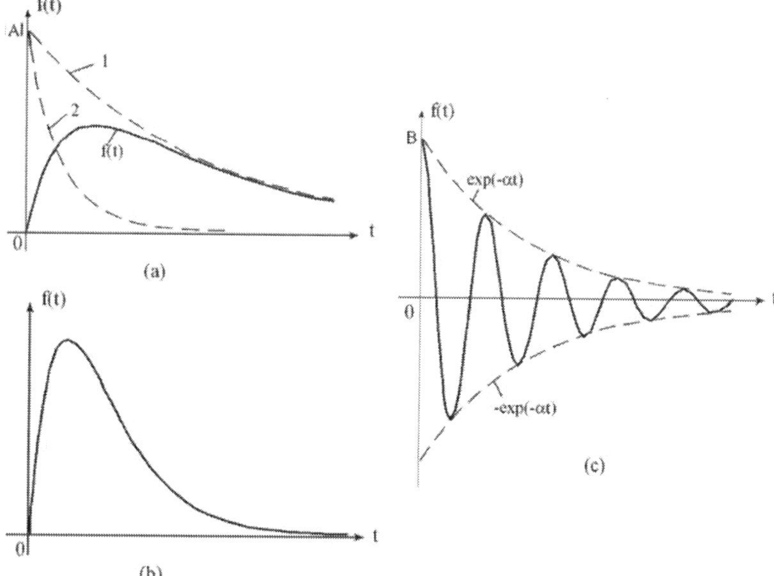

Figure 1.14 An overdamped response (a), a critical response (b) and an underdamped response (c).

which is shown in Fig. 1.14(c). Here term α is the rate of decay and ω_n is the angular frequency of the oscillations.

Now the critical damping may be interpreted as the boundary case between the overdamped and underdamped responses. It should be noted however that the critical damping is of a more theoretical than practical interest, since the exact satisfaction of the critical damping condition $\alpha = \omega_d$ in a circuit, which has a variety of parameters, is of very low probability. Therefore, the transient response in a second order circuit will always be of an exponential or oscillatory form. Let us now consider a numerical example.

Example 1.2

The circuit shown in Fig 1.15 represents an equivalent circuit of a one-phase transformer and has the following parameters: $L_1 = 0.06$ H, $L_2 = 0.02$ H, $M = 0.03$ H, $R_1 = 6\,\Omega$, $R_2 = 1\,\Omega$. If the transformer is loaded by an inductive load, whose parameters are $L_{ld} = 0.005$ H and $R_{ld} = 9\,\Omega$, a) determine the characteristic equation of a given circuit and b) find the roots and write the expression of a natural response.

Solution

Using mesh analysis, we may write a set of two algebraic equations (which represent two differential equations in operational form)

$$(R_1 + sL_1)i_1 - sM\,i_2 = 0$$

$$-sM\,i_1 + (R_2 + sL_2 + R_{ld} + sL_{ld})i_2 = 0.$$

The determinant of this set of two equations is

$$\det = \begin{vmatrix} R_1 + sL_1 & -sM \\ -sM & (R_2 + R_{ld}) + s(L_2 + L_{ld}) \end{vmatrix}$$

$$= (L_1 L_2' - M^2)s^2 + (R_1 L_2' + R_2' L_1)s + R_1 R_2',$$

where, to shorten the writing, we assigned $L_2' = L_2 + L_{ld}$ and $R_2' = R_2 + R_{ld}$.

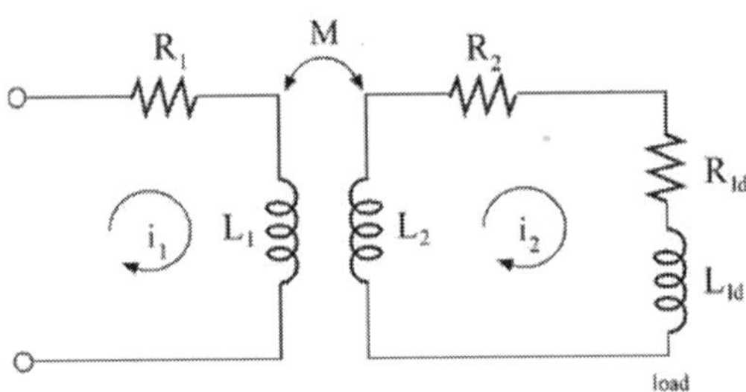

Figure 1.15 A given circuit for example 1.2.

Letting det $=0$, we obtain the characteristic equation in the form

$$s^2+\frac{R_1L_2'+R_2'L_1}{L_1L_2'-M^2}s+\frac{R_1R_2'}{L_1L_2'-M^2}=0.$$

Substituting the given values, we have

$$s^2+\frac{6\cdot 0.025+10\cdot 0.06}{0.06\cdot 0.025-0.03^2}s+\frac{6\cdot 10}{0.06\cdot 0.025-0.03^2}=0,$$

or

$$s^2+12.5\cdot 10^2 s+10\cdot 10^4=0.$$

The roots of this equation are:

$$s_1=\left[-\frac{12.5}{2}+\sqrt{\left(\frac{12.5}{2}\right)^2-10}\right]\cdot 10^2=-0.860\cdot 10^2\ \mathrm{s}^{-1}$$

$$s_2=\left[-\frac{12.5}{2}-\sqrt{\left(\frac{12.5}{2}\right)^2-10}\right]\cdot 10^2=-11.60\cdot 10^2\ \mathrm{s}^{-1},$$

which are two different negative real numbers. Therefore the natural response is:

$$i_n(t)=A_1e^{-86t}+A_2e^{-1160t},$$

which consists of two exponential functions and is of the overdamped kind.

It should be noted that in second order circuits, which contain two energy-storing elements of the same kind (two L-s, or two C-s), the transient response cannot be oscillatory and is always exponential overdamped. It is worthwhile to analyze the roots of the above characteristic equation. We may then obtain

$$s_{1,2}=\frac{1}{2(L_1L_2'-M^2)}\left[(R_1L_2'+R_2'L_1)\pm\sqrt{(R_1L_2'+R_2'L_1)^2-4(L_1L_2'-M^2)R_1R_2'}\right] \tag{1.34}$$

The expression under the square root can be simplified to the form: $(R_1L_2'+R_2'L_1)^2+4R_1R_2'M^2>0$, which is always positive, i.e., both roots are negative real numbers and the transient response of the overdamped kind. These results once again show that in a circuit, which contains energy-storing elements of the same kind, the transient response cannot be oscillatory.

In conclusion, it is important to pay attention to the fact that all the real roots of the characteristic equations, under study, were negative as well as the real part of the complex roots. This very important fact follows the physical reality that the natural response and transient-state cannot exist in infinite time. As we already know, the natural response takes place in the circuit free of sources and must vanish due to the energy losses in the resistances. Thus, natural responses, as exponential functions e^{st}, must be of a negative power $(s<0)$ to decay with time.

1.7 INDEPENDENT AND DEPENDENT INITIAL CONDITIONS

From now on, we will use the term "switching" for any change or interruption in an electrical circuit, planned as well as unplanned, i.e. different kinds of faults or other sudden changes in energy distribution.

1.7.1 Two switching rules (laws)

The principle of a gradual change of energy in any physical system, and specifically in an electrical circuit, means that the energy stored in magnetic and electric fields cannot change instantaneously. Since the magnetic energy is related to the magnetic flux and the current through the inductances (i.e., $w_m = \lambda i_L/2$), both of them **must not be allowed to change instantaneously**. In transient analysis it is common to assume that the switching action takes place at an instant of time that is defined as $t = 0$ (or $t = t_0$) and occurs instantaneously, i.e. in **zero time**, which means **ideal switching**. Henceforth, we shall indicate two instants: the instant just prior to the switching by the use of the symbol 0_-, i.e. $t = 0_-$, and the instant just after the switching by the use of the symbol 0_+, i.e. $t = 0_+$, (or just 0), as shown in Fig. 1.16. Using mathematical language, the value of the function $f(0_-)$, is the "limit from the left", as t approaches zero from the left and the value of the function $f(0_+)$ is the "limit from the right", as t approaches zero from the right.

Keeping the above comments in mind, we may now formulate two switching rules.

(a) First switching law (or first switching rule)

The first switching rule/law determines that **the current (magnetic flux) in an inductance just after switching $i_L(0_+)$ is equal to the current (flux) in the same inductance just prior to switching**

$$i_L(0_+) = i_L(0_-) \tag{1.35a}$$

$$\lambda(0_+) = \lambda(0_-). \tag{1.35b}$$

Equation 1.35a determines the initial value of the inductance current and enables

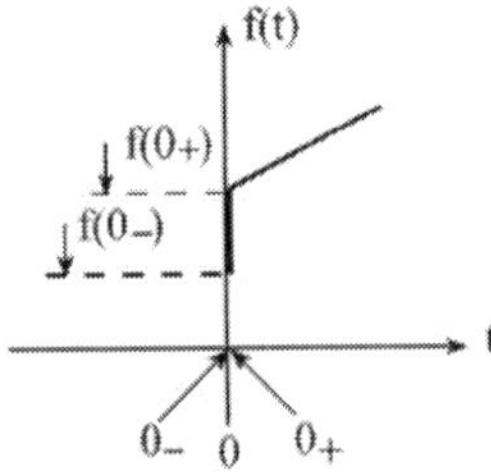

Figure 1.16 The instants: prior to switching (0_-), switching (0) and after switching (0_+).

us to find the integration constant of the natural response in circuits containing inductances. If the initial value of the inductance current is zero (zero initial conditions), the inductance at the instant $t = 0$ (and only at this instant) is equivalent to an open circuit (open switch) as shown in Fig. 1.17(a). If the initial value of the inductance current is not zero (non-zero initial conditions) the inductance is equivalent at the instant $t = 0$ (and only at this instant) to a current source whose value is the initial value of the inductance current $I_s = i_L(0)$, as shown in Fig. 1.17(b). Note that this equivalent, current source may represent the inductance in a most general way, i.e., also in the case of the zero initial current. In this case, the value of the current source is zero, and inner resistance is infinite (which means just an open circuit).

(b) Second switching law (or second switching rule)

The *second* switching rule/law determines that **the voltage (electric charge) in a capacitance just after switching $v_C(0_+)$ is equal to the voltage (electric charge) in the same capacitance just prior to switching**

$$v_C(0_+) = v_C(0_-) \tag{1.36a}$$

$$q(0_+) = q(0_-). \tag{1.36b}$$

Equation 1.36a determines the initial value of the capacitance voltage and

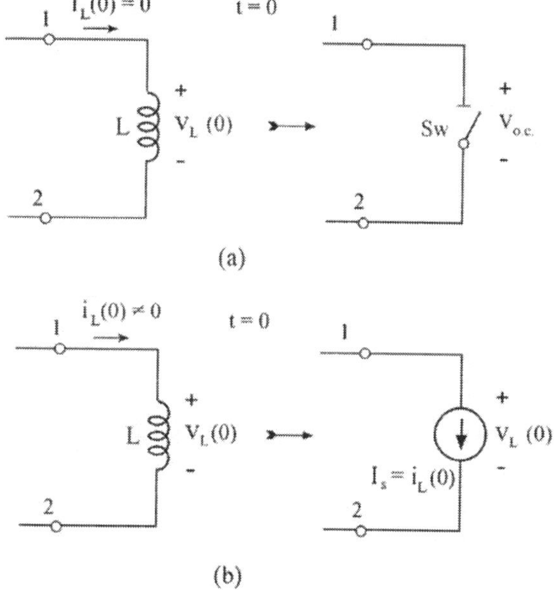

Figure 1.17 An equivalent circuit for an inductance at $t = 0$, with a zero initial current (a) and with current $i_L(0)$ (b).

enables us to find the integration constant of the natural response in circuits containing capacitances. If the initial value of the voltage across a capacitance is zero, zero initial conditions, the capacitance at the instant $t = 0$ (and only at this instant) is equivalent to a short-circuit (closed switch) as shown in Fig. 1.18(a). If the initial value of the capacitance voltage is not zero (non-zero initial conditions), the capacitance, at the instant $t = 0$ (and only at this instant), is equivalent to the voltage source whose value is the initial capacitance voltage $V_s = v_C(0)$, as shown in Fig. 1.18(b). Note that this equivalent, voltage source may represent the capacitance in a most general way, i.e., also in the case of the zero initial voltage. In this case, the value of the voltage source is zero, and inner resistance is zero (which means just a short-circuit).

In a similar way, as a current source may represent an inductance with a zero initial current, we can also use the voltage source as an equivalent of the capacitance with a zero initial voltage. Such a source will supply zero voltage, but its zero inner resistance will form a short-circuit.

If the initial conditions are zero, it means that the current through the inductances and the voltage across the capacitances will start from zero value, whereas if the initial conditions are non-zero, they will continue with the same values, which they possessed prior to switching.

The initial conditions, given by equations 1.35 and 1.36, i.e., the currents

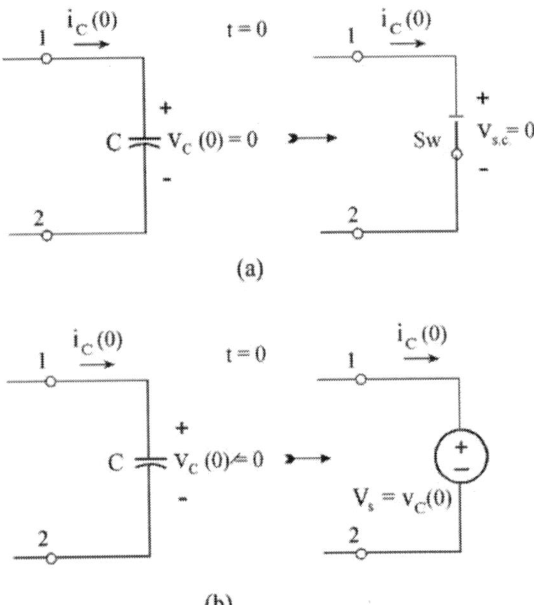

Figure 1.18 An equivalent circuit for a capacitance, at $t = 0$, with zero initial voltage (a) and with non-zero initial voltage $v_C(0)$ (b).

through the inductances and voltages across the capacitances, are called **independent initial conditions**, since they do not depend either on the circuit sources or on the status of the rest of the circuit elements. It does not matter how they had been set up, or what kind of switching or interruption took place in the circuit.

The rest of the quantities in the circuit, i.e., the currents and the voltages in the resistances, the voltages across the inductances and currents through the capacitances, can change abruptly and their values at the instant just after the switching ($t = 0_+$) are called **dependent initial conditions.** They depend on the independent initial conditions and on the status of the rest of the circuit elements. The determination of the dependent initial conditions is actually the most arduous part of the classical method. In the next sections, methods of determining the initial conditions will be introduced. We shall first, however, show how the independent initial conditions can be found.

1.7.2 Methods of finding independent initial conditions

For the determination of independent initial conditions the given circuit/network shall be inspected at its steady-state operation prior to the switching. Let us illustrate this procedure in the following examples.

Example 1.3

In the circuit in Fig. 1.19, a transient-state occurs due to the closing of the switch (*Sw*). Find the expressions of the independent initial values, if prior to the switching the circuit operated in a d.c. steady-state.

Solution

By inspection of the given circuit, we may easily determine 1) the current through the inductance and 2) the voltages across two capacitances.

1) Since the two capacitances in a d.c. steady-state are like an open switch the

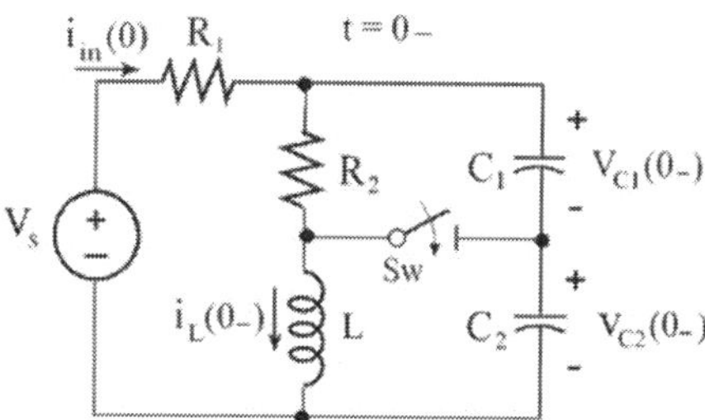

Figure 1.19 The circuit of example 1.3 at instant time $t = 0_-$.

inductance current is

$$i_L(0_-) = \frac{V_s}{R_1 + R_2}.$$

2) Since the voltage across the inductance in a d.c. steady-state is zero (the inductance provides a closed switch), the voltage across the capacitances is

$$v_C(0_-) = R_2 i_L(0_-).$$

This voltage is divided between two capacitors in inverse proportion to their values (which follows from the principle of their charge equality, i.e., $C_1 v_{C1} = C_2 v_{C2}$), which yields:

$$v_{C1}(0_-) = R_2 i_L(0_-) \frac{C_2}{C_1 + C_2}$$

$$v_{C2}(0_-) = R_2 i_L(0_-) \frac{C_1}{C_1 + C_2}.$$

Example 1.4

Find the independent initial conditions $i_L(0_-)$ and $v_C(0_-)$ in the circuit shown in Fig. 1.20, if prior to opening the switch, the circuit was under a d.c. steady-state operation.

Solution

1) First, we find the current i_4 with the current division formula (no current is flowing through the capacitance branch)

$$i_4 = I_s \frac{R_5}{R_5 + R_4 + R_3//R_1} = I_s \frac{R_5(R_1 + R_3)}{R_1 R_3 + R_1 R_4 + R_1 R_5 + R_3 R_4 + R_3 R_5}.$$

Using once again the current division formula, we obtain the current through the inductance

$$i_L(0_-) = i_4 \frac{R_3}{R_3 + R_1} = i_s \frac{R_3 R_5}{R_1 R_3 + R_1 R_4 + R_1 R_5 + R_3 R_4 + R_3 R_5}.$$

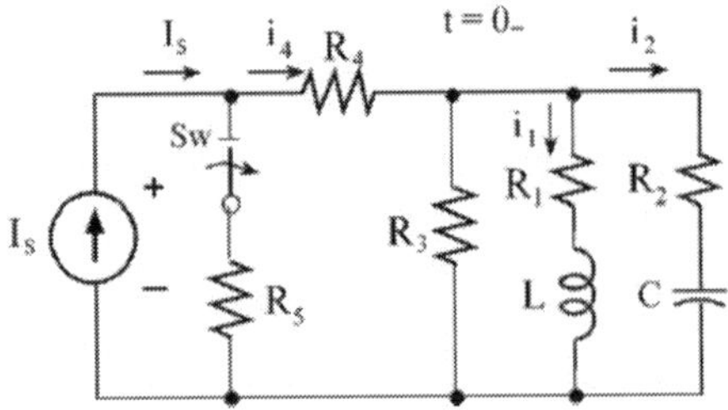

Figure 1.20 The circuit prior to the switching $t = 0_-$ of example 1.4.

2) The capacitance voltage can now be found as the voltage drop in resistance R_1

$$v_C(0_-) = R_1 i_L(0_-).$$

The examples given above show that in order to determine the independent initial conditions, i.e., the initial values of inductance currents and/or capacitance voltages, we must consider the circuit under study prior to the switching, i.e. at instant $t = 0_-$. It is usual to suppose that the previous switching took place a long time ago so that the transient response has vanished. We may apply all known methods for the analysis of circuits in their steady-state operation. Our goal is to choose the most appropriate method based on our experience in order to obtain the quickest answer for the quantities we are looking for.

1.7.3 Methods of finding dependent initial conditions

As already mentioned the currents and voltages in resistances, the voltages across inductances and the currents through capacitances can change abruptly at the instant of switching. Therefore, the initial values of these quantities should be found in the circuit just after switching, i.e., at instant $t = 0_+$. Their new values will depend on the new operational conditions of the circuit, which have been generated after switching, as well as on the values of the currents in the inductances and voltages of the capacitances. For this reason we will call them **dependent initial conditions**.

As we have already observed, the natural response in the circuit of the second order is, for instance, of form equation 1.31. Therefore, two arbitrary constants A_1 and A_2, **called integration constants**, have to be determined to satisfy the two initial conditions. One is the initial value of the function and the other one, as we know from mathematics, is the initial value of its first derivative. Thus, for circuits of the second order or higher the initial values of derivatives at $t = 0_+$ must also be found. We also consider the initial values of these derivatives as **dependent initial conditions**.

In order to find the dependent initial conditions we must consider the analyzed circuit, which has arisen after switching and in which all the inductances and capacitances are replaced by current and voltage sources (or, with zero initial conditions, by an open and/or short-circuit). Note that this circuit fits only at the instant $t = 0_+$. For finding the desirable quantities, we may use all the known methods of steady-state analysis. Let us introduce this technique by considering the following examples.

Example 1.5

Consider once again the circuit in Fig. 1.20. We now however need to find the initial value of current $i_2(0_+)$, which flows through the capacitance and therefore can be changed instantaneously.

Solution

We start the solution by drawing the equivalent circuit for instant $t = 0_+$, i.e. just after switching, Fig. 1.21. The inductance and capacitance in this circuit are replaced by the current and voltage sources, whose values have been found in Example 1.4 and are assigned as I_{LO} and V_{CO}.

The achieved circuit has two nodes and the most appropriate method for its solution is node analysis. Thus,

$$-I_s + G_3 V_{ab} + I_{LO} + i_2(0) = 0,$$

where $G_3 = 1/R_3$. Substituting $V_{ab} = V_{CO} + R_2 i_2(0)$ for V_{ab} we may obtain

$$i_2(0)(1 + G_3 R_2) = I_s - I_{LO} - G_3 V_{CO},$$

or

$$i_2(0) = \frac{I_s - I_{LO} - G_3 V_{CO}}{1 + G_3 R_2}.$$

Example 1.6

Let us say that we are interested in finding the initial value of the input current in the circuit of Example 1.3, shown in Figure 1.19.

Solution

Since the current we are looking for is a current in a resistance, which can change abruptly, we shall consider the circuit at instant $t = 0_+$. This circuit is shown in Fig. 1.22 where the inductance is replaced by a current source and the capacitances are replaced by voltage sources.

The quickest way to find $i_{in}(0_+)$ is by using the superposition principle. For this purpose, we shall consider two circuits: in the first one only the voltage sources are in action (circuit (b) in Fig. 1.22) and in the other one only the current source is in action (circuit (c) in Fig. 1.22). By inspection of the first circuit and by applying Kirchhoff's voltage law to the outer loop, we have

$$i'_{in}(0_+) = \frac{V_s - V_{C1} - V_{C2}}{R_1}.$$

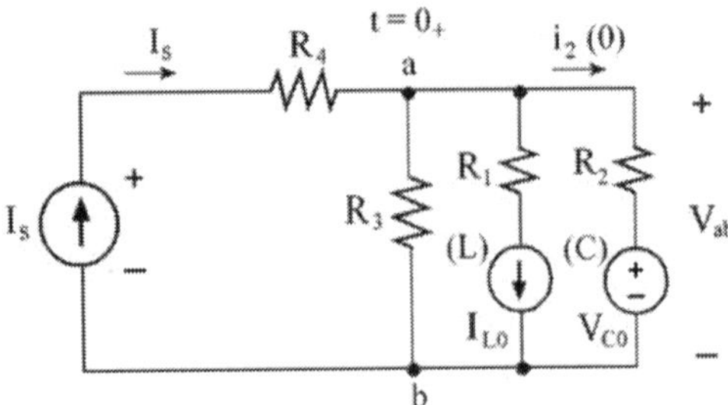

Figure 1.21 An equivalent circuit for Example 1.5.

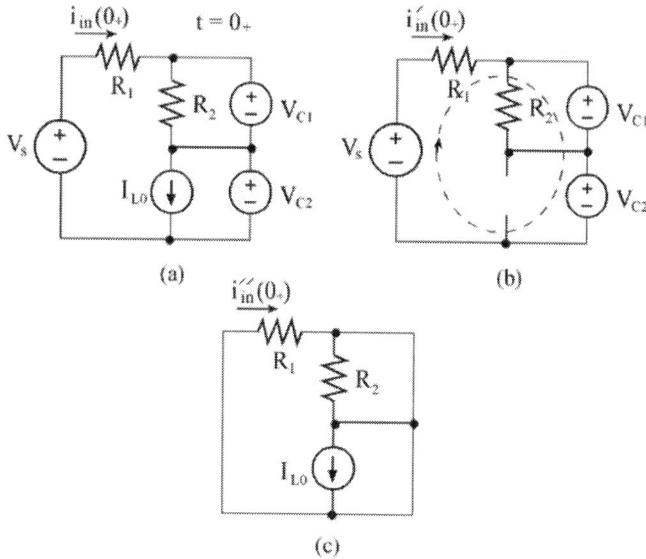

Figure 1.22 The circuit for finding $i_{in}(0_+)$ (a), the subcircuit with voltage sources (b) and the subcircuit with a current source (c).

By inspection of the second circuit in which the current source is short-circuited, we have

$$i''_{in}(0_+) = 0.$$

Therefore, finally

$$i_{in}(0_+) = i'_{in}(0_+) = \frac{V_s - V_{C1} - V_{C2}}{R_1}.$$

Example 1.7

As a numerical example, let us consider the circuit in Fig. 1.23. Suppose that we wish to find the initial value of the output voltage, just after switch *Sw* instantaneously changes its position from "1" to "2". The circuit parameters are: $L = 0.1$ H, $C = 0.1$ mF, $R_1 = 10\,\Omega$, $R_2 = 20\,\Omega$, $R_{ld} = 100\,\Omega$, $V_{s1} = 110$ V and $V_{s2} = 60$ V.

Solution

In order to answer this question, we must first find the independent initial conditions, i.e., $i_L(0_+)$ and $v_C(0_+)$. By inspection of the circuit for instant $t = 0_-$, Fig. 1.23(a), we have

$$i_L(0_-) = \frac{V_{s1}}{R_1 + R_{ld}} = \frac{110}{100 + 10} = 1 \text{ A},$$

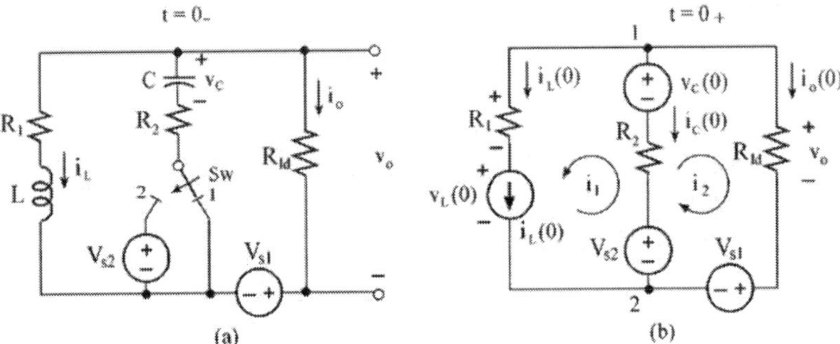

Figure 1.23 A given circuit for Example 1.7(a) and its equivalent at $t = 0_+$(b).

and

$$v_C(0_-) = V_{s1}\frac{R_1}{R_1 + R_{ld}} = 110\frac{10}{100+10} = 10\ \text{V}.$$

With two switching rules we have

$$i_L(0_+) = i_L(0_-) = 1\ \text{A}$$

$$v_C(0_+) = v_C(0_-) = 10\ \text{V},$$

and we can now draw the equivalent circuit for instant $t = 0_+$, Fig. 1.23(b). By inspection, using KCL (Kirchhoff's current law), we have

$$R_{ld}i_2 + R_2(i_2 + i_1) = -V_{s1} + V_{s2} + v_C(0). \tag{1.37}$$

Keeping in mind that $i_2 = i_o$ and $i_1 = i_L(0)$, we obtain

$$i_2(0) = \frac{-V_{s1} + V_{s2} + v_C(0) - R_2 i_L(0)}{R_2 + R_{ld}} = \frac{-110 + 60 + 10 - 20\cdot 1}{20 + 100} = -0.5\ \text{A}.$$

Thus the initial value of the output current is -0.5 A. Note that, prior to switching, the value of the output current was -1 A, therefore, with switching the current drops to half of its previous value.

The circuit of this example is of the second order and, as earlier mentioned, its natural response consists of two unknown constants of integration. Therefore, we shall also find the derivative of the output current at instant $t = 0_+$. By differentiating equation 1.37 with respect to time, and taking into consideration that V_{s1} and V_{s2} are constant, we have

$$(R_2 + R_{ld})\frac{di_o}{dt} + R_2\frac{di_L}{dt} = \frac{dv_C}{dt},$$

and, since $\dfrac{di_L}{dt} = \dfrac{1}{L}v_L$ and $\dfrac{dv_C}{dt} = \dfrac{1}{C}i_C$,

$$\left.\frac{di_o}{dt}\right|_{t=0} = \frac{1}{R_2 + R_{ld}}\left[\frac{1}{C}i_C(0) - \frac{R_2}{L}v_L(0)\right].$$

By inspection of the circuit in Fig. 1.23(b) once again, we may find

$$v_L(0) = V_{s1} + R_{ld} i_o(0) - R_1 i_L(0) = 110 + 100(-0.5) - 10 \cdot 1 = 40\,\mathrm{V}.$$

$$i_C(0) = -i_o(0) - i_L(0) = 0.5 - 1 = -0.5\,\mathrm{A}.$$

Thus,

$$\left.\frac{di_o}{dt}\right|_{t=0} = \frac{1}{20 + 100}\left(\frac{-0.5}{0.1 \cdot 10^{-3}} - \frac{10}{0.1} 40\right) = -75\,\mathrm{A\,s^{-1}}.$$

1.7.4 Generalized initial conditions

Our study of initial conditions would not be complete without mention of the so-called *incorrect initial conditions*, i.e. by which it looks as though the two switching laws are disproved.

(a) Circuits containing capacitances

As an example of such a "disproval", consider the circuit in Fig. 1.24(a). In this circuit, the voltage across the capacitance prior to switching is $v_C(0_-) = 0$ and after switching it should be $v_C(0_+) = V_s$, because of the voltage source. Thus,

$$v_C(0_+) \neq v_C(0_-),$$

and the second switching law is disproved.

This paradox can be explained by the fact that the circuit in Fig 1.24(a) is not a physical reality, but only a mathematical model, since it is built of two ideal elements: an ideal voltage source and an ideal capacitance. However, every electrical element in practice has some value of resistance, and generally speaking some value of inductance (but this inductance is very small and in our future discussion it will be neglected). Because, in a real switch, the switching process takes some time (even very small), during which the spark appears, the latter is also usually approximated by some value of resistance. By taking into consideration just the resistances of the connecting wires and/or the inner resistance of the source or the resistance of the spark, connected in series, and a resistance, which represents the capacitor insulation, connected in parallel, we obtain the

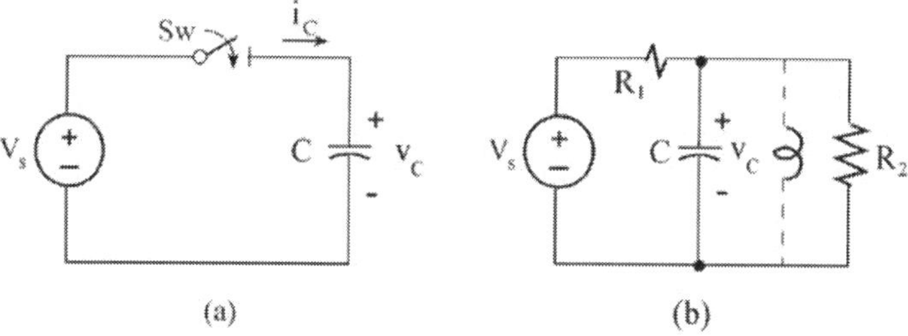

Figure 1.24 An incorrect circuit model of a source and a capacitor (a) and its corrected version (b).

circuit shown in Fig. 1.24(b). In this circuit, the second switching law is correct and we may write

$$v_C(0_+) = v_C(0_-).$$

Now, at the instant of switching, i.e., at $t = 0$, the magnitude of the voltage drop across this resistance will be as large as the source value. As a result the current of the first moment will be very large, however not unlimited, like it is supposed to be in Fig. 1.24(a). In order to illustrate the transient behavior in the circuit discussed, let us turn to a numerical example. Suppose that a 1.0 nF condenser is connected to a 100 V source and let the resistance of the connecting wires be about one hundredth of an ohm. In such a case, the "spike" of the current will be $I_\delta = 100/0.01 = 10{,}000$ A, which is a very large current in a 100 V source circuit (but it is not infinite). This current is able to charge the above condenser during the time period of about 10^{-11} s, since the required charge is $q = CV = 10^{-9} \cdot 10^2 = 10^{-7}$ C and $\Delta t \cong \Delta q/\Delta i = 10^{-7}/10^4 = 10^{-11}$ s. This period of time is actually equal to the time constant of the series RC circuit, $\tau = RC = 10^{-2} \cdot 10^{-9} = 10^{-11}$ s.

From another point of view, the amount of the charge, which is transferred by an exponentially decayed current, is equal to the product of its initial value, I_0 and the time constant. Indeed, from Fig. 1.25, we have

$$q = \int i\, dt = I_0 \int_0^\infty e^{-t/\tau} dt = I_0(-\tau)e^{-t/\tau}\Big|_0^\infty = I_0\tau, \tag{1.38}$$

i.e., $q = 10{,}000 \cdot 10^{-11} = 10^{-7}$ C, as estimated earlier. This result (equation 1.38) justifies using an impulse function δ (see further on) for representing very large (approaching infinity) magnitudes applying very short (approaching zero) time intervals, whereas their product stays finite, as shown in Fig. 1.25.

Note that the second resistance R_2 is very large (hundreds of mega ohms), so that the current through this resistance, being very small (less than a tenth of a microampere), can be neglected.

In conclusion, when a capacitance is connected to a voltage source, a very large current, tens of kiloamperes, charges the capacitance during a vanishing time interval, so that we may say that the capacitance voltage changes from zero to its final value, practically immediately. However, of course, none of the

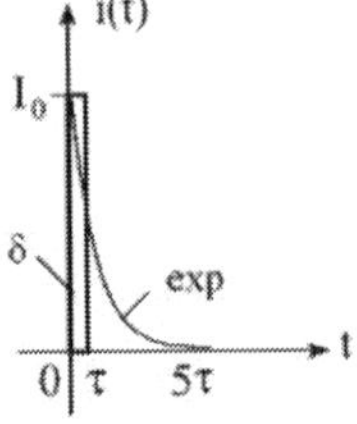

Figure 1.25 A large and fast decaying exponent and an equivalent impulse.

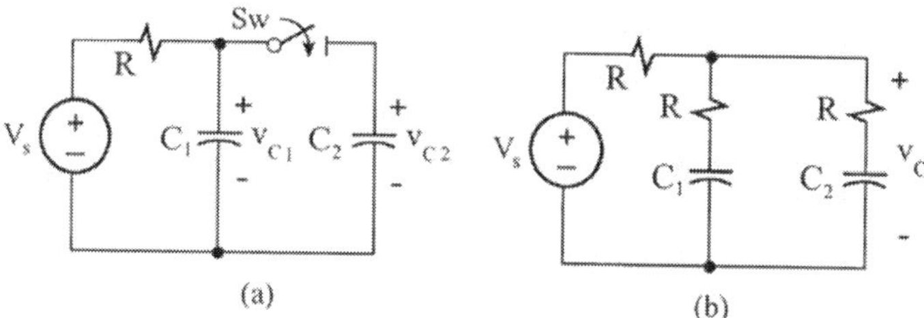

Figure 1.26 A circuit in which the second switching law is "disproved": prior switching (a) and after switching (b).

physical laws, neither the switching law nor the law of energy conservation, has been disproved.

As a second example, let us consider the circuit in Fig. 1.26(a). At first glance, applying the second switching law, we have

$$v_{C1}(0_+) = v_{C1}(0_-) = V_s$$
$$v_{C2}(0_+) = v_{C2}(0_-) = 0. \tag{1.39}$$

But after switching, at $t = 0$, the capacitances are connected in parallel, Fig. 1.26(b), and it is obvious that

$$v_{C1}(0_+) = v_{C2}(0_-) \tag{1.40}$$

which is in contrast to equation 1.39.

To solve this problem we shall divide it into two stages. In the first one, the second capacitance is charged practically immediately in the same way that was explained in the previous example. During this process, part of the first capacitance charge is transferred by a current impulse to the second capacitance, so that the entire charge is distributed between the two capacitances in reciprocal proportion to their values. The common voltage of these two capacitances, connected in parallel, after the switching at instant $t = 0$, is reduced to a new value lower than the applied voltage V_s.

In the second stage of the transient process in this circuit, the two capacitances will be charged up so that the voltage across the two of them will increase up to the applied voltage V_s. To solve this second stage problem we have to know the new initial voltage in equation 1.40. We shall find it in accordance with equation 1.36b which, as was mentioned earlier, expresses the physical principal of continuous electrical charges, i.e. the latter cannot change instantaneously. This requirement is general but even more stringent than the requirement of continuous voltages, and therefore is called the **generalized second switching law**. Thus,

$$q_\Sigma(0_+) = q_\Sigma(0_-) = C_1 v_{C1}(0_-). \tag{1.41}$$

This law states that: **the total amount of charge in the circuit cannot change instantaneously and its value prior to switching is equal to its value just after the switching, i.e., the charge always changes gradually.**

Since the new equivalent capacitance after switching is $C_{eq} = C_1 + C_2$, we may write

$$q_\Sigma(0_+) = (C_1 + C_2)v_{C1}(0_+) = C_1 v_{C1}(0_-).$$

Since, in this example, $v_{C1}(0_-) = V_s$, we finally have

$$v_{C1}(0_+) = \frac{C_1}{C_1 + C_2} v_{C1}(0_-) = \frac{C_1}{C_1 + C_2} V_s. \tag{1.42}$$

With this initial condition, the integration constant can easily be found.

It is interesting to note that by taking into consideration the small resistances (wires, sparks, etc.) the circuit becomes of second order and its characteristic equations will have two roots (different real negative numbers). One of them will be very small, determining the first stage of transients, and the second one, relatively large, will determine the second stage.

Let us now check the energy relations in this scheme, Fig. 1.26, before and after switching. The energy stored in the electric field of the first capacitance (prior to switching) is $w_e(0_-) = \frac{1}{2} C_1 V_{C1}^2(0_-) = \frac{1}{2} C_1 V_s^2$ and the energy stored in the electric field of both capacitances (after switching) is $w_e(0_+) = \frac{1}{2}(C_1 + C_2)v_C^2(0_+)$. Thus, the energy "lost" is

$$\Delta w_e = w_e(0_-) - w_e(0_+) = \frac{C_1 V_s^2}{2} - \frac{C_1 + C_2}{2}\left(\frac{C_1 V_s}{C_1 + C_2}\right)^2 = \frac{C_1 C_2 V_s^2}{2(C_1 + C_2)}. \tag{1.43}$$

This energy actually dissipates in the above-discussed resistances.

When two capacitances, connected in series, switch to the voltage source, as shown in Fig. 1.27(a), the transients will also consist of two stages. In the first stage, the current impulse will charge two capacitances equally to the same charge

$$q(0_+) = V_s \frac{C_1 C_2}{C_1 + C_2}, \tag{1.44}$$

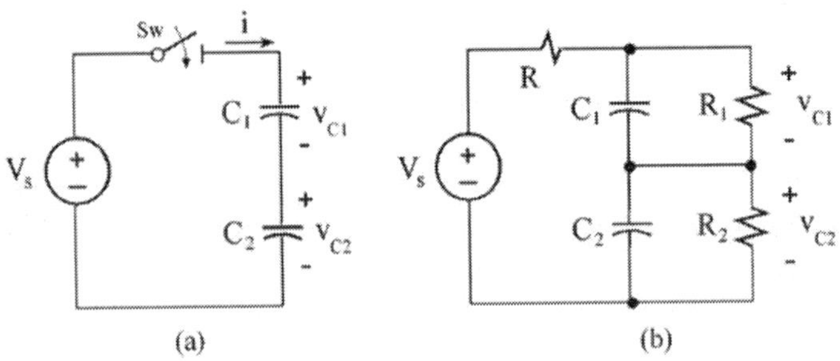

Figure 1.27 Two capacitances in series are connected to the voltage source: incorrect (a) and correct (b) circuits.

but to different voltages, in reciprocal proportion to their values:

$$v_{C1}(0_+) = V_s \frac{C_2}{C_1 + C_2}, \quad v_{C2}(0_+) = V_s \frac{C_1}{C_1 + C_2}. \tag{1.45}$$

However, in accordance with the correct equivalent circuit in Fig. 1.27(b), the final steady-state voltages (at $t \to \infty$) across two capacitances must be determined by the voltage division in proportion to their resistances:

$$v_{C1}(\infty) = V_s \frac{R_1}{R_1 + R_2}, \quad v_{C2}(\infty) = V_s \frac{R_2}{R_1 + R_2}. \tag{1.46}$$

This change in voltages, from equation 1.45 to equation 1.46, takes place during the second stage with the time constant $\tau = (C_1 + C_2)/(G_1 + G_2)$ (proof of this expression is left to the reader as an exercise).

Finally it should be noted that the very fast charging of the capacitances by the flow of *very large currents* (*current impulses*) results in relatively *small energy dissipation*, so that usually no damage is caused to the electrical equipment. Indeed, with the numerical data of our first example, we may calculate

$$w_d = RI_\delta^2 \int_0^\infty e^{-\frac{2t}{\tau}} = RI_\delta^2 \frac{\tau}{2} = 10^{-2}(10^4)^2 \cdot 10^{-11} \cdot 0.5 = 0.5 \cdot 10^{-5}\ \text{J},$$

which is negligibly small. Checking the law of energy conservation, we may find that the energy being delivered by the source is

$$w_s = \int_0^\infty V_s i\, dt = V_s \int_0^\infty C \frac{dv_C}{dt} dt = CV_s \int_0^{V_s} dv_C = CV_s^2,$$

and the energy being stored into the capacitances is $w_e = \frac{1}{2}CV_s^2$, i.e., half of the energy delivered by the source is dissipated in the resistances. Calculating this energy yields

$$\Delta w_s = \frac{CV_s^2}{2} = \frac{10^{-9} \cdot 10^4}{2} = 0.5 \cdot 10^{-5}\ \text{J},$$

as was previously calculated.

(b) Circuits containing inductances

We shall analyze the circuits containing inductances keeping in mind that such circuits are dual to those containing capacitances and using the results, which have been obtained in our previous discussion.

Consider the circuit shown in Fig. 1.28 in which the current prior to switching is $i_L(0_-) = I_0$ and after switching is supposed to be $i_L(0_+) = 0$, so that the first switching law is disproved

$$i_L(0_+) \neq i_L(0_-).$$

However, by taking into consideration the small parameters G, R_L, and C, we

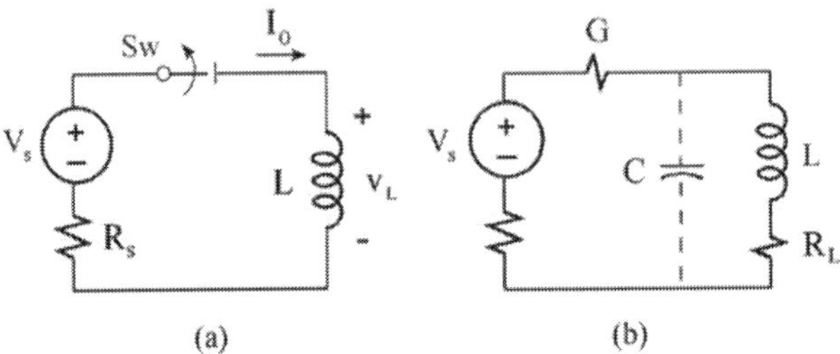

Figure 1.28 An incorrect circuit containing a disconnected inductance (a) and its improved equivalent (b).

may obtain the correct circuit, shown in Fig 1.28(b), in which all the physical laws are proven.

In this circuit, the open switch is replaced by a very small conductance G (very big resistance), so that we can now write $i_L(0_+) = i_L(0_-)$, but because of the vanishingly small time constant $\tau = GL$, the current decays almost instantaneously.

From another point of view the almost abrupt change of inductance current results in a very large voltage induced in inductance, $v_L = L(di/dt)$, which is applied practically all across the switch, and causes an arc, which appears between the opening contacts of the switch. Let us estimate the magnitude of such an overload across the coil in Fig. 1.28(a), having 0.1 H and 20 Ω;, which disconnects almost instantaneously from the voltage source, and the current through the coil prior to switching was 5 A. Assume that the time of switching is $\Delta t = 10$ μs (note that this time, during which the current changes from the initial value to zero, can be achieved if the switch is replaced by a resistor of at least 50 kΩ, as shown in Fig. 1.28(b)), then the overvoltage will be $V_{\max} \cong L(\Delta i/\Delta t) = 0.1 \cdot 5 \cdot 10^5 = 50$ kV.

Such a high voltage usually causes an arc, which appears between the opening contacts of the switch. This transient phenomenon is of great practical interest since in power system networks the load is mostly of the inductance kind and any disconnection of the load and/or short-circuited branch results in overvoltages and arcs. However, the capacitances associated with all the electric parts of power systems affect its transient behavior and usually result in reducing the overvoltages. (We will analyze this phenomenon in more detail also taking into consideration the capacitances, see Chapter 2).

Consider next the circuit in Fig. 1.29, which is dual to the circuit in Fig. 1.26. (It should be noted that the duality between the two circuits above, Figs 1.28 and 1.29, and the corresponding capacitance circuits, in Figs 1.24 and 1.26, is not full. For full duality the voltage sources must be replaced by current sources. However, the quantities, the formulas, and the transient behavior are similar.) In this circuit, prior to switching $i_{L1}(0_-) = I_0$ and $i_{L2}(0_-) = 0$. Applying the first

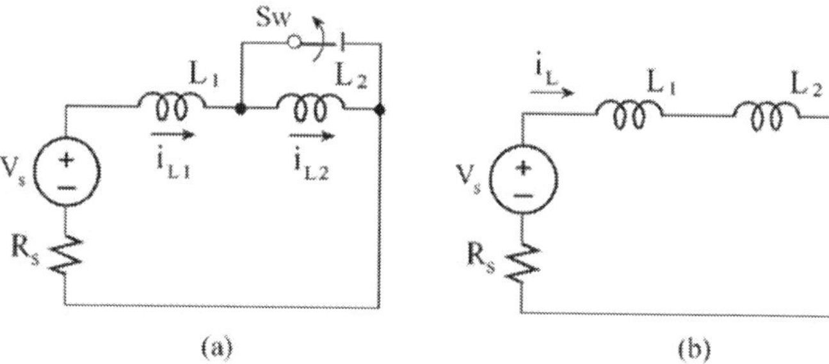

Figure 1.29 A circuit containing two inductances, in which the first switching law is "disproved": prior to switching (a) and after switching (b).

switching law we shall write

$$\begin{aligned} i_{L1}(0_+) &= i_{L1}(0_-) = I_0 \\ i_{L2}(0_+) &= i_{L2}(0_-) = 0. \end{aligned} \tag{1.47}$$

After switching the two inductances are connected in series, Fig 1.29(b), therefore

$$i_{L1}(0_+) = i_{L2}(0_-), \tag{1.48}$$

which is obviously contrary to equation 1.47. However, we may consider the transient response of this circuit as similar to that in capacitance and conclude that it is composed of two stages. In the first stage, the currents change almost instantaneously, in a very short period of time $\Delta t \to 0$, so that voltage impulses appear across the inductances. In the second stage, the current in both inductances changes gradually from its initial value up to its steady-state value. In order to find the initial value of the common current flowing through both inductances connected in series (just after switching and after accomplishing the first stage) we may apply the so-called **first generalized switching law** (equation 1.35b). This law states that: **the total flux linkage in the circuit cannot change instantaneously and its value prior to switching is equal to its value just after switching, i.e. the flux linkage always changes gradually.**

If an electrical circuit contains only one inductance element, then

$$Li_L(0_+) = Li_L(0_-) \quad \text{or} \quad i_L(0_-) = i_L(0_+),$$

and the first switching law regarding flux linkages (equation 1.35b) is reduced to a particular case with regard to the currents. For this reason the first switching law, regarding flux linkages, is more general.

Applying the first generalized law to the circuit in Fig. 1.29, we have

$$L_1 i_{L1}(0_-) + L_2 i_{L2}(0_-) = L_1 i_{L1}(0_+) + L_2 i_{L2}(0_+), \tag{1.49}$$

or since $i_{L1}(0_+) = i_{L2}(0_+) = i_L(0_+)$ we have

$$i_L(0_+) = \frac{L_1 i_{L1}(0_-) + L_2 i_{L2}(0_-)}{L_1 + L_2}.$$

Substituting $i_{L2}(0_-) = 0$ and $i_{L1}(0_-) = I_0$ the above expression becomes

$$i_L(0_+) = \frac{L_1}{L_1 + L_2} I_0. \tag{1.50}$$

This equation enables us to determine the initial condition of the inductance current in the second stage of a transient response.

The energy stored in the magnetic field of two inductances prior to switching is

$$w_m(0_-) = \frac{L_1 i_{L1}^2(0_-)}{2} + \frac{L_2 i_{L2}^2(0_-)}{2}, \tag{1.50a}$$

and after switching

$$w_m(0_+) = \frac{(L_1 + L_2) i_L^2(0_+)}{2}. \tag{1.50b}$$

Then the amount of energy dissipated in the first stage of the transients, i.e., in circuit resistances and in the arc, with equations 1.50a and 1.50b will be

$$\Delta w_m = w_m(0_-) - w_m(0_+) = \frac{1}{2} \frac{L_1 L_2}{L_1 + L_2} [i_{L1}(0_-) - i_{L2}(0_-)]^2. \tag{1.51}$$

(Developing this formula is left to the reader as an exercise.) For the circuit under consideration the above equation 1.51 becomes

$$\Delta w_m = \frac{1}{2} \frac{L_1 L_2}{L_1 + L_2} I_0^2. \tag{1.52}$$

It is interesting to note that this expression is similar to formula 1.43 for a capacitance circuit. Let us now consider a numerical example.

Example 1.8

In the circuit in Fig. 1.30(a) the switch opens at instant $t = 0$. Find the initial current $i(0_+)$ in the second stage of the transient response and the energy

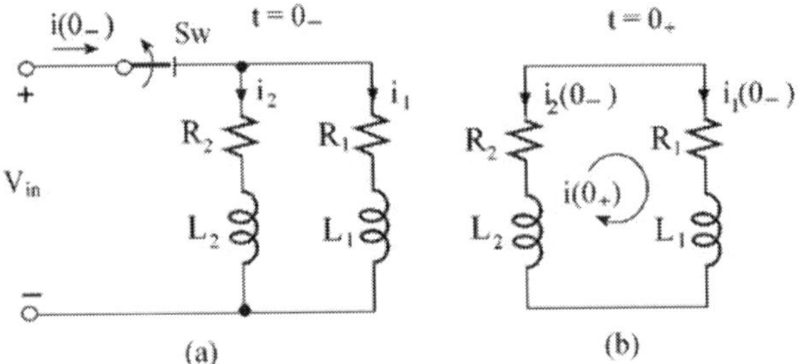

Figure 1.30 A circuit for Example 1.8: prior to switching (a) and after switching (b).

dissipated in the first stage if the parameters are: $R_1 = 50\,\Omega$, $R_2 = 40\,\Omega$, $L_1 = 160$ mH, $L_2 = 40$ mH, $V_{in} = 200$ V.

Solution

The values of the two currents in circuit (a) are

$$i_{L1}(0_-) = \frac{V_{in}}{R_1} = \frac{200}{50} = 4\text{ A}$$

and

$$i_{L2}(0_-) = \frac{V_{in}}{R_2} = \frac{200}{40} = 5\text{ A}.$$

Thus, the initial value of the current in circuit (b), in accordance with equation 1.49, is

$$i_L(0_+) = \frac{L_1 i_{L1}(0_-) - L_2 i_{L2}(0_-)}{L_1 + L_2} = \frac{160\cdot 4 - 40\cdot 5}{160 + 40} = 2.2\text{ A}.$$

Note that for the calculation of the initial current $i(0_+)$ in circuit (b), we took into consideration that the current $i_{L2}(0_-)$ is negative since its direction is opposite to the direction of $i(0_+)$, which has been chosen as the positive direction. The dissipation of energy, in accordance with equation 1.51, is

$$\Delta w_m = \frac{L_1 L_2 [i_{L1}(0_-) - i_{L2}(0_-)]^2}{2(L_1 + L_2)} = \frac{160\cdot 40\cdot 10^{-3}(4+5)^2}{2(160+40)} \cong 1.3\text{ J}.$$

As a final example, consider the circuit in Fig. 1.31. This circuit of two inductive branches in parallel to a current source is a complete dual to the circuit in Fig. 1.27, in which two capacitances in series are connected to a voltage source.

Prior to switching the inductances are short-circuited, so that both currents $i_{L1}(0_-)$ and $i_{L2}(0_-)$ are equal to zero. The current of the current source flows through the switch. (In the dual circuit, the voltages across the capacitances prior to switching are also zero.) At the instant of switching the currents through

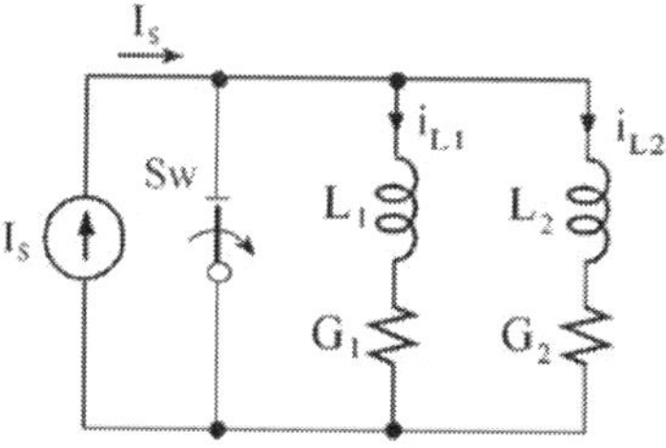

Figure 1.31 A circuit of two parallel inductances and a current source, which is a complete dual to the circuit with two series capacitances and a voltage source.

the inductances change almost instantaneously, so that their sum should be I_s. This abrupt change of currents results in a voltage impulse across the opening switch. Since this voltage is much larger than the voltage drop on the resistances, we may neglect these drops and assume that the inductances are connected in parallel. As we know, the current is divided between two parallel inductances in inverse proportion to the value of the inductances. Thus,

$$i_{L1}(0_+) = I_s \frac{L_2}{L_1 + L_2} \quad \text{and} \quad i_{L2}(0_+) = I_s \frac{L_1}{L_1 + L_2}. \tag{1.53}$$

These expressions enable us to determine the initial condition in the second stage of the transient response. The steady-state values of the inductance currents will be directly proportional to the conductances G_1 and G_2. Hence, the induced voltages across the inductances will be zero (the inductances are now short-circuited) and the resistive elements are in parallel (note that in the capacitance circuit of Fig. 1.27 the voltages across the capacitances in steady state are also directly proportional, but to the resistances, which are parallel to the capacitances). Thus,

$$i_{L1}(\infty) = I_s \frac{G_1}{G_1 + G_2} \quad \text{and} \quad i_{L2}(\infty) = I_s \frac{G_2}{G_1 + G_2}.$$

Knowing the initial and final values, the complete response can be easily obtained (see the next chapter).

1.8 METHODS OF FINDING INTEGRATION CONSTANTS

From our previous study, we know that the natural response is formed from a sum of exponential functions:

$$f_n(t) = A_1 e^{s_1 t} + A_2 e^{s_2 t} + \cdots = \sum_1^n A_k e^{s_k t}. \tag{1.54}$$

where the number of exponents is equal to the number of roots of a characteristic equation. In order to determine the integration constants $A_1, A_2, \ldots A_n$ it is necessary to formulate n equations, which must obey the instant of switching, $t = 0$ (or $t = t_0$). By differentiation of the above expression $(n-1)$ times, we may obtain

$$\begin{aligned} A_1 + A_2 + \cdots &= \sum_1^n A_k = f_n(0) \\ s_1 A_1 + s_2 A_2 + \cdots &= \sum_1^n s_k A_k = f'_n(0) \\ &\cdots \\ s_1^{n-1} A_1 + s_2^{n-1} A_2 + \cdots &= \sum_1^n s_k^{n-1} A_k = f_k^{(n-1)}(0), \end{aligned} \tag{1.55}$$

where it has been taken into consideration that

$$\begin{aligned} &A_k e^{s_k t}\big|_{t=0} = A_k \\ &\frac{d}{dt} A_k e^{s_k t}\bigg|_{t=0} = s_k A_k \\ &\dots \\ &\frac{d^{(n-1)}}{dt^{(n-1)}} A_k e^{s_k t}\bigg|_{t=0} = s_k^{n-1} A_k. \end{aligned} \tag{1.56}$$

The initial values of the natural responses are found as

$$\begin{aligned} &f_n(0) = f(0) - f_f(0) \\ &f'_n(0) = f'(0) - f'_f(0) \\ &\dots \\ &f_n^{(n-1)}(0) = f^{(n-1)}(0) - f_f^{(n-1)}(0) \end{aligned} \tag{1.57}$$

Thus, for the formulation in equation 1.55 of its left side quantities, we must know:

(1) the initial values of the complete transient response $f(0)$ and its $(n-1)$ derivatives, and
(2) the initial values of the force response $f_f(0)$ and its $(n-1)$ derivatives.

The technique of finding the initial values of the complete transient response in (1) has been discussed in the previous section. In brief, according to this technique: a) we have to determine the independent initial condition (currents through the inductances at and voltages across the capacitances at $t = 0_-$), and b) by inspection of the equivalent circuit which arose after switching, i.e., at $t = 0$, we have to find all other quantities by using Kirchhoff's two laws and/or any known method of circuit analysis. For determining the initial values in (2), the forced response must also be found. Let us now introduce the procedure of finding integration constants in more detail.

Consider a first order transient response and assume, for instance, that the response we are looking for is a current response. Then its natural response is

$$i_n(t) = Ae^{st}.$$

Knowing the current initial value $i(0_+)$ and its force response $i_f(t)$ we may find

$$A = i(0_+) - i_f(0). \tag{1.58}$$

If the response is of the second order and the roots of the characteristic equation are real, then

$$i_n(t) = A_1 e^{s_1 t} + A_2 e^{s_2 t}, \tag{1.59}$$

and after differentiation, we obtain

$$i'_n(t) = s_1 A_1 e^{s_1 t} + s_2 A_2 e^{s_2 t}. \tag{1.59a}$$

Suppose that we found $i(0)$ and $i'(0)$, and also $i_f(0)$ and $i'_f(0)$, then with equation 1.57

$$i_n(0) = i(0) - i_f(0)$$
$$i'_n(0) = i'(0) - i'_f(0), \tag{1.60}$$

and in accordance with equation 1.55 we have two equations for determining two unknowns: A_1 and A_2

$$A_1 + A_2 = i_n(0)$$
$$s_1 A_1 + s_2 A_2 = i'_n(0). \tag{1.61}$$

The solution of equation 1.61 yields

$$A_1 = \frac{i'_n(0) - s_2 i_n(0)}{s_1 - s_2}$$
$$A_2 = \frac{i'_n(0) - s_1 i_n(0)}{s_2 - s_1}. \tag{1.61a}$$

If the roots of the characteristic equation are complex-conjugate, $s_{1,2} = \alpha \pm j\omega_n$, then A_1 and A_2 are also complex-conjugate, $\mathbf{A}_{1,2} = Ae^{\pm j\vartheta}$ and the natural response (equation 1.59) may be written in the form

$$i_n(t) = Ae^{+j\vartheta}e^{-\alpha t}e^{+j\omega_n t} + Ae^{-j\vartheta}e^{-\alpha t}e^{-j\omega_n t} = Be^{-\alpha t}\sin(\omega_n t + \beta), \tag{1.62}$$

where $B = 2A$ and $\beta = \vartheta + 90°$. Taking a derivative of equation 1.62 we will have

$$i'_n(t) = -B\alpha\, e^{-\alpha t}\sin(\omega_n t + \beta) + B\omega_n e^{-\alpha t}\cos(\omega_n t + \beta). \tag{1.63}$$

Equations 1.62 and 1.63 for instant $t = 0$, with the known initial conditions (equation 1.60), yield

$$B\sin\beta = i_n(0),$$
$$-B\alpha\sin\beta + B\omega_n\cos\beta = i'_n(0). \tag{1.64}$$

By division of the second equation by the first one, we have

$$\omega_n \cot\beta = \frac{i'_n(0)}{i_n(0)} + \alpha,$$

and the solution is

$$\beta = \tan^{-1}\left[\frac{\omega_n i_n(0)}{i'_n(0) + \alpha i_n(0)}\right] \tag{1.65a}$$

$$B = \frac{i_n(0)}{\sin\beta}. \tag{1.65b}$$

The natural response (equation 1.62) might be written in a different form (which is preferred in some textbooks)

$$i_n(0) = e^{-\alpha t}(M \sin \omega_n t + N \cos \omega_n t), \tag{1.66}$$

where

$$M = B \cos \beta \quad \text{and} \quad N = B \sin \beta. \tag{1.67}$$

Then, by differentiating equation 1.66 and with the known initial conditions, the two equations for determining two unknowns, M and N, may be written as

$$\begin{aligned} N &= i_n(0), \\ M\omega_n - \alpha N &= i'_n(0), \end{aligned} \tag{1.68a}$$

and

$$M = \frac{i'_n(0) + \alpha i_n(0)}{\omega_n}. \tag{1.68b}$$

Knowing M and N we can find B and β and vice versa. Thus for instance

$$\beta = \tan^{-1}\frac{N}{M} \quad \text{and} \quad B = \sqrt{M^2 + N^2}$$

(substituting M and N from equation 1.68 into these expressions yields equation 1.65).

If the characteristic equation is of an order higher than two, the higher derivatives shall be found and the solution shall be performed in accordance with equation 1.55.

Example 1.9

Using the results of Example 1.7 (Fig. 1.23), find the two integration constants of the natural response of current i_o. The circuit of Example 1.7 after switching is shown here in Fig. 1.32(a).

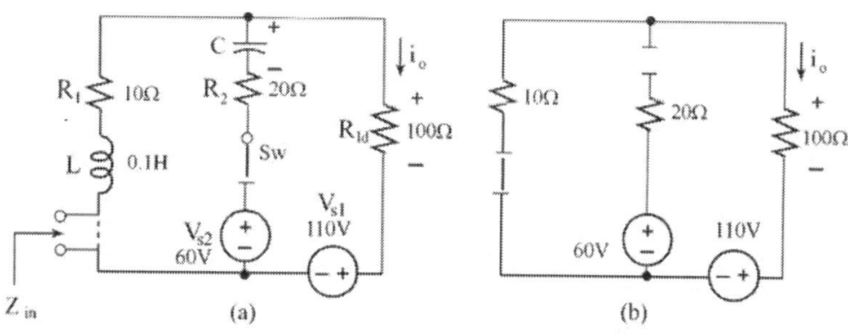

Figure 1.32 A given circuit for Example 1.9: prior to switching (a) and its equivalent in steady-state operation (b).

Solution

From Example 1.7 it is known that $i_o(0) = -0.5$ A and $i'_o(0) = -75$ A s^{-1}. To find the two constants of the integration we have to know: 1) the two roots of the second order characteristic equation and 2) the forced response.

1) In order to determine the characteristic equation we must short-circuit the voltage sources and find the input impedance by opening, for instance, the inductance branch, Fig. 1.32(a),

$$Z_{in} = R_1 + sL + \frac{(R_2 + 1/sC)R_{ld}}{R_2 + R_{ld} + 1/sC}.$$

Equaling zero and substituting the numerical values, we obtain the characteristic equation

$$s^2 = 350s + 9.17 \cdot 10^4 = 0,$$

and the roots are a complex-conjugate pair $s_{1,2} = -175 \pm j247$ s^{-1}.

2) By inspection of the circuit in the steady-state operation, Fig. 1.32(b), we have

$$i_{o,f} = \frac{-110}{100 + 10} = -1 \text{ A}.$$

(Note that this current is negative, since it flows opposite to the positive direction, assigned by a solid arrow). Now we can find the initial values of the natural response. With equation 1.60 and noting that $i'_{o,f} = 0$, we have

$$i_{o,n}(0) = i_o(0) - i_{o,f}(0) = -0.5 - (-1) = 0.5 \text{ A}$$

$$i'_{o,n}(0) = -75 - 0 = -75 \text{ A s}^{-1}.$$

Since the roots are complex numbers, we shall use equation 1.65 (or equation 1.68):

$$\beta = \tan^{-1} \frac{0.5 \cdot 247}{-75 + 0.5 \cdot 175} = 84.2^\circ$$

$$B = \frac{0.5}{\sin 84.2^\circ} = 0.502.$$

(With equation 1.68 $N = i_n(0) = 0.5$ and $M = (-75 + 175 \cdot 0.5)/247 = 0.0506$ and $\beta = \tan^{-1}(0.5/0.0506) = 84.2^\circ$).

Chapter #2

TRANSIENT RESPONSE OF BASIC CIRCUITS

2.1 INTRODUCTION

In this chapter, we shall proceed with transient analysis and apply the classical approach technique, which was introduced in the previous chapter, for a further and intimate understanding of the transient behavior of different kinds of circuits. It will be shown that by applying the so-called **five-step solution** we may greatly simplify the transient analysis of any circuit, upon any interruption and under any supply, so that the determination of transient responses becomes a simple procedure.

Starting with relatively simple *RC* and *RL* circuits, we will progress to more complicated *RLC* circuits, wherein their transient analysis is done under both kinds of supplies, d.c. and a.c. The emphasis is made on the treatment of *RLC* circuits, in the sense that these circuits are more general and are more important when the power system networks are analyzed via different kinds of interruptions. All three kinds of transients in *RLC* circuit, overdamped, underdamped and critical damping, are analyzed in detail.

In power system networks, when interrupted, different kinds of resonances, on a fundamental or system frequency, as well as on higher or lower frequencies, may occur. Such resonances usually cause excess voltages and/or currents. Thus, the transients in an *RLC* circuit under this resonant behavior are also treated and the conditions for such overvoltages and overcurrents have been defined.

It is shown that using the superposition principle in transient analysis allows the simplification of the entire solution by bringing it to zero initial conditions and to only one supplied source. The theoretical material is accompanied by many numerical examples.

2.2 THE FIVE STEPS OF SOLVING PROBLEMS IN TRANSIENT ANALYSIS

As we have seen in our previous study of the classical method in transient analysis, there is no general answer, or ready-made formula, which can be applied to every kind of electrical circuit or transient problem. However, we can formulate a **five-step solution**, which will be applicable to any kind of circuit or problem. Following these five steps enables us to find the complete response in transient behavior of an electrical circuit after any kind of switching (turning on or off different kinds of sources, short and/or open-circuiting of circuit elements, changing the circuit configuration, etc.). We shall summarize the five-step procedure of solving transient problems by the classical approach as follows:

1) *Determination of a characteristic equation and evaluation of its roots.* Formulate the input impedance as a function of s by inspection of the circuit, which arises after switching, at instant $t=0_+$. Note that all the independent voltage sources should be short-circuited and the current sources should be open-circuited. Equate the expression of $Z_{in}(s)$ to zero to obtain the characteristic equation $Z_{in}(s)=0$. Solve the characteristic equation to evaluate the roots.

The input impedance can be determined in a few different ways: a) As seen from a *voltage source*; b) Via any branch, which includes one or more energy storing elements L and/or C (by opening this branch). The characteristic equation can also be obtained using: c) an input admittance as seen from a *current source* or d) with the determinant of a matrix (of circuit parameters) written in accordance with mesh or node analysis.

Knowing the roots s_k the expression of a natural response (for instance, of current) may be written as

$$i_n(t)=\sum_k A_k e^{s_k t}, \quad \text{for real roots (see 1.31)}$$

or

$$i_n(t)=\sum_k B_k \sin(\omega_{n,k} t+\beta_k), \quad \text{for complex roots (see 1.33)}$$

2) *Determination of the forced response.* Consider the circuit, which arises after switching, for the instant time $t\to\infty$, and find the steady-state solution for the response of interest. Note that any of the appropriate methods (which are usually studied in introductory courses) can be applied to evaluate the solution $i_f(t)$.

3) *Determination of the independent initial conditions.* Consider the circuit, which existed prior to switching at instant $t=0_-$. Assuming that the circuit is operating in steady state, find all the currents through the inductances $i_L(0_-)$ and all the voltages across the capacitances $v_C(0_-)$. By applying two switching laws (1.35) and (1.36), evaluate the independent initial conditions

$$i_L(0_+)=i_L(0_-), \quad v_C(0_+)=v_C(0_-). \tag{2.1}$$

4) *Determination of the dependent initial conditions.* When the desirable response is current or voltage, which can change abruptly, we need to find their initial values, i.e. at the first moment after switching. For this purpose the inductances

must be replaced by current sources, having the values of the currents through these inductances at the moment prior to switching $i_L(0_-)$ and the capacitances should be replaced by voltage sources, having the values of the voltages across these capacitances prior to switching $v_C(0_-)$. If the current through an inductance prior to switching was zero, this inductance should be replaced by an open circuit (i.e., open switch), and if the voltage across a capacitance prior to switching was zero, this capacitance should be replaced by a short circuit (i.e., closed switch). By inspecting and solving this equivalent circuit, the initial values of the desirable quantities can be found. If the characteristic equation is of the second or higher order, the initial values of the derivatives must also be found. This can be done by applying Kirchhoff's two laws and using the other known initial conditions.

5) *Determination of the integration constants.* With all the known initial conditions apply equations (1.58), (1.61) or (1.65), (1.68), and by solving them find the constants of the integration (see section 1.8). The number of constants must be the same as the order of the characteristic equation. For instance, if the characteristic equation is of the first order, then only one constant of integration has to be calculated as

$$A = i(0_+) - i_f(0), \tag{2.2a}$$

and the complete response will be

$$i(t) = i_f(t) + [i(0_+) - i_f(0)]e^{st}. \tag{2.2b}$$

Keeping the above-classified rules in mind, we shall analyze (in the following sections) the transient behavior of different circuits.

2.3 FIRST ORDER *RL* CIRCUITS

2.3.1 *RL* circuits under d.c. supply

Let us start with a simple *RL* series circuit, which is connected to a d.c. voltage source, to illustrate how to determine its complete response by using the 5-step solution method. This circuit, shown in Fig, 2.1(a), has been previously analyzed (in its short-circuiting behavior) by applying a mathematical approach.

1) Determining the input impedance and equating it to zero yields

$$Z_{in}(s) = R + sL = 0. \tag{2.3a}$$

The root of these equations is

$$s = -\frac{R}{L}. \tag{2.3b}$$

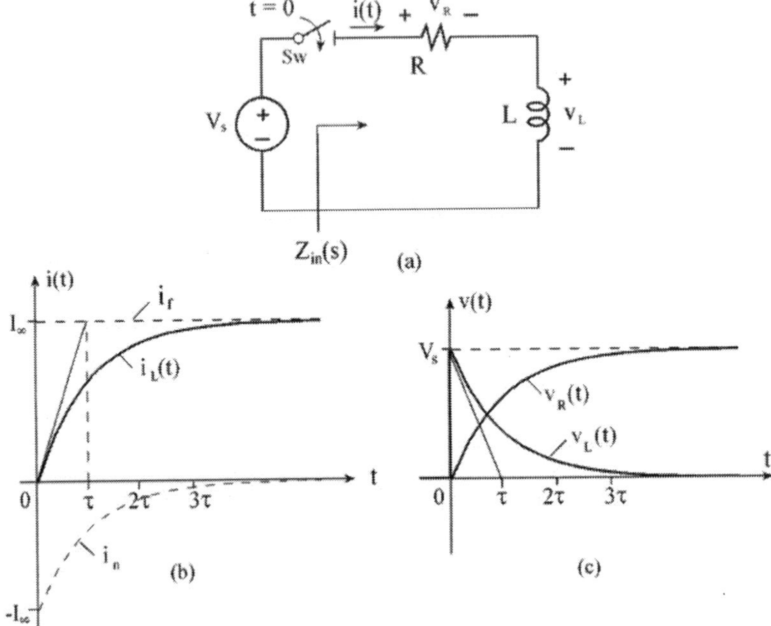

Figure 2.1 A series *RL* circuit switching at $t = 0$ (a), the current plot after switching (b) and the voltages $v_L(t)$ and $v_R(t)$ (c).

Thus, the natural response will be

$$i_n(t) = Ae^{-\frac{R}{L}t}. \tag{2.3c}$$

2) The forced response, i.e. the steady-state current (after the switch is closed, at $t \to \infty$, the inductance is equivalent to a short circuit) will be

$$i_f(t) = \frac{V_s}{R} = I_\infty. \tag{2.4}$$

3) Because the current through the inductance, prior to closing the switch, was zero, the independent initial condition is

$$i_L(0_+) = i_L(0_-) = 0.$$

4) Since no dependent initial conditions are required, we proceed straight to the 5th step.

5) With equation 2.2a we have

$$A = 0 - \frac{V_s}{R} = -I_\infty,$$

and

$$i(t) = I_\infty - I_\infty e^{-\frac{R}{L}t} = I_\infty\left(1 - e^{-\frac{R}{L}t}\right). \tag{2.5}$$

This complete response and its two components, natural and forced responses, are shown in Fig. 2.1(b). Note that the natural response, at $t = 0$, is exactly equal to the steady-state response, but is opposite in sign, so that the whole current at the first moment of the transient is zero (in accordance with the initial conditions). It should once again be emphasized that the natural response appears to insure the initial condition (at the beginning of the transients) and disappears at the steady state (at the end of the transients). It is logical therefore, to conclude that in a particular case, when the steady state, i.e., the forced response at $t = 0$, equals the initial condition, the natural response will not appear at all(*).

The time constant in this example is

$$\tau = \frac{L}{R} \quad \text{or in general} \quad \tau = \frac{1}{|s|}.$$

The time constant, in this example, is also found graphically as a line segment on the asymptote, i.e. on the line of a steady-state value, determined by the intercept of a tangent to the curve $i(t)$ at $t = 0$ and the asymptote, as shown in Fig. 2.1(b).

Knowing the current response, we can now easily find the voltages across the inductance, v_L and the resistance, v_R:

$$v_L = L\frac{di}{dt} = L\frac{d}{dt}[I_\infty(1 - e^{-(R/L)t})] = L\frac{V_s}{R}\left(-\frac{R}{L}\right)(-e^{-(R/L)t}) = V_s e^{-(R/L)t},$$

and

$$v_R = Ri = V_s(1 - e^{-(R/L)t}),$$

where $V_s = RI_\infty$.

Both these curves are shown in Fig 2.1(c). As we can see at the first moment the whole voltage is applied to the inductance and at the end of the transient it is applied to the resistance. This voltage exchange between two circuit elements occurs gradually during the transient.

Before we turn our attention to more complicated RL circuits, consider once again the circuit of Fig 1.8, which is presented here (for the reader's convenience) in Fig. 2.2(a). The time constant of this circuit has been found (see (1.20)) and is the same as in a series RL circuit. Therefore the natural response (step 1) is $Ae^{-(R/L)t}$. The forced response (step 2) here is $i_{L,f} = I_s$ and the initial value (step 3) is zero. Hence, the integration constant subsequently (step 5) is $A = 0 - I_s =$

(*)This statement is only true in first order circuits.

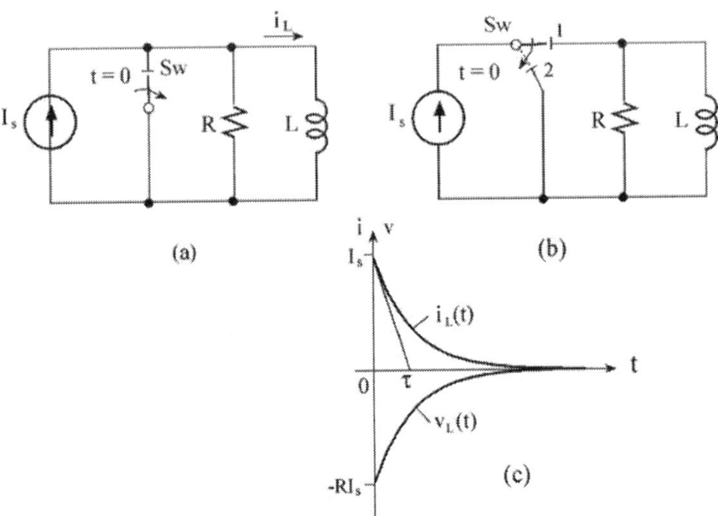

Figure 2.2 An *RL* parallel circuit (a), the circuit in which the inductance discharges through a resistance (b) and the plots of the discharging current and voltage (c).

$-I_s$. Thus, the complete response will be $i_L = I_s(1 - e^{-(R/L)t})$, which is in the same form as in the *RL* series circuit.

To complete our analysis of a simple *RL* series circuit, consider the circuit in Fig. 2.2(b), in which the switch changes its position from "1" to "2" instantaneously and the inductance "discharges" through the resistance. In this case, the natural response, obviously, is the same as in the circuit (a), but the forced response is zero. Therefore, we have $i_L = Ae^{-(R/L)t} = I_s e^{-(R/L)t}$, where $A = I_s$ since the initial value of the inductance current (prior to switching) is I_s. This response and the voltage across the inductance and the resistance are shown in Fig. 2.2(c). Verifying the voltage response is left to the reader.

Let us illustrate the 5-step method by considering more complicated circuits in the following numerical examples.

Example 2.1

In the circuit, Fig. 2.3(a), find current $i_2(t)$ after opening the switch. The circuit parameters are $V_1 = 20\,\text{V}$, $V_2 = 4\,\text{V}$, $R_1 = 8\,\Omega$, $R_2 = 2\,\Omega$, $R_3 = R_4 = 16\,\Omega$ and $L = 1\,\text{mH}$.

Solution

1) We start our solution by expressing the impedance *Z(s)* of the circuit that arises after switching, at the instant $t = 0_+$. We shall determine $Z_{in}(s)$ as seen from source V_2. (However, the impedance $Z_{in}(s)$ can be found in a few different ways, as will be shown further on.) By inspecting the circuit in Fig. 2.3(b) we

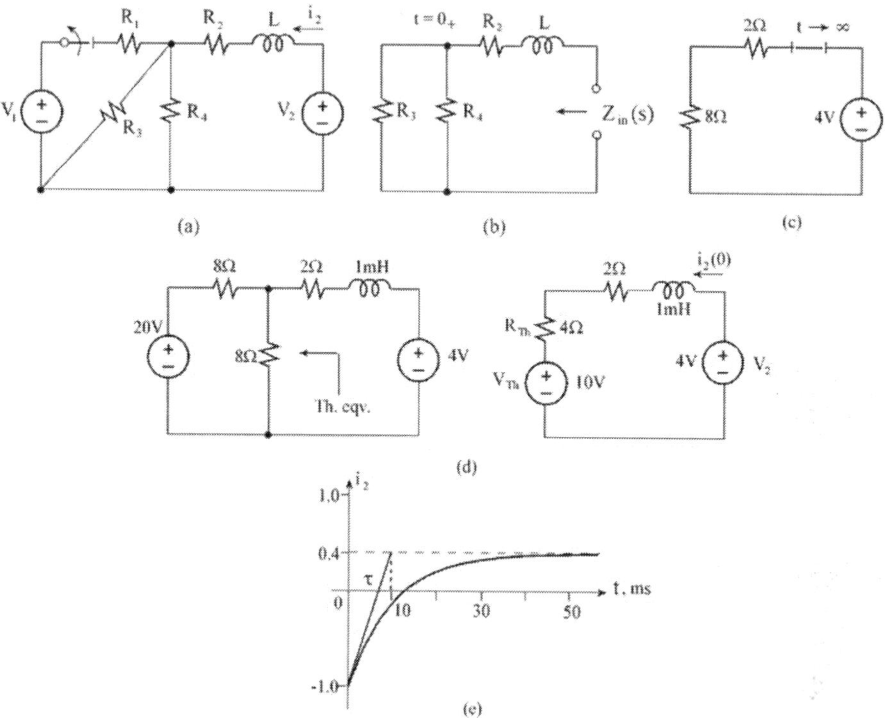

Figure 2.3 The given circuit (a), its equivalent for $t = 0$ (b), its equivalent for $t \to \infty$ (c), its equivalent for $t < 0$ (d) and the curve of current $i_2(t)$ (e).

have

$$Z_{in}(s) = sL + R_2 + \frac{R_3 R_4}{R_3 + R_4}.$$

Substituting the numerical values and equating the expression to zero yields

$$10^{-3}s + 2 + 8 = 0.$$

This equation has the root

$$s = -100\ \mathrm{s}^{-1} \quad \text{and} \quad \tau = 0.01\ \mathrm{s},$$

and the natural response will be

$$i_{2,n} = Ae^{-100t}.$$

2) The forced response, i.e., the steady-state current $i_{2,f}$, is found in the circuit, Fig. 2.2(c) that is derived from the given circuit after the switching, at $t \to \infty$,

while the inductance behaves as a short circuit

$$i_{2,f}=\frac{V_2}{R_{eq}}=\frac{4}{10}=0.4\text{ A}.$$

3) The independent initial condition, i.e., $i_L(0_-)$ is found in the circuit prior to switching, shown in Fig. 2.3(d). Using Thévenin's equivalent for the left part of the circuit, as shown in (d), we have

$$i_2(0_+)=i_2(0_-)=\frac{V_2-V_{Th}}{R_2+R_{Th}}=\frac{4-10}{2+4}=-1\text{ A}.$$

4) None of the dependent initial conditions is needed.

5) In order to evaluate constant A, we use equation 2.2a: $A=i_2(0_+)-i_f(0)=-1-0.4=-1.4$ A. Thus the complete response is $i_2(t)=0.4-1.4e^{-100t}$ A, which is sketched in Fig. 2.3(e).

Example 2.2

For the circuit shown in Fig. 2.4(a) find the current response $i_1(t)$ after closing the switch. The circuit parameters are: $R_1=R_2=20\ \Omega$, $L_1=0.1$ H, $L_2=0.4$ H, $V_s=120$ V.

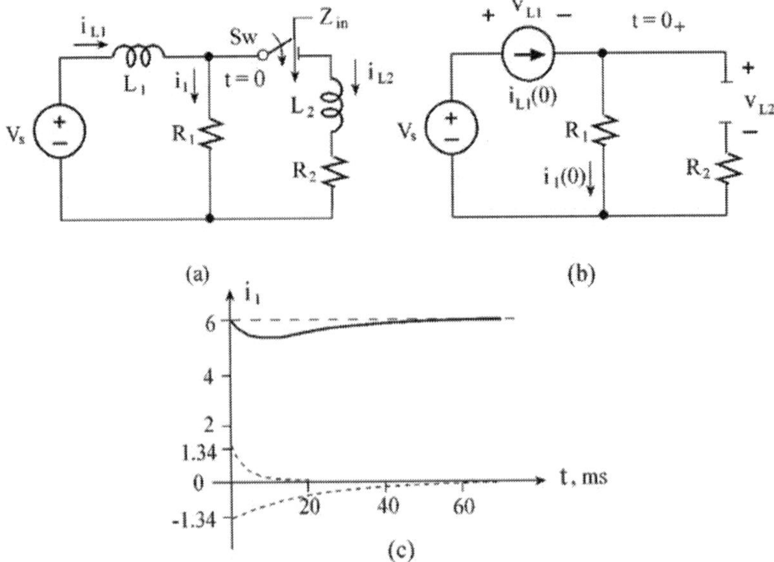

Figure 2.4 A given circuit for Example 2.2(a), its equivalent at time $t=0_+$ (b) and the plot of current $i_1(t)$ and its components (c).

Solution

1) The input impedance is found as seen from the L_2 branch (we just "measure" it from the open switch point of view), with the voltage source short-circuited

$$Z_{in}(s) + sL_2 + R_2 + \frac{R_1 sL_1}{R_1 + sL_1}.$$

Equating this expression to zero and after simplification, we get the characteristic equation

$$s^2 + \frac{R_1 L_1 + R_2 L_1 + R_1 L_2}{L_1 L_2} + \frac{R_1 R_2}{L_1 L_2} = 0,$$

or by substituting the numerical data

$$s^2 + 3 \cdot 10^2 s + 10^4 = 0.$$

Thus, the roots of this equation are

$$s_1 = -38.2\ \mathrm{s}^{-1}, \quad s_2 = -262\ \mathrm{s}^{-1},$$

and the natural response is

$$i_{1,n} = A_1 e^{-38.2t} + e^{-262t}.$$

2) By inspecting the circuit after the switch is closed, at $t \to \infty$, we may determine the forced response

$$i_{1,f} = \frac{V_s}{R_1} = \frac{120}{20} = 6\ \mathrm{A}.$$

3) By inspection of the circuit prior to switching we observe that $i_{L1}(0_-) = 120/20 = 6\ \mathrm{A}$ and $i_{L2}(0_-) = 0$. Therefore, the independent initial conditions are

$$i_{L1}(0_+) = 6\ \mathrm{A}, \quad i_{L2}(0_+) = 0.$$

4) Since the characteristic equation is of the second order, and the desired response, which is a current through a resistance, can be changed abruptly, we need its two dependent initial conditions, namely:

$$i_1(0) \quad \text{and} \quad \left.\frac{di}{dt}\right|_{t=0}.$$

By inspection of the circuit in Fig. 2.4(b) for instant $t = 0_+$, we may find $t_1(0) = 6\ \mathrm{A}$. (Note that in this specific case the current i_1 does not change abruptly and, therefore, its initial value equals its steady-state value, but because the circuit is of the second order, the transient response of the current is expected.)

By applying KCL we have $i_1 = i_{L1} - i_{L2}$ and after the differentiation and evaluation of $t = 0$ we obtain

$$\left.\frac{di}{dt}\right|_{t=0} = \left.\frac{di_1}{dt}\right|_{t=0} - \left.\frac{di_2}{dt}\right|_{t=0} = \frac{1}{L_1} v_{L1}(0) - \frac{1}{L_2} v_{L2}(0).$$

Since, Fig. 2.4(b), $v_{R1}(0) = V_s$, then $v_{L1}(0) = 0$ and $v_{R1}(0) = v_{L2}(0) = 120$ V. Therefore, we have

$$\left.\frac{di}{dt}\right|_{t=0} = 0 - \frac{120}{0.4} = -300,$$

and we may obtain two equations

$$A_1 + A_2 = i(0) - i_f(0) = 6 - 6 = 0$$

$$s_1 A_1 + s_2 A_2 = \left.\frac{di}{dt}\right|_{t=0} - \left.\frac{di_f}{dt}\right|_{t=0} = -300 - 0 = -300.$$

Solving these two equations yields $A_1 = -1.34$, $A_2 = 1.34$ and the answer is

$$i_1(t) = 6 - 1.34e^{-38.2t} + 1.34e^{-262t}\ \text{A}.$$

This current and its components are plotted in Fig. 2.4(c).

Example 2.3

Consider the circuit of the transformer of Example 1.2, which is shown here in Fig. 2.5 in a slightly different form. For measuring purposes, the transformer is connected to a 120 V d.c.-source. Find both current i_1 and i_2 responses.

1) The characteristic equation and its roots have been found in Example 1.2: $s_1 = -86\ \text{s}^{-1}$, $s_2 = -1160\ \text{s}^{-1}$. Therefore, the natural responses are

$$i_{1,n} = A_1 e^{-86t} + A_2 e^{-1160t}$$

$$i_{2,n} = B_1 e^{-86t} + B_2 e^{-1160t}.$$

2) The forced responses are found by inspection of the circuit after switching $(t \to \infty)$:

$$i_{1,f} = \frac{V_s}{R_1} = \frac{120}{6} = 20\ \text{A}, \quad i_{2,f} = 0.$$

3) The independent initial conditions are zero, since prior to switching no currents are flowing through the inductances: $i_1 = (0_+) = i_1(0_-) = 0$, $i_2 = (0_+) = i_2(0_-) = 0$.

4) In order to determine the integration constant we need to evaluate the current derivatives. By inspection of the circuit in Fig. 2.5(b), we have $v_{L1}(0) = 120$ V, $v_{L2}(0) = 0$, and

$$L_1 \left.\frac{di_1}{dt}\right|_{t=0} + M \left.\frac{di_2}{dt}\right|_{t=0} = 120$$

$$L_2 \left.\frac{di_2}{dt}\right|_{t=0} + M \left.\frac{di_1}{dt}\right|_{t=0} = 0.$$

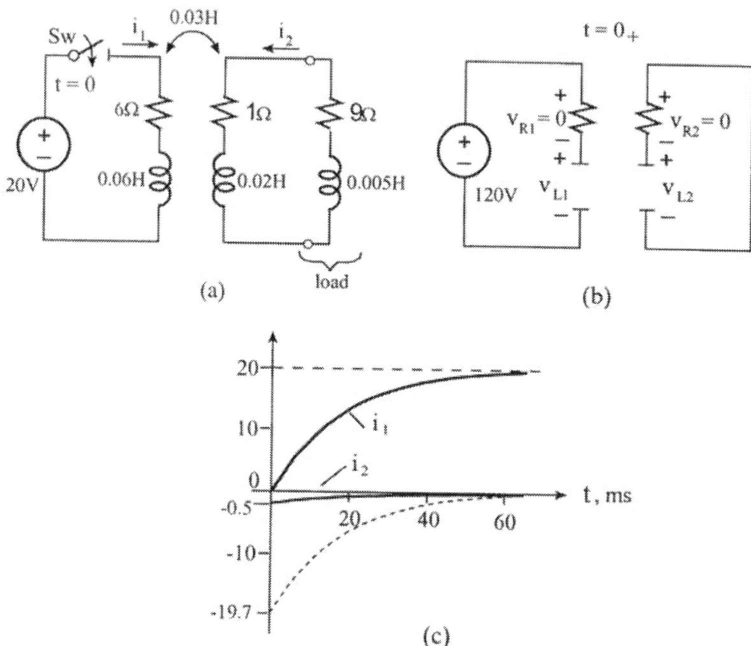

Figure 2.5 The circuit of a transformer (a), its equivalent for $t = 0_+$ (b) and the plots of two currents (c).

Solving these two relatively simple equations yields

$$\left.\frac{di_1}{dt}\right|_{t=0} = 5000, \quad \left.\frac{di_2}{dt}\right|_{t=0} = -6000.$$

5) With the initial value of $i_{1,n}(0) = i_1(0) - i_{1,f}(0) = 0 - 20 = -20$ and the initial value of its derivative

$$\left.\frac{di_{1,n}}{dt}\right|_{t=0} = \left.\frac{di_1}{dt}\right|_{t=0} - \left.\frac{di_{1,f}}{dt}\right|_{t=0} = 5000 - 0 = 5000,$$

we obtain two equations in the two integration constants of current i_1

$$A_1 + A_2 = -20$$

$$s_1 A_1 + s_2 A_2 = 5000,$$

for which the solution is: $A_1 = -19.7$, $A_2 = -0.3$. In a similar way, the two equations in the two integration constants of current i_2

$$B_1 + B_2 = 0$$
$$s_1 B_1 + s_2 B_2 = -6000,$$

for which the solution is $B_1 = -0.52$, $B_2 = 0.52$.

Therefore, the current responses are

$$i_1 = 20 - 19.7e^{-86t} - 0.3e^{-1160t}$$

$$i_2 = -0.52e^{-86t} + 0.52e^{-1160t}.$$

These two currents are sketched in Fig. 2.5(c). Note that the second exponential parts decay much faster than the first ones and are not shown in Fig. 2.5(c). Note also that the second exponential term in i_1 is relatively small and might be completely neglected.

Example 2.4

As a final example of inductive circuits let us consider the "inductance" node circuit, which is shown in Fig. 2.6(a). Find the currents i_1 and i_2 after switching, if the circuit parameters are: $L_1 = L_2 = 0.05$ H, $L_3 = 0.15$ H, $R_1 = R_2 = R_3 = 1\ \Omega$ and $V_s = 15$ V.

Solution

1) Let us determine the characteristic equation by using mesh analysis. The

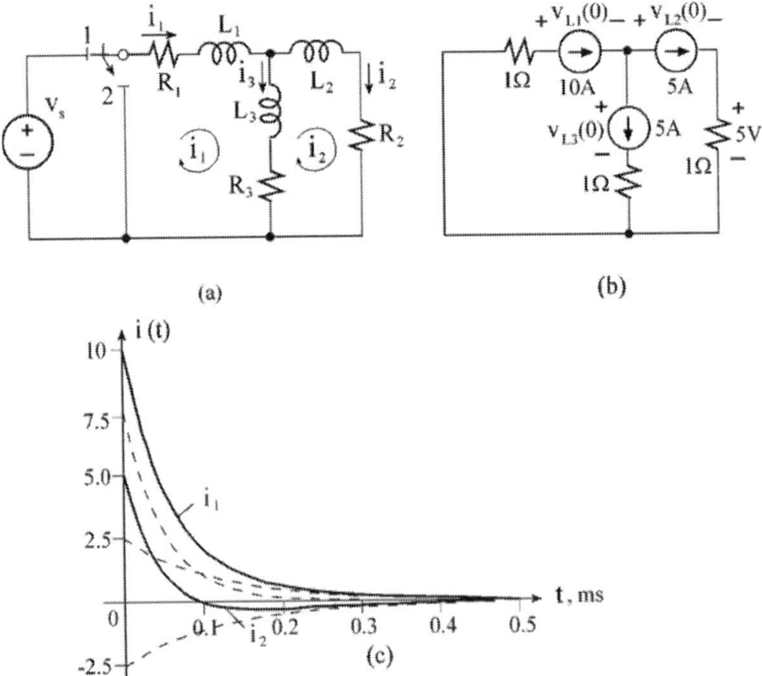

Figure 2.6 A circuit containing an "inductance node" (a), its equivalent at $t = 0$ (b) and the plots of the currents and their components (c).

impedance matrix is

$$\begin{bmatrix} s(L_1+L_3)+R_1+R_3 & -(sL_3+R_3) \\ -(sL_3+R_3) & s(L_2+L_3)+R_2+R_3 \end{bmatrix} = \begin{bmatrix} 0.2s+2 & -(0.15s+1) \\ -(0.15s+1) & 0.2s+2 \end{bmatrix}$$

Equating the determinant to zero and after simplification, we obtain the characteristic equation

$$0.0175s^2 + 0.5s + 3 = 0,$$

for which the roots are

$$s_1 = -8.6\ \mathrm{s}^{-1}, \quad s_2 = -20\ \mathrm{s}^{-1}.$$

Thus, the natural responses of the currents are

$$i_{1,n} = A_1 e^{-8.6t} + A_2 e^{-20t}$$

$$i_{2,n} = B_1 e^{-8.6t} + B_2 e^{-20t}.$$

2) The steady-state values of the currents are zero, since after switching the circuit is source free.

3) The independent initial conditions can be found by inspection of the circuit in Fig. 2.6(a) prior to switching and keeping in mind that all the inductances are short-circuited

$$i_1(0) = i_1(0_-) = \frac{V_s}{R_1 + R_2//R_3} = \frac{15}{1.5} = 10\ \mathrm{A}$$

$$i_2(0) = i_2(0_-) = \frac{10}{2} = 5\ \mathrm{A}.$$

Note that only two initial independent currents can be found (although the circuit contains three inductances), since the third current is dependent on two others. However, because the circuit is of the second order, the two initial values are enough for solving this problem.

4) Next, we have to find the initial values of the current derivatives for which we must find the voltage drops in the inductances $v_{L1}(0)$ and $v_{L2}(0)$ for the instant of switching, i.e., $t = 0$. By inspection of the circuits in Fig. 2.6(b), we have

$$v_{L1}(0) + v_{L2}(0) = -15, \quad v_{L1}(0) + v_{L3}(0) = -15, \tag{2.6a}$$

$$v_{L2}(0) = v_{L3}(0). \tag{2.6b}$$

With KCL we may write $i_1 = i_2 + i_3$ and by differentiation

$$\frac{di_1}{dt} = \frac{di_2}{dt} + \frac{di_3}{dt}, \quad \text{or} \quad \frac{v_{L1}}{L_1} = \frac{v_{L2}}{L_2} + \frac{v_{L3}}{L_3}.$$

With equation 2.6b we have

$$\frac{1}{L_1}v_{L1}=\left(\frac{1}{L_2}+\frac{1}{L_3}\right)v_{L2}, \quad \text{or} \quad v_{L2}=\frac{L_2L_3}{(L_2+L_3)L_1}v_{L1}=60.75v_{L1},$$

and with equation 2.6a $v_{L1}=-8.57$ V and $v_{L2}=-6.43$ V. Therefore,

$$\left.\frac{di_1}{dt}\right|_{t=0}=\frac{v_{L1}}{L_1}=-\frac{8.57}{0.05}=-171.4\ \mathrm{A\,s^{-1}}$$

$$\left.\frac{di_2}{dt}\right|_{t=0}=\frac{v_{L2}}{L_2}=-\frac{6.43}{0.05}=-128.6\ \mathrm{A\,s^{-1}}.$$

5) We may now obtain a set of equations to evaluate the integration constant

$$A_1+A_2=10$$

$$s_1A_1+s_2A_2=-171.4.$$

for which the solution is $A_1\cong 2.5$, $A_2\cong 7.5$. In a similar way we can obtain

$$B_1+B_2=5$$

$$s_1B_1+s_2B_2=-128.6,$$

and the solution is $B_1\cong-2.5$, $B_2\cong 7.5$. Therefore, two current responses are

$$i_1=2.5e^{-8.6t}+7.5e^{-20t}$$

$$i_2=-2.5e^{-8.6t}+7.5e^{-20t}.$$

The plots of these currents and their components are shown in Fig. 2.6 (c).

2.3.2 *RL* circuits under a.c. supply

As we already know, the natural response does not depend on the source function, and therefore the first step of the solution, i.e. determining the characteristic equation and evaluating its roots, is the same as in previous cases. This is also understandable from the fact that the natural response arises from the solution of the homogeneous differential equation, which has zero on the right side. The forced response can be determined from the steady-state solution of the given circuit. The symbolic, or phasor, method should be used for this solution.

To illustrate the above principles, let us consider the circuit shown in Fig. 2.7. The solution will be completed by applying the five steps as previously done. In the first step, we have to determine the characteristic equation and its root. However, for such a simple circuit it is already known that $s=-R/L$. Therefore the natural response is

$$i_n=Ae^{-t/\tau}, \quad \text{where} \quad \tau=L/R. \tag{2.7}$$

In the next step, our attention turns to obtaining the steady-state current.

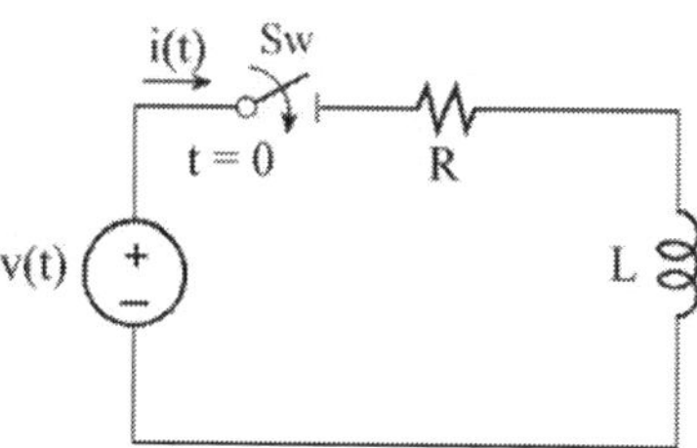

Figure 2.7 A series RL circuit switching to an a.c. source.

Applying the phasor method we have

$$\tilde{I}_m = \frac{\tilde{V}_m}{Z} = \frac{V_m}{\sqrt{R^2 + (\omega L)^2}} \angle (\psi_v - \varphi),$$

where $\tilde{V}_m = V_m e^{j\psi_v}$ and $\tilde{I}_m = I_m e^{j\psi_i}$ are voltage and current phasors respectively and $\varphi = \psi_v - \psi_i = \tan^{-1}(\omega L/R)$ is the phase angle difference between the voltage and current phasors. Thus,

$$i_f = I_m \sin(\omega t + \psi_i), \tag{2.8}$$

where

$$I_m = \frac{V_m}{\sqrt{R^2 + (\omega L)^2}}, \quad \psi_i = \psi_v - \varphi.$$

In the next two steps, 3 and 4, we shall determine the only initial condition, which is necessary to find the current through the inductance. Since prior to switching this current was zero, we have $i(0_+) = i(0_-) = 0$. In the final step, with this initial value we may obtain the integration constant

$$A = i(0) - i_f(0) = -I_m \sin \psi_i. \tag{2.9}$$

Thus, the complete response of an RL circuit to applying an a.c. voltage source is

$$i = i_f + i_n = I_m \sin(\omega t + \psi_i) - I_m \sin \psi_i e^{-t/\tau}. \tag{2.10}$$

This current and its components are plotted in Fig. 2.8(a).

Note that the initial values of $i_f(0)$ and $i_n(0)$ are equal and opposite in sign, so that with zero initial conditions the actual current $i(t)$ always starts with the zero value. If switching occurs at the instant that $\psi_i = \pm\pi/2$, then the total response reaches its maximum value at the point of one-half a period. This extreme value of the current may increase up to twice that of the amplitude of the steady-state current and occurs if the time constant of the circuit is much greater than the period of the a.c. current so that the natural response current decays relatively slowly. Thus, if $\tau \gg T$, where T is a period of an a.c. current, then $i_{\max} \to 2I_m$. This is shown in Fig 2.8(b). If, however, the time constant of the circuit is small compared to the period of the a.c. current, the natural current

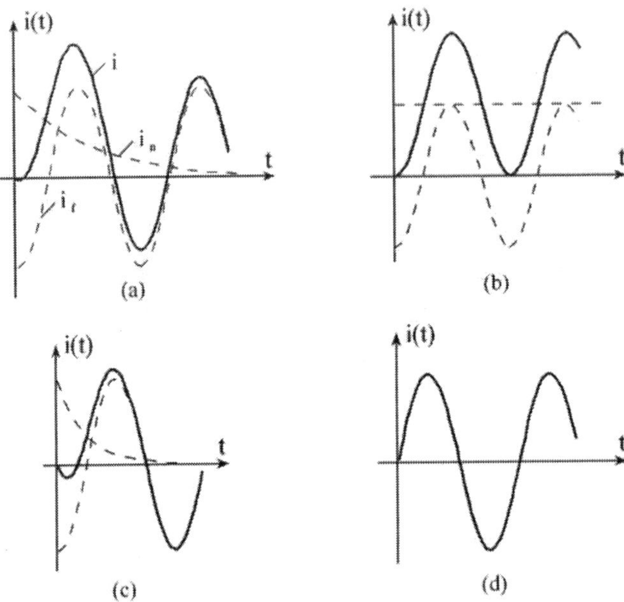

Figure 2.8 The transient response of a series RL circuit when switching to an a.c. voltage source (a) and maximal (b), minimal (c) and zero (d) responses.

decreases quickly during the first half period and no considerable excess current can develop, as shown in Fig. 2.8(c). If the phase angle ψ_i is zero, which means that the forced (steady-state) current passes through zero at the instant of switching, no transient current (equation 2.9) occurs, so that the a.c. current immediately starts in its normal way, Fig. 2.8(d).

In highly inductive circuits, which are common for industrial networks, the displacement angle between the voltage and current is nearly 90°. Thus the favorable case, Fig. 2.8(d), corresponds to the switching on at the maximum instantaneous voltage, which usually occurs in high voltage circuit breakers. The switching-on process in such breakers is initiated by a discharged spark between the breaker contacts, wherein the contacts approach each other relatively slowly compared with the a.c. frequency, and when the voltage passes its maximum.

We shall now illustrate the transients in a.c. circuits by the following numerical examples.

Example 2.5

In an RL circuit of Fig. 2.7, the switch closes at $t = 0$. Find the complete current response and sketch its plot, if $r = 10\,\Omega$, $L = 0.01$ H, and $v_s = 120\sqrt{2}\sin(1000t + 15^\circ)$ V.

Solution

1) The time constant of the circuit is

$$\tau = \frac{L}{R} = \frac{0.01}{10} = 10^{-3} = 1\ \mathrm{ms}$$

and the natural response is

$$i_n = Ae^{-1000t}.$$

2) The steady-state current is calculated by phasor analysis. The impedance of the circuit is $Z(j\omega) = R + j\omega L = 10 + j10 = \sqrt{2}10\angle 45°\ \Omega$, the voltage source phasor is $\tilde{V}_{s,m} = 100\sqrt{2}e^{j15°}$. Thus, the current phasor will be

$$\tilde{I}_f = \frac{\tilde{V}_{s,m}}{Z} = \frac{100\sqrt{2}\angle 15°}{10\sqrt{2}\angle 45°} = 10\angle -30°\ \mathrm{A},$$

and the current versus time is

$$i_f = 10\sin(1000t - 30°)\ \mathrm{A}.$$

3) The initial condition is zero, i.e., $i(0_+) = i(0_-)) = 0$.

4) Non-dependent initial conditions are needed.

5) The integration constant can now be found $A = i(0) - i_f(0) = 0 - 10\sin(-30°) = 5$ and the complete response is $i = 10\sin(1000t - 30°) + 5e^{-1000t}$ A, which is sketched in Fig. 2.9.

Example 2.6

At the receiving end of the transmission line in a no-load operation, a short-circuit fault occurs. The impedance of the line is $Z = (1 + j5)\ \Omega$ and the a.c.

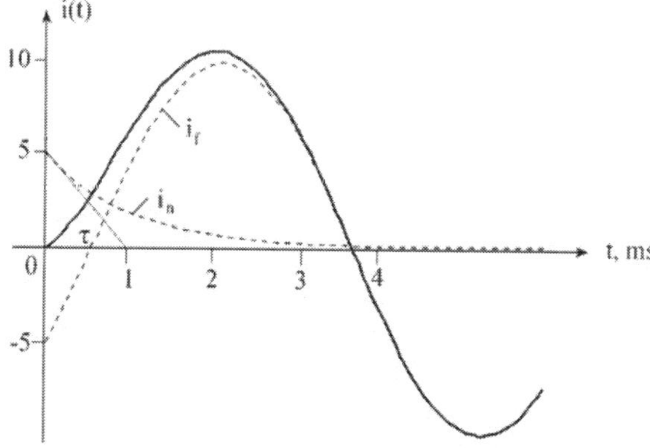

Figure 2.9 A current response in a series *RL*, of example 2.5, circuit switched to the a.c. source.

voltage at the sending end is 10 kV at 60 Hz. a) Find the transient short-circuit current if the instant of short-circuiting is when the voltage phase angle is 1) $-\pi/4+\varphi$; 2) $-\pi/2+\varphi$ and b) estimate the maximal short-circuit current and the applied voltage phase angle under the given conditions.

Solution

a) First we shall evaluate the line inductance $L = x/\omega = 5/2\pi 60 = 0.01326 \cong 13.3$ mH. The voltage at the sending end versus time is $v_s = 10\sqrt{2}\sin(\omega t + \psi_v)$.

1) The time constant of the line (which is represented by RL in series) is $\tau = L/R = 13.3/1 = 13.3$ ms or $s = -1/\tau = -75.2\ \text{s}^{-1}$ and the natural current is

$$i_n = Ae^{-75.2t}$$

2) The steady-state short current (r.m.s.) is found using phasor analysis:

$$\tilde{I}_f = \frac{10\angle\psi_v}{1+j5} = \frac{10\angle\psi_v}{5.1\angle 78.7^\circ} = 1.96\angle\psi_v - 78.7^\circ.$$

Thus

$$i_f = I_m\sin(377t + \psi_v - 78.7^\circ),$$

where

$$I_m = 1.96\sqrt{2}\ \text{A} \quad \text{and} \quad \omega = 2\pi 60 = 377\ \text{rad/s}.$$

3) Because of the zero initial condition, $i(0_+) = i(0_-) = 0$.

5) We omit step 4) (since no dependent initial conditions are needed) and evaluate constant A for two cases:

(1) $\psi_v = -180^\circ/4 + 78.7^\circ = 33.7^\circ$
and $A = i(0) - i_f(0) = 0 - I_m\sin(33.7^\circ - 78.7^\circ) = (\sqrt{2}/2)I_m$.

Therefore, the complete response is

$$I_{sc} = I_m\sin(\omega t - \pi/4) + (\sqrt{2}/2)I_m e^{-75.2t}.$$

(2) $\psi_v = -180^\circ/2 + 78.7^\circ = -11.3^\circ$
and $A = i(0) - i_f(0) = 0 - I_m\sin(-11.3^\circ - 78.7^\circ) = I_m$.

Therefore, the complete response is

$$i_{sc} = I_m\sin(\omega t - \pi/2) + I_m e^{-75.2t}.$$

b) The maximal value of the short-circuit current is dependent on the initial phase angle of the applied voltage and will appear if the natural response is the largest possible one as in (2), i.e., when $A = I_m$. The instant at which the current reaches its peak is about half of the period after switching. To find the exact time we have to equate the current derivative to zero. Thus,

$$\frac{di_{sc}}{dt} = \frac{di_f}{dt} + \frac{di_n}{dt} = 0, \quad \text{or} \quad \frac{di_f}{dt} = -\frac{di_n}{dt}.$$

Performing this procedure we may find

$$I_m\omega\cos(\omega t + \psi_v - \varphi) = I_m\frac{1}{\tau}\sin(\psi_v - \varphi)e^{-\frac{t}{\tau}},$$

or in accordance with (2)

$$\cos(\omega t - \pi/2) = \frac{1}{\omega\tau}e^{-\frac{\omega t}{\omega\tau}}.$$

Taking into consideration that

$$\omega\tau = \frac{x}{L}\cdot\frac{L}{R} = 5$$

we may solve the above transcendental equation finding

$$\omega t_{(\max)} \cong 3.03 \text{ rad}.$$

Therefore, the short-circuit current, of the form $i_{sc} = I_m\sin(\omega t - \pi/2) - I_m e^{-t/\tau}$, will reach its maximal value at $\omega t_{(\max)} \cong 3.03$ rad, Fig. 2.10, and this value will be

$$I_{\max} = I_m(\sin(3.03 - \pi/2) + e^{-3.03/5}) \cong 1.54 I_m.$$

Example 2.7

The switch in the circuit of Fig. 2.11 closes at $t = 0$, after being open for a long

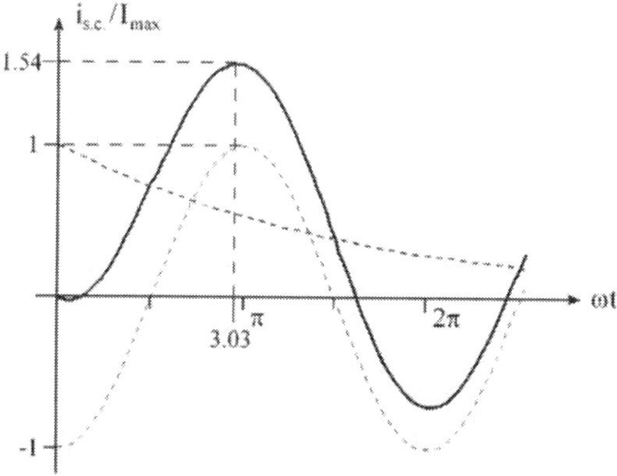

Figure 2.10 A plot of the short-circuit current in which it reaches its maximal value.

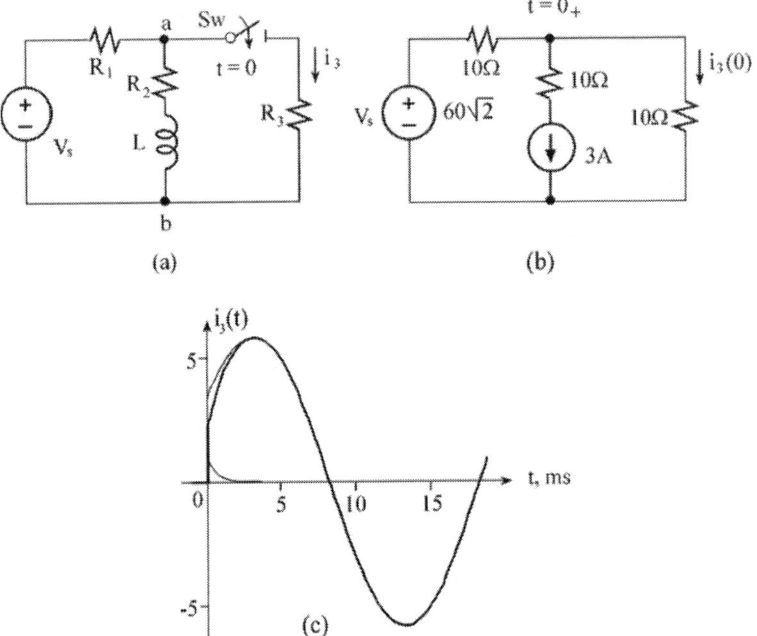

Figure 2.11 A given circuit of Example 2.7 (a), its equivalent at $t = 0$ (b) and the current plot (c).

time. Find the transient current $i_3(t)$, if $R_1 = R_2 = R_3 = 10\ \Omega$, $L = 0.01$ H and $V_{sm} = 120\sqrt{2}$ V at $f = 50$ Hz and $\psi_v = 30°$.

Solution

1) The simplest way to determine the characteristic equation is by observing it from the inductive branch

$$Z(s) = sL + R_2 + R_1//R_2 = 0.$$

With the given data we have

$$0.01s + 15 = 0, \quad \text{or} \quad s = -1500\ \text{s}^{-1},$$

and

$$i_{3,n} = Ae^{-1500t}.$$

2) The forced response of the current will be found by nodal analysis

$$\tilde{V}_a = \frac{\tilde{V}_s}{R_1\left[\dfrac{1}{R_1} + \dfrac{1}{R_2 + jx_L} + \dfrac{1}{R_3}\right]} = \frac{120 \angle 30°}{2 + \dfrac{1}{1 + j0.314}} = 41.1 \angle 35.6°,$$

where $x_L = \omega t = 314 \cdot 0.01 = 3.14\ \Omega$. Thus

$$\tilde{I}_3 = \frac{\tilde{V}_a}{R_3} = 4.11 \angle 35.6^\circ \quad \text{and} \quad i_{3,f} = 4.11\sqrt{2}\sin(\omega t + 35.6^\circ).$$

3) The independent initial condition may be obtained from the circuit prior to switching:

$$\tilde{I}_L = \frac{\tilde{V}_s}{R_1 + R_2 + jx_L} = 5.92 \angle 21.1^\circ.$$

Therefore, $i_L(0_-) = 5.92\sqrt{2}\sin 21.1^\circ = 3.0$ A.

4) With the superposition principle being applied to the circuit in Fig. 2.11(b), we obtain

$$i_3(0) = i_3'(0) + i_3''(0) = \frac{60\sqrt{2}}{20} - \frac{3}{2} = 2.74 \text{ A}.$$

Note that the current i_3 is a resistance current and it changes abruptly.

5) The integration constant is now found as

$$A = i_3(0) - i_{3,f}(0) = 2.74 - 4.11\sqrt{2}\sin 35.6^\circ = -0.64.$$

Therefore,

$$i_3(t) = 4.11\sqrt{2}\sin(\omega t + 35.6^\circ) - 0.64e^{-1500t} \text{ A},$$

which is plotted in Fig. 2.11(c).

Example 2.8

As our next example consider the circuit in Fig. 2.12 and find the current through the switch, which closes at $t = 0$ after being open for a long time. The circuit parameters are: $R_1 = 2\ \Omega$, $x_1 = 10\ \Omega$, $R_2 = 20\ \Omega$, $x_2 = 50\ \Omega$ and $V_m = 15$ V at $f = 50$ Hz and $\psi_v = -15^\circ$.

Solution

1) After short-circuiting, the circuit is divided into two parts, so that each of

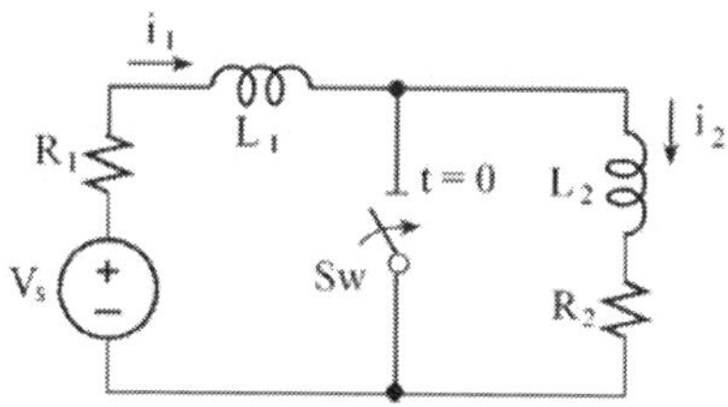

Figure 2.12 A given circuit for Example 2.8.

them has two different time constants:

$$\tau_1 = \frac{L_1}{R_1} = \frac{x_1}{\omega R_1} = \frac{10}{314\cdot 2} = 15.9\ \text{ms}, \quad \text{or} \quad s_1 = -1/\tau_1 = -62.9\ \text{s}^{-1},$$

$$\tau_2 = \frac{L_2}{R_2} = \frac{x_2}{\omega R_2} = \frac{50}{314\cdot 20} = 7.96\ \text{ms}, \quad \text{or} \quad s_2 = -1/\tau_2 = -125\ \text{s}^{-1}.$$

Thus, the natural response of the current contains two parts:

$$i_{sw,n} = A_1 e^{-62.9t} + A_2 e^{-125t}.$$

2) The right loop of the circuit is free of sources, so that only the left side current will contain the forced response:

$$i_{1,f} = \frac{15}{\sqrt{2^2 + 10^2}} \sin(314t - 15^\circ - \tan^{-1}10/2) = 1.47 \sin(314t - 93.7^\circ)\ \text{A}.$$

3) The independent initial conditions, i.e., the currents into two inductances prior to switching, are the same:

$$i_L(0_+) = i_{L1}(0_-) = i_{L2}(0_-) = \frac{15}{\sqrt{22^2 + 60^2}} \sin(-15^\circ - \tan^{-1}60/22) = -0.234\ \text{A}.$$

4–5) Since non-dependent initial conditions are required, we may now evaluate the integration constants:

$$A_1 = i_L(0) - i_{1,f}(0) = -0.234 - 1.47 \sin(-93.7) = 1.23,$$

$$A_2 = i_L(0) - 0 = -0.234.$$

Therefore, the answer is:

$$i_{sw} = i_1 - i_2 = 1.47 \sin(314t - 93.7^\circ) + 1.23 e^{-62.9t} + 0.234 e^{-125t}\ \text{A}.$$

Example 2.9

Our final example of *RL* circuits will be the circuit shown in Fig. 2.13, in which both kinds of sources, d.c. and a.c., are presented. Consider the above circuit and find the transient current through resistance R_1. The circuit parameters are: $R_1 = R_2 = 5\ \Omega$, $L = 0.01$ H, $I_s = 4$ A d.c. and $v_s(t) = 100\sqrt{2}\sin(1000t + 15^\circ)$ V.

Solution

1) The characteristic equation for this circuit may be determined as

$$Z(s) = R_1 + R_2 + sL = 0 \quad \text{or} \quad 0.01s + 10 = 0,$$

which gives

$$s = -1000\ \text{s}^{-1} \quad \text{or} \quad \tau = 1\ \text{ms}.$$

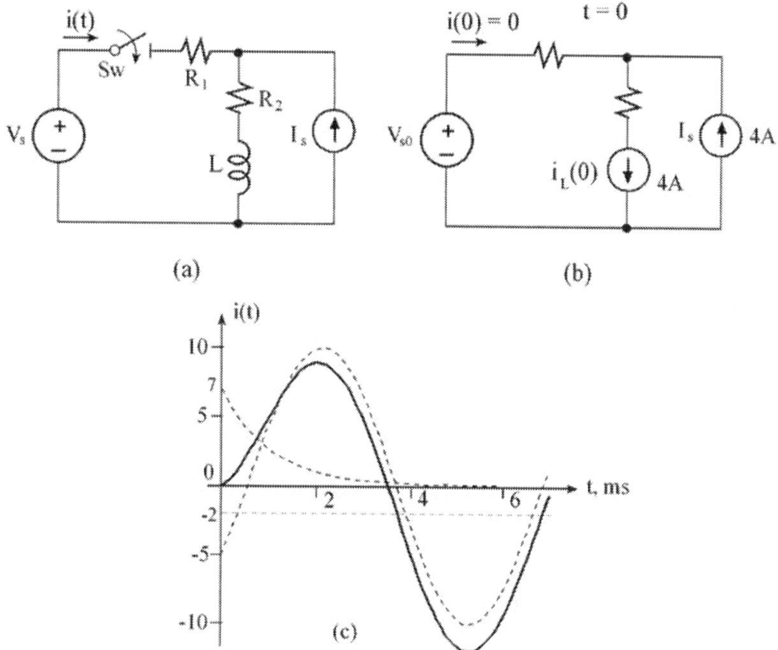

Figure 2.13 A given circuit of Example 2.9 (a), its equivalent for $t=0$ (b) and the plots of the current and its components (c).

Thus,

$$i_{1,n} = Ae^{-1000t}.$$

2) The forced response (using the superposition principle) is

$$i_f = i_{(Is)} + i_{(vs)} = -2 + \frac{100\sqrt{2}}{\sqrt{10^2+10^2}}\sin(1000t + 15^\circ - 45^\circ)$$
$$= -2 + 10\sin(1000t - 30^\circ)\text{ A}.$$

3) The inductance current prior to (and after) switching is $i_L(0) = i_L(0_-) = I_s = 4$ A.

4) The initial value of the current through R_1 (the dependent initial condition) is found in the circuit of Fig. 2.13(b). By inspection of this circuit, we shall conclude that this current is zero (since both branches with current sources, which possess an infinite inner resistance, behave as an open circuit for the voltage source, and the two equal current sources are connected in the right loop in series without sending any current to the left loop). Thus, $i_1(0) = 0$.

5) The integration constant, therefore, is obtained as $A = i_1(0) - i_{1,f}(0) = 0 + 2 - 10\sin(-30°) = 7$ A. Hence,

$$i_1(t) = -2 + 10\sin(1000t - 30°) + 7e^{-1000t} \text{ A}.$$

This current is plotted in Fig. 2.13(c).

2.3.3 Applying the continuous flux linkage law to *L*-circuits

As we have observed earlier (see Section 1.7.4), when an *RL* circuit is disconnected from a source, say for instance a d.c. source, by the rapid opening of a switch a very high voltage appears across the switch, which may result in a breakdown of the circuit insulation. In this section, we shall review this phenomenon by introducing a number of examples in which the problem is solved using the continuous flux linkage principle. Let us consider the circuit in Fig. 2.14(a). The current prior to switching is $i(0_-) = I_0 = V_0/R$ and, according to the switching law, at the first moment after switching it remains the same

$$i(0_+) = I_0.$$

Because the resistance of an open switch is infinite $R_{sw} \to \infty$, the voltage across the switch will also be infinite $v_{sw} \to \infty$. In reality an infinite voltage will not be reached, since the resistance of the actual switches in the open position is very high, but not infinite. Another reason that the voltage cannot reach infinity is that the spark appears between the switch contacts, and the stored energy is

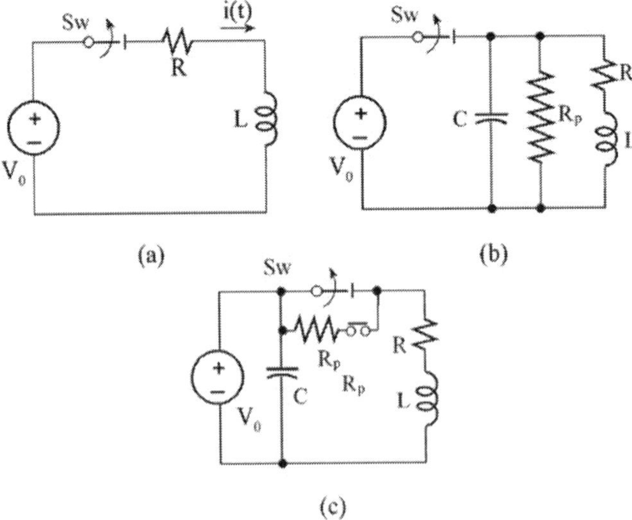

Figure 2.14 A series *RL* circuit switched off instantaneously (a), an *RL* circuit with a parallel resistance and capacitance (b) and an *RL* circuit with a resistance and capacitance parallel to the switch (c).

dissipated in ionizing the air surrounding the contacts. This phenomenon is used in special inductance coils for generating high voltage peaks (for instance, in the ignition system of an automobile such a coil is used to initialize the arcs across the spark plugs to ignite the gasoline in the engine cylinders).

In power circuits, such excess voltages are detrimental and must be avoided. It is useful to connect a substantial resistance parallel to the circuit, Fig. 2.14(b), or, which is even better, parallel to the switch (or breaker), Fig. 2.14(c). In these figures, C represents the stray capacitance shunting the breaker. The presence of an inductance and capacitance raises the differential equation to one of the second order, which will be examined in the following sections. Let us next consider a few examples of the switching phenomenon in first order RL circuits.

Example 2.10

In the circuit of Fig. 2.15, which contains two coils, the switch opens almost instantaneously and coil L_2, whose current prior to switching was different from that of coil L_1, is connected in series with coil L_1. (a) Find the transient current and (b) Estimate the voltage drop between the switch contacts, if the estimated switching time is $\Delta t \approx 10\ \mu\text{s}$. The circuit parameters are: $L_1 = 20\ \text{mH}$, $L_2 = 80\ \text{mH}$, $R_1 = 2\ \Omega$, $R_2 = R_3 = 4\ \Omega$ and $V_s = 12\ \text{V}$ (see also Example 1.8).

Solution (a)

1) By inspection of the circuit after switching, we observe that two coils are connected in series, thus

$$s = \frac{R_1 + R_2}{L_1 + L_2} = \frac{6}{(20 + 80)10^{-3}} = 60\ \text{s}^{-1},$$

and the natural response is

$$i_n = Ae^{-60}.$$

2) The forced response is:

$$i_L = \frac{V_s}{R_1 + R_2} = \frac{12}{2 + 4} = 2\ \text{A}.$$

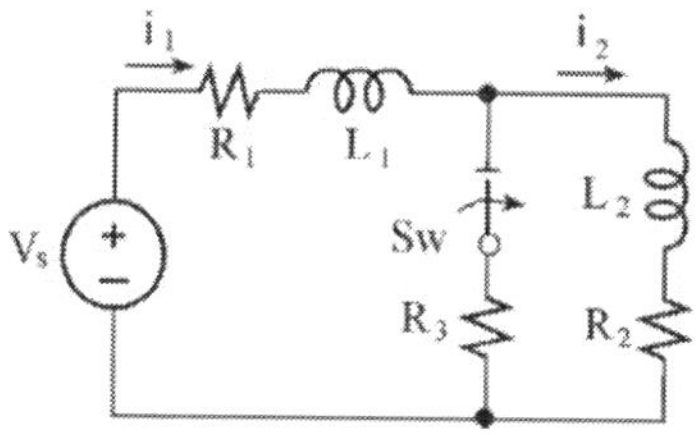

Figure 2.15 A given circuit for Example 2.10.

3) The currents in each coil prior to switching are:

$$i_1(0_-) = \frac{V_s}{R_1 + R_3//R_2} = \frac{12}{2+2} = 3\text{ A} \quad \text{and} \quad i_2(0_-) = \frac{1}{2} i_1(0_-) = 1.5\text{ A}.$$

4) Using the first generalized switching law regarding flux linkages (1.35b), we may write

$$L_1 i_1(0_-) + L_2 i_2(0_-) = (L_1 + L_2) i(0_+).$$

Therefore the common current $i(0_+)$ in both coils after switching is

$$i(0_+) = \frac{20 \cdot 3 + 80 \cdot 1.5}{20 + 80} = 1.8\text{ A}.$$

5) The integrating constant is

$$A = i(0_+) - i_f(0) = 1.8 - 2 = -0.2,$$

and the complete constant is

$$i_f + i_n = 2 - 0.2e^{-60t}\text{ A}.$$

Solution (b)

To approximate the voltage drop we use the expression

$$v_{sw}(0) \cong L_2 \frac{|\Delta i_2|}{\Delta t}.$$

Since the current rise is $\Delta i_2 = i_2(0_-) - i(0_+) = 1.5 - 1.8 = -0.3$ A, therefore

$$v_{sw} \cong 80 \cdot 10^{-3} \frac{0.3}{10 \cdot 10^{-6}} \cong 2.4 \cdot 10^3 = 2.4\text{ kV}.$$

Example 2.11

In the circuit of Fig. 2.16, with $L_1 = L_2 = 24$ mH, $M = 12$ mH and $R_1 = R_2 = 1\ \Omega$, the switch opens practically instantaneously after being closed for a long time. Find the current i_2 and estimate the voltage drop in the switch, if $\Delta t_{sw} \cong 1\ \mu$s.

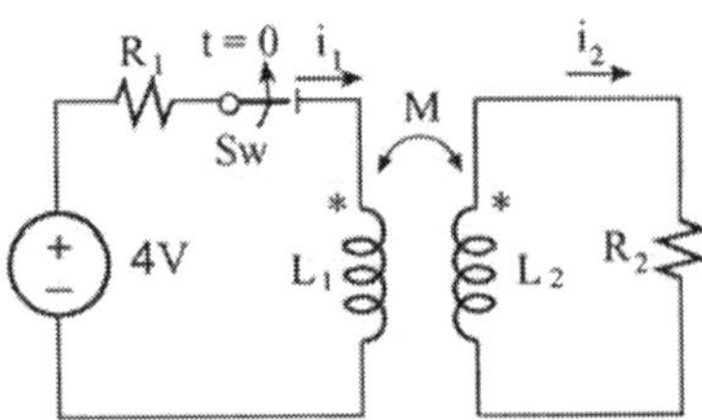

Figure 2.16 A given circuit of Example 2.11.

Solution

1) The time constant of the secondary circuit is

$$\tau = \frac{L_2}{R_2} = \frac{24 \cdot 10^{-3}}{1} = 24\ \text{ms} \quad \text{or} \quad s = -41.7\ \text{s}^{-1}$$

and the natural response is

$$i_{2,n} = Ae^{-41.7t}.$$

2) Since the circuit after switching is source free, the forced response is zero: $i_{2,f} = 0$.

3) The initial value of the current in the transformer secondary may be found in accordance with the principle of flux linkage continuance (first generalized switching law), i.e.,

$$(\lambda_L + \lambda_M)_{t=0_+} = (\lambda_L + \lambda_M)_{t=0_-},$$

or

$$L_1 i_1(0_+) - Mi_2(0_+) + L_2 i_2(0_+) - Mi_1(0_+) = L_1 i_1(0_-) - Mi_2(0_-) + L_2 i_2(0_-) - Mi_1(0_-).$$

Since $i_2(0_-) = 0$, $i_1(0_-) = 4\ \text{A}$ and $i_1(0_+) = 0$, we have $(L_2 - M)i_2(0_+) = (L_1 - M)i_1(0_-)$ and since $L_1 = L_2$, we have $i_2(0_+) = i_1(0_-) = 4\ \text{A}$.

5) Omitting step 4 (since non-dependent initial values are needed) we obtain

$$A = i_2(0) - i_{2,f}(0) = 4,$$

and

$$i_2 = 4e^{-41.7t}.$$

The voltage drop across the switch will be

$$v_{sw} = \left| L\frac{-\Delta i_1}{\Delta t} - M\frac{\Delta i_2}{\Delta t} \right| = \frac{(24 \cdot 4 + 12 \cdot 4)10^{-3}}{10^{-6}} = 144 \cdot 10^3 = 144\ \text{kV}.$$

Checking the energy preservation, we may find:

The energy prior to switching:

$$\frac{L_1 i_1^2(0_-)}{2} + \frac{L_2 i_2^2(0_-)}{2} + Mi_1(0_-)i_2(0_-) = \frac{L_1 i_1^2(0_-)}{2} = \frac{24 \cdot 10^{-3} \cdot 4^2}{2} = 192\ \text{mJ},$$

and the energy after switching:

$$\frac{L_1 i_1^2(0_+)}{2} + \frac{L_2 i_2^2(0_+)}{2} + Mi_1(0_+)i_2(0_+) = \frac{L_2 i_2^2(0_+)}{2} = \frac{24 \cdot 10^{-3} \cdot 4^2}{2} = 192\ \text{mJ},$$

which are the same.

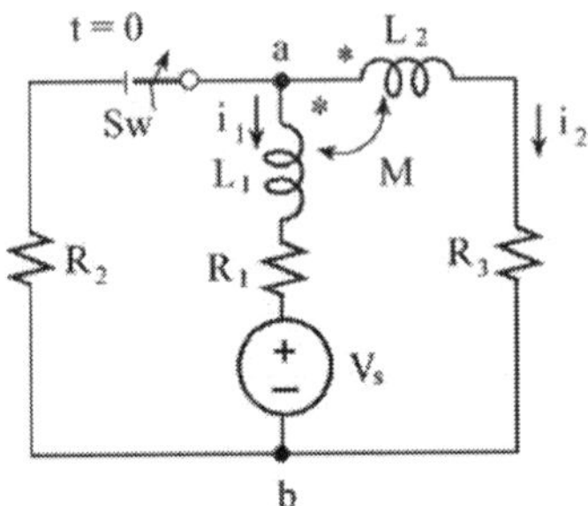

Figure 2.17 A given circuit for Example 2.12.

Example 2.12

In the circuit of Fig. 2.17 containing the mutual inductance, the switch opens practically instantaneously, after being closed for a long time. Find the transient response of current i_1 for two cases: (a) Both dotted terminals are connected to the common node "a" and (b) Only one dotted terminal is connected to the common node "a". The circuit parameters are $R_1 = 5\ \Omega$, $R_2 = R_3 = 10\ \Omega$, $L_1 = 0.1$ H, $L_2 = 0.2$ H, $M = 0.05$ H and $V_s = 60$ V.

Solution (a)

1) The characteristic equation will be determined by writing the KVL equation for the right loop and equating it to zero (note that after switching $i_1 = i_2 = i$ and all the elements are connected in series):

$$[R_1 + R_2 + s(L_1 + L_2 - 2M)]i = 0.$$

Thus,

$$(0.1 + 0.2 - 0.1)s + 5 + 10 = 0 \quad \text{and} \quad s = -75\ \text{s}^{-1}.$$

Therefore, the natural response is

$$i_n = Ae^{-75t}.$$

2) The steady-state current is

$$i_f = \frac{V_s}{R_1 + R_2} = \frac{60}{5 + 10} = 4\ \text{A}.$$

3) The initial value of the current $i(0_+)$ shall now be found using the first generalized law, i.e.,

$$i_1(0_+)(L_1 + L_2 - 2M) = i_1(0_-)L_1 + i_2(0_-)L_2 - [i_1(0_-) + i_2(0_-)]M,$$

or

$$i_1(0_+) = \frac{i_1(0_-)L_1 + i_2(0_-)L_2 - [i_1(0_-) + i_2(0_-)]M}{(L_1 + L_2 - 2M)}$$

$$= \frac{6 \cdot 0.1 + 3 \cdot 0.2 - 9 \cdot 0.05}{0.1 + 0.2 - 0.1} = 3.75 \text{ A},$$

where the currents prior to switching are (by inspection of the circuit in Fig. 2.17): $i_1(0_-) = 6$ A and $i_2(0_-) = 3$ A.

4–5) The integration constant can now be evaluated as

$$A = i(0) - i_f(0) = 3.75 - 4 = -0.25,$$

and the complete response is

$$i(t) = 4 - 0.25e^{-75t} \text{ A}.$$

Solution (b)

1) The exchange of the position of the dotted terminals results in a positive sign connection of the mutual inductance. Therefore,

$$[R_1 + R_2 + s(L_1 + L_2 + 2M)]i = 0,$$

or

$$15 + 0.4s = 0 \quad \text{and} \quad s = -37.5 \text{ s}^{-1}.$$

Thus,

$$i_n = Ae^{-37.5t}.$$

2) The forced response is not influenced by the dotted terminal exchange and remains the same $i_f = 4$ A.

3) The initial condition is now found as

$$i_1(0_+) = \frac{i_1(0_-)L_1 + i_2(0_-)L_2 + [i_1(0_-) + i_2(0_-)]M}{(L_1 + L_2 + 2M)}$$

$$= \frac{6 \cdot 0.1 + 3 \cdot 0.2 + 9 \cdot 0.05}{0.1 + 0.2 + 0.1} = 4.125 \text{ A}.$$

4–5) The integration constant is

$$A = 4.125 - 4 = 0.125,$$

and the complete response in this case is

$$i(t) = 4 + 0.125e^{-37.5t} \text{ A}.$$

Example 2.13

Our last example in this section will be the circuit shown in Fig. 2.18(a). This

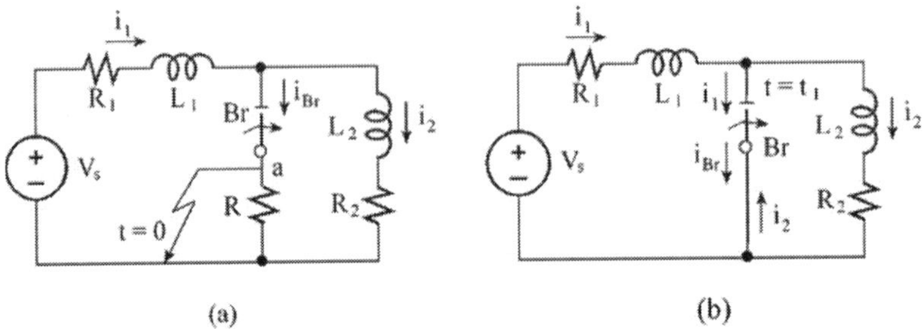

Figure 2.18 A given circuit for Example 2.13(a) and the circuit after the short-circuiting for the second stage of the transients (b).

circuit represents the equivalent of a d.c. supply network. At the instant of time $t = 0$, the short-circuit fault occurs at node "a" and when the short-circuit current i_{sc} through the breaker reaches the value $I = 500$ A, the circuit breaker opens practically instantaneously. Find the transient response of current i_2 after the fault. The circuit parameters are $R_1 = 1\ \Omega$, $R = R_2 = 9\ \Omega$, $L_1 = 0.01$ H, $L_2 =$ 0.45 H and $V_s = 1100$ V.

Solution

First stage (the period between a short circuit $t = 0$ and opening the circuit breaker, BR, $t = t_1$).

1) Since the circuit is divided into two sub circuits: the left one with current i_1 and the right one with current i_2, we shall obtain two time constants and two natural responses:

$$(1)\quad \tau_1 = \frac{L_1}{R_1} = \frac{0.01}{1} = 0.01\text{ s},\quad \text{or}\quad s_1 = -100\text{ s}^{-1}\quad \text{and}\quad i_{1,n} = A_1 e^{-100t},$$

$$(2)\quad \tau_2 = \frac{L_2}{R_2} = \frac{0.45}{9} = 0.05\text{ s},\quad \text{or}\quad s_2 = -20\text{ s}^{-1}\quad \text{and}\quad i_{2,n} = A_2 e^{-20t}.$$

2) The forced responses in these circuits are:

$$(1)\quad i_{1,f} = \frac{V_s}{R_1} = \frac{1100}{1} = 1100\text{ A},\qquad (2)\quad i_{2,f} = 0.$$

3) The initial conditions of the above two currents may be obtained by inspection of the given circuit prior to short-circuiting

$$(1)\quad i_1(0_-) = \frac{V_s}{R_1 + R_2//R_3} = \frac{1100}{1 + 4.5} = 200\text{ A},\qquad (2)\quad i_2(0_-) = \frac{1}{2} i_1(0_-) = 100\text{ A}.$$

4–5) The integration constants are

$$(1)\quad A_1 = i_1(0) - i_{1,f} = 200 - 1100 = -900,$$

$$(2)\quad A_2 = i_2(0) - i_{2,f} = 100 - 0 = 100.$$

The complete response in each of the circuits is

$$(1)\quad i_1(t) = 1100 - 900e^{-100t}\ \mathsf{A}, \qquad (2)\quad i_2(t) = 100e^{-20t}\ \mathsf{A}.$$

In order to determine the instant of time, at which the breaker opens, we must solve the equation

$$i_{Br} = i_1 - i_2|_{t=t_1} = 500,$$

or

$$1100 - 900e^{-100t_1} - 100e^{-20t_1} = 500.$$

This transcendental equation can now be solved by the iteration approach. Since the time constant of the second circuit is relatively large, we assume that current i_2 is a constant. Thus, the first estimation of time t_1 will be found as

$$900e^{-100t_1} = 1100 - 600, \quad \text{and} \quad -100t_1 = \ln\frac{500}{900} \quad \text{or} \quad t_1^{(1)} = 5.6\ \mathsf{ms}$$

For the second estimation we assume that current $i_2 = 100e^{-20\cdot 5.6\cdot 10^{-3}} = 89.4\ \mathsf{A}$, therefore, now

$$900e^{-100t_1} = 511,$$

to which the solution is

$$t_1^{(2)} = -\frac{\ln(511/900)}{100} = 0.566\cdot 10^{-2} \cong 5.7\ \mathsf{ms}.$$

Since this result is very close to the previous one, no more estimations are needed, and the value $t_1 = 5.7$ ms is taken as the answer.

Second stage (the period of time after the breaker opens $t > 5.7$ ms)

1) In this stage the circuit consists of only one loop, whose characteristic equation is

$$R_1 + R_2 + s(L_1 + L_2) = 0.$$

Upon substitution of the numerical data the time constant becomes

$$\tau = \frac{0.46}{10} = 0.046\ \mathsf{s}, \quad \text{or} \quad s = 21.7\ \mathsf{s}^{-1}.$$

2) The forced response is

$$i_{2,f} = \frac{1100}{1+9} = 110\ \mathsf{A}.$$

3) The current values prior to switching are

$$i_1(0_-) = i_1(t_1) = 1100 - 900e^{-100\cdot 5.7\cdot 10^{-3}} = 591\ \text{A},$$

$$i_2(0_-) = i_2(t_1) = 100e^{-20\cdot 5.7\cdot 10^{-3}} = 89\ \text{A}.$$

The initial value of i_2 after switching (note that both currents are now equal), in accordance with the first generalized law, is

$$i_2(0_+) = \frac{i_1(0_-)L_1 + i_2(0_-)L_2}{L_1 + L_2} = \frac{591\cdot 0.01 + 89\cdot 0.45}{0.46} \cong 100\ \text{A}.$$

4–5) The integration constant can now be found $A = i_2(0) - i_f = 100 - 110 = -10$.

Therefore, the complete response of current i_2 after the short-circuit fault is

$$i_2(t) = 100e^{-20t} \qquad \text{for} \quad 0 < t < 5.7$$

$$i_2(t) = 110 - 10e^{-21.7(t-t_1)} \quad \text{for} \quad 5.7 < t < \infty.$$

Note that at the moment $t = 5.7$ ms the current changes rapidly (however, the total magnetic flux of both inductances remains the same).

2.4 *RC* CIRCUITS

We shall approach the transient analysis of *RC* circuits keeping in mind the principle of duality. As we have noted the *RC* circuit is dual to the *RL* circuit. This means that we may use all the achievements and results we obtained in the previous section regarding the inductive circuit for capacitance circuit analysis. For instance, the time constant of a simple *RL* circuit has been obtained as $\tau_L = L/R$, for a simple *RC* circuit it must be $\tau_C = C/G$ (i.e., L is replaced by C and R by G, which are dual elements). Since $G = 1/R$, the time constant of an *RC* circuit can, of course, be written as $\tau_C = RC$. In the following sections, more examples of such duality will be presented.

2.4.1 Discharging and charging a capacitor

Consider once again the *RC* circuit (also see section 1.3.1) shown in Fig. 2.19(a), in which R and C are connected in parallel. Prior to switching the capacitance was charged up to the voltage of the source V_s. After opening the switch, the capacitance discharges through the resistance. The time constant of the circuits is $\tau = RC$ and the initial value of the capacitance voltage is $V_{C0} = V_s$. The forced response component of the capacitance voltage is zero, since the circuit after switching is source free. Thus,

$$v_C(t) = V_{C0}e^{-\frac{t}{RC}}. \tag{2.11}$$

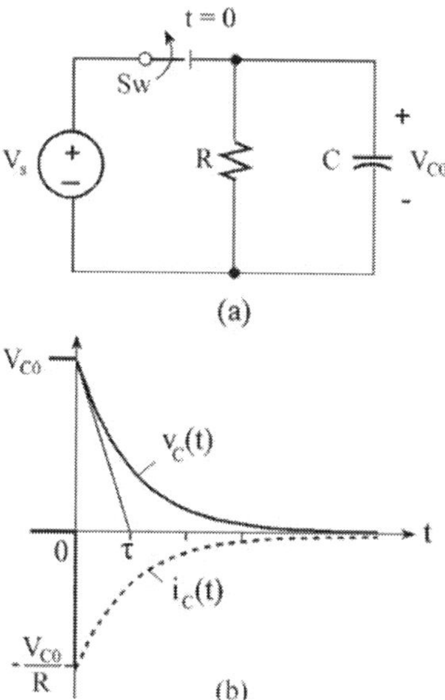

Figure 2.19 A circuit of a parallel connection of resistance and capacitance (a) and the plots of the discharging voltage and current (b).

The current response will be

$$i_C = C\frac{dv_C}{dt} = -\frac{V_{CO}}{R}e^{-\frac{t}{RC}}. \tag{2.12}$$

Note that 1) the current changes abruptly at $t = 0$ from zero (prior to switching) to V_{CO}/R and 2) its direction is opposite to the charging current. This current and the capacitance voltage are plotted in Fig. 2.19(b). Also note that the voltage curve in Fig. 2.19(b) is similar to the current curve in the RL circuit, and inversely the current curve is similar to the voltage curve in the RL circuit, as shown in Fig. 2.2(c). This fact is actually another example of duality.

Let us now show that the energy stored in the electric field of the capacitance completely dissipates in the resistance, converging into heat, during the transients. The energy stored is

$$w_e = \frac{CV_{CO}^2}{2}. \tag{2.13}$$

The energy dissipated is

$$w_R = \int_0^\infty \frac{v_C^2}{R}dt = \frac{V_{CO}^2}{R}\int_0^\infty e^{-\frac{2t}{RC}}dt = -\frac{RCV_{CO}^2}{2R}e^{-\frac{2t}{RC}}\bigg|_0^\infty = \frac{CV_{CO}^2}{2}. \tag{2.14}$$

Hence, the energy conservation law has been conformed to.

Consider next the circuit of Fig. 2.20, in which the capacitance is charging

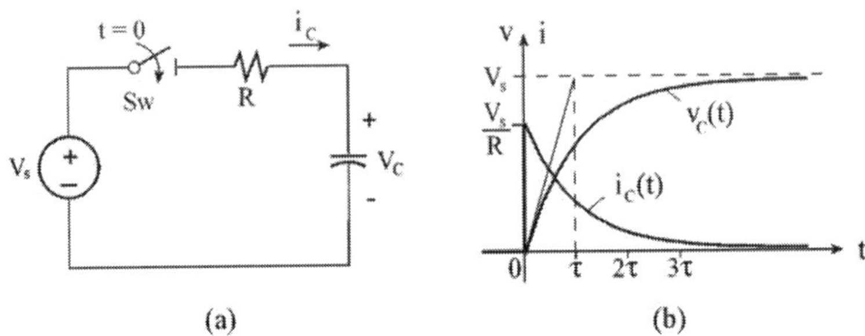

Figure 2.20 An *RC* circuit in which the capacitance is charging (a) and the plots of the voltage and current responses (b).

through the resistance after closing the switch. The natural response of this circuit is similar to the previous circuit, i.e.,

$$v_{C,n} = Ae^{-t/RC}.$$

However, because of the presence of a voltage source, the forced response (step 2) will be $v_{C,f} = V_s$, since in the steady-state operation the current is zero (the capacitance is fully charged), and the voltage across the capacitance is equal to the source voltage.

Next, we realize that the initial value of the capacitance voltage, prior to switching (step 3), is zero, and the constant of integration (step 5) is obtained as $A = 0 - V_s = -V_s$.

The complete response, therefore is

$$v_C = V_s - V_s e^{-t/RC} = V_s(1 - e^{-t/RC}). \tag{2.15}$$

The current response can now be found as

$$i_C = C\frac{dv_C}{dt} = \frac{V_s}{R}e^{-t/RC}. \tag{2.16}$$

Both responses, voltage and current, are plotted in Fig. 2.20(b). Note again that these two curves are similar to the current and voltage response curves respectively in the *RL* circuit as shown in Fig. 2.1(b) and (c).

2.4.2 *RC* circuits under d.c. supply

Let us now consider more complicated *RC* circuits, fed by a d.c. source. If, for instance, in such circuits a few resistances are connected in series/parallel, we may simplify the solution by determining R_{eq} and reducing the circuit to a simple *RC*-series, or *RC*-parallel circuit. An example of this follows.

Example 2.14

Consider the circuit of Fig. 2.21 with $R_1 = R_2 = R_3 = R_4 = 50\ \Omega$, $C = 100\ \mu\text{F}$ and

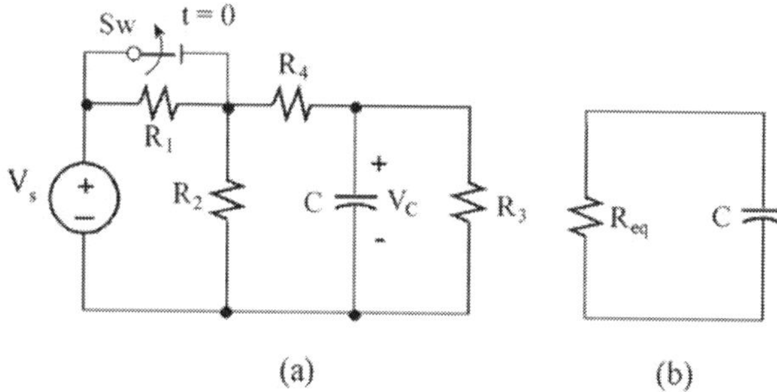

Figure 2.21 A given circuit of Example 2.14 (a) and its simplified equivalent (b).

$V_s = 250$ V. Find the voltage across the capacitance after the switch opens at $t = 0$.

Solution

After the voltage source is "killed" (short-circuited), we may determine the equivalent resistance, which is in series/parallel to the capacitance, Fig. 2.21(b): $R_{eq} = (R_1//R_2 + R_4)//R_3$, which, upon substituting the numerical values, results in $R_{eq} = 30\,\Omega$. Thus, the time constant (step 1) is

$$\tau = R_{eq}C = 30 \cdot 100 \cdot 10^{-6} = 3 \text{ ms}, \quad \text{and} \quad v_{C,n} = Ae^{-t/3}, \quad (t \text{ is in ms}).$$

By inspection of the circuit in its steady-state operation ($t \to \infty$) the voltage across the capacitor (the forced response) can readily be found (step 2): $v_{C,f} =$ 50 V. The initial value of the capacitance voltage (step 3) must be determined prior to switching:

$$v_C(0_+) = v_C(0_-) = V_s \frac{R_3}{R_3 + R_4} = 250 \frac{50}{100} = 125 \text{ V}.$$

Hence, the integration constant (step 5) is found to be $A = v_C(0) - v_{C,f} = 125 - 50 = 75$, and the complete response is

$$v_C(t) = 50 + 75e^{-t/3}.$$

With the above expression of the integration constant (see step 5), the complete response in the first order circuit can be written in accordance with the following formula (given here in its general notation, for either voltage or current):

$$f(t) = f_f + f_n = f_f + (f_0 - f_{f,0})e^{-t/\tau}, \tag{2.17}$$

where f_0 and $f_{f,0}$ are the initial values of the complete and the forced responses respectively. Or in the form

$$f(t) = f_f(1 - e^{-t/\tau}) + f_0 e^{-t/\tau}, \tag{2.18}$$

and for zero initial conditions ($f_0 = 0$)

$$f(t) = f_f(1 - e^{-t/\tau}). \tag{2.19}$$

In the following examples, we shall consider more complicated *RC* circuits.

Example 2.15

At the instant $t = 0$ the capacitance is interswitched between two voltage sources, as shown in Fig. 2.22(a). The circuit parameters are $R_1 = 20\ \Omega$, $R_2 = 10\ \Omega$, $R_3 = R_4 = 100\ \Omega$, $C = 0.01$ F, and the voltage sources are $V_{s1} = 60$ V and $V_{s2} = 120$ V. Find voltage $v_C(t)$ and current $i_2(t)$ for $t > 0$.

Solution

1) The input impedance, Fig. 2.22(b), is:

$$Z_{in}(s) = \frac{1}{sC} + R_2//R_3//R_4.$$

Upon substitution of the numerical data and equating it to zero yields

$$\frac{1}{s}10^2 + \frac{50}{6} = 0, \quad \text{or} \quad s = -12\ \text{s}^{-1},$$

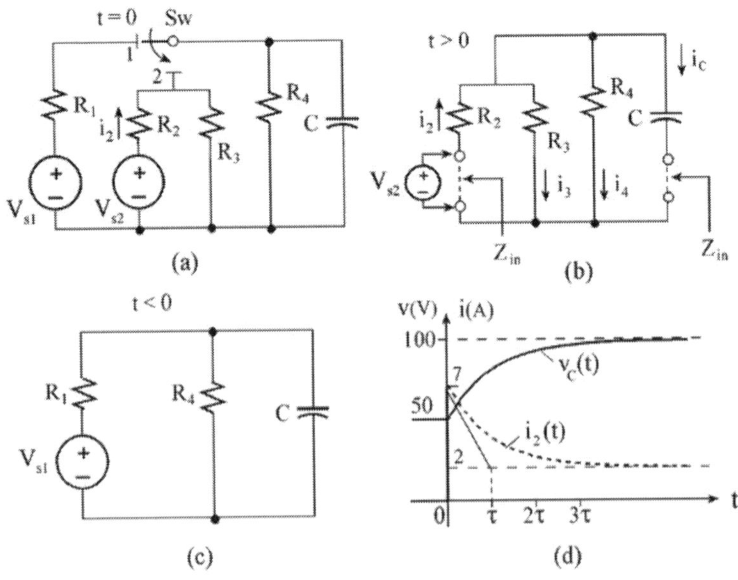

Figure 2.22 A given *RC* circuit of Example 2.15 (a), a circuit for determining the input impedance and the forced response (b), a circuit for determining the initial value (c) and the curves of the voltage and current responses (d).

and the natural response becomes

$$v_{C,n} = Ae^{-12t}.$$

2) The forced response is found as the voltage drop in two parallel resistances $R_{3,4} = 50\ \Omega$. With the voltage division formula, we obtain

$$v_{C,f} = V_s \frac{R_{3,4}}{R_2 + R_{3,4}} = 120 \frac{50}{10 + 50} = 100\ \text{V}.$$

3) The initial value of the capacitance voltage must be determined from the circuit prior to switching, as shown in Fig. 2.22(c). Applying the voltage division once again, we have

$$v_C(0_+) = v_C(0_-) = 60 \frac{100}{20 + 100} = 50\ \text{V}.$$

5) (Step 4 is omitted, as it is unnecessary). In accordance with equation 2.17 we obtain

$$v_C(t) = 100 + (50 - 100)e^{-12t} = 100 - 50e^{-12t}\ \text{V}.$$

Current i_2 can now be easily found as, Fig. 2.22(b),

$$i_2(t) = i_R + i_C = \frac{v_C}{R_{3,4}} + C\frac{dv_C}{dt} = 2 - 1e^{-12t} + 0.01(-50)(-12)e^{-12t}$$

$$= 2 + 5e^{-12t}\ \text{A}.$$

Both curves, of v_C and i_2, are plotted in Fig. 2.22(d). Note that the current i_2 changes abruptly from zero to 7 A. Our next example will be a second order *RC* circuit.

Example 2.16

Consider the second order *RC* circuit shown in Fig. 2.23(a), having $R_1 = R_3 = 200\ \Omega$, $R_2 = R_4 = 100\ \Omega$, $C_1 = C_2 = 100\ \mu\text{F}$ and two sources $V_s = 300$ V and $I_s = 1$ A. The switch opens at $t = 0$ after having been closed for a long time. Find current $i_2(t)$ for $t > 0$.

Solution

1) We shall determine the characteristic equation by using mesh analysis for the circuit in Fig. 2.23(a) after opening the switch and with "killed" sources

$$\left(\frac{1}{sC_1} + \frac{1}{sC_2} + R_1 + R_2\right)i_2 - \frac{1}{sC_2}i_3 = 0$$

$$-\frac{1}{sC_2}i_2 + \left(\frac{1}{sC_2} + R_3\right)i_3 = 0.$$

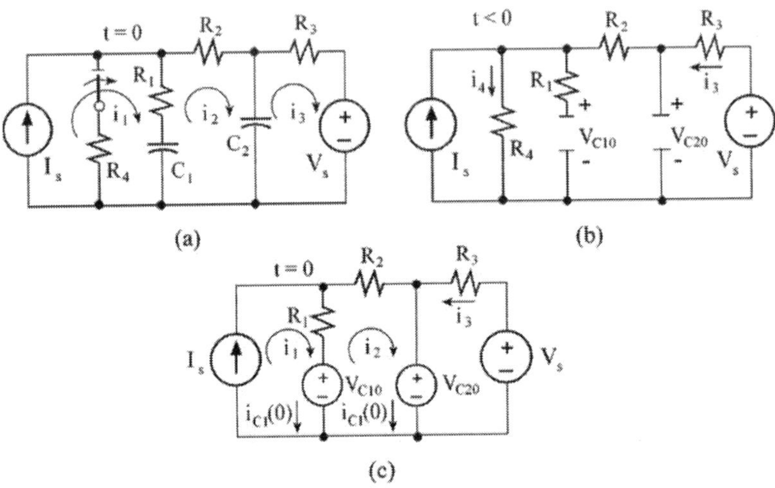

Figure 2.23 A second order *RC* circuit of Example 2.16 (a), an equivalent circuit for the calculation of the independent initial conditions (b) and an equivalent circuit for the calculation of the dependent initial conditions.

Equating the determinant for this set of equations to zero, we may obtain the characteristic equation (note that $C_1 = C_2 = C$)

$$\left(\frac{2}{sC} + R_1 + R_2\right)\left(\frac{1}{sC} + R_3\right) - \left(\frac{1}{sC}\right)^2 = 0.$$

Upon substituting the numerical data the above becomes $6s^2 + 700s + 10^4 = 0$, and the roots are $s_1 = -16.7\ \mathrm{s}^{-1}$ and $s_2 = -100\ \mathrm{s}^{-1}$. Therefore, the natural response becomes

$$i_{2,n} = A_1 e^{-16.7t} + A_2 e^{-100t}\ \mathrm{A}.$$

2) By inspection of the circuit in Fig. 2.23(a), in its steady-state operation (after the switch had been open for a long time), we may conclude that the only current flowing through resistance R_2 is the current of the current source, i.e., $i_{2,f} = I_s = 1\ \mathrm{A}$.

3) In order to determine the independent initial condition, i.e. the capacitance voltages at $t = 0$, we shall consider the circuit equivalent for this instant of time, shown in Fig. 2.23(b). Using the superposition principle, we may find the current through resistance R_3 as

$$i_3 = \frac{V_3}{R_2 + R_3 + R_4} - I_s \frac{R_4}{R_2 + R_3 + R_4} = \frac{300}{400} - 1\frac{100}{400} = 0.5\ \mathrm{A},$$

and the voltage across capacitance C_2 as $v_{C2} = V_{C20} = V_s - R_3 i_3 = 300 -$

$200 \cdot 0.5 = 200$ V. In a similar way

$$i_4 = \frac{V_3}{R_2 + R_3 + R_4} + I_s \frac{R_2 + R_3}{R_2 + R_3 + R_4} = \frac{300}{400} + 1 \frac{100 + 200}{400} = 1.5 \text{ A},$$

and

$$v_{C1}(0) = V_{C10} = R_4 i_4 = 100 \cdot 1.5 = 150 \text{ V}.$$

4) Since the response that we are looking for is the current in a resistance, it can change abruptly. For this reason, and also since the response is of the second order, we must determine the dependent initial conditions, namely $i_2(0)$ and $di_2/dt|_{t=0}$. This step usually has an abundance of calculations. (This is actually the reason why the transformation methods, in which there is no need to determine the dependent initial conditions, are preferable). However, let us now perform these calculations in order to complete the classical approach.

In order to determine $i_2(0)$ we must consider the equivalent circuit, which fits instant $t=0$, Fig. 2.23(c). With the mesh analysis we have $R_1[i_2(0) - I_s] + R_2 i_2(0) = V_{C10} - V_{C20}$, or

$$i_2(0) = \frac{V_{C10} - V_{C20} + R_1 I_s}{R_1 + R_2} = \frac{150 - 200 + 200 \cdot 1}{200 + 100} = 0.5 \text{ A}.$$

For the following calculations, we also need the currents through the capacitances, i.e., through the voltage sources, which represent the capacitances. First, we find current i_3:

$$i_3 = (V_s - V_{C20})/R_3 = (300 - 200)/200 = 0.5 \text{ A},$$

then

$$i_{C1}(0) = I_s - i_2(0) = 1 - 0.5 = 0.5 \text{ A}$$

$$i_{C2}(0) = i_2(0) + i_3(0) = 0.5 + 0.5 = 1.0 \text{ A}.$$

In order to determine the derivative of i_2, we shall write the KVL equation for the middle loop (Fig. 2.23(c)):

$$-v_{C1} - R_1(I_s - i_2) + R_2 i_2 + v_{C2} = 0.$$

After differentiation we have

$$(R_1 + R_2)\frac{di_2}{dt} = \frac{dv_{C1}}{dt} - \frac{dv_{C2}}{dt} = \frac{1}{C}(i_{C1} - i_{C2}),$$

or

$$\left.\frac{di_2}{dt}\right|_{t=0} = \frac{1}{(R_1 + R_2)C}(i_{C1} - i_{C2}) = \frac{1}{300 \cdot 10^{-4}}(0.5 - 1) = -16.7.$$

5) In accordance with equation 1.61 we can now find the integration constants

$$A_1 + A_2 = i_2(0) - i_{2,f}(0) = 0.5 - 1$$

$$s_1 A_1 + s_2 A_2 = \left.\frac{di_2}{dt}\right|_{t=0} - \left.\frac{di_{2,f}}{dt}\right|_{t=0} = -16.7 - 0,$$

or

$$A_1 + A_2 = -0.5$$
$$-16.7A_1 - 100A_2 = -16.7,$$

to which the solution is

$$A_1 = \frac{-0.5(-100) + 16.7}{-100 + 16.7} = -0.8$$

$$A_2 = -0.5 - A_1 = -0.5 + 0.8 = 0.3.$$

Thus the complete response is

$$i_2(t) = 1 - 0.8e^{-16.7t} + 0.3e^{-100t}\ \text{A}.$$

2.4.3 *RC* circuits under a.c. supply

If the capacitive branch switches to the a.c. supply of the form $v_s = V_{sm}\sin(\omega t + \psi_v)$, as shown in Fig, 2.24(a), the forced response of the capacitance

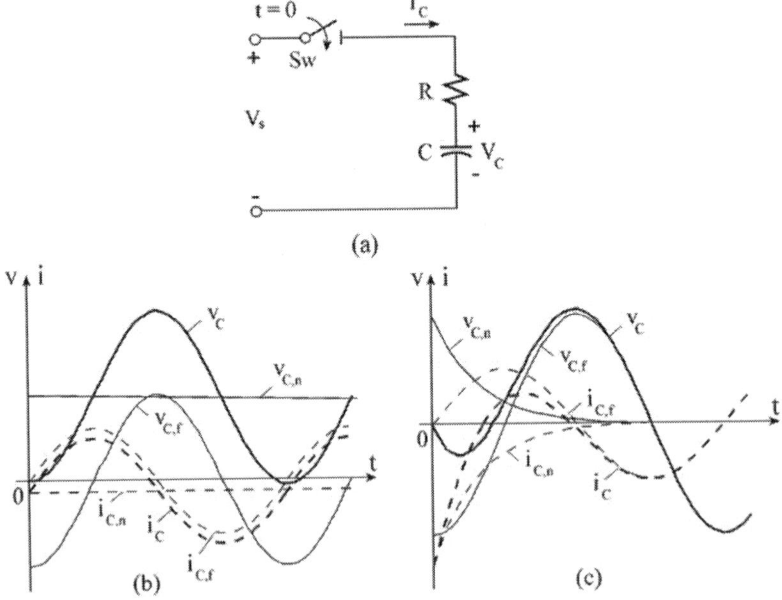

Figure 2.24 An *RC* circuit under an a.c. supply (a), the transient response when the overvoltage occurs (b) and the transient response with the current peak (c).

voltage will be

$$v_{C,f} = V_{Cm}\sin(\omega t + \psi_v - \varphi - \pi/2). \tag{2.20}$$

Here phase angle ψ_v (switching angle), is appropriate to the instant of switching $t=0$

$$V_{cm} = \frac{1}{\omega C}\frac{V_{sm}}{\sqrt{R^2 + (1/\omega C)^2}} \tag{2.21a}$$

and

$$\varphi = \tan^{-1}(-1/R\omega C) \tag{2.21b}$$

Since the natural response does not depend on the source, it is

$$v_{C,f} = Ae^{-t/RC}.$$

With zero initial conditions, i.e., $v_C(0)=0$, the integration constant becomes

$$A = v_C(0) - v_{C,f}(0) = -V_{Cm}\sin(\psi_v - \varphi - \pi/2). \tag{2.22}$$

Thus, the complete response of the capacitance voltage will be

$$v_C(t) = V_{Cm}[\sin(\omega t + \psi_v - \varphi - \pi/2) - \sin(\psi_v - \varphi - \pi/2)e^{-t/RC}], \tag{2.23}$$

and of the current

$$i_C(t) = C\frac{dv_C}{dt} = I_m\left[\sin(\omega t + \psi_v - \varphi) + \frac{1}{\omega RC}\sin(\psi_v - \varphi - \pi/2)e^{-t/RC}\right], \tag{2.24}$$

where

$$I_m = \omega C V_{Cm} = \frac{V_{sm}}{R\sqrt{1 + (1/\omega RC)^2}} \tag{2.25}$$

and

$$A = \frac{I_m}{\omega RC}\sin(\psi_v - \varphi - \pi/2). \tag{2.26}$$

Since, during the transient behavior, the natural response is added to the forced response of the voltage and current, it may happen that the complete responses will exceed their rated amplitudes. The maximal values of overvoltages and current peaks depend on the switching angle and time constant. If switching occurs at the moment when the forced voltage equals its amplitude value, i.e. when the switching angle $\psi_v = \varphi$ and with a large time constant, the overvoltage may reach the value of an almost double amplitude, $2V_{Cm}$. This is shown in Fig. 2.24(b). It should be noted that the current in this case will almost be its regular value, since at the switching moment its forced response equals zero, and the initial value of the natural response (equation 2.26) is small because of the large resistance due to the large time constant, Fig. 2.24(b). (Compare with

Figs. 2.9 and 2.10 of the current response in an RL circuit under an a.c. supply). On the other hand, if the time constant is small due to the small resistance R, the current peak, at $t = 0$, may reach a very high level, many times that of its rated amplitude, Fig. 2.24(c). However the overvoltage will not occur.

We shall now consider a few numerical examples.

Example 2.17

In the circuit of Fig. 2.25(a), with $R_1 = R_2 = 5\,\Omega$, $C = 500\,\mu\text{F}$ and $v_s = 100\sqrt{2}\sin(\omega t + \pi/2)$, find current $i(t)$ after switching.

Solution

There are two ways of finding the current: 1) straightforwardly and 2) first to find the capacitance voltage and then to perform the differentiation $i = C(dv_C/dt)$. We will present both ways.

1) The time constant (step 1) is $\tau = RC = 5\cdot 500\cdot 10^{-6} = 2.5\cdot 10^{-3}$, therefore $s = -1/\tau = -400\ \text{s}^{-1}$ and the natural response is $i_n = Ae^{-400t}$. The forced response (step 2) is

$$i_f = I_m \sin(\omega t + \pi/2 - \varphi) = 17.5\sin(\omega t + 141.8^\circ),$$

where

$$I_m = \frac{100\sqrt{2}}{\sqrt{5^2 + (1/314\cdot 5\cdot 10^{-4})^2}} = 17.5\ \text{A}$$

and

$$\varphi = \tan^{-1}\frac{-1/(314\cdot 5\cdot 10^{-4})}{5} = -51.8^\circ.$$

The initial value of the capacitance voltage (the initial independent condition, step 3) must be found in the circuit of Fig. 2.25(a) prior to switching

$$v_C(0_-) = \frac{100\sqrt{2}\cdot 6.37}{\sqrt{10^2 + 6.37^2}}\sin\left(\frac{\pi}{2} - \tan^{-1}\frac{-6.37}{10} - \frac{\pi}{2}\right) = 40.8\ \text{V},$$

where

$$x_C = \frac{1}{314\cdot 5\cdot 10^{-4}} = 6.37\ \Omega.$$

The initial value of the current, which is the dependent initial condition (step 4) may be found from the equivalent circuit, for the instant of switching, $t = 0$, which is shown in Fig. 2.25(b):

$$i(0) = \frac{100\sqrt{2} - 40.8}{5} = 20.1\ \text{A}.$$

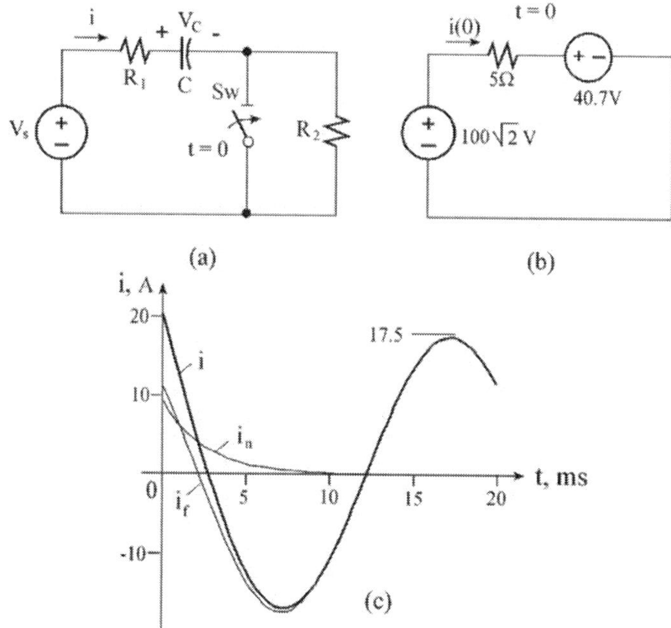

Figure 2.25 A given circuit for Example 2.15 (a), its equivalent for calculating $i(0)$ (b) and the plot of the current (c).

The integration constant and complete response (step 5) will then be

$$A_i = i(0) - i_f(0) = 20.1 - 17.5 \sin 141.8^\circ = 9.3 \text{ A},$$

and

$$i(t) = 17.5 \sin(314t + 141.8^\circ) + 9.3e^{-400t} \text{ A}.$$

2) The difference in the calculation according to way 2) is that we do not need Step 4. Step 1 is the same; therefore, the natural response of the capacitance voltage is $v_{C,n} = Ae^{-400t}$, and we continue with Step 2:

$$v_{C,f} = \frac{100\sqrt{2}\cdot 6.37}{\sqrt{5^2 + 6.37^2}} \sin(314t + \pi/2 + 51.8^\circ - \pi/2) = 111.3 \sin(314t + 51.8^\circ) \text{ V}.$$

Step 3 has already been performed so we can calculate the complete response as

$$v_C(t) = 111.3 \sin(314t + 51.8^\circ) - 46.7e^{-400t} \text{ V},$$

where

$$A_v = v_C(0) - v_{C,f}(0) = 40.8 - 111.3 \sin 51.8^\circ = -46.7.$$

The current can now be evaluated as

$$i = C\frac{dv_C}{dt} = 17.5\sin(314t + 51.8^\circ + \pi/2) + 9.3e^{-400t}\ \mathsf{A},$$

where

$$I_m = 5\cdot 10^{-4}\cdot 314\cdot 111.3 = 17.5 \quad \text{and} \quad A_i = 5\cdot 10^{-4}(-400)(-46.7) = 9.3\ \mathsf{A},$$

which is the same as previously obtained. The plot of current i is shown in Fig. 2.25(c).

Example 2.18

In the circuit of Fig. 2.26(a), the switch closes at $t = 0$. Find the current in the switching resistance R_3. The circuit parameters are: $R_1 = R_2 = R_3 = 10\ \Omega$, $C = 250\ \mu\mathsf{F}$ and $v_s = 100\sqrt{2}\sin(\omega t + \psi_v)$ at $f = 60\ \mathsf{Hz}$. To determine the switching angle ψ_v, assume that at the instant of switching $v_s = 0$ and its derivative is positive.

Solution

The voltage is zero if ψ_v is 0° or 180°. Since the derivative of the sine wave at 0° is positive (and at 180° it is negative), we should choose $\psi_v = 0°$.

1) To determine the time constant (step 1) we shall first find the equivalent resistance $R_{eq} = R_2 + R_1//R_3 = 10 + 5 = 15\ \Omega$. Thus, $\tau = R_{eq}C = 15\cdot 250\cdot 10^{-6} = 3.75\ \mathsf{ms}$ and $s = -1/\tau = -267\ \mathsf{s}^{-1}$. Therefore, the natural response is

$$i_{3,n} = Ae^{-267t}\ \mathsf{A}.$$

2) The forced response shall be found by using node analysis

$$\frac{\tilde{V}_a - \tilde{V}_s}{R_1} + \frac{\tilde{V}_a}{R_2 - jx_C} + \frac{\tilde{V}_a}{R_3} = 0.$$

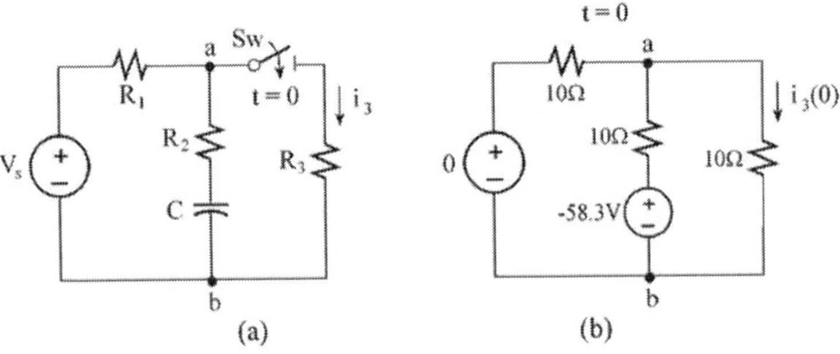

Figure 2.26 A given *RC* circuit for Example 2.16 (a) and its equivalent for determining the initial value of the current $i_3(0)$ (b).

Upon substituting $1/377 \cdot 2.5 \cdot 10^{-4} = 10.6$ for x_C, $141 \angle 0°$ for $\tilde{V}_s$ and 10 for R_3 and R_1 the above equation becomes

$$\frac{\tilde{V}_a - 141}{10} + \frac{\tilde{V}_a}{10 - j10.6} + \frac{\tilde{V}_a}{10} = 0,$$

to which the solution is

$$\tilde{V}_a = 55.9 \angle -11.42° \quad \text{and} \quad \tilde{I}_3 = \frac{\tilde{V}_a}{R_3} = \frac{55.9 \angle -11.42°}{10} = 5.59 \angle -11.42°.$$

The forced response, therefore, is

$$i_{3,f} = 5.59 \sin(377t - 11.42°) \text{ A}.$$

3) The initial value of the capacitance voltage is found by inspection of the circuit prior to switching. By using the voltage division formula we have

$$\tilde{V}_C = \frac{\tilde{V}_s(-jx_C)}{R_1 + R_2 - jx_C} = \frac{141(-j10.6)}{20 - j10.6} = 66.0 \angle -62.07°.$$

Therefore,

$$v_C(0) = 66.0 \sin(-62.07°) = -58.3 \text{ V}.$$

4) The initial value of the current may now be found by inspection of the circuit in Fig. 2.26(b), which fits the instant of switching, $t = 0$. At this moment, the value of the voltage source is $v_s(0) = 0$ and the capacitance voltage is $v_C(0) = -58.3$ V. Using nodal analysis again, we have

$$\frac{V_a}{10} + \frac{V_a + 58.3}{10} + \frac{V_a}{10} = 0,$$

to which the solution is $V_a = -19.4$ V and the initial value of current is

$$i_3(0) = \frac{V_a}{R_3} = \frac{-19.4}{10} = -1.94 \text{ A}.$$

5) The integration constant will be $A = i_3(0) - i_{3,f}(0) = -1.94 - 5.59 \sin(-11.42°) = -0.83$, and the complete response is

$$i_3 = 5.59 \sin(377t - 11.42°) - 0.83 e^{-267t} \text{ A}.$$

Example 2.19

As a last example in this section, consider the circuit in Fig. 2.27(a), in which $R = 100\ \Omega$, $C = 10\ \mu\text{F}$ and two sources are $v_s = 1000\sqrt{2} \sin(1000t + 45°)$ V and $I_s = 4$ A d.c. Find the response of the current through the voltage source after opening the switch and sketch it.

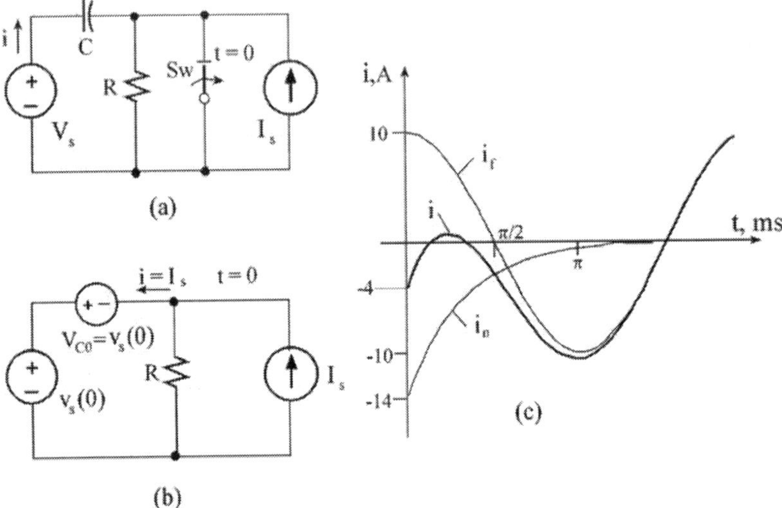

Figure 2.27 A given RC circuit of Example 2.17 (a), its equivalent for instant $t=0$ (b) and the plot of current i (c).

Solution

The time constant (step 1) is $\tau = RC = 100\cdot 10^{-5} = 10^{-3} = 1\ \text{ms}$ or $s = -1000\ \text{s}^{-1}$ and $i_n = Ae^{-1000t}$. The forced current (step 2) is found as a steady-state current in Fig. 2.27(a) after opening the switch

$$\tilde{I} = \frac{\tilde{V}_s}{R + jx_C} = \frac{1000\sqrt{2}\angle 45^\circ}{100 - j100} = 10\angle 90^\circ$$

in which

$$x_C = 1/\omega C = 1/10^3\cdot 10^{-5} = 100\ \Omega.$$

Thus,

$$i_f = 10\sin(1000t + 90^\circ)\ \text{A}.$$

The initial value of the capacitance voltage (step 3) must be evaluated in the circuit 2.27(a) prior to switching. By inspecting this circuit, and noting that the resistance and the current source are short-circuited, we may conclude that this voltage is equal to source voltage $v_C(0) = v_s(0)$.

By inspection of the circuit in Fig. 2.27(b), we shall find the initial value of current i (step 4), which is equal to the current source flowing in a negative direction, i.e., $i(0) = -4$ A. (Note that two voltage sources are equal and opposed to each other.)

Finally the complete response (step 5) in accordance with equation 2.17 will be:

$$i = i_f + [i(0) - i_f(0)]e^{st} = 10\sin(1000t + 90°) - 14e^{-1000t}\ \text{A},$$

where $i_f(0) = 10\sin 90° = 10$ A. The plot of this current is shown in Fig. 2.27(c). Note that the period of the forced current is

$$T = \frac{2\pi}{1000} = 2\pi 10^{-3}\ \text{s} \quad \text{or} \quad T = 2\pi\ \text{ms}.$$

2.4.4 Applying a continuous charge law to C-circuits

As we have observed earlier (see section 1.7.4) switching on circuits containing capacitances may result in very high pulses of current. (This phenomenon is dual to overvoltages in circuits containing inductances when switching them off as studied in section 2.3.3). When trying to solve these kinds of circuits the second switching law for capacitance voltages is usually disproved. However, as we already know, the problem might be solved by the principle of physics that electric charges are always continuous and cannot be abruptly changed. In this section we shall continue analyzing these kinds of circuits by introducing more numerical examples.

Example 2.20

Consider the circuit shown in Fig. 2.28(a), in which capacitance C_3 switches on in parallel to capacitance C_2. The resistances of the wires are very low and are presented by two resistors $R_2 = R_3 \approx 0.1\ \Omega$. The rest of the parameters are $R_1 = 40$ kΩ, $C_1 = 4$ μF, $C_2 = 1$ μF, $C_3 = 3$ μF and $V_s = 100$ V. (a) Assuming that the voltage change of two capacitances C_2 and C_3 occurs abruptly, find the charging current i, and the voltage v_2 across the capacitances, connected in parallel, in the second stage of transients. (b) Find the time and the charge interchanging between these two capacitances and the current pulse.

Solution

(a) After switching, capacitances C_2 and C_3 are connected in parallel with $C_{2,3} = C_2 + C_3 = 4$ μF, as shown in the circuit of Fig. 2.28(b). (The resistances R_1 and R_2 are neglected in comparison with R_1). The time constant of this circuit (step 1) is $\tau = R_1 C_{eq} = 40\cdot 10^3 \cdot 2\cdot 10^{-6} = 80\cdot 10^{-3} = 80$ ms, where

$$C_{eq} = \frac{C_1 C_{2,3}}{C_1 + C_{2,3}} = \frac{4\cdot 4}{4+4} = 2\ \mu\text{F},$$

and the root of the characteristic equation is $s = -1/\tau = -12.5\ \text{s}^{-1}$. The natural responses of the current and voltage will be:

$$i_n = Ae^{-12.5t}$$

$$v_2 = Be^{-12.5t}.$$

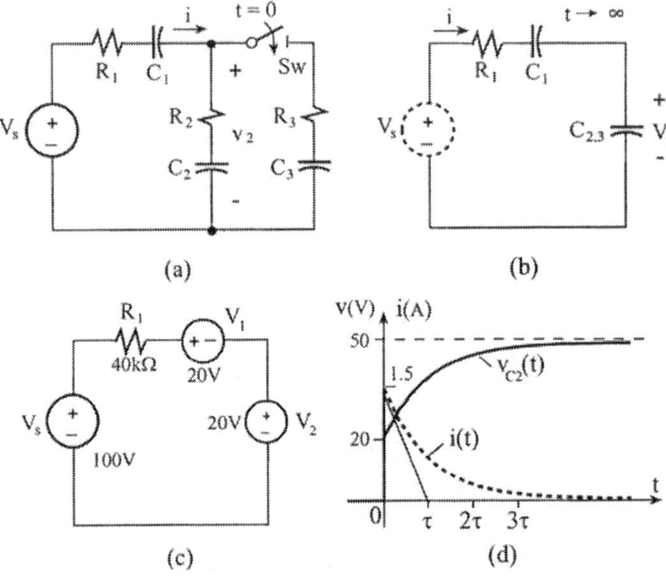

Figure 2.28 A given circuit for Example 2.20 (a), a circuit of the steady-state operation ($t \to \infty$) (b), an equivalent circuit of the instant of switching $t = 0$ (c) and the plots of the current and voltage (d).

The forced response (step 2) of the current is equal to zero as a steady-state current through the capacitance at the d.c. supply. However, the forced response of the capacitance voltages becomes half of the supply voltages, as divided between two equal capacitances C_1 and $C_{2,3}$. Thus, $i_f = 0$ and $v_{2,f} = 50\,\text{V}$.

Next, we shall find the initial value of the voltage of the two capacitances in parallel (step 3). With the generalized second switching law (1.36b), or the principle of continuous charges, we have

$$v_2(0) = \frac{C_2 v_{C2}(0_-) + C_3 v_{C3}(0_-)}{C_2 + C_3}.$$

Here $v_{C3}(0_-)$ should be zero and $v_{C2}(0_-)$ can be found with the voltage division formula

$$v_{C2}(0_-) = V_s \frac{C_1}{C_1 + C_2} = 100 \frac{4}{4+1} = 80\ \text{V}.$$

Thus,

$$v_2(0) = \frac{1 \cdot 80 + 3 \cdot 0}{1+4} = 20\ \text{V}.$$

In the next step (step 4) we shall find the initial value of the current, as the dependent initial condition. By inspection of the equivalent circuit fitting instant

$t=0$, Fig. 2.28(c), we obtain

$$i(0)=\frac{V_s-v_{C1}(0)-v_{C2}(0)}{R_1}=\frac{100-20-20}{40\cdot 10^3}=1.5\ \text{mA}.$$

The integration constants (step 5) are

$$A=i(0)-i_f(0)=1.5-0=1.5\ \text{mA}$$

$$B=v_2(0)-v_{2,f}(0)=20-50=-30\ \text{V}.$$

The complete response of the current and capacitance voltage can now be written

$$i(t)=1.5e^{-12.5t}\ \text{A}$$

$$v_2(t)=50-30e^{-12.5t}\ \text{V}.$$

Both curves are sketched in Fig. 2.28(d).

(b) In order to find the time of the first stage of the transients we must take into consideration the wire resistances. Thus, after switching, the time constant of the right loop of the circuit, in which the first stage of the transients takes place, may be estimated as $\tau=2R_1C_{eq}=0.2\cdot 0.75\cdot 10^{-6}=0.15\ \mu\text{s}$. Here:

$$C_{eq}=\frac{C_2C_3}{C_2+C_3}=\frac{1\cdot 3}{1+3}=0.75\ \mu\text{F}.$$

The time of the first stage is estimated as $T\cong 5\tau=0.75\ \mu\text{s}$, which is about $2\cdot 10^{-6}$ times shorter than the second stage.

The charge of C_2, $q_2(0_-)=C_2v_2(0_-)=1\cdot 10^{-6}\cdot 80=80\ \mu\text{C}$, prior to switching decreases, during the first stage to $q_2(0_+)=C_2v_2(0_+)=1\cdot 10^{-6}\cdot 20=20\ \mu\text{C}$. Thus, the interchange of the charges between two capacitances is $\Delta q=80-20=60\ \mu\text{C}$. The current peak will be $I_\delta=v_{C2}(0)/2R_2=80/0.2=400\ \text{A}$, which results in transferring the charge $\Delta q=I_\delta\tau=400\cdot 0.15\cdot 10^{-6}=60\ \mu\text{C}$, as previously calculated.

Example 2.21

In the circuit of Fig. 2.29(a) the capacitance C_2 has been charged prior to switching up to voltage -6 V. Find current i and voltage v_C after switching, if $R_1=300\ \Omega$, $R_2=600\ \Omega$, $C_1=300\ \mu\text{F}$, $C_2=200\ \mu\text{F}$ and $V_s=36$ V.

Solution

In order to find the time constant and the root of the characteristic equation (step 1), we must find the equivalent resistance and capacitance:

$$R_{eq}=R_1//R_2=\frac{300\cdot 600}{300+600}=200\ \Omega,$$

$$C_{eq}=C_1//C_2=C_1+C_2=300+200=500\ \mu\text{F}.$$

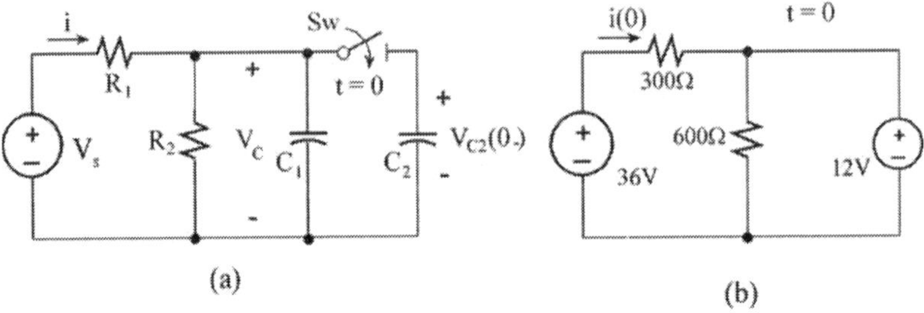

Figure 2.29 A circuit of Example 2.21 (a) and its equivalent at instant $t=0$ (b).

Thus, $\tau = R_{eq}C_{eq} = 200 \cdot 500 \cdot 10^{-6} = 0.1$ s and $s = -1/\tau = -10\ \mathrm{s}^{-1}$, and the natural responses are

$$i_n = Ae^{-10t}, \quad v_{C,n} = Be^{-10t}.$$

The forced responses (step 2) are:

$$i_f = \frac{V_s}{R_1 + R_2} = \frac{36}{300+600} = 0.04\ \mathrm{A} = 40\ \mathrm{mA},$$

$$v_{C,f} = V_s \frac{R_2}{R_1 + R_2} = 36 \frac{600}{300+600} = 24\ \mathrm{V}.$$

The initial value of the voltage of these two capacitances (step 3) shall be found using the second generalized law and by taking into consideration that (in the circuit prior to switching) $v_{C2}(0_-)$ is negative, i.e., $v_{C2}(0_-) = -6$ V, and $V_{C1}(0_-) =$ 24 V. Hence,

$$v_C(0) = \frac{C_1 v_{C1}(0) + C_2 v_{C2}(0)}{C_1 + C_2} = \frac{300 \cdot 24 + 200(-6)}{300 + 200} = 12\ \mathrm{V}.$$

The initial value of the current, which is a dependent initial condition (step 4), is found in the circuit of Fig. 2.29(b) for instant $t = 0$. By inspection, we find:

$$i(0) = \frac{V_s - v_{C2}}{R_1} = \frac{36-12}{300} = 0.08\ \mathrm{A} = 80\ \mathrm{mA}.$$

Now the integration constants (step 5) can be found

$$A = i(0) - i_f(0) = 80 - 40 = 40\ \mathrm{mA}$$

$$B = v_c(0) - v_{C,f}(0) = 12 - 24 = -12\ \mathrm{V}.$$

Thus, the complete responses are

$$i(t) = 40 + 40e^{-10t} = 40(1 + e^{-10t})\ \mathrm{mA}$$

$$v_C(0) = 24 - 12e^{-10t}\ \mathrm{V}.$$

Example 2.22

As a last example for this section, consider the circuit shown in Fig. 2.30(a), in

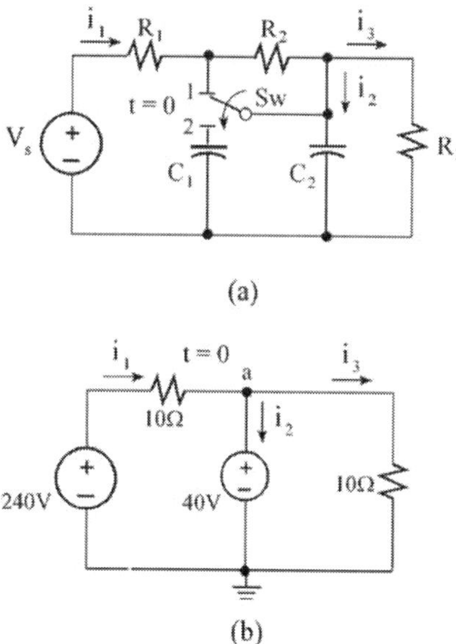

Figure 2.30 A given circuit of Example 2.22 (a) and its equivalent at the instant of switching, $t = 0$ (b).

which upon switching, the configuration of the circuit has been changed, namely, resistor R_2 connects in series to resistor R_1 and capacitance C_1 being uncharged connects in parallel to capacitance C_2. Assume that the switching occurs instantaneously and find all the current responses, $i_1(t)$, $i_2(t)$ and $i_3(t)$ after switching. The circuit parameters are: $R_1 = R_2 = 5\ \Omega$, $R_3 = 10\ \Omega$, $C_1 = 750\ \mu\text{F}$, $C_2 = 250\ \mu\text{F}$ and $V_s = 240\ \text{V}$.

Solution

The time constant (step 1) may be easily found after determining the equivalent resistance and capacitance:

$$R_{eq} = (R_1 + R_2)//R_3 = (5 + 5)//10 = 5\ \Omega,$$

$$C_{eq} = C_1 + C_2 = 250 + 750 = 1000\ \mu\text{F}.$$

Therefore, the time constant is $\tau = R_{eq}C_{eq} = 5 \cdot 10^{-3} = 5\ \text{ms}$ and $s = -1/\tau = -200\ \text{s}^{-1}$. The natural responses of the currents, therefore, are

$$i_{1,n} = A_1 e^{-200t}, \quad i_{3,n} = A_3 e^{-200t}$$

and

$$i_2 = i_1 - i_3 = (A_1 - A_3)e^{-200t}.$$

The forced responses (step 2) are found in the circuit after switching in its steady-state operation:

$$i_{1,f} = i_{3,f} = \frac{V_s}{R_1 + R_2 + R_3} = \frac{240}{5+5+10} = 12\ \text{A},$$

$$i_{2,f} = i_C = 0.$$

The initial value of the voltage across the two capacitances (in parallel) (step 3) may be found using the principle of continuous charge (the second generalized law):

$$v_C(0) = \frac{C_1 v_{C1}(0_-) + C_2 v_{C2}(0_-)}{C_1 + C_2} = \frac{0 + 250 \cdot 160}{750 + 250} = 40\ \text{V},$$

where $v_{C1}(0_-) = 0$ and $v_{C2}(0_-) = 240 \cdot 10/(5+10) = 160$ V. The initial values of the currents, which are dependent initial conditions (step 4), can be obtained in the equivalent circuit of Fig. 2.30(b). Since the potential of node "a" is 40 V, we have:

$$i_1(0) = \frac{240 - 40}{10} = 20\ \text{A},$$

$$i_3(0) = \frac{40}{10} = 4\ \text{A}, \quad i_2(0) = i_1(0) - i_3(0) = 20 - 4 = 16\ \text{A}.$$

The integration constants (step 5) are:

$$A_1 = i_1(0) - i_{1,f}(0) = 20 - 12 = 8\ \text{A}$$

$$A_2 = i_2(0) - i_{2,f}(0) = 16 - 0 = 16\ \text{A}$$

$$A_3 = i_3(0) - i_{3,f}(0) = 4 - 12 = -8\ \text{A},$$

and the complete responses of the three currents are:

$$i_1(t) = 12 + 8e^{-200t}\ \text{A}$$

$$i_2(t) = 16e^{-200t}\ \text{A}$$

$$i_3(t) = 12 - 8e^{-200t}\ \text{A}.$$

Note that current $i_2(t)$ might also be found as the difference between i_1 and i_3, i.e., $i_2(t) = i_1(t) - i_3(t) = 16e^{-200t}$ A, which is the same as was found earlier.

It is worthwhile to calculate current i_2, which is actually the current through two parallel capacitances, also as $i_2 = C_{eq}(dv_C/dt)$. In order to do this we first have to find the capacitance voltage. Since its forced value is $240 \cdot 10/(5+5+10) = 120$ V, we have

$$v_C(t) = 120 + (40 - 120)e^{-200t} = 120 - 80e^{-200t}\ \text{V},$$

and

$$i_2(t) = C_{eq}\frac{dv_C}{dt} = 10^{-3}(-80)(-200)e^{-200t} = 16e^{-200t}\ \text{A},$$

which again is the same as was calculated earlier.

2.5 THE APPLICATION OF THE UNIT-STEP FORCING FUNCTION

The reason that any transient responses at all appear in electrical circuits is because of the discontinuity or switching actions which take place at an instant of time that is defined as $t = 0$ (or $t = t_0$). In studying transient responses, it is convenient, in many cases, to use a special function, which represents this kind of discontinuous or switching action, and is called a **unit-step function**. Thus, the operation of a switch in series with a voltage source is equivalent to a forcing function, which is zero up to the instant that the switch is closed and is equal to the value of the voltage source thereafter. This change of voltage occurs abruptly (since we are considering the switch as an ideal device working instantaneously), expressing a discontinuity of the voltage at the instant the switch is closed. Such kinds of functions, which are discontinuous or have discontinuous derivatives, are called **singularity functions**. The two most important of them are the unit-step function and the unit-impulse function (see further on). Thus, the mathematical definition of the unit-step forcing function is

$$u(t) = \begin{cases} 0 & t < 0 \\ 1 & t > 0. \end{cases} \tag{2.27a}$$

or

$$u(t - t_0) = \begin{cases} 0 & t < t_0 \\ 1 & t > t_0. \end{cases} \tag{2.27b}$$

Therefore, the unit-step function is zero for all negative values of its argument ($t < 0$) and is unity for all positive values ($t > 0$). This is shown in Fig. 2.31. Note that at the instant of time $t = 0$ is not defined: but it is zero as a left limit and unity as a right limit. In accordance with equation 2.27a, a switching action takes place at the instant $t = 0$ and in accordance with equation 2.27b at instant $t = t_0$ ($t_0 \neq 0$, since, if $t_0 = 0$ we get equation 2.27a). To indicate that any voltage source of the value V is switching at $t = 0$ (or $t = t_0$) to a general network, we write $v(t) = Vu(t)$, which is illustrated in Fig. 2.32(a). Such representation of switching on sources by using a unit forcing function instead of a switch by itself is common and useful in transient analysis. (Note that the unit-step function is itself dimensionless.)

Because of the wide use of the unit-step function in transient analysis, we

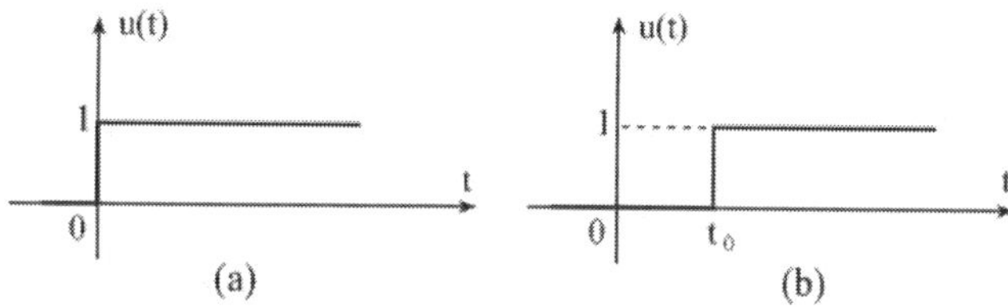

Figure 2.31 A unit-step function applied at $t = 0$ (a) and applied at $t = t_0$ (b).

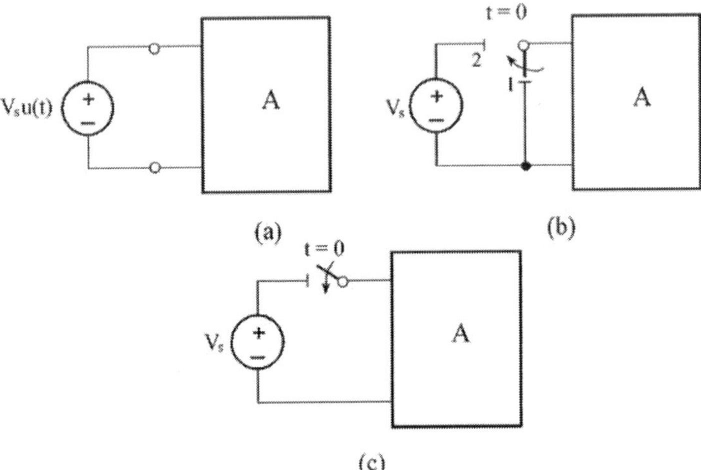

Figure 2.32 A circuit in which the voltage source is applied at $t = 0$ (a) and its switching equivalent drawn correctly (b) and incorrectly (c).

shall explain the features of this function in more detail. In the circuit of Fig. 2.32(a) the ideal voltage source possesses a zero internal impedance, so that circuit A is short-circuited the entire time, also prior to $t = 0$, even when the applied voltage equals zero. We have the same conditions in the circuit (b), which is therefore the correct equivalent of the circuit with the discontinuous forcing function (a). (Note that the switch in this circuit is an ideal instantaneously operating switch.) The circuit in (c) cannot be the correct equivalent of (a) since prior to switching circuit A is open-circuited. However, after switching, $t \geq 0$, the circuits in (c) and (a) are equivalent, and if this is the only time interval we are interested in, and if the initial currents which flow from the two circuits, A in (a) and in (c), are identical at $t = 0$, then Fig. 2.32(c) becomes a useful equivalent of Fig. 2.32(a).

The circuit with a discontinuous current source is a dual of the circuit with a discontinuous voltage source and is shown in Fig. 2.33. The above explanation regarding the voltage source may be easily understood from this figure. Using two unit step functions, we can obtain the rectangular pulse of a forcing function, as is shown in Fig. 2.34.

To show an application of the unit-step function in transient analysis, let us consider a numerical example in which a pulse current is applied.

Example 2.23

In the circuit, shown in Fig. 2.35(a), find the output voltage, if the current pulse of amplitude $I = 2$ A and duration $t_0 = 0.01$ s, shown in Fig. 2.35(b), is applied to this circuit.

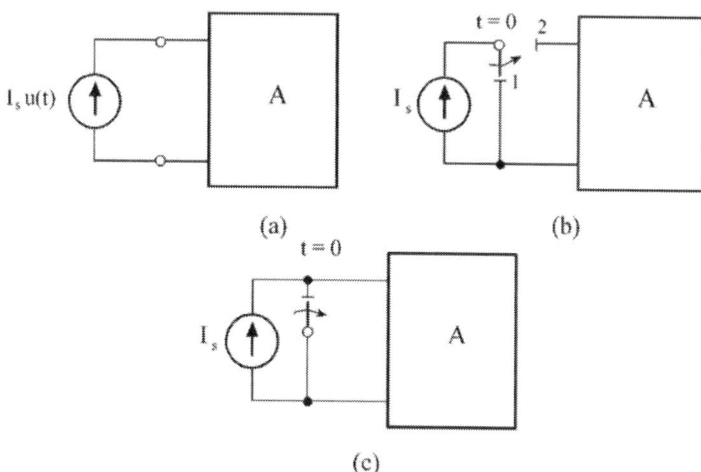

Figure 2.33 A circuit in which the current source is applied at $t = 0$ (a) and its switching equivalent drawn correctly (b) and incorrectly (c).

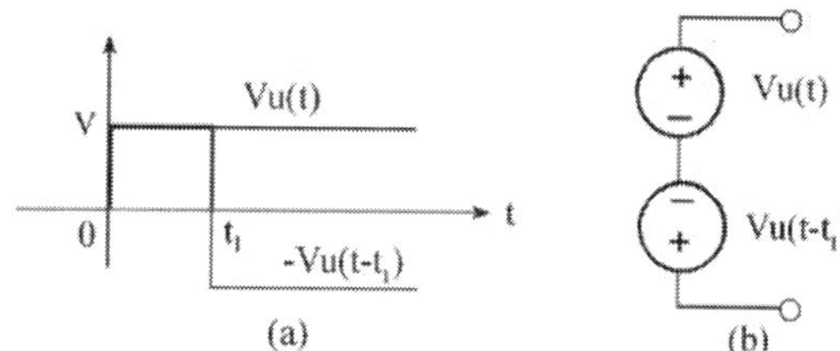

Figure 2.34 A rectangular forcing function (a) and a combined source, which yields the rectangular pulse (b).

Solution

The output voltage can be represented as

$$v_o(t) = v_o'(t) + v_o''(t),$$

where $v_o'(t)$ is the part of the total response due to the positive current source acting alone and $v_o''(t)$ is the part due to the negative current source acting alone. Starting with the first part of the output voltage and following the five steps, we must do as follows:

1) To obtain the characteristic equation we shall equal the input admittance to zero, since the current source possesses an infinite impedance (an open circuit):

$$Y_{in} = \frac{1}{15} + \frac{1}{10 + 1/(1/100 + 1/2s)} = \frac{1}{15} + \frac{100 + 2s}{1000 + 220s},$$

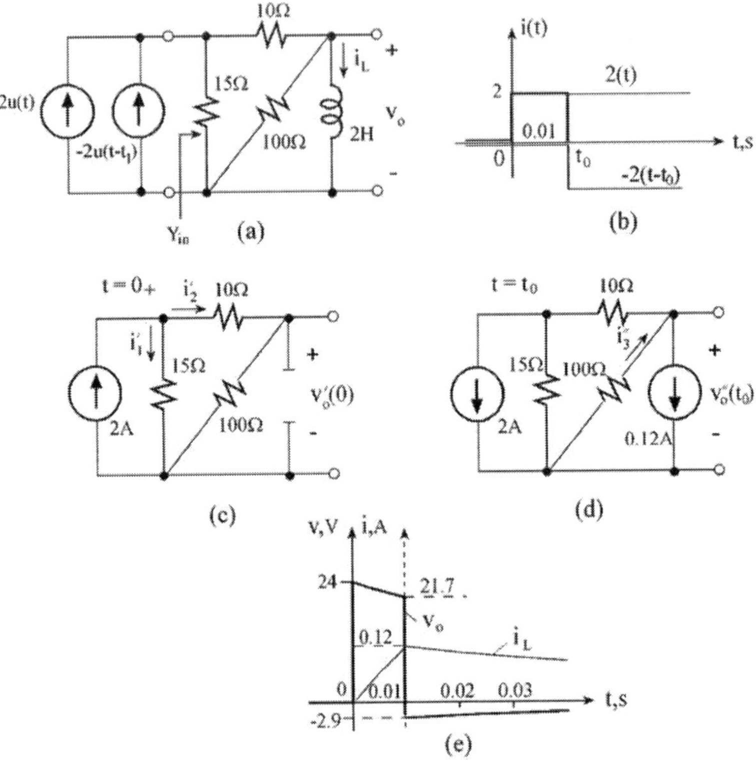

Figure 2.35 A circuit of Example 2.23 (a), the input current pulse (b), an equivalent circuit for determining $v_o'(0)$ (c), an equivalent circuit for determining $v_o''(t_0)$ (d) and the output voltage response (e).

or

$$250s + 2500 = 0 \quad \text{and} \quad s = -10\ \text{s}^{-1}.$$

(Alternatively, we may obtain the characteristic equation by equaling the admittance, seen from the inductance branch, to zero, which is left for the reader as an exercise.) Thus,

$$v_o'(t) = A'e^{-10t}\ \text{V}.$$

2) The forced response is zero as a voltage drop across an inductance in a d.c. circuit, thus, $v_{o,f} = 0$.

3) Inspecting the circuit for $t < 0$, we have $i_L(0_-) = 0$, so that $i_L(0_+) = 0$.

4) In the circuit drawn for $t = 0$, Fig. 2.35(c), we have

$$v_o'(0) = 100 i_2'(0) = 100 \cdot 2 \frac{15}{110 + 15} = 24\ \text{V}.$$

5) The integration constant, therefore, is $A' = v_o'(0) - v_{o,f}' = 24 - 0 = 24$. and the first part of the voltage response is $v_o'(t) = 24e^{-10t}$ V.

To find the second part of the voltage $v_o''(t)$ we start from step 3, since the root of the characteristic equation and the forced response have already been found, i.e., $s = -10\,\text{s}^{-1}$ and $v_{o,f}'' = v_{o,f}' = 0$.

3) The independent initial condition for the inductance current at the instance of second commutation, t_0, is

$$i_L''(t_{0-}) = \frac{1}{L}\int_0^{t_0} v_o' dt = \frac{24}{2}\int_0^{0.01} e^{-10t} = \frac{12}{-10} e^{-10t}\Big|_0^{0.01} = 0.12\ \text{A}.$$

4) To find the initial condition of v_o'' in the second transient interval we must consider the given circuit for $t = t_0$, in which the inductance is represented by a current source of 0.12 A.

$$i_3'' = 2\frac{15}{110 + 15} + 0.12\frac{25}{100 + 25} = 0.264\ \text{A}$$

and $v_o''(t_{0+}) = -100 \cdot 0.264 = -26.4\,\text{V}$.

5) We can now find the constant of integration: $A'' = v_o''(0) - v_{o,f} = -24.6 - 0 = -24.6\,\text{V}$, and

$$v_o''(t) = -24.6e^{-10t}\ \text{V} \quad \text{for} \quad t > t_0.$$

Then

$$v_o = v_o' + v_o'' = 24e^{-10t} - 24.6e^{-10(t - t_0)}\ \text{V} \quad \text{for} \quad t > t_0.$$

To simplify this expression we designate $t' = t - t_0$ or $t = t' + t_0$, then

$$v_o = 24e^{-10t_0}e^{-10t'} - 24.6e^{-10t'} = -2.9e^{-10t'},$$

which means that the y-axis has been moved to the new origin at t_0, i.e., now $t_0 = 0$.

The output voltage and inductance current are shown in Fig. 2.35(e). Note that the output voltage form is almost a rectangular pulse, i.e. similar to the input current pulse. In other words, the current pulse is transferred to the voltage pulse. Note also that this is correct in the case that $t_0 \ll \tau$ or $t_0/\tau \ll 1$, where $\tau = L/R_{eq}$.

2.6 SUPERPOSITION PRINCIPLE IN TRANSIENT ANALYSIS

In this section, we shall show how the property of superposition can be used for solving problems in transient analysis. Suppose that a new branch connects to a general active network A after closing the switch and, suppose that we are looking for the transient current in any other branch of the network, say i_1, as shown in Fig. 2.36(a). The transient behavior of the entire circuit can be written as a superposition of two regimes: 1) the previous one, which existed prior to switching and 2) an additional one, which is a result of the switching. Therefore, the unknown current i_1 will be the sum of the two currents. The first one, i_1' is the current which flowed in branch "1" before switching, figure (b), and the second one, i_1'', is the additional current which appears in circuit P, figure (c). This circuit arises from the original circuit A, in which all the sources have been "killed" and the switch is replaced by a voltage source, which is oppositely equal to the voltage across the open switch in circuit A, as shown in Fig. 2.36(c). (Remember that "to kill" a source means that the source is replaced by its inner resistance/impedance, or that the ideal voltage source is simply short-circuited and the current source is simply open-circuited.) Hence,

$$i_1 = i_1' + i_1'' \quad \text{or} \quad i_1 = i_{1pr} + i_{1ad}, \tag{2.28}$$

where i_1' is the previous current and i_1'' is the additional one. It is very important to note that, in the additional circuit, the independent conditions are zero.

If any branch in a general network is disconnected, as shown in Fig. 2.37(a), we may apply the principle of duality, which means that the switch in the passive circuit must be replaced by a *current source* that is oppositely equal to the current through the closed switch in circuit A, as shown in Fig. 2.37(c). The required current i_1 will then again be the sum of the previous current, which flows in the circuit with the switch closed, Fig. 2.37(b), and the additional current, which will flow in the passive circuit with the source current, Fig. 2.37(c), as indicated in equation 2.28. The above technique is illustrated in the following examples.

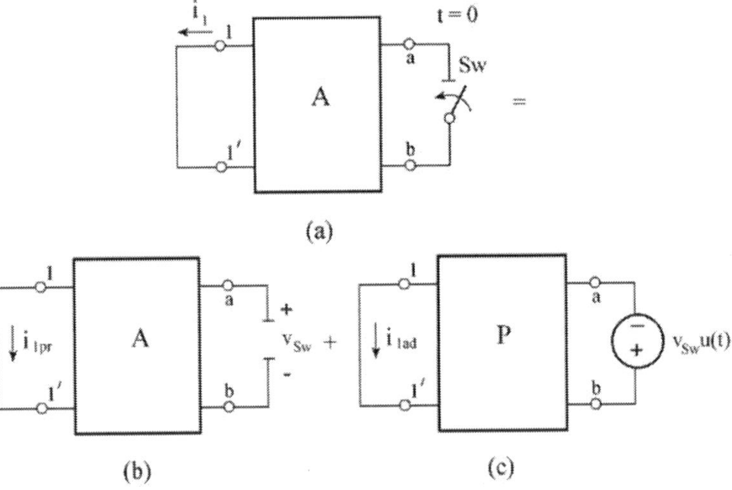

Figure 2.36 A given circuit (a), a previous circuit prior to switching (b) and the additional passive circuit with a voltage source (c).

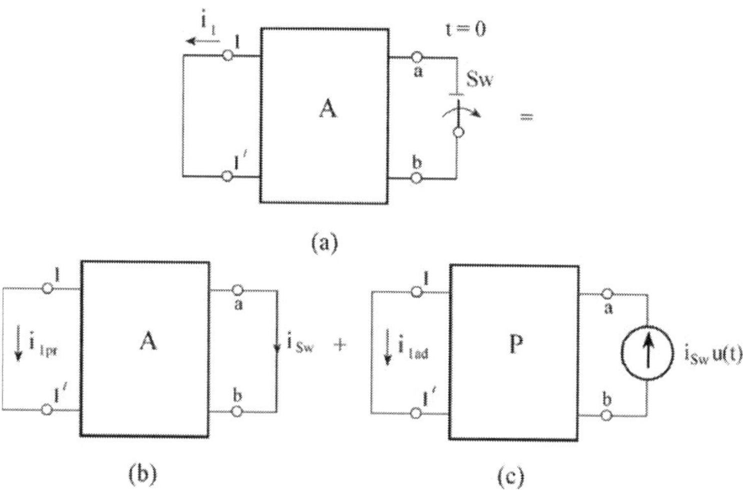

Figure 2.37 A given circuit (a), a previous circuit prior to switching (b) and the additional passive circuit with a current source (c).

Example 2.24

In the circuit, shown in Fig. 2.38(a), find current i after opening the switch, using the principle of superposition. The parameters of the circuit are: $v_s = 100 \sin \omega t$ at $f = 60$ Hz, $\omega L = 10\ \Omega$ and $R_1 = R_2 = 10\ \Omega$.

Solution

First, we find the currents in the circuit of Fig. 2.38(a) prior to switching

$$i' = \frac{100}{\sqrt{5^2 + 10^2}} \sin(\omega t + \psi_i) = 8.94 \sin(\omega t - 63.4°)\ \text{A},$$

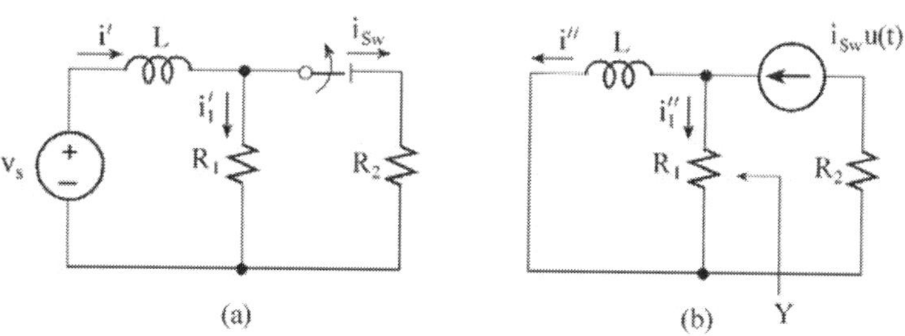

Figure 2.38 A circuit prior to switching (a) and the additional circuit (b).

where the current phase angle is $\psi_i = 0 - \tan^{-1}(10/5) = -63.4^\circ$, and

$$i_{sw} = \frac{i'}{2} = 4.47\sin(\omega t - 63.4^\circ).$$

Now we shall find the transient current in the circuit of Fig. 2.38(b), in which the initial value of the inductance current is zero. The characteristic equation is

$$Y = \frac{1}{sL} + \frac{1}{R_1} = 0 \quad \text{or} \quad sL + R_1 = 0,$$

and its root is

$$s = \frac{R_1}{L} = -\frac{10}{10/\omega} = -377\ \mathrm{s}^{-1}.$$

Hence, the natural response is

$$i''_n = Ae^{-377t}.$$

The forced response of current i''_f is found with phasor analysis:

$$\tilde{I}'_f = \tilde{I}_{sw}\frac{R_1}{R_1 + j\omega L} = (4.47\angle -63.4^\circ)\frac{10}{10 + j10} = 3.16\angle -108.4^\circ.$$

Therefore,

$$i''_f(t) = 3.16\sin(\omega t - 108.4^\circ)\ \mathrm{A}.$$

Since the initial value of this current is zero (zero initial conditions), we have

$$A = 0 - i''_f(0) = -3.16\sin(-108.4^\circ) = 3.0\ \mathrm{A}.$$

Thus, the total response of current i is

$$i = i' - i'' = 8.49\sin(\omega t - 63.4^\circ) - 3.16\sin(\omega t - 108.4^\circ) - 3e^{-377t}\ \mathrm{A}. \quad \text{(a)}$$

The initial value of this current at $t = 0$ is $i(0) = -8$ A, which is the current through the inductance prior to switching ($i'(0) = 8.94\sin(-63.4^\circ) \cong -8$ A).

Note that the same current can be found as current i_1 through resistance R_1, since in the original circuit both currents are equal. The forced response of this current is determined as

$$\tilde{I}''_{1,f} = \tilde{I}_{sw}\frac{j\omega L}{R_1 + j\omega L} = 4.47\angle -63.4^\circ\frac{j10}{10 + j10} = 3.16\angle -18.4^\circ,$$

or, as versus time,

$$i''_{1,f} = 3.16\sin(\omega t - 18.4^\circ).$$

Since $i''_L(0) = 0$, the initial value of the current through resistance R_1 is

$$i''_1(0) = 4.47\sin(-63.4^\circ) = -4\ \mathrm{A},$$

and the integration constant for this current is $A = -4 - 3.16\sin(-18.4°) = -3$. Hence the total current is

$$i_1 = i_1' + i_1'' = 4.47\sin(\omega t - 63.4°) + 3.16\sin(\omega t - 18.4°) - 3e^{-377t}\ \mathsf{A}. \tag{b}$$

This current at $t = 0$ yields $i_1(0) = -8$ A, which is again the value of the inductance current prior to switching. Both results (a) and (b) can be simplified to the same expression

$$i(t) = 5\sqrt{2}\sin(\omega t - 45°) - 3e^{-377t}\ \mathsf{A}.$$

Example 2.25

In the circuit, having all R's of 10 Ω, $C = 1$ μF and $V_s = 60$ V, shown in Fig. 2.39(a), the switch closes at time $t = 0$. Find current i_1 using the superposition theorem.

Solution

First, we shall find the previous current i_1' and the voltage across the open switch. By inspection of the circuit in Fig. 2.39(a), we may find

$$i_1' = \frac{V_s}{R_1 + R_3 + R_4} = \frac{60}{3 \cdot 10} = 2\ \mathsf{A} \quad \text{and} \quad V_{sw} = R_4 i_1' = 10 \cdot 2 = 20\ \mathsf{V}.$$

To find the time constant of the circuit in Fig. 2.39(b), we must determine R_{eq}

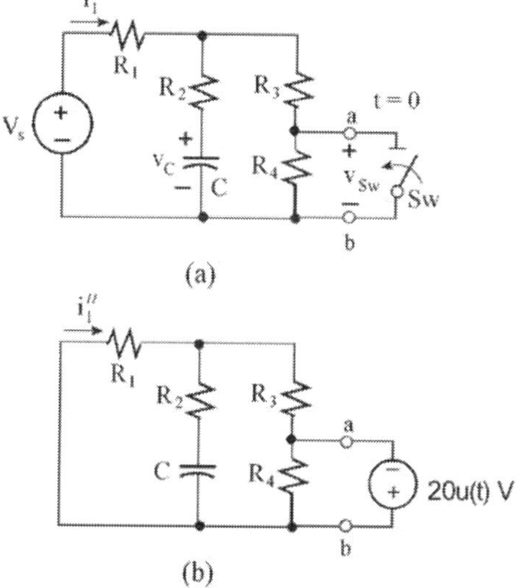

Figure 2.39 A circuit of Example 2.25 prior to switching (a) and an additional circuit for finding the transient response (b).

seen from the capacitance (note that R_4 is short-circuited by the voltage source): $R_{eq} = R_2 + R_1//R_3 = 15\,\Omega$ and $\tau = R_{eq}C = 15 \cdot 10^{-6}$ s, or $s = -1/s = -66.7 \cdot 10^3\ \mathrm{s}^{-1}$.

The forced response of the current in Fig. 2.39(b) will be found as:

$$i''_{1,f} = \frac{V_{sw}}{R_1 + R_3} = \frac{20}{10 + 10} = 1\ \mathrm{A}.$$

The initial value of current i''_1, since the capacitance voltage is zero (which means that the capacitance is short-circuited, i.e., zero initial conditions), will be

$$i'_1(0) = \frac{1}{2}\frac{V_{sw}}{R_3 + R_1//R_2} = \frac{1}{2}\frac{20}{10 + 5} = 0.667\ \mathrm{A}.$$

Therefore, the arbitrary constant will be: $A = i''_1(0) - i''_{1,f} = 0.667 - 1 = -0.133$ A. The additional current now is

$$i''_1 = 1 - 0.333e^{-66.7 \cdot 10^3 t}\ \mathrm{A},$$

and the total current will be:

$$i_1 = i'_1 + i''_1 = 3 - 0.333e^{-66.6 \cdot 10^3 t}\ \mathrm{A}.$$

2.7 *RLC* CIRCUITS

This section is devoted to analyzing very important circuits containing three basic circuit elements: *R*, *L*, and *C*. These circuits are considered important because the networks involved in many practical transient problems in power systems can be reduced to one or to a number of simple circuits made up of these three elements. In particular, the most important are series or parallel *RLC* circuits, with which we shall start our analysis.

From our preceding study, we already know that the transient response of a second order circuit contains two exponential terms and the natural component of the complete response might be of three different kinds: overdamped, underdamped or critical damping. The kind of response depends on the roots of the characteristic equation, which in this case is a quadratic equation. We also know that in order to determine *two* arbitrary integration constants, A_1 and A_2, we must find two initial conditions: 1) the value of the function at the instant of switching, $f(0)$, and 2) the value of its derivative, $(df/dt)|_{t=0}$.

In the following section, we shall deepen our knowledge of the transient analysis of second order circuits in their practical behavior and by solving several practical examples

2.7.1 *RLC* circuits under d.c. supply

We shall start our practical study of transients in second order circuits by considering examples in which the d.c. sources are applied. At the same time, we must remember that only the forced response is dependent on the sources. Natural responses on the other hand depend only on the circuit configuration and its parameter and do not depend on the sources. Therefore, by determining the natural responses we are actually practicing solving problems for both kinds of sources, d.c. and a.c. However, it should be mentioned that the natural response depends on from which source the circuit is fed: the voltage source or the current source. These two sources possess different inner resistances (impedances) and therefore they determine whether the source branch is short-circuited or open, which influences of course the equivalent circuit.

In our next example, we shall elaborate on the methods of determining characteristic equations and show how the kind of source (voltage or current) and the way it is connected may influence the characteristic equation. Let us determine the characteristic equation of the circuit, shown in Fig. 2.40, depending on the kind of source: voltage source or current source and on the place of its connection: (1) in series with resistance R_1, (2) in series with resistance R_2, (3) between nodes m–n.

(1) *Source connected in series with resistance* R_1

If a voltage source is connected in series with resistance R_1, Fig. 2.40(b), we may use the input impedance method for determining the characteristic equation. This impedance as seen from the source is

$$Z(s) = R_1 + (R_2 + 1/sC)//(R_3 + sL).$$

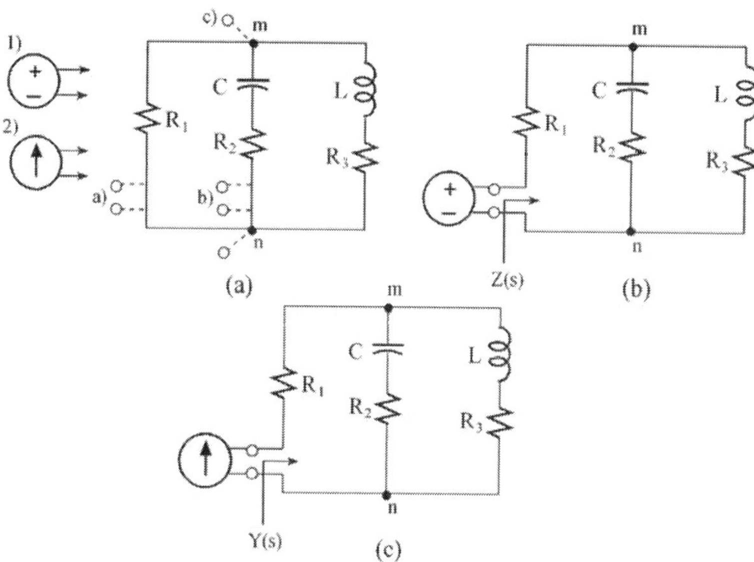

Figure 2.40 A given circuit (a), a circuit in which a voltage source is connected to the branch of R_1 (b) and a circuit in which a current source is connected to the branch of R_1 (c).

Performing the above operation and upon simplification and equating $Z(s)$ to zero we obtain

$$(R_1+R_2)LCs^2+(\sum R_iR_jC+L)s+(R_1+R_3)=0. \tag{2.29}$$

where $\sum R_iR_j=R_1R_2+R_1R_3+R_2R_3$, and the roots of (2.29) are

$$s_{1,2}=-\frac{1}{2}\left(\frac{R_{eq}}{L}+\frac{1}{R_{12}C}\right)\pm\sqrt{\frac{1}{4}\left(\frac{R_{eq}}{L}+\frac{1}{R_{12}C}\right)^2-\varepsilon\frac{1}{LC}},$$

where

$$R_{eq}=\frac{\sum R_iR_j}{R_1+R_2},\quad R_{12}=R_2+R_2\quad\text{and}\quad\varepsilon=\frac{R_1+R_3}{R_1+R_2}.$$

If a current source is connected in series with resistance R_1 we may use the input admittance method. By inspection of Fig. 2.40(c), and noting that the branch with resistance R_1 is opened ($Y_1=0$), we have

$$Y(s)=0+\frac{1}{R_2+1/sC}+\frac{1}{R_3+sL}=0,$$

or, after simplification,

$$LCs^2+(R_2+R_3)Cs+1=0, \tag{2.30}$$

and the roots of (2.30) are

$$s_{1,2}=-\frac{1}{2}\frac{R_{23}}{L}\pm\sqrt{\frac{1}{4}\left(\frac{R_{23}}{L}\right)^2-\frac{1}{LC}},\quad\text{where}\quad R_{23}=R_2+R_3.$$

Since the characteristic equations 2.29 and 2.30 are completely different, and therefore their roots are also different, we may conclude that the transient response in the same circuit, but upon applying different kinds of sources, will be different.

(2) We leave this case to the reader to solve as an exercise.

(3) Source is connected between nodes *m–n*.

If a voltage source is connected between nodes m–n, the circuit is separated into three independent branches: 1) a branch with resistance R_1, in which no transients occur at all; 2) a branch with R_2 and C in series, for which the characteristic equation is $R_2Cs+1=0$; and 3) a branch with R_3 and L in series, for which the characteristic equation is $Ls+R_3=0$.

If a current source is connected between nodes m–n, by using the rule $Y_{in}(s)=0$ we may obtain

$$Y_{mn}=\frac{1}{R_1}+\frac{1}{R_2+1/sC}+\frac{1}{R_3+sL}=0.$$

Performing the above operations and upon simplification, we obtain

$$(R_1 + R_2)LCs^2 + (\sum R_i R_j C + L)s + (R_1 + R_3) = 0, \tag{2.31}$$

where $\sum R_i R_j$ is like in equation 2.29. Note that this equation (2.31) is the same as (2.29), which can be explained by the fact that connecting the sources in these two cases does not influence the configuration of the circuit: the voltage source in (1) keeps the branch short-circuited and the current source in (3) keeps the entire circuit open-circuited. In all the other cases the sources change the circuit configuration.

In the following analysis we shall discuss three different kinds of responses: overdamped, underdamped, and critical damping, which may occur in *RLC* circuits. Let us start with a free source simple *RLC* circuit.

(a) Series connected RLC circuits

Consider the circuit shown in Fig. 2.41. At the instant $t = 0$ the switch is moved from position "1" to "2", so that the capacitor, which is precharged to the initial voltage V_0, discharges through the resistance and inductance. Let us find the transient responses of $v_C(t)$, $i(t)$ and $v_L(t)$. The characteristic equation is

$$R + sL + \frac{1}{sC} = 0, \quad \text{or} \quad s^2 + \frac{R}{L}s + \frac{1}{LC} = 0. \tag{2.32}$$

The roots of this equation are

$$s_{1,2} = -\frac{R}{2L} \pm \sqrt{\left(\frac{R}{2L}\right)^2 - \frac{1}{LC}} \tag{2.33a}$$

or as previously assigned (see section 1.6.2)

$$s_{1,2} = -\alpha \pm \sqrt{\alpha^2 - \omega_d^2}, \tag{2.33b}$$

where $\alpha = R/2L$ is the exponential damping coefficient and $\omega_d = 1/\sqrt{LC}$ is the resonant frequency of the circuit.

An overdamped response: Assume that the roots (equation 2.32) are real (or more precisely negative real) numbers, i.e., $\alpha > \omega_d$ or $R > 2\sqrt{L/C}$. The natural

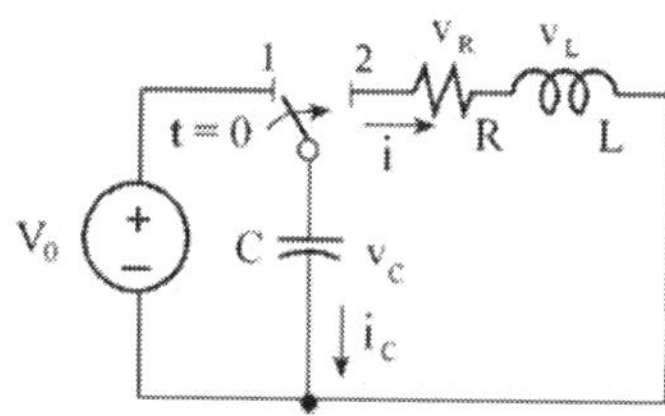

Figure 2.41 A series connected *RLC* circuit.

response will be the sum of two decreasing exponential terms. For the capacitance voltage it will be

$$v_{C,n} = A_1 e^{s_1 t} + A_2 e^{s_2 t}.$$

Since the absolute value of s_2 is larger that that of s_1, the second term, containing this exponent, has the more rapid rate of decrease.

The circuit in Fig. 2.41 after switching becomes source free; therefore, no forced response will occur and we continue with the evaluation of the initial conditions. For the second order differential equation, we need two initial conditions. The first one, an independent initial condition, is the initial capacitance voltage, which is V_0. The second initial condition, a dependent one, is the derivative dv_C/dt, which can be expressed as a capacitance current divided by C

$$\left.\frac{dv_{C,n}}{dt}\right|_{t=0} = \frac{1}{C} i_{C,n}(0) = 0. \tag{2.34}$$

This derivative equals zero, since in a series connection $i_C(0) = i_L(0)$ and the current through an inductance prior to switching is zero. Now we have two equations for determining two arbitrary constants

$$\begin{aligned} A_1 + A_2 &= V_0, \\ s_1 A_1 + s_2 A_2 &= 0. \end{aligned} \tag{2.35}$$

The simultaneous solution of equations 2.35 yields

$$A_1 = \frac{V_0 s_2}{s_2 - s_1} \quad \text{and} \quad A_2 = \frac{V_0 s_1}{s_1 - s_2}. \tag{2.36}$$

Therefore, the natural response of the capacitance voltage is

$$v_{C,n} = \frac{V_0}{s_2 - s_1}(s_2 e^{s_1 t} - s_1 e^{s_2 t}). \tag{2.37}$$

The current may now be obtained by a simple differentiation of the capacitance voltage, which results in

$$i_n(t) = C\frac{dv_C}{dt} = CV_0 \frac{s_1 s_2}{s_2 - s_1}(e^{s_1 t} - e^{s_2 t}) = \frac{V_0}{L(s_2 - s_1)}(e^{s_1 t} - e^{s_2 t}). \tag{2.38}$$

(The reader can easily convince himself that $s_1 s_2 = 1/LC$.) Finally, the inductance voltage is found as

$$v_{L,n}(t) = L\frac{di_n}{dt} = \frac{V_0}{s_2 - s_1}(s_1 e^{s_1 t} - s_2 e^{s_2 t}). \tag{2.39}$$

The plots of $v_{C,n}$, i_n, and $v_{L,n}$ are shown in Fig. 2.42(a). As can be seen from the inductance voltage plot, at $t = 0$ it abruptly changes from zero to $-V_0$, at the instant $t = t_1$ it equals zero and after that, the inductance voltage remains

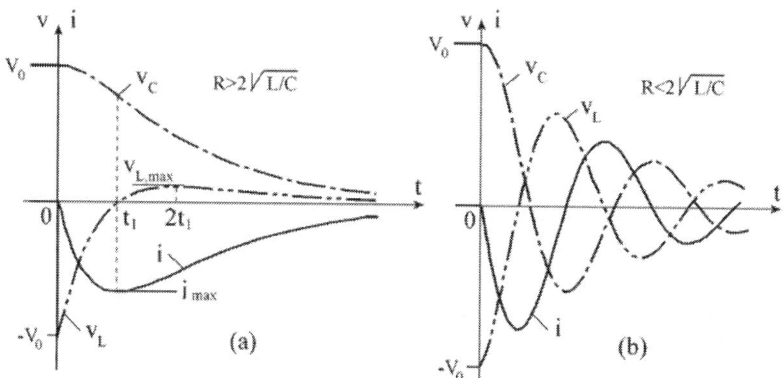

Figure 2.42 The natural responses of $v_{C,n}$, $v_{L,n}$ and i_n in a series connected RLC circuit: overdamped (a), and underdamped (b).

positive. The instant of t_1 can be found from the equation $s_1e^{s_1t_1} - s_2e^{s_2t_1} = 0$, to which the solution is

$$t_1 = \frac{\ln(s_2/s_1)}{s_1 - s_2}. \tag{2.40}$$

At this instant of time, the current reaches its maximum. By equating $dv_{L,n}/dt$ to zero it can be readily found that $v_{L,n}$ has its maximal value at $t = 2t_1$. The overdamped response is also called an *aperiodical response.* The energy exchange in such a response can be explained as follows. The energy initially stored in the capacitance decreases continuously with the decrease of the capacitance voltage. This energy is stored in the inductance throughout the period that the current increases. After $t = t_1$, the current decreases and the energy stored in the inductance decreases. Throughout the entire transient response, all the energy dissipates into resistance, converting into heat.

An underdamped response: Assume now that the roots of equation 2.32 are complex conjugate numbers, i.e., $\alpha < \omega_d$ or $R < 2\sqrt{L/C}$, and $s_{1,2} = -\alpha \pm j\omega_n$, where $\omega_n = \sqrt{\omega_d^2 - \alpha^2}$ is the frequency of the natural response, or *natural frequency*, and $\alpha = R/2L$ is, as previously, the exponential *damping coefficient*. As we have observed earlier (see section 1.6.2), the natural response of, for instance, the capacitance voltage in this case becomes a damped sinusoidal function of the form (1.33):

$$v_{C,n}(t) = Be^{-\alpha t}\sin(\omega_n t + \beta), \tag{2.41a}$$

where the arbitrary constants B and β can be found as was previously by solving two simultaneous equations

$$B\sin\beta = V_0,$$

$$-\alpha\sin\beta + \omega_n\cos\beta = 0,$$

to which the solution is (also see (1.65)):

$$B = \frac{V_0}{\sin\beta} \quad \text{and} \quad \beta = \tan^{-1}\frac{\omega_n}{\alpha}.$$

By using trigonometrical identities we may also obtain:

$$\sin\beta = \frac{\tan\beta}{\sqrt{1+\tan^2\beta}} = \frac{\omega_n}{\sqrt{\alpha^2+\omega_n^2}}$$

$$B = \frac{V_0\sqrt{\alpha^2+\omega_n^2}}{\omega_n} = V_0\sqrt{\frac{\alpha^2}{\omega_n^2}+1} \quad \text{or} \quad B = V_0\frac{\omega_d}{\omega_n},$$

where

$$\omega_d = \sqrt{\alpha^2+\omega_n^2}.$$

We may also look for the above response in the form of two sinusoids as in (1.66):

$$v_{C,n}(t) = e^{-\alpha t}(M\sin\omega_n t + N\cos\omega_n t). \tag{2.41b}$$

In this case, the arbitrary constants can be found, as in (1.68), with

$$(dv_{C,n}/dt)|_{t=0} = 0$$

and $v_{C,n}(0) = V_0$:

$$N = v_{C,n}(0) = V_0$$

$$M = \frac{\alpha}{\omega_n}V_0.$$

This results in

$$\beta = \tan^{-1}\frac{N}{M} = \frac{\omega_n}{\alpha} \quad \text{and} \quad B = \sqrt{M^2+N^2} = V_0\sqrt{\frac{\alpha^2}{\omega_n^2}+1} = V_0\frac{\omega_d}{\omega_n},$$

which is as was previously found. Therefore,

$$v_{C,n}(t) = e^{-\alpha t}\left(\frac{\alpha}{\omega_n}V_0\sin\omega_n t + V_0\cos\omega_n t\right), \tag{2.42a}$$

or

$$v_{C,n}(t) = V_0\frac{\omega_d}{\omega_n}e^{-\alpha t}\sin(\omega_n t+\beta). \tag{2.42b}$$

The current becomes

$$i_n(t) = C\frac{dv_{C,n}}{dt} = V_0\frac{\omega_d}{\omega_n}Ce^{-\alpha t}[-\alpha\sin(\omega_n t+\beta) + \omega_n\cos(\omega_n t+\beta)]$$

$$= \frac{V_0}{\omega_n L}e^{-\alpha t}\sin(\omega_n t+\beta+v),$$

where $\tan \nu = \omega_n/(-\alpha)$ and, since $\tan \beta = \omega_n/\alpha$, $\beta + \nu = 180°$. Therefore,

$$i_n(t) = -\frac{V_0}{\omega_n L} e^{-\alpha t} \sin \omega_n t. \tag{2.43}$$

The inductance voltage may now be found as

$$v_{L,n}(t) = L\frac{di_n}{dt} = -\frac{V_0}{\omega_n} e^{-\alpha t}[-\alpha \sin \omega_n t + \omega_n \cos \omega_n t]$$

$$= -\frac{V_0 \omega_d}{\omega_n} e^{-\alpha t} \sin(\omega_n t + \nu) = \frac{V_0 \omega_d}{\omega_n} e^{-\alpha t} \sin(\omega_n t - \beta). \tag{2.44}$$

The plots of $v_{C,n}$, i_n, and $v_{L,n}$ are shown in Fig. 2.42(b). This kind of response is also called *an oscillatory or periodical response.*

The energy, initially stored in the capacitance, during this response is interchanged between the capacitance and inductance and is accompanied by energy dissipation into the resistance. The transients will finish, when the entire capacitance energy $CV_0/2$ is completely dissipated.

Critical damping response: If the value of a resistance is close to $2\sqrt{L/C}$, i.e., $R \to 2\sqrt{L/C}$, the natural frequency $\omega_n = \sqrt{1/LC - R^2/4L^2} \to 0$ and the ratio in equation 2.43 $\sin \omega_n t/\omega_n \to \frac{0}{0}$ is indefinite. Applying l'Hopital's rule, gives

$$\lim_{\omega_n \to 0} \left(\frac{\sin \omega_n t}{\omega_n}\right) = \left.\frac{d/d\omega_n(\sin \omega_n t)}{d/d\omega_n(\omega_n)}\right|_{\omega_n \to 0} = \left.\frac{t \cos \omega_n t}{1}\right|_{\omega_n \to 0} = t.$$

Therefore in this critical response the current will be

$$i_n(t) = -\frac{V_0}{L} t e^{-\alpha t}, \tag{2.45}$$

which is also aperiodical. The capacitance voltage can now be found as

$$v_{C,n}(t) = \frac{1}{C}\int i_n(t)dt = \frac{1}{C}\left(-\frac{V_0}{L}\right)\frac{1}{\alpha^2} e^{-\alpha t}(-\alpha t - 1),$$

or since $\alpha^2 = 1/LC$,

$$v_{C,n} = V_0(1 + \alpha t)e^{-\alpha t}. \tag{2.46}$$

Finally, the inductive voltage is

$$v_{L,n}(t) = L\frac{di_n}{dt} = -V_0(e^{-\alpha t} - \alpha t e^{-\alpha t}) = -V_0(1 - \alpha t)e^{-\alpha t}. \tag{2.47}$$

It is also worthwhile to introduce here the graphical representation of the roots of the characteristic equation. On the complex plane the roots, which

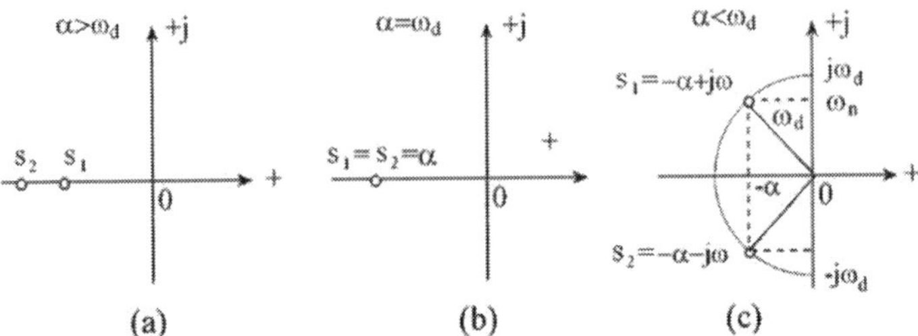

Figure 2.43 The location of the roots of the characteristic equation on the complex plane: overdamped response (a), critical damping response (b), and underdamped response (c).

define the three different cases of the transient response, are located as shown in Fig. 2.43.

The position of the roots on the complex plane, Fig. 2.43 (in other words the dependency of a specific kind of natural response on the relationship between the circuit parameters), is related to the quality factor of a resonant *RLC* circuit. Indeed, by rewriting the critical damping condition as $R/2L = 1/\sqrt{LC}$ we have $\frac{1}{2} = \sqrt{L/C}/R = Q$, this in terms of the resonant circuit is the quality factor. (In our future study, we shall call $Z_c = \sqrt{L/C}$ a surge or natural impedance.) Hence, if $Q < 1/2$ (the position of the roots is shown in (a)), the natural response is overdamped, if $Q > 1/2$ it is underdamped (c) and if $Q = 1/2$ the response is critical damping (b). Hence, the natural response becomes an underdamped oscillatory response, if the resistance of the *RLC* circuit is relatively low compared to the natural impedance.

In (a), two negative real roots are located on the negative axis (in the left half of the complex plane), which indicates the *overdamped* response. Note that $|s_2| > |s_1|$ and therefore $e^{s_2 t}$ decreases faster than $e^{s_1 t}$. In (b), two equal negative roots $s_1 = s_2 = -\alpha$, which indicate the *critical* damping, are still located on the real axis at the boundary point, i.e., no real roots are possible to the right of this point. In (c), the two roots become complex-conjugate numbers, located on the left half circle whose radius is the resonant frequency ω_d. This case indicates an *underdamped* response, having an oscillatory waveform of natural frequency. Note that the two frequencies $\pm j\omega_d$ represent a dissipation-free oscillatory response since the damping coefficient α is zero. This is, of course, a theoretical response: however there are very low resistive circuits in which the natural response could be very close to the theoretical one. Finally, in Fig. 2.44 the change of the form of the natural response with regard to decreasing the damping coefficient is shown.

(b) Parallel connected RLC *circuits*

The circuit containing an *RLC* in parallel is shown in Fig. 2.45. At the instant of $t = 0$ the switch is moved from position "1" to position "2", so that the initial value of the inductance current is I_0. In such a way, this circuit is a full dual

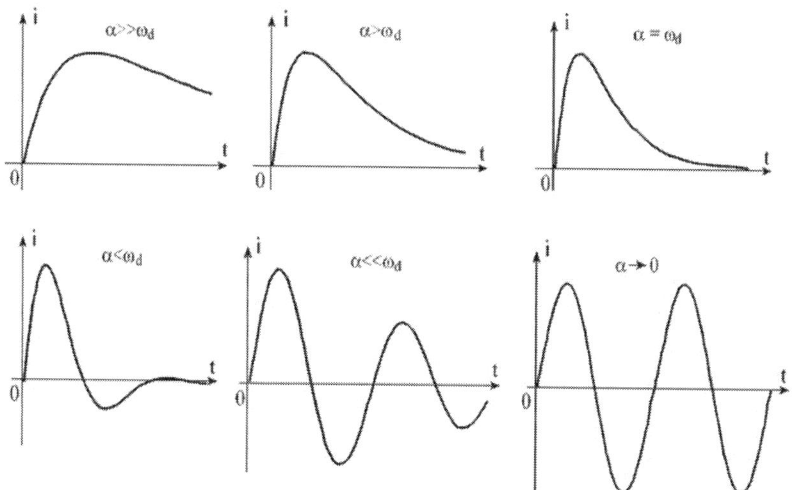

Figure 2.44 The transformation of the natural response in an *RLC* circuit by decreasing the damping coefficient α.

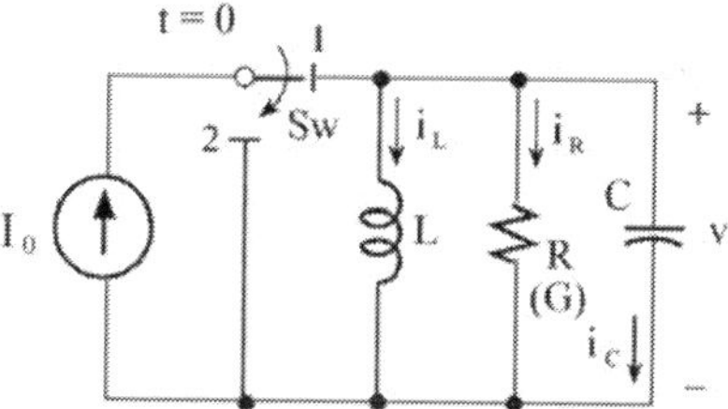

Figure 2.45 A parallel-connected *RLC* circuit.

of the circuit containing an *RLC* in series with an initial capacitance voltage. In order to perform the transient analysis of this circuit we shall apply the principle of duality. As a reminder of the principle of duality: the mathematical results for *RLC* in series are appropriate for *RLC* in parallel after interchanging between the dual parameters ($R \rightarrow G$, $L \rightarrow C$, $C \rightarrow L$), and then the solutions for currents are appropriate for voltages and vice versa. The roots of the characteristic equation will be of the same form: $s_{1,2} = -\alpha \pm \sqrt{\alpha^2 - \omega_d^2}$, but the meaning of α is different: $\alpha = G/2C$ (instead of $\alpha = R/2L$ for a series circuit), however, it is more common to write the above expression as $\alpha = 1/2RC$. The resonant frequency $\omega_d = 1/\sqrt{LC}$ remains the same, since the interchange between L and C does not change the expression.

Underdamped response: The common voltage of all three elements is appropriate to the common current in the series circuit, therefore (see equation 2.38).

$$v_n(t) = \frac{I_0}{C(s_2 - s_1)}(e^{s_1 t} - e^{s_2 t}). \tag{2.48}$$

The inductor current is appropriate to the capacitor voltage in the series circuit, therefore (see (2.37))

$$i_{L,n}(t) = \frac{I_0}{s_2 - s_1}(s_2 e^{s_1 t} - s_1 e^{s_2 t}). \tag{2.49}$$

In a similar way, we shall conclude that the capacitor current is appropriate to the inductance voltage (see equation 2.39)

$$i_{C,n}(t) = \frac{I_0}{s_2 - s_1}(s_1 e^{s_1 t} - s_2 e^{s_2 t}). \tag{2.50}$$

In order to check these results we shall apply the KCL for the common node of the parallel connection and by noting that $i_{R,n}(t) = v_n(t)/R$, we may obtain

$$i_{L,n} + i_{C,n} + i_{R,n} = 0,$$

or

$$\frac{I_0}{s_2 - s_1}\left(s_2 e^{s_1 t} - s_1 e^{s_2 t} + s_1 e^{s_1 t} - s_2 e^{s_2 t} + \frac{1}{RC}e^{s_1 t} - \frac{1}{RC}e^{s_2 t}\right)$$

$$= \frac{I_0}{s_2 - s_1}\left(s_2 + s_1 + \frac{1}{RC}\right)(e^{s_1 t} - e^{s_2 t}) = 0,$$

since $s_2 + s_1 = -2\alpha = -1/RC$.

Overdamped response: The analysis of the overdamped response in a parallel circuit can be performed in a similar way to an underdamped response, i.e., by using the principle of duality. This is left for the reader as an exercise.

(c) Natural response by two nonzero initial conditions

Our next approach in the transient analysis of an *RLC* circuit shall be the more general case in which both energy-storing elements *C* and *L* are previously charged. For this reason, let us consider the current in Fig. 2.46. In this circuit prior to switching, the capacitance is charged to voltage V_{C0} and there is current I_{L0} flowing through the inductance. Therefore, this circuit differs from the one

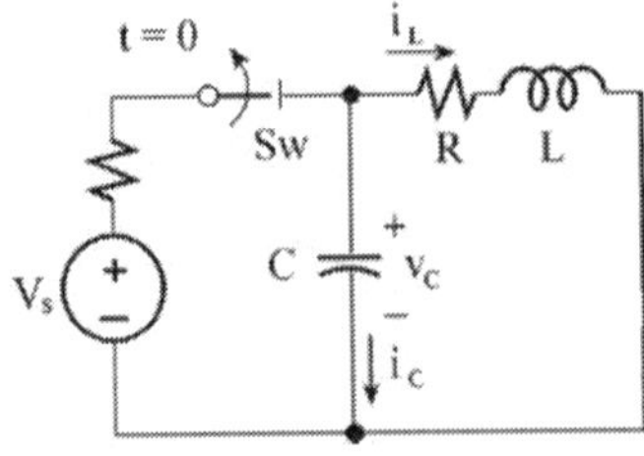

Figure 2.46 *RLC* circuit with a non-zero initial condition.

in Fig. 2.41 in that the initial condition of the inductor current is now $i_L(0_-) = I_{L0}$, but not zero. The capacitance current is now, after switching, $i_C(0) = -i_{L0}(0) = -I_{L0}$. By determining the initial value of the capacitance voltage derivative in equation 2.34, we must substitute $-I_{L0}$ for $i_C(0)$. Therefore,

$$\left.\frac{dv_C}{dt}\right|_{t=0} = -\frac{1}{C} I_{L0},$$

and the set of equations for determining the constants of integration becomes

$$\begin{aligned} A_1 + A_2 &= V_{C0} \\ s_1 A_1 + s_2 A_2 &= -(1/C) I_{L0}, \end{aligned} \tag{2.51}$$

to which the solution is

$$A_1 = \left(V_{C0} + \frac{I_{L0}}{s_2 C}\right)\frac{s_2}{s_2 - s_1} \quad \text{and} \quad A_2 = \left(V_{C0} + \frac{I_{L0}}{s_1 C}\right)\frac{s_1}{s_1 - s_2}. \tag{2.52}$$

The natural responses of an *RLC* circuit will now be

$$v_{C,n}(t) = \left(V_{C0} + \frac{1}{s_2 C} I_{L0}\right)\frac{s_2}{s_2 - s_1} e^{s_1 t} + \left(V_{C0} + \frac{1}{s_1 C} I_{L0}\right)\frac{s_1}{s_1 - s_2} e^{s_2 t}, \tag{2.53}$$

or in a slightly different way

$$v_{C,n}(t) = \frac{V_{C0}}{s_2 - s_1}(s_2 e^{s_1 t} - s_1 e^{s_2 t}) + \frac{I_{L0}}{C(s_2 - s_1)}(e^{s_1 t} - e^{s_2 t}), \tag{2.54}$$

which differs from equation 2.37 by the additional term due to the initial value of the current I_{L0}.

The current response will now be

$$i_n(t) = \frac{V_{C0}}{L(s_2 - s_1)}(e^{s_1 t} - e^{s_2 t}) + \frac{I_{L0}}{s_2 - s_1}(s_1 e^{s_1 t} - s_2 e^{s_2 t}), \tag{2.55}$$

and the inductance voltage

$$v_{L,n}(t) = \frac{V_{C0}}{s_2 - s_1}(s_1 e^{s_1 t} - s_2 e^{s_2 t}) + \frac{L I_{L0}}{s_2 - s_1}(s_1^2 e^{s_1 t} - s_2^2 e^{s_2 t}). \tag{2.56}$$

The above equations 2.54–2.56 can also be *written in terms of hyperbolical functions*. Such expressions are used for transient analysis in some professional books.(*) We shall first write roots s_1 and s_2 in a slightly different form

$$s_{1,2} = -\alpha \pm \gamma, \quad \text{where} \quad \gamma = \sqrt{\alpha^2 - \omega_d^2}, \tag{2.57a}$$

(*)Greenwood, A. (1991) *Electrical Transients in Power Systems*. Wiley, New York, Chichester, Brisbane, Toronto, Singapore.

then

$$s_2 - s_1 = -2\gamma, \quad s_1 s_2 = \alpha^2 - \gamma^2 = \omega_d^2 = 1/LC,$$

and

$$e^{s_{1,2}t} = e^{-\alpha t} e^{\pm\gamma t} = e^{-\alpha t}(e^{\gamma t} + e^{-\gamma t}) = e^{-\alpha t}[\cosh \gamma t \pm \sinh \gamma t]. \qquad (2.57b)$$

With the substitution of equation 2.57(a) for $s_{1,2}$ and taking into account the above relationships, after a simple mathematical rearrangement, one can readily obtain

$$v_{C,n}(t) = \left[V_{CO}\left(\cosh \gamma t + \frac{\alpha}{\gamma} \sinh \gamma t \right) + \frac{I_{LO}}{\gamma C} \sinh \gamma t \right] e^{-\alpha t}, \qquad (2.58)$$

and

$$i_n(t) = \left[-\frac{V_{CO}}{\gamma L} \sinh \gamma t + I_{LO}\left(\cosh \gamma t - \frac{\alpha}{\gamma} \sinh \gamma t \right) \right] e^{-\alpha t}. \qquad (2.59)$$

It should be noted that $1/\gamma C$ and γL (like $1/\omega C$ and ωL) are some kinds of resistances in units of Ohms.

For the overdamped response

$$s_{1,2} = -\alpha \pm j\omega,$$

which means that γ must be substituted by $j\omega$ and the hyperbolic *sine* and *cosine* turn into trigonometric ones

$$v_{C,n}(t) = \left[V_{CO}\left(\cos \omega_n t + \frac{\alpha}{\omega_n} \sin \omega_n t \right) + \frac{I_{LO}}{\omega_n C} \sin \omega_n t \right] e^{-\alpha t},$$

or

$$v_{C,n}(t) = e^{-\alpha t} \left[\left(\frac{I_{LO}}{\omega_n C} + \frac{V_{CO}\alpha}{\omega_n} \right) \sin \omega_n t + V_{CO} \cos \omega_n t \right]. \qquad (2.60)$$

(Which, by assumption $I_{LO} = 0$, turns into the previously obtained one in equation 2.42a.)

At this point we shall once more turn our attention to the *energy relations* in the *RLC* circuit upon its natural response. As we have already observed, the energy is stored in the magnetic and electric fields of the inductances and capacitances, and dissipates in the resistance. To obtain the relation between these processes in a general form we shall start with a differential equation describing the above circuit:

$$L\frac{di}{dt} + v_C + Ri = 0.$$

Multiplying all the terms of the equation by $i = C(dv_C/dt)$, we obtain

$$Li\frac{di}{dt} + Cv_C\frac{dv_C}{dt} + Ri^2 = 0.$$

Taking into consideration that

$$f\frac{df}{dt}=\frac{1}{2}\frac{d}{dt}(f^2)$$

we may rewrite

$$\frac{d}{dt}\left(\frac{Li^2}{2}\right)+\frac{d}{dt}\left(\frac{Cv_C^2}{2}\right)+Ri^2=0,$$

or

$$\frac{d}{dt}\left(\frac{Li^2}{2}+\frac{Cv_C^2}{2}\right)=-Ri^2. \qquad (2.61)$$

The term inside the parentheses gives the sum of the stored energy and, therefore, the derivative of this energy is always negative (if, of course, $i\neq 0$), or, in other words, the total stored energy changes by decreasing. The change of each of the terms inside the parentheses can be either positive or negative (when the energy is exchanged between the inductance and capacitance), but it is impossible for both of them to change positively or increase. This means that the total stored energy decreases during the transients and the rate of decreasing is equal to the rate of its dissipating into resistance (Ri^2).

At this point, we will continue our study of transients in *RLC* circuits by solving numerical examples.

Example 2.26

In the circuit of Fig. 2.47 the switch is changed instantaneously from position "1" to "2". The circuit parameters are: $R_1=2\,\Omega$, $R_2=10\,\Omega$, $L=0.1$ H, $C=0.8$ mF and $V_s=120$ V. Find the transient response of the inductive current.

Solution

The given circuit is slightly different from the previously studied circuit in that the additional resistance is in series with the parallel-connected inductance and capacitance branches.

In order to determine the characteristic equation and its roots (step 1), we

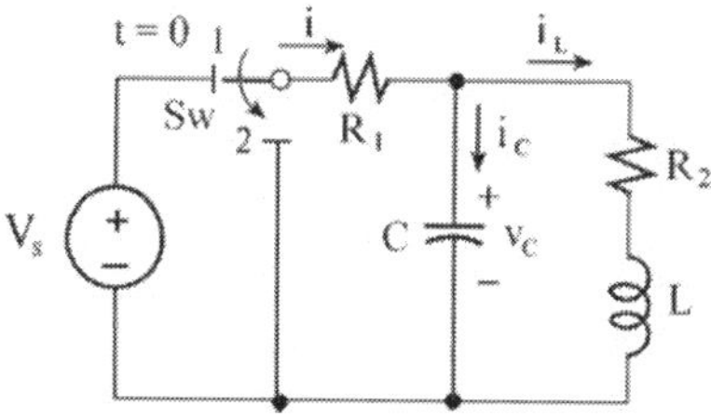

Figure 2.47 A given circuit for Example 2.26.

must indicate the input impedance (seen from the inductance branch)

$$Z(s) = (R_2 + sL) + R_1//(1/sC),$$

which results in

$$s^2 + \left(\frac{R_2}{L} + \frac{1}{R_1 C}\right)s + \frac{R_1 + R_2}{R_1}\frac{1}{LC} = 0,$$

or

$$s^2 + 725s + 7.5\cdot 10^4 = 0,$$

where

$$2\alpha = \left(\frac{R_2}{L} + \frac{1}{R_1 C}\right) = \frac{10}{0.1} + \frac{10^3}{2\cdot 0.8} = 725 \text{ s}^{-1}$$

$$\omega_d = \frac{R_1 + R_2}{R_1}\frac{1}{LC} = \frac{2+10}{2}\frac{10^3}{0.1\cdot 0.8} = 7.5\cdot 10^4 \text{ rad/s}.$$

Thus,

$$s_{1,2} = (-3.625 \pm \sqrt{3.625^2 - 7.5})\cdot 10^2 = -125, \ -600 \text{ s}^{-1}$$

and

$$v_{C,n}(t) = A_1 e^{-125t} + A_2 e^{-600t}.$$

Since the circuit after switching is source free, no forced response (step 2) is expected.

The initial conditions (step 3) are:

$$v_C(0) = v_C(0\_) = V_s\frac{R_2}{R_1 + R_2} = 120\frac{10}{2+10} = 100 \text{ V}$$

$$i_L(0) = i_L(0\_) = \frac{V_s}{R_1 + R_2} = \frac{120}{2+10} = 10 \text{ A}.$$

The initial value of the current derivative (step 4) is found as

$$\left.\frac{di_L}{dt}\right|_{t=0} = \frac{v_L(0)}{L} = \frac{v_C(0) - R_2 i_L(0)}{L} = \frac{100 - 10\cdot 10}{L} = 0.$$

By solving the two equations below (step 5)

$$A_1 + A_2 = 10,$$

$$s_1 A_1 + s_2 A_2 = 0,$$

we have (see equation 2.36)

$$A_1 = \frac{I_{L0} s_2}{s_2 - s_1} = \frac{10(-600)}{-600+125} = 12.6, \quad A_2 = \frac{I_{L0} s_1}{s_1 - s_2} = \frac{10(-125)}{-125+600} = -2.6.$$

Thus,

$$i_L(t) = 12.6e^{-125t} - 2.6e^{-600t}\ \text{A}.$$

In the next example, we will consider an RLC circuit, having a zero independent initial condition, which is connected to a d.c. power supply.

Example 2.27

In the circuit with $R = 100\ \Omega$, $R_1 = 5\ \Omega$, $R_2 = 3\ \Omega$, $L = 0.1$ H, $C = 100\ \mu$F and $V_s = 100$ V, shown in Fig. 2.48, find current $i_L(t)$ for $t > 0$. The voltage source is applied at $t = 0$, due to the unit forcing function $u(t)$.

Solution

The input impedance seen from the inductive branch is

$$Z_{12}(s) = R_1 + sL + \left(R_2 + \frac{1}{sC}\right) // R,$$

or, after performing the algebraic operations and equating it to zero, we obtain

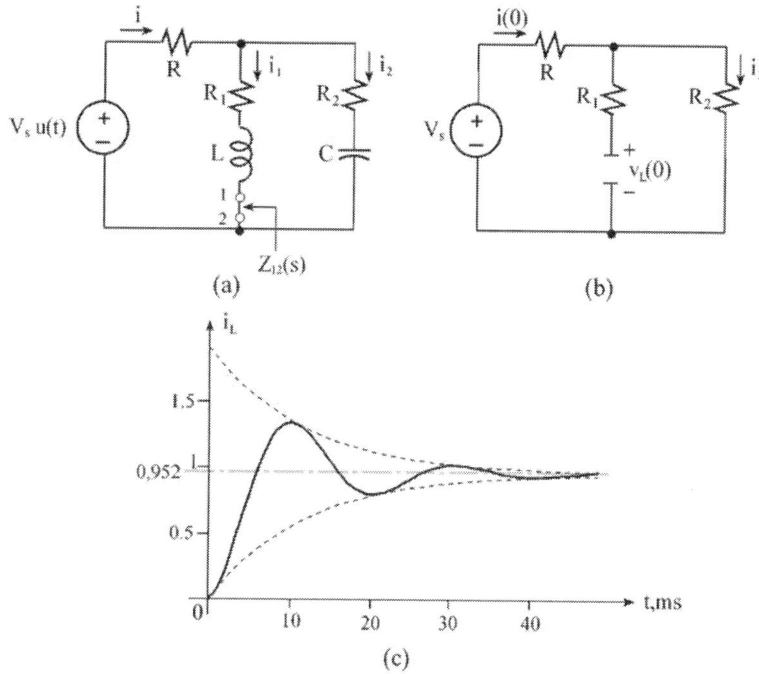

Figure 2.48 A given circuit for Example 2.27 (a), an equivalent circuit for instant $t = 0$ (b) and the current plot (c).

the characteristic equation

$$s^2+\left(\frac{R_{eq}}{L}+\frac{1}{(R+R_2)C}\right)s+\frac{R+R_1}{R+R_2}\frac{1}{LC}=0,$$

where

$$R_{eq}=(RR_1+RR_2+R_1R_2)/(R+R_2).$$

Substituting the numerical values yields

$$s^2+176.2s+10.2\cdot 10^4=0,$$

to which the roots are:

$$s_{1,2}=-88.1\pm j307\ \mathrm{s}^{-1}.$$

Since the roots are complex numbers, the natural response is

$$i_{L,n}(t)=Be^{-88.1t}\sin(307t+\beta).$$

The forced response is

$$i_{L,f}=\frac{V_s}{R+R_1}=\frac{100}{100+5}=0.952\ \mathrm{A}.$$

The independent initial conditions are zero, therefore

$$v_C(0)=0 \quad \text{and} \quad i_L(0)=0.$$

The dependent initial condition is found in circuit (b), which is appropriate to the instant of switching $t=0$:

$$\left.\frac{di_L}{dt}\right|_{t=0}=\frac{v_L(0)}{L}=\frac{i(0)R_2}{L}=\frac{V_sR_2}{(R+R_2)L}=\frac{100\cdot 3}{(100+3)0.1}=29.2.$$

The integration constant can now be found from

$$B\sin\beta=i(0)-i_f(0)=0-0.952=-0.952$$

$$-\alpha B\sin\beta+\omega_n B\cos\beta=\left.\frac{di}{dt}\right|_{t=0}-\left.\frac{di_f}{dt}\right|_{t=0}=29.2-0=29.2,$$

to which the solution is

$$\beta=\tan^{-1}\frac{307}{-(29.2/0.952)+88}=79.4^\circ \quad \text{and} \quad B=-\frac{0.952}{\sin 79.4^\circ}=-0.968.$$

Therefore, the complete response is

$$i_L=i_{L,f}+i_{L,n}=0.952-0.968e^{-88t}\sin(307t+79.4^\circ)\ \mathrm{A}.$$

To plot this curve we have to estimate the time constant of the exponent, $\tau=1/88\cong 11$ ms, and the period of sine, $T=2\pi/307\cong 20$ ms. The plot of the current is shown in Fig. 2.48(c).

Example 2.28

In the circuit with $R_1 = R_2 = 10\,\Omega$, $L = 5$ mH, $C = 10\,\mu$F and $V_s = 100$ V, in Fig. 2.49, find current $i_2(t)$ after the switch closes.

Solution

The input impedance seen from the source is $Z_{in}(s) = R_1 + sL + R_2//(1/sC)$. Then the characteristic equation becomes

$$s^2 + \left(\frac{R_1}{L} + \frac{1}{R_2 C}\right)s + \frac{R_1 + R_2}{R_2}\frac{1}{LC} = 0.$$

Substituting the numerical values and solving this characteristic equation, we obtain the roots:

$$s_{1,2} = (-6 \pm j2)10^3\ \mathrm{s}^{-1}.$$

The natural response becomes

$$i_{2,n} = Be^{-6\cdot 10^3 t}\sin(2\cdot 10^3 t + \beta).$$

The forced response is

$$i_{2,f} = \frac{V_s}{R_1 + R_2} = \frac{100}{10 + 10} = 5\ \mathrm{A}.$$

The independent initial conditions are

$$i_1(0) = i_L(0) = \frac{V_s}{R_1 + R_2} = \frac{100}{10 + 10} = 5\ \mathrm{A} \quad \text{and} \quad v_C(0) = 0.$$

In order to determine the initial conditions for current i_2, which can change abruptly, we must consider the given circuit at the moment of $t = 0$. Since the capacitance at this moment is a short-circuit, the current in R_2 drops to zero, i.e., $i_2 = 0$. With the KVL for the right loop we have

$$R_2 i_2 - v_C = 0 \quad \text{or} \quad R_2 i_2 = v_C,$$

and

$$\left.\frac{di_2}{dt}\right|_{t=0} = \frac{1}{R_2}\frac{1}{C}i_3(0) = \frac{1}{10\cdot 10^{-5}}5 = 5\cdot 10^4.$$

Here $i_3(0) = i_1(0) = 5$ A, since $i_2(0) = 0$.

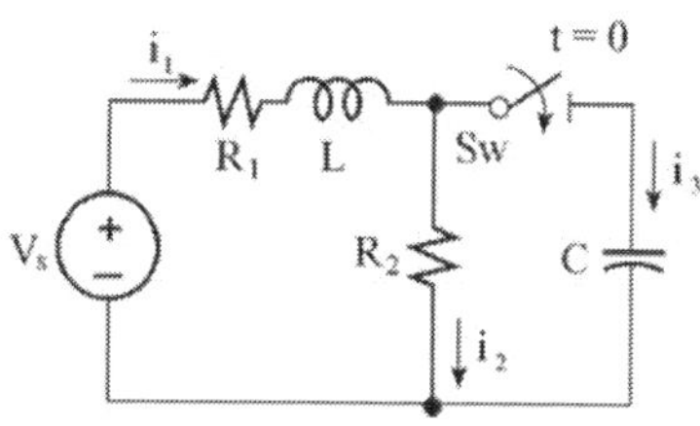

Figure 2.49 A given circuit for Example 2.28.

Our last step is to find the integration constants. We have

$$B\sin\beta = i_2(0) - i_{2,f}(0) = -5$$

$$-6\cdot 10^3 \sin\beta + 2\cdot 10^3 \cos\beta = \left.\frac{di_2}{dt}\right|_{t=0} = 5\cdot 10^4,$$

to which the solution is

$$\beta = \tan^{-1}\frac{-5\cdot 2\cdot 10^3}{50\cdot 10^3 - 30\cdot 10^3} = -26.6^\circ, \quad B = \frac{-5}{\sin(-26.6^\circ)} = 11.2.$$

Therefore, the complete response is

$$i_2 = 5 + 11.2e^{-6000t}\sin(2000t - 26.6^\circ)\ \mathrm{A}.$$

Example 2.29

Consider once again the circuit shown in Fig. 2.40, which is redrawn here, Fig. 2.50. This circuit has been previously analyzed and it was shown that the natural response is dependent on the kind of applied source, voltage or current. We will now complete this analysis and find the transient response a) of the current $i(t)$ when a voltage source of 100 V is connected between nodes m–n, Fig. 2.50(a); and (b) of the voltage $v(t)$ when a current source of 11 A is connected between nodes m–n, Fig. 2.50(b). The circuit parameters are $R_1 = R_2 = 100\ \Omega$, $R_3 = 10\ \Omega$, $L = 20$ mH and $C = 2\ \mu$F.

Solution

(a) In this case an ideal voltage source is connected between nodes m and n.

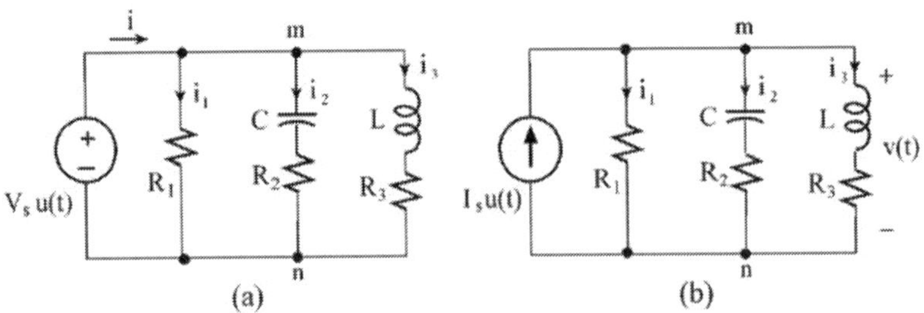

Figure 2.50 A circuit for Example 2.29 driven by a voltage source (a) and by a current source (b).

Therefore each of the three branches operates independently, and we may find each current very simply.

$$i_{1,f} = \frac{V_s}{R_1} = \frac{100}{100} = 1 \text{ A} \quad \text{(no natural response)}$$

$$i_2 = i_{2,f} + i_{2,n} = 0 + \frac{V_s}{R_2} e^{s_2 t} = 1 e^{-5000t} \text{ A},$$

where $s_2 = -\frac{1}{RC} = \frac{1}{100 \cdot 2 \cdot 10^{-6}} = 5000 \text{ s}^{-1}$,

$$i_3 = i_{3,f} + i_{3,n} = \frac{V_s}{R_3} - \frac{V_s}{R_3} e^{s_3 t} = 10 - 10e^{-500t} \text{ A},$$

where $s_3 = -\frac{R_3}{L} = \frac{10}{20 \cdot 10^{-3}} = -500 \text{ s}^{-1}$.

Therefore, the total current is

$$i = i_1 + i_2 + i_3 = 11 + e^{-5000t} - 10e^{-500t} \text{ A}.$$

(b) In this case, in order to find the transient response we shall, as usual, apply the five-step solution. The characteristic equation (step 1) for this circuit has already been determined in equation 2.29. With its simplification, we have

$$s^2 + \left[\frac{R_{eq}}{L} + \frac{1}{(R_1 + R_2)C}\right] s + \frac{R_1 + R_3}{R_1 + R_2} \frac{1}{\sqrt{LC}} = 0,$$

where $R_{eq} = (R_1 R_2 + R_1 R_3 + R_2 R_3)/(R + R_2)$.

Upon substituting the numerical data the solution is

$$s_{1,2} = (-2.75 \pm j2.5)10^3 \text{ s}^{-1}.$$

Thus the natural response will be

$$v_n = Be^{-2.75 \cdot 10^3 t} \sin(2.5 \cdot 10^3 t + \beta) \text{ V}.$$

The forced response (step 2) is

$$v_f = I_s \frac{R_1 R_3}{R_1 + R_3} = 11 \frac{100 \cdot 10}{100 + 10} = 100 \text{ V}.$$

The independent initial conditions (step 3) are zero, i.e., $v_C(0) = 0$, $i_L(0) = 0$.

Next (step 4) we shall find the dependent initial condition, which will be used to determine the voltage derivative:

the voltage drop in the inductance, which is open circuited

$$v_L(0) = I_s(R_1 // R_2) = 11 \cdot 50 = 550 \text{ V};$$

the capacitance current, since the capacitance is short-circuited

$$i_C(0) = I_s \frac{R_1}{R_1 + R} = 11 \frac{100}{200} = 5.5 \text{ A};$$

the initial value of the node voltage (which is the voltage across the inductance) $v(0) = v_L(0) = 550$ V. In order to determine the voltage derivative we shall apply Kirchhoff's two laws

$$i_R + i_C + i_L = I_s, \quad v = R_1 i_R = R_2 i_C + v_C,$$

and, after differentiation, we have

$$\frac{di_R}{dt} + \frac{di_C}{dt} + \frac{di_L}{dt} = 0$$

$$\frac{dv}{dt} = R_2 \frac{di_C}{dt} + \frac{dv_C}{dt}.$$

By taking into consideration that

$$\left.\frac{dv_C}{dt}\right|_{t=0} = \frac{i_C(0)}{C}, \quad \left.\frac{di_L}{dt}\right|_{t=0} = \frac{v_L(0)}{L} \quad \text{and} \quad i_R = \frac{v}{R_1},$$

the solution for $(dv/dt)|_{t=0}$ becomes

$$\left.\frac{dv}{dt}\right|_{t=0} = \frac{R_1}{R_1 + R_2}\left(\frac{i_C(0)}{C} - \frac{R_2}{L} v_L(0)\right),$$

which, upon substitution of the data, results in $(dv/dt)|_{t=0} = 0$.

The integration constant, can now be found by solving the following set of equations

$$B \sin \beta = v(0) - v_f(0) = 550 - 100 = 450$$

$$-2.75 \cdot 10^3 \sin \beta + 2.5 \cdot 10^3 \cos \beta = \left.\frac{dv}{dt}\right|_{t=0} - \left.\frac{dv_f}{dt}\right|_{t=0} = 0.$$

The solution is

$$\beta = \tan^{-1} \frac{2 \cdot 5}{2.75} = 42.3^\circ$$

$$B = \frac{450}{\sin 42.3^\circ} = 669.$$

Therefore, the complete response is

$$v(t) = 100 + 669 e^{-2.75 \cdot 10^3 t} \sin(2.5 \cdot 10^3 t + 42.3^\circ) \text{ V}.$$

Note that this response is completely different from the one achieved in circuit (a). However, the forced response here, i.e., the node voltage, is 100 V, which is the same as the node voltage in circuit (a) due to the 100 V voltage source.

2.7.2 *RLC* circuits under a.c. supply

The analysis of an *RLC* circuit under a.c. supply does not differ very much from one under d.c. supply, since the natural response does not depend on the source and the five-step solution may again be applied. However, the evaluation of the forced response is different and somehow more labor consuming, since phasor analysis (based on using complex numbers) must be applied. Let us now illustrate this approach by solving numerical examples.

Example 2.30

Let us return to the circuit shown in Fig. 2.51 of Example 2.26 and suppose that the switch is moved from position "2" to "1", connecting this circuit to the a.c. supply: $v_s = V_m \sin(\omega t + \psi_v)$, having $V_m = 540$ V at $f = 50$ Hz and $\psi_v = 0°$. Find the current of the inductive branch, i_L, assuming that the circuit parameters are: $R_1 = 2\ \Omega$, $R_2 = 10\ \Omega$, $L = 0.1$ H and the capacitance $C = 100\ \mu$F, whose value is chosen to improve the power factor.

Solution

The characteristic equation of the circuit has been found in Example 2.26, in

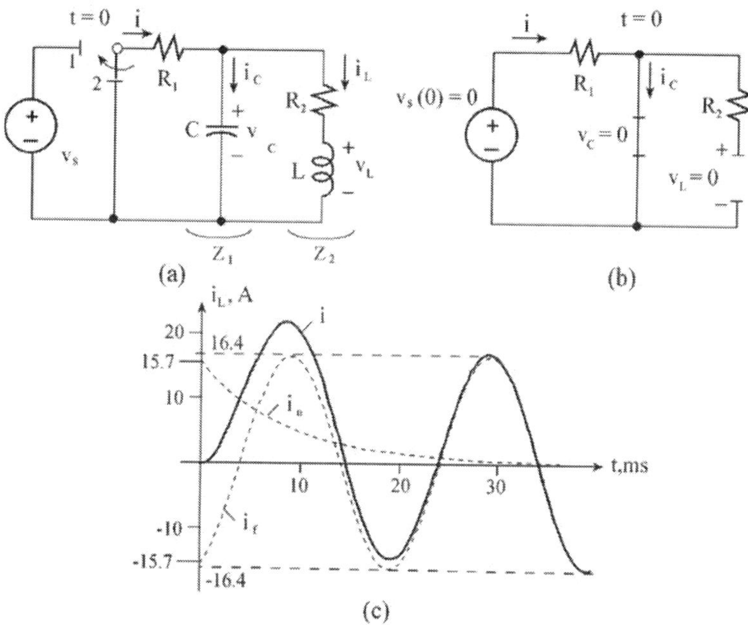

Figure 2.51 A given circuit of Example 2.30 (a), circuit equivalent at $t = 0$ (b) and the plot of current i_L (c).

which

$$\alpha = \frac{1}{2}\left(\frac{R_2}{L} + \frac{1}{R_1 C}\right) = \frac{1}{2}\left(\frac{10}{0.1} + \frac{10^4}{2}\right) = 2.55 \cdot 10^3$$

and

$$\omega_d^2 = \frac{R_1 + R_2}{R_1}\frac{1}{LC} = \frac{2+10}{2}\frac{10^4}{0.1} = 0.6 \cdot 10^6.$$

Thus, the roots are

$$s_{1,2} = (-2.55 \pm \sqrt{2.55^2 - 0.6})10^3 \cong (-0.12, -5.0)10^3 \text{ s}^{-1}$$

and the natural response is

$$i_{L,n} = A_1 e^{-120t} + A_2 e^{-5000t}.$$

The next step is to find the forced response. By using the phasor analysis method we have

$$\tilde{I}_{L,m} = \frac{\tilde{V}}{Z_{in}}\frac{Z_1}{Z_2 + Z_1} = \frac{540}{105\angle -17.4^\circ}\frac{31.8\angle -90^\circ}{10} = 16.4\angle -72.6^\circ \text{ A},$$

where

$$Z_2 = R_2 + j\omega L = 10 + j31.4 = 32.9\angle 72.3^\circ, \quad Z_1 = -j1/\omega C = -j31.8$$

and

$$Z_{in} = R_1 + Z_2//Z_1 = 105\angle -17.4^\circ.$$

The forced response is

$$i_{L,f} = 16.4 \sin(314t - 72.6^\circ).$$

Since no initial energy is stored either in the capacitance or in the inductance, the initial conditions are zero: $v_C(0) = 0$ and $i_L(0) = 0$. By inspection of the circuit for the instant $t = 0$, Fig. 2.51(b), in which the capacitance is short-circuited, the inductance is open-circuited and the instant value of the voltage source is zero, we may conclude that $v_L(0) = 0$. Therefore, the second initial condition for determining the integration constant is

$$\left.\frac{di_L}{dt}\right|_{t=0} = \frac{v_L(0)}{L} = 0.$$

Thus, we have

$$A_1 + A_2 = i_L(0) - i_{L,f}(0) = 0 - 16.4 \sin(-72.6^\circ) = 15.65$$

$$s_1 A_1 + s_2 A_2 = \left.\frac{di_L}{dt}\right|_{t=0} - \left.\frac{di_{L,f}}{dt}\right|_{t=0} = 0 - 16.4 \cdot 314 \cos(-72.6^\circ) = -1540,$$

and

$$A_1 = \frac{(-5 \cdot 15.65 + 1.54)10^3}{(-5 + 0.12)10^3} = 15.72, \quad A_2 = 15.65, \quad A_1 = -0.07 \cong 0.$$

Therefore, the complete response is

$$i_L(t) = 16.4 \sin(314t - 72.6^\circ) + 15.7e^{-120t} \text{ A}.$$

This response is plotted in Fig. 2.51(c), whereby the time constant of the exponential term is $\tau = 1/120 = 8.3$ ms.

Example 2.31

A capacitance of 200 μF is switched on at the end of a 1000 V, 60 Hz transmission line with $R = 10\ \Omega$ and load $R_1 = 30\ \Omega$ and $L = 0.1$ H, Fig. 2.52. Find the transient current i and sketch it, if the instant of switching the voltage phase angle is zero, $\psi_v = 0$.

Solution

The characteristic equation is obtained by equating the input impedance to zero

$$s^2 + 2\alpha s + \omega_d^2 = 0,$$

Here

$$\alpha = \frac{1}{2}\left(\frac{R_1}{L} + \frac{1}{RC}\right) = \frac{1}{2}\left(\frac{30}{0.1} + \frac{10^4}{10 \cdot 2}\right) = 4 \cdot 10^2$$

and

$$\omega_d^2 = \frac{R + R_1}{R}\frac{1}{LC} = \frac{10 + 30}{10}\frac{10^4}{0.1 \cdot 2} = 20 \cdot 10^4,$$

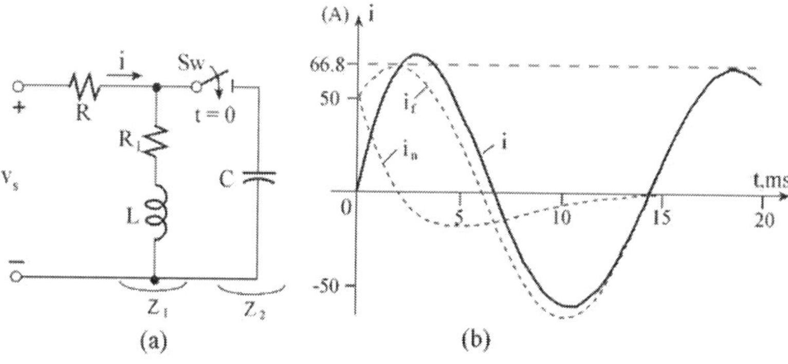

Figure 2.52 A given circuit of Example 2.31 (a) and the current plot (b).

which results in the roots

$$s_{1,2} = -\alpha \pm \sqrt{\alpha^2 - \omega_d^2} = (-4 \pm \sqrt{16 - 20})10^2 = (-4 \pm j2)10^2 \ \mathrm{s}^{-1}.$$

Thus, the natural response is

$$i_n(t) = Be^{-400t} \sin(200t + \beta).$$

The forced response is found by phasor analysis

$$\tilde{I} = \frac{\tilde{V}_s}{Z_{in}} = \frac{1000}{21.2 \angle -50^\circ} = 47.2 \angle 50^\circ,$$

where

$$Z_1 = R_1 + j\omega L = 30 + j37.7 = 48.2 \angle 51.5^\circ,$$

$$Z_2 = -j\frac{1}{\omega C} = -j\frac{10}{377 \cdot 2} = -j13.3$$

and

$$Z_{in} = R + \frac{Z_1 Z_2}{Z_1 + Z_2} = 10 + \frac{48.2 \angle 51.5^\circ \cdot 13.3 \angle -90^\circ}{30 + j37.7 - j13.3} = 21.2 \angle -50^\circ.$$

Therefore,

$$i_f = 47.2\sqrt{2}\sin(377t + 50^\circ) = 66.8 \sin(377t + 50^\circ)\ \mathrm{A}.$$

The independent initial conditions are $v_C(0) = 0$, $i_L(0) = 25.7 \sin 43.3^\circ = 17.6$ A, since prior to switching:

$$\tilde{I}_{L,m} = \frac{V_s\sqrt{2}}{\sqrt{(R + R_1)^2 + x_L^2}} = \frac{1000\sqrt{2}}{\sqrt{40^2 + 37.7^2}} = 25.7 \quad \text{and} \quad \varphi = \tan^{-1}\frac{37.7}{40} = 43.3^\circ.$$

The next step is to determine the initial values of $i(0)$ and $(di/dt)|_{t=0}$. Since the input voltage at $t = 0$ is zero and the capacitance voltage is zero, we have $i(0) = [v_s(0) - v_C(0)]/R = 0$. The initial value of the current derivative is found with Kirchhoff's voltage law applied to the outer loop

$$-v_s + Ri + v_C = 0,$$

and, after differentiation, we have

$$\left.\frac{di}{dt}\right|_{t=0} = \frac{1}{R}\left(\frac{dv_s}{dt} - \frac{dv_C}{dt}\right)_{t=0} = \frac{1}{10}[533 - (-88)]10^3 = 62.1 \cdot 10^3.$$

Here

$$\left.\frac{dv_s}{dt}\right|_{t=0} = 1000\sqrt{2} \cdot 377 \cos\psi_v = 533 \cdot 10^3$$

and

$$\left.\frac{dv_C}{dt}\right|_{t=0} = \frac{1}{C} i_C = \frac{10^4}{2}(-17.6) = -88 \cdot 10^3,$$

because $i_C(0) = -i_L(0) = -17.6$ (note that $i(0) = 0$). Hence, we now have two simultaneous equations for finding the integration constants

$$B \sin \beta = i(0) - i_f(0) = 0 - 47.2\sqrt{2} \sin 50° = -51.1$$

$$-4 \cdot 10^2 B \sin \beta + 2 \cdot 10^2 B \cos \beta = \left.\frac{di}{dt}\right|_{t=0} - \left.\frac{di_f}{dt}\right|_{t=0}$$

$$= 62.1 \cdot 10^3 - 47.2\sqrt{2} \cdot 377 \cos 50° = 45.9 \cdot 10^3,$$

for which the solution is

$$\beta = \tan^{-1} \frac{2 \cdot 10^2}{45.9 \cdot 10^3/(-51.1) + 4 \cdot 10^2} = 158.2° \quad \text{and} \quad B = \frac{-51.1}{\sin 158.2°} = -137.6.$$

Thus, the complete response is

$$i = 66.8 \sin(377t + 50°) - 137.6 e^{-400t} \sin(200t + 158.2°).$$

The plot of this curve can be seen in Fig. 2.52(c).

2.7.3 Transients in *RLC* resonant circuits

An *RLC* circuit whose quality factor Q is high (at least as large as 1/2) is considered a resonant circuit and, when interrupted, the transient response will be oscillatory. If the natural frequency of such oscillations is equal or close to some of the harmonics inherent in the system voltages or currents, then the resonant conditions may occur. In power system networks, the resonant circuit may arise in many cases of its operation.

In transmission and distribution networks, resonance may occur if an extended underground cable (having preponderant capacitance) is connected to an overhead line or transformer (having preponderant induction). The natural frequency of such a system may be close to the lower harmonics of the generating voltage. When feeder cables of high capacitances are protected against short-circuit currents by series reactors of high inductances, the resonance phenomenon may also arise. Banks of condensers, used, for example, for power factor correction, and directly connected under full voltage with the feeding transformer, may form a resonance circuit, i.e., where no sufficient damping resistance is present. Such circuits contain relatively small inductances and thus the frequency of the transient oscillation is extremely high.

Very large networks of high voltage may have such a great capacitance of the transmission lines and the inductance of the transformers, that their natural frequency approaches the system frequency. This may happen due to line-to-ground fault and would lead to significant overvoltages of fundamental frequency. More generally, it is certain that, for every alteration in the circuits and/or variation of the load, the capacitances and inductances of an actual network change substantially. In practice it is found, therefore, that the resonance during the transients in power systems, occur if and when the natural system frequency is equal or close to one of the generalized frequencies. During the resonance some harmonic voltages or currents, inherent in the source or in the load, might be amplified and cause dangerous overvoltages and/or overcurrents.

It should be noted that in symmetrical three-phase systems all higher harmonics of a mode divisible by 2 or 3 vanish, the fifth and seventh harmonics are the most significant ones due to the generated voltages and the eleventh and the thirteenth are sometimes noticeable due to the load containing electronic converters.

We shall consider the transients in the *RLC* resonant circuit in more detail assuming that the resistances in such circuits are relatively low, so that $R \ll Z_C$, where $Z_C = \sqrt{L/C}$, which is called a *natural or characteristic* impedance (or resistance); it is also sometimes called a *surge impedance.*

(a) *Switching on a resonant* RLC *circuit to an a.c. source*

The natural response of the current in such a circuit, Fig. 2.53 (see sections 1.62 and 2.72) may be written as

$$i_n = I_n e^{-\alpha t} \sin(\omega_n t + \beta). \tag{2.62}$$

The natural response of the capacitance voltage will then be

$$v_{C,n} = \frac{1}{C}\int i_n dt = I_n \frac{e^{-\alpha t}}{C(\alpha^2 + \omega_n^2)}[-\alpha \sin(\omega_n t + \beta) - \omega_n \cos(\omega_n t + \beta)],$$

upon simplification, combining the sine and cosine terms to a common sine term with the phase angle $(90° + \delta)$,

$$v_{C,n} = V_{C,n} e^{-\alpha t} \sin[\omega_n t + \beta - (90° + \delta)], \tag{2.63}$$

where

$$V_{C,n} = I_n \sqrt{\frac{L}{C}}, \tag{2.64a}$$

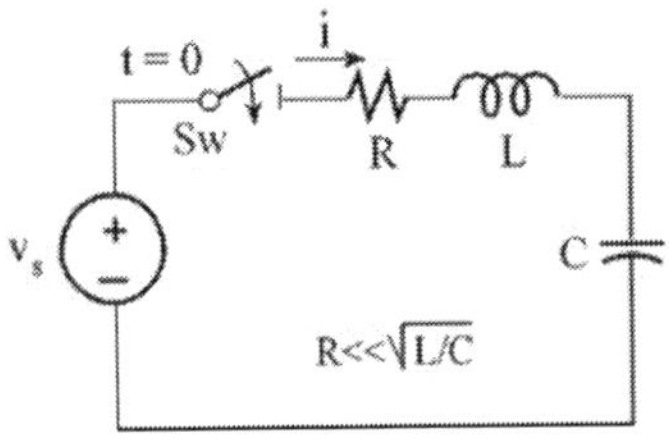

Figure 2.53 A resonant *RLC* circuit.

$$\delta = \tan^{-1}\left(\frac{\alpha}{\omega_n}\right), \tag{2.64b}$$

and

$$(\alpha^2 + \omega_n^2) = \omega_d^2 + \frac{1}{LC}. \tag{2.65}$$

The natural response of the inductive voltage may be found simply by differentiation:

$$v_{L,n} = L\frac{di_{i,n}}{dt} = LI_n e^{-\alpha t}\,[-\alpha\sin(\omega_n t + \beta) + \omega_n\cos(\omega_n t + \beta)],$$

or after simplification, as was previously done, we obtain

$$v_{L,n} = I_n\sqrt{\frac{L}{C}}e^{-\alpha t}\sin[\omega_n t + \beta + (90^\circ + \delta)]. \tag{2.66}$$

It is worthwhile to note here that by observing equation 2.63 and equation 2.66 we realize that $v_{C,n}$ is lagging slightly more and $v_{L,n}$ is leading slightly more than 90° with respect to the current. This is in contrast to the steady-state operation of the *RLC* circuit, in which the inductive and capacitive voltages are displaced by exactly $\pm 90^\circ$ with respect to the current. The difference, which is expressed by the angle δ, is due to the exponential damping. This angle is analytically given by equation 2.64b and indicates the *deviation* of the *displacement angle* between the current and the inductive/capacitance voltage from 90°. Since the resistance of the resonant circuits is relatively small, we may approximate

$$\omega_n = \sqrt{\frac{1}{LC} - \left(\frac{R}{2L}\right)^2} \cong \frac{1}{\sqrt{LC}} \quad \text{and} \quad \tan\delta \cong R/2\sqrt{L/C}. \tag{2.67}$$

For most of the parts of the power system networks resistance R is much smaller than the natural impedance $\sqrt{L/C}$ so that the angle δ is usually small and can be neglected.

By switching the *RLC* circuit, Fig. 2.53, to the voltage source

$$v_s = V_m\sin(\omega t + \psi_v) \tag{2.68}$$

the steady-state current will be

$$i_f = I_f\sin(\omega t + \psi_i), \tag{2.69}$$

the amplitude of which is

$$I_f = \frac{V_m}{\sqrt{R^2 + (\omega L - 1/\omega C)^2}}, \tag{2.70}$$

and the phase angle is

$$\psi_i = \psi_v - \varphi, \quad \varphi = \tan^{-1}\frac{\omega L - 1/\omega C}{R}. \tag{2.71}$$

The steady-state capacitance voltage is

$$v_{C,f} = \frac{I_f}{\omega C}\sin(\omega t + \psi_i - 90^\circ). \tag{2.72}$$

For the termination of the arbitrary constant, β, we shall solve a set of equations, written for $i_n(0)$ and $v_{C,n}(0)$ in the form:

$$i_n(0) = i(0) - i_f(0)$$
$$v_{C,n}(0) = v_C(0) - v_{C,f}(0).$$

Since the independent initial conditions for current and capacitance voltage are zero and the initial values of the forced current and capacitance voltage are $i_f(0) = I_f \sin\psi_i$, and $v_{C,f}(0) = -(I_f/\omega C)\cos\psi_i$ we have

$$\begin{aligned} I_n \sin\beta &= 0 - I_f \sin\psi_i \\ -I_n\sqrt{\frac{L}{C}}\cos\beta &= 0 + \frac{I_f}{\omega C}\cos\psi_i. \end{aligned} \tag{2.73}$$

The simultaneous solution of these two equations, by dividing the first one by the second one, results in

$$\tan\beta = \frac{\omega}{\omega_n}\tan\psi_i. \tag{2.74}$$

Whereby the phase angle β of the natural current can be determined and, with its value, the first equation in 2.73 give the initial amplitude of the transient current

$$I_n = -I_f\frac{\sin\psi_i}{\sin\beta} = -I_f\sin\psi_i\sqrt{1 + \frac{1}{\tan^2\psi_i}\left(\frac{\omega_n}{\omega}\right)^2},$$

or

$$I_n = -I_f\sqrt{\sin^2\psi_i + \left(\frac{\omega_n}{\omega}\right)^2\cos^2\psi_i}. \tag{2.75}$$

The initial amplitude of the transient capacitance voltage can also be found with equation 2.64(a)

$$V_{C,n} = I_n\sqrt{\frac{L}{C}} = -I_f\sqrt{\frac{L}{C}}\sqrt{\sin^2\psi_i + \left(\frac{\omega_n}{\omega}\right)^2\cos^2\psi_i},$$

or, with the expression $V_{C,f} = I_f/\omega C$ (see equation 2.72), we may obtain

$$V_{C,n} = -V_{C,f}\sqrt{\left(\frac{\omega}{\omega_n}\right)^2 \sin^2\psi_i + \cos^2\psi_i}. \tag{2.76}$$

From the obtained equations 2.74, 2.75 and 2.76 we can understand that the phase angle, β, and the amplitudes, I_n and $V_{C,n}$, of the transient current and capacitance voltage depend on two parameters, namely, the instant of switching, given by the phase angle ψ_i of the steady-state current and the ratio of the natural, ω_n to the a.c. source frequency, ω. Using the obtained results let us now discuss a couple of practical cases.

(b) *Resonance at the fundamental (first) harmonic*

In this case, with $\omega_n = \omega$ the above relationships become very simple. According to equation 2.74

$$\tan\beta = \tan\psi_i \quad \text{and} \quad \beta = \psi_i, \tag{2.77}$$

i.e., the initial phase angles of the natural and forced currents are equal. According to equations 2.75 and 2.76

$$I_n = -I_f \quad \text{and} \quad V_{C,n} = -V_{C,f}, \tag{2.78}$$

which means that the amplitudes of the natural current and capacitance voltages are negatively equal (in other words they are in the opposite phase) to their steady-state values. Since the frequencies ω and ω_n are equal, we can combine the sine function of the forced response (the steady-state value) and the natural response, and therefore the complete response becomes

$$\begin{aligned} i &= I_f(1 - e^{-\alpha t})\sin(\omega t + \psi_i) \\ v_C &= V_{C,f}(1 - e^{-\alpha t})\sin(\omega t + \psi_i - 90^\circ). \end{aligned} \tag{2.79}$$

The plot of the transient current (equation 2.79) is shown in Fig. 2.54. As can be seen, in a resonant circuit the current along with the voltages reach their maximal values during transients after a period of 3–5 times the time constant of the exponential term. Since the time constant here is relatively low, due to

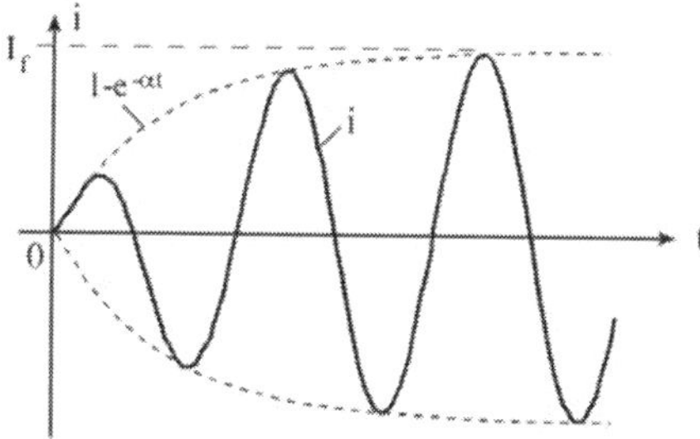

Figure 2.54 A current plot after switching in a resonant circuit.

the small resistances of the resonant circuits the current and voltages reach their final values after very many cycles. It should be noted that these values of current and voltages at resonance here, are much larger than in a regular operation.

(c) *Frequency deviation in resonant circuits*

In this case, equations 2.77 and 2.78 can still be considered as approximately true. However since the natural and the system frequencies are only approximately (and not exactly) equal, we can no longer combine the natural and steady-state harmonic functions and the complete response will be of the form

$$
\begin{aligned}
i &= I_f[\sin(\omega t + \psi_i) - e^{-\alpha t}\sin(\omega_n t + \psi_i)] \\
v_C &= V_{C,f}[\sin(\omega t + \psi_i - 90°) - e^{-\alpha t}\sin(\omega_n t + \psi_i - 90°)].
\end{aligned}
\tag{2.80}
$$

Since the natural current/capacitance voltage now has a slightly different frequency from the steady-state current/capacitance voltage, they will be displaced in time soon after the switching instant. Therefore, they will no longer subtract as in equation 2.79, but will gradually shift into such a position that they will either add to each other or subtract, as shown in Fig. 2.55. As can be seen with increasing time the addition and subtraction of the two components occur periodically, so that *beats* of the total current/voltage appear. These beats then diminish gradually and are decayed after the period of the 3–5 time constant, τ. It should also be noted that, as seen in Fig. 2.55, the current/capacitance voltage soon after switching rises up to nearly twice its large final value; so that in this case switching the circuit to an a.c. supply will be more dangerous than in the case of resonance proper. By combining the trigonometric functions in equation 2.80 (after omitting the damping factor $e^{-\alpha t}$ and the phase angle ψ_i,

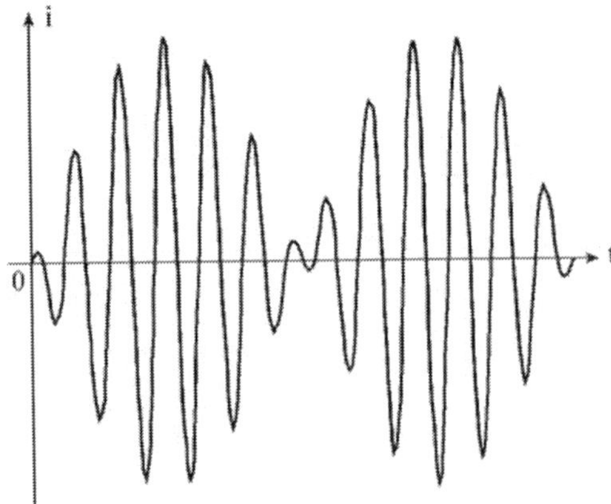

Figure 2.55 A plot of the current when the natural and fundamental frequencies are approximately equal.

i.e., supposing that the switching occurs at $\psi_i = 0$) we may obtain

$$i = 2I_f \sin\frac{\omega - \omega_n}{2} t \cdot \cos\frac{\omega + \omega_n}{2} t$$
$$v_C = -2V_{C,f} \sin\frac{\omega - \omega_n}{2} t \cdot \cos\frac{\omega + \omega_n}{2} t. \tag{2.81}$$

These expressions represent, however, the circuit behavior only a short time after the switching-on, as long as the damping effect is small. In accordance with the above expressions, and by observing the current change in Fig. 2.55, we can conclude that two oscillations are presented in the above current curve. One is a rapid oscillation of high frequency, which is a mean value of ω and ω_n, and the second one is a sinusoidal variation of the amplitude of a much lower frequency, which is the difference between ω and ω_n, and represents the beat frequency.

(d) *Resonance at multiple frequencies*

In this case, the transient phenomena are largely dependent on the instant of switching, i.e. on the angle ψ_i. Two extreme cases are of particular interest: 1) $\psi_i = 0$ and 2) $\psi_i = 90°$.

If the switching occurs the moment at which $\psi_i = 90°$, i.e., at the instant at which the steady-state current is maximal, while the capacitance voltage passes through zero, then the natural phase angle (equation 2.74) will also be

$$\beta = 90°.$$

Then, see equations 2.75 and 2.76,

$$I_n = -I_f \quad \text{and} \quad V_{C,n} = -\frac{\omega}{\omega_n} V_{C,f}, \tag{2.82}$$

and the total current and voltage become

$$i = I_f[\sin(\omega t + 90°) - e^{-\alpha t}\sin(\omega_n t + 90°)]$$
$$v_C = V_{C,f}\left(\sin\omega t - \frac{\omega}{\omega_n} e^{-\alpha t}\sin\omega_n t\right). \tag{2.83}$$

For the cases in which the natural frequency ω_n is higher than the forced frequency ω, the current rises, at the instant half a cycle after the instant of switching, to almost twice the amount of the steady-state current, which is shown in Fig. 2.56. The excess capacitance voltage in this case, however, is relatively small due to the small ratio of the frequencies in the second term of the capacitance voltage, which lowers its natural response.

If the switching occurs, the moment at which $\psi_i = 0$, i.e., at the instant the steady-state current passes through zero and the capacitance voltage reaches its maximum, the natural current phase angle (equation 2.74) will also be zero

$$\beta = 0,$$

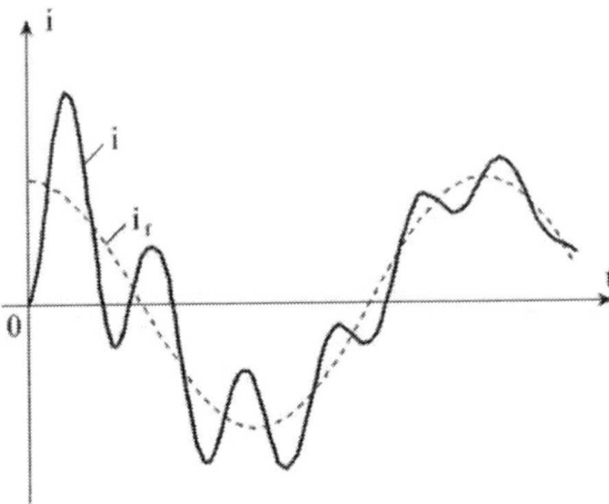

Figure 2.56 A plot of the current when $\psi_i = 90°$.

and, in accordance with equations 2.75 and 2.76,

$$I_n = -\frac{\omega_n}{\omega} I_f \quad \text{and} \quad V_{C,n} = -V_{C,f}. \tag{2.84}$$

Now the total current waveform and the total capacitance voltage waveform become

$$i = I_f \left(\sin \omega t - e^{-\alpha t} \frac{\omega_n}{\omega} \sin \omega_n t \right)$$

$$v_C = V_{C,f} \left[\sin(\omega t - 90°) - e^{-\alpha t} \sin(\omega_n t - 90°) \right],$$

which is almost inversely what is was in the former case. The plots of both the current and voltage are shown in Fig. 2.57. As can be seen, half a natural period after the switching moment the capacitance voltage is nearly doubled. The total current in this case may reach enormously high values due to the large ratio of

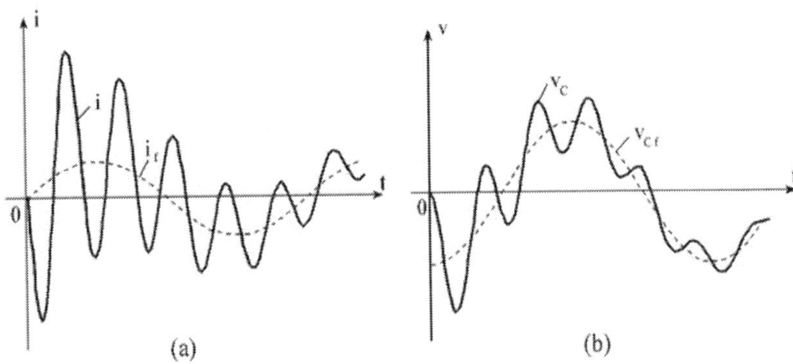

Figure 2.57 The plots of the current (a) and the capacitance voltage (b) for the case where $\psi_i = 0$.

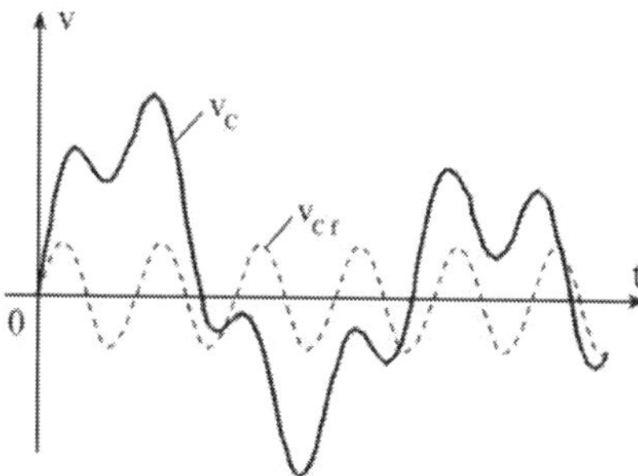

Figure 2.58 The plot of the capacitance voltage for $\psi_i = 90°$ and $\omega_n < \omega$.

the frequencies, which determine the natural component initial amplitude, when the natural frequency is higher than the system frequency.

For cases in which the natural frequency is lower than the force frequency, the transient phenomenon is significantly changed. Hence, here the most dangerous case is where the switching on occurs at the initial phase $\psi_i = 90°$ and the capacitance voltage (equation 2.83) rises to almost as much as the ratio of the frequencies ω/ω_n times the amount of its final value, Fig. 2.58.

In conclusion, as was previously mentioned, some parts of power system networks, particularly the windings of electrical machines and transformers, predominantly possess inductances, while other parts, particularly underground cables and high-voltage overhead lines, predominantly possess capacitances. Hence, the possibility of resonant conditions always exists, and by switching-on in such circuits the resonant phenomena may appear. The magnitude of the transient currents and voltages is dependent on the natural frequency and its ratio to the forced frequency as well as the instant of switching. Since it can never be predicted at what exact instant the switching occurs, we must always expect and analyze the most unfavorable cases.

2.7.4 Switching-off in *RLC* circuits

We have seen in sections 1.74 and 2.3.3 that very high voltages may develop if a current is suddenly interrupted. However, the presence of capacitances, which are associated with all electric circuit elements, as shown in Fig. 2.14, may change the transient behavior of such circuits. Thus, the raised voltages charge all these capacitances and thereby the actual voltages will be lower. To show this, consider a very simple approximation of such an arrangement by the parallel connection of L and C, as shown in Fig. 2.59. After instantaneously opening the switch, the current of the inductance flows through the capacitance charging it up to the voltage of V_C. The magnetic energy stored in the inductance, $W_m = \frac{1}{2}LI_L^2$, where I_L is the current through the inductance prior to switching, will be changed into the electric energy of the capacitance $W_e = \frac{1}{2}CV_C^2$. Since both amounts of energy, at the first moment after switching,

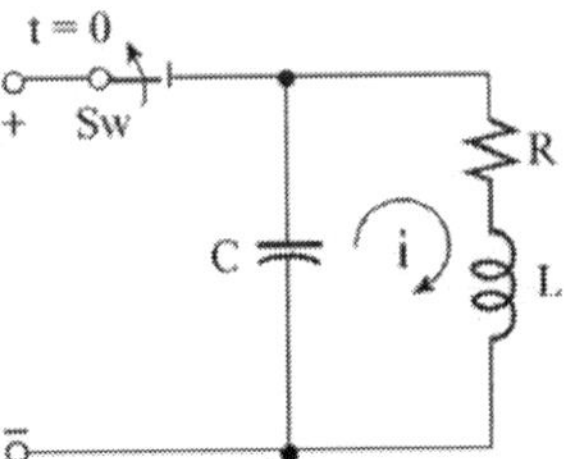

Figure 2.59 A circuit in which a coil with a parallel capacitance is disconnected from the voltage source.

are equal (by neglecting the energy dissipation due to low resistances), we have

$$\frac{CV_C^2}{2} = \frac{LI_L^2}{2},$$

and the maximal transient overvoltage appearing across the switch is

$$V_C = \sqrt{\frac{L}{C}}\, I_L. \tag{2.85}$$

Recalling from section 1.74, Fig. 1.28(a), that the overvoltage, by interrupting the coil of 0.1 H with the current of 5 A, was 50 kV, we can now estimate it more precisely. Assuming that the equivalent capacitance of the coil and the connecting cable is $C = 6$ nF, and is connected in parallel to the coil, as shown in Fig. 2.59,

$$V_C = \sqrt{\frac{0.1}{6 \cdot 10^{-9}}}\, 5 = 20.4 \text{ kV}.$$

Hence, for reducing the overvoltages, capacitances should be used. Subsequently, by connecting the additional condensers of large magnitudes, the overvoltage might be reduced to moderate values.

For a more exact calculation, we shall now also consider the circuit resistances. By using the results obtained in the previous section, we shall take into consideration that when the circuit is disconnected, the forced response is absent. However, the independent initial values are not zero, hence the initial value of the transient (natural) current through the inductive branch is found as

$$I_0 = i_L(0) - 0, \tag{2.86a}$$

and similarly for the capacitance voltage

$$V_0 = v_C(0) - 0. \tag{2.86b}$$

With the current derivatives

$$\left.\frac{di_L}{dt}\right|_{t=0} = (1/L)v_L(0) = \frac{V_0 - Ri_L(0)}{L} \quad \text{and} \quad \left.\frac{di_{L,n}}{dt}\right|_{t=0} = \left.\frac{di_L}{dt}\right|_{t=0} - 0,$$

we have two equations for determining two integration constants

$$I_n \sin \beta = I_0, \tag{2.87a}$$

$$I_n(-\alpha \sin \beta + \omega_n \cos \beta) = \frac{V_0 - RI_0}{L}. \tag{2.87b}$$

By dividing equation 2.87b by equation 2.87a, and substituting $R/2L$ for α and

$$\sqrt{1/LC - (R/2L)^2}$$

for ω_n upon simplification we obtain

$$\tan \beta = \frac{\sqrt{L/C - (R/2)^2}}{V_0/I_0 - R/2}. \tag{2.88a}$$

For circuits having small resistances, namely if $R/2 \ll \sqrt{L/C}$, the above equation becomes

$$\tan \beta = \frac{\sqrt{L/C}}{V_0/I_0 - R/2}. \tag{2.88b}$$

Using equation 2.88 with equation 2.87a, we may obtain (the details of this computation are left for the reader to convince himself of the obtained results)

$$I_n = \sqrt{I_0^2 + \frac{(V_0 - RI_0/2)^2}{L/C - (R/2)^2}} \cong \sqrt{I_0^2 + \frac{C}{L}(V_0 - RI_0/2)^2}, \tag{2.89}$$

and with equation 2.64a

$$V_{C,n} = \sqrt{\frac{L}{C}} I_n = \sqrt{\frac{L}{C} I_0^2 + \frac{(V_0 - RI_0/2)^2}{1 - (C/L)(R/2)^2}} \cong \sqrt{\frac{L}{C} I_0^2 + \left(V_0 - \frac{1}{2} RI_0\right)^2} \tag{2.90}$$

The above relationships express, in an exact and approximate way, the amplitudes of transient oscillations of the current and capacitance voltage. They are valid for switching-off in any d.c. as well as in any a.c. circuit.

Example 2.32

Assume that, for reducing the overvoltage, which arises across the switch, by disconnecting the previously considered coil of 0.1 H inductance and 20 Ω inner resistance, the additional capacitance of 0.1 μF is connected in parallel to the coil, Fig. 2.59. Find the transient voltage across the switch, if the applied voltage is 100 V dc.

Solution

We shall first find the current phase angle. Since $(R/2 = 10) \ll (\sqrt{L/C} = 10^3)$, using equation 2.88b and taking into consideration that $V_0 = V_s$ and $I_0 = V_s/R$, we have

$$\tan\beta = \frac{\sqrt{L/C}}{R - R/2} = \frac{\sqrt{1/LC}\,L}{R/2} = \frac{\omega_n}{\alpha}.$$

The damping coefficient and the natural frequency are

$$\alpha = \frac{R}{2L} = \frac{20}{2\cdot 0.1} = 100\ \mathrm{s}^{-1}, \quad \omega_n = \frac{1}{\sqrt{LC}} = \frac{1}{\sqrt{0.1\cdot 0.1\cdot 10^{-6}}} = 10^4\,\frac{\mathrm{rad}}{\mathrm{s}},$$

therefore,

$$\beta = \tan^{-1}\frac{\omega_n}{\alpha} = \tan^{-1}100 = 89.4^\circ.$$

In accordance with the approximate expression (equation 2.90), we have

$$V_{C,n} = \sqrt{\frac{L}{C}\frac{V_s^2}{R^2} - \left(V_s - \frac{1}{2}V_s\right)^2} \cong V_s\frac{\sqrt{L/C}}{R} = 100\,\frac{1000}{20} = 5\cdot 10^3\ \mathrm{V}.$$

(Note that this value is less than the previous estimation.) The capacitance voltage versus time (equation 2.63) therefore, is

$$v_{C,n}(t) = -V_{C,n}e^{-\alpha t}\sin(\omega t + \beta - \delta - 90^\circ) \cong -5e^{-100t}\sin(10^4 t - 2\delta)\ \mathrm{kV}.$$

Where δ is a *displacement angle* (equation 2.64b): $\delta = \tan^{-1}(\alpha/\omega_n) \cong 0.6^\circ$ (note that $\beta \cong 90^\circ - \delta = 89.4^\circ$ as calculated above). The negative sign of the capacitance voltage indicates the discharging process.

The voltage across the switch can now be found as the difference between the voltages of the source and the capacitance. Thus,

$$v_{sw} = V_s - v_C(t) = 100 + 5\cdot 10^3 e^{-100t}\sin(10^{4t} - 1.2^\circ)\ \mathrm{V},$$

which for $t = 0$ gives v_{sw} zero. Instead, $v_{sw}(0) = 100 + 5\cdot 10^3\sin(-1.2^\circ) \cong 0$.

The plot of v_{sw} is shown in Fig. 2.60 (the source voltage here is unproportionally enlarged relative to the transient voltage to clarify the relation between

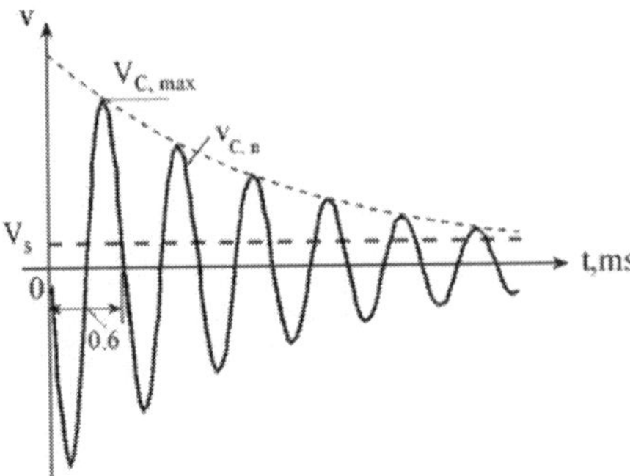

Figure 2.60 A plot of the voltage across the switch in the circuit of Fig. 2.59 after opening the switch.

these two voltages). As can be seen this voltage does not suddenly jump to its maximal value, but rises as a sinusoidal and reaches the peak after one-quarter of the natural period (which in this example is about 1.57 ms). Within this time, the contacts of the switch (circuit breaker) must have separated enough to avoid any sparking or an arc formation.

The circuit in Fig. 2.61(a) represents a very special resonant circuit, in which $R_1 = R_2 = \sqrt{L/C}$. As is known, the resonant frequency of such a circuit may be any frequency, i.e., the resonance conditions take place in this circuit, when it is connected to an a.c. source of any frequency. If such a circuit is interrupted, for instance by being switched off, the two currents i_C and i_L are always oppositely equal. In addition, since the time constants of each branch are equal ($\tau_L = L/R_1 = R_2C = \tau_C$), both currents decay equally, as shown in Fig. 2.61(b). Therefore, no current will flow through the switch when interrupted, providing its sparkless operation.

(a) *Interruptions in a resonant circuit fed from an a.c. source*

Finally, consider a resonant *RLC* circuit when disconnected from an a.c. source. The initial condition in such a circuit may be found from its steady-state operation prior to switching. Let the driving voltage be $v_s = V_m \sin(\omega t + \psi_v)$, then the current and the capacitance voltage (see, for example, the circuit in Fig. 2.59) are

$$\begin{aligned} i &= I_m \sin(\omega t + \psi_i) \\ v_C &= V_m \sin(\omega t + \psi_v), \end{aligned} \tag{2.91}$$

where

$$I_m = \frac{V_m}{\sqrt{R^2 + (\omega L)^2}} \quad \text{and} \quad \varphi = \tan^{-1}\frac{\omega L}{R} \quad (\psi_i = \psi_v - \varphi). \tag{2.92}$$

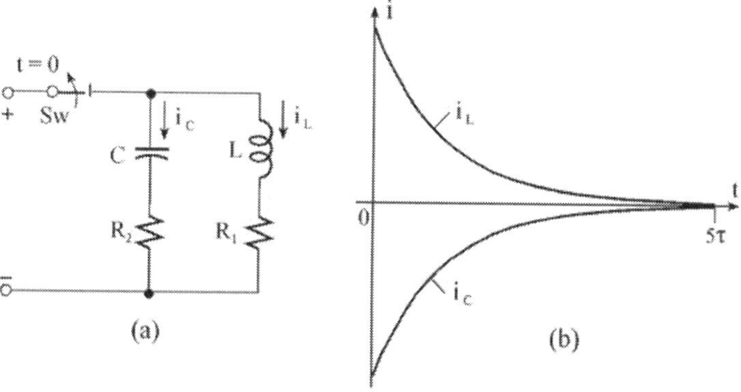

Figure 2.61 A special resonant circuit (a) and two transient currents after switching (b).

The initial conditions may now be found as

$$i(0) = I_m \sin \psi_i = I_0$$
$$v_C(0) = V_m \sin \psi_v = V_0. \tag{2.93}$$

Since the forced response in the switched-off circuit is zero, the initial values (equation 2.93) are used as the initial conditions for determining the integration constants in equation 2.87. Therefore, by substituting equation 2.93 in equations 2.88–2.90, and upon simplification and approximation for very small resistances, we obtain

$$\tan \beta = \sqrt{\frac{L}{C}} \frac{I_m \sin \psi_i}{V_m \sin \psi_v} = \frac{\omega_n}{\omega} \frac{\sin \psi_i}{\sin \psi_v \sin \varphi}, \tag{2.94}$$

where it is taking into account that the ratio

$$\frac{V_m}{I_m} = \frac{\omega L}{\sin \varphi},$$

and

$$I_n = I_m \sqrt{\sin^2 \psi_i + \left(\frac{\omega}{\omega_n \sin \varphi}\right)^2 \sin \psi_v} \cong I_m \sqrt{\sin^2 \psi_i + \left(\frac{\omega}{\omega_n}\right)^2 \cos^2 \psi_i} \tag{2.95}$$

$$V_{C,n} = V_s \sqrt{\left(\frac{\omega_n \sin \varphi}{\omega}\right)^2 \sin^2 \psi_i + \sin^2 \psi_v} \cong V_s \sqrt{\left(\frac{\omega_n}{\omega}\right)^2 \sin^2 \psi_i + \cos^2 \psi_i}, \tag{2.96}$$

where the second approximation (the right hand term) is done for $\varphi \cong 90°$, i.e., $\sin \varphi \cong 1$ and $\sin \psi_v = \sin(\psi_i + 90°) = \cos \psi_i$.

As can be seen from the above expressions, the natural current and capacitance voltage magnitudes are dependent on the phase displacement angle φ (or the power factor of the circuit), on the ratio of the natural frequency ω_n and the system frequency ω, and on the current phase angle ψ_i, which is given by the instant of switching. Therefore, in *RLC* circuits with a natural frequency higher than the system frequency (which usually happens in power networks), the transient voltage across the capacitance may attain its maximal value, which is as large as the ratio of the frequencies. This occurs in highly inductive circuits with $\varphi \cong 90°$ due to the interruption of the current while passing through its amplitude, i.e., when $\psi_i = 90°$. However, the switching-off practically occurs at the zero passage of the current, i.e., when $\psi_i = 0$. In this very favorable case the transient voltage amplitude, with equation 2.96, will now be equal to the voltage before the interruption. The voltage across the switch contacts reaches a maximum, which, with small damping, is twice the value of the source amplitude

$$v_{sw,\max} = 2V_s,$$

and then decays gradually. The initial angle of the transient response in this case, with equation 2.94, will be

$$\beta \cong 0.$$

Suppose that the circuit in Fig. 2.59, which has been analyzed, represents, for instance, the interruption at the sending end of the underground cable or overhead line having a significant capacitance to earth, while the circuit in Fig. 2.62 may represent the interruption at the receiving end of such a cable or overhead line. One of such interruptions could be a short-circuit fault and its following switching-off. The analysis of this circuit is rather similar to the previous one. The difference, though, is that here the initial capacitance voltage is zero and the forced response is present. Therefore, the initial conditions for the transient (natural) response will be

$$\begin{aligned} i_{L,n}(0) &= i_L(0) - i_f(0) = I_0 \\ v_{C,n}(0) &= 0 - v_{C,f}(0) = V_0, \end{aligned} \tag{2.97}$$

and for the current derivative, we have

$$\left.\frac{di_{L,n}}{dt}\right|_{t=0} = \frac{1}{L}v_L(0) - \left.\frac{di_f}{dt}\right|_{t=0} = \frac{v_s(0) - Ri_L(0)}{L} - \left.\frac{di_f}{dt}\right|_{t=0}.$$

The current through the inductance prior to switching might be found as a short-circuit current

$$i_{sc} = \frac{V_m}{\sqrt{R^2 + (\omega L)^2}} \sin(\omega t + \psi_v - \varphi_{sc}), \tag{2.98a}$$

where

$$\varphi_{sc} = \tan^{-1}\frac{\omega L}{R}, \tag{2.98b}$$

and ψ_v is a voltage source phase angle at switching instant $t = 0$.

Since switching in a.c. circuits usually occurs at the moment when the current passes zero, we shall assume that $I_0 = 0$ and $\psi_i(0) = \psi_v - \varphi_{sc} = 0$ (or the voltage phase angle at the switching moment is equal to the short-circuit phase angle).

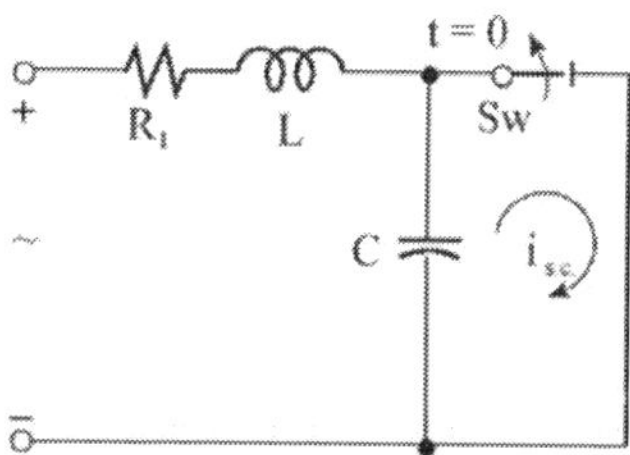

Figure 2.62 An *RLC* circuit, which arises after having been short-circuited.

Thus,

$$I_0 = 0 \quad \text{and} \quad \psi_v = \varphi_{sc}. \tag{2.99}$$

The forced response of the current and the capacitance voltage are found in the circuit after the disconnection of the short-circuit current, i.e. in the open-circuit, in which the cable or the line is disconnected (no load operation). In this regime the entire circuit is highly capacitive ($1/\omega C \gg \omega L$). Therefore, we have

$$I_f \cong V_m \omega C \quad \text{and} \quad \varphi_f \cong -90°. \tag{2.100}$$

Now, the two equations for finding the integration constant are

$$\begin{aligned} I_n \sin\beta &= 0 - i_f(0) \cong -I_f \sin(\psi_v + 90°) = -I_f \cos\psi_v \\ I_n(-\alpha \sin\beta + \omega_n \cos\beta) &= \frac{V_m \sin\psi_v}{L} - \omega_n I_f \sin\psi_v, \end{aligned} \tag{2.101}$$

for which the solution is

$$\tan\beta = \frac{-\omega_n \omega}{(\omega_n^2 + \omega^2)\tan\psi_v + \alpha\omega}. \tag{2.102}$$

Since in power system circuits the natural frequency usually is much higher than the system frequency, the above expression might be simplified for low resistive circuits to

$$\tan\beta \cong -\frac{\omega}{\omega_n \tan\psi_v}. \tag{2.102a}$$

Thus, the oscillation amplitudes of the natural current and capacitance voltage are

$$I_n = \frac{I_f \cos\psi_v}{\sin\beta} = I_f\sqrt{1 + \cot^2\beta} \cong I_f \frac{\omega_n}{\omega} \sin\psi_v, \tag{2.103}$$

$$V_{C,n} = \sqrt{\frac{L}{C}} I_n \cong \sqrt{\frac{L}{C}} I_f \frac{\omega_n}{\omega} \sin\psi_v = V_m \sin\psi_v, \tag{2.104}$$

where $I_f = \omega C V_m$. Let us illustrate this case in the following example.

Example 2.33

Determine the maximum voltage across the breaker and the transient current after it opens, disconnecting the system's short-circuit fault, as shown in Fig. 2.63(a). The system is fed by an underground cable, through a reactor (whose purpose is to reduce the short-circuit current). The parameters of the reactor and the cable are $L_1 = 6.13$ mH, $R_0 = 0.2\ \Omega$/km, $L_0 = 0.318$ mH/km and $C_0 = 0.267\ \mu$F/km. The system voltage is 10 kV (rms) at 60 Hz and the fault occurs at 13.5 km from the sending end. Suppose that the arc, which appears

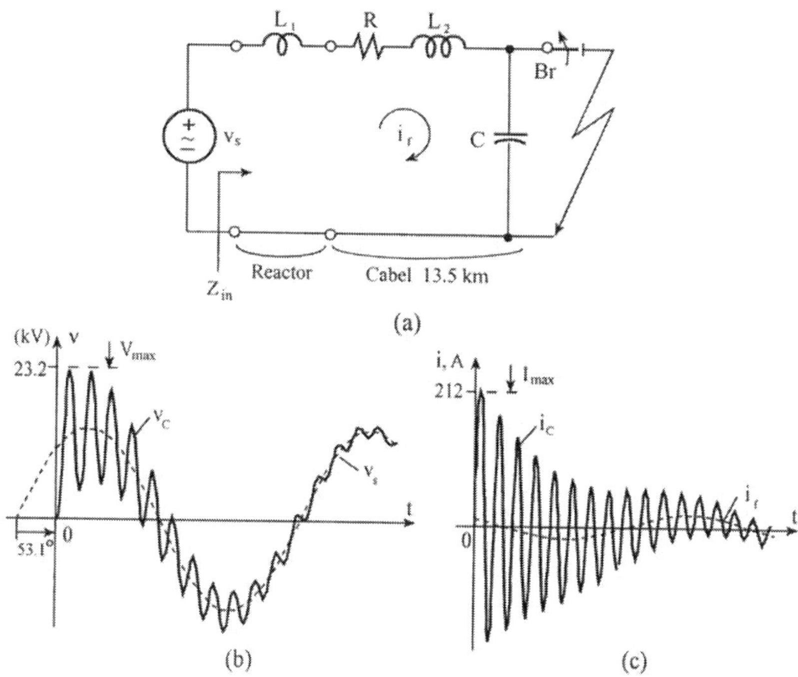

Figure 2.63 A given circuit for Example 2.33 (a) and the plots of the capacitance voltage (b) and transient current (c).

at the first moment of switching, is extinguished at a current pause, i.e. at its zero value.

Solution

The total circuit parameters are: $L = L_1 + L_0 l = 6.13 + 0.318 \cdot 13.5 = 10.4$ mH, $C = C_0 l = 0.267 \cdot 13.5 = 3.6$ μF, $R = R_0 l = 0.2 \cdot 13.5 = 2.7\ \Omega$.

The natural frequency and damping coefficient are

$$\omega_n \cong \frac{1}{\sqrt{LC}} = \frac{1}{\sqrt{10.4 \cdot 10^{-3} \cdot 3.6 \cdot 10^{-9}}} = 5.17 \cdot 10^3 \text{ rad/s}$$

$$\alpha = \frac{R}{2L} = \frac{2.7}{2 \cdot 10.4 \cdot 10^{-3}} = 130 \text{ s}^{-1},$$

and the characteristic impedance is

$$R_c = \sqrt{\frac{L}{C}} = \sqrt{\frac{10.4 \cdot 10^{-3}}{3.6 \cdot 10^{-6}}} = 53.7\ \Omega.$$

The forced current amplitude and phase angle are

$$I_f = \frac{V_s}{Z_{in}} = \frac{10\sqrt{2}}{733} = 13.6\sqrt{2}\ \mathsf{A} \quad \text{and} \quad \varphi_{in} \cong -90^\circ,$$

where

$$Z_{in} = \sqrt{R^2 + (\omega L - 1/\omega C)^2}$$
$$= \sqrt{2.7^2 + (377\cdot 10.4\cdot 10^{-3} - 1/377\cdot 3.6\cdot 10^{-6})^2} \cong 733\ \Omega.$$

Since the short current switching-off occurs at $\psi_i = 0$, the forced voltage initial angle should be (2.99)

$$\psi_v = \varphi_{sc} = \tan^{-1}\frac{\omega L}{R} = \tan^{-1}\frac{3.9}{2.7} = 53.1^\circ,$$

and the forced current phase angle will be

$$\psi_i = \psi_v - \varphi_{in} = 53.1^\circ - (-90^\circ) = 143.1^\circ.$$

Now we can find the phase angle of the natural current (equation 2.102a)

$$\tan\beta \cong \frac{-\omega}{\omega_n \tan\psi_v} = \frac{-377}{5.17\cdot 10^3 \tan 53.1^\circ} = -54.7\cdot 10^{-3} \quad \text{and} \quad \beta = -3.13^\circ.$$

The magnitude of the transient capacitance voltage is (equation 2.104)

$$V_{C,n} \cong V_m \sin\psi_v = 10\sqrt{2}\sin 53.1^\circ = 8.0\sqrt{2}\ \mathsf{kV},$$

and the complete capacitance voltage is

$$v_C(t) = 10\sqrt{2}\sin(\omega t + 53.1^\circ) + 8.0\sqrt{2}e^{-130t}\sin(5.17\cdot 10^3 t - 93.1^\circ)\ \mathsf{kV}.$$

The transient current oscillation amplitude (equation 2.103) is

$$I_n \cong I_f\frac{\omega_n}{\omega}\sin\psi_v = 13.6\sqrt{2}\frac{5.17\cdot 10^3}{377}\sin 53.1^\circ = 149.2\sqrt{2}\ \mathsf{A},$$

and the complete current response is

$$i(t) = 13.6\sqrt{2}\sin(\omega t + 143.1^\circ) + 149.2\sqrt{2}e^{-130t}\sin(5.17\cdot 10^3 t - 3.13^\circ)\ \mathsf{A}.$$

Checking for $t = 0$ yields

$$i(0) = 13.6\sqrt{2}\sin(143.1^\circ) + 149.2\sqrt{2}\sin(-3.13^\circ) \cong 0$$

(since the switching occurs at the zero current), and

$$v_C(0) = 10\sqrt{2}\sin 53.1^\circ + 8.0\sqrt{2}\sin(-93.1^\circ) \cong 0$$

(since the cable was short-circuited prior to switching).

The voltage across the breaker is equal to the capacitance voltage and its maximum will occur at the moment when the forced response reaches its first maximum and the natural response is positive. Thus,

$$t_{\max,v} = \frac{(90^\circ - 53.1^\circ)/57.3^\circ}{\omega} = \frac{0.644}{377} \cong 1.71 \text{ ms},$$

(note that 1 rad = 57.3°) and the maximum voltage is

$$V_{sw,\max} \cong 10\sqrt{2} + 8.0\sqrt{2}e^{-130t_{\max,v}} = 16.4\sqrt{2} = 23.2 \text{ kV}.$$

The current maximum will occur at the moment when the natural response reaches its first maximum, i.e., at the time

$$t_{\max,i} = \frac{(90^\circ + 3.13^\circ)/57.3^\circ}{5.17 \cdot 10^{-3}} = 0.314 \text{ ms},$$

and the maximum current is

$$I_{\max} \cong [13.6 \sin(\omega t_{\max,i} + 143.1^\circ) + 149.2 \cdot e^{-130t_{\max,i}})\sqrt{2} = 212 \text{ A}.$$

The plots of the capacitance voltage and the current are shown in Fig. 2.63(b) and (c).

Chapter #3

TRANSIENT ANALYSIS USING THE LAPLACE TRANSFORM TECHNIQUES

3.1 INTRODUCTION

In the introductory courses of circuit analysis the transient response is usually examined for relatively simple circuits of one or two energy storage elements. This analysis is based on general (or classical) techniques, involves writing the differential equations for the network, and proceeds to use them to obtain the differential equation in terms of one variable. Then the complete solution, including the natural and forced responses, has to be obtained. The tedium and complexity of using this technique is in determining the initial conditions of the unknown variables and their derivatives and then evaluating the arbitrary constants by utilizing those initial conditions. This procedure usually requires a great amount of work, which increases with the complexity of the network. Therefore, we now focus our attention on more effective methods of transient analysis.

A simplification of solving different problems can be achieved by using mathematical transformation. We are already familiar with one kind of mathematical transformation: the phasor transform technique, which allows simplifying the solution of the circuit steady-state response to sinusoidal sources. As we have seen, this very useful technique transforms the trigonometrical equations describing a circuit in the time domain into the algebraic equations in the frequency domain. Then the solution for the desirable variable (being actually manipulated by complex numbers) is transformed back to the time domain.

In this chapter a very powerful tool for the transient analysis of circuits, i.e., the *Laplace transform techniques*, will be introduced. This method enables us to convert the set of integro-differential equations describing a circuit in its transient behavior in the time domain to the set of linear algebraic equations in the complex frequency domain. Then using an algebraic operation, one may solve them for the variables of interest. Finally, with the help of the inverse transform, the desired solution can be expressed in terms of time. The paramount benefit of applying the Laplace transform to circuit analysis is in "automatically" taking the initial conditions into account: they appear when a derivative or integral is transformed.

Moreover, the concept of the frequency-domain equivalent circuit, based on the Laplace transform analysis, will be introduced. These circuits can be analyzed using techniques such as nodal and mesh analysis, Thévenin's and Norton's theorems, source transformations and so on, as described in earlier chapters.

So, the transform method in general can be represented by the expression

$$f(t) \leftrightarrow F(s),$$

which shows the one-to-one correspondence between the time-domain function $f(t)$ and its frequency domain transform $F(s)$, where $s = \sigma + j\omega$ is the *complex frequency.*

3.2 DEFINITION OF THE LAPLACE TRANSFORM

The so called *two-sided or bilateral Laplace transform* of $F(t)$ is defined as(*)

$$F(s) = \int_{-\infty}^{\infty} e^{-st} f(t)dt. \tag{3.1}$$

In circuit analysis problems the forcing and response functions do not usually exist endlessly in time, but rather they are initiated at some specific instant of time selected as $t = 0$. Thus, such functions that do not exist for $t < 0$ can be described with the help of unit step functions as $f(t)u(t)$ (see sections 2.5 and 3.3.1). For these functions the Laplace transform defining integral is taken with the lower limit at $t = 0_-$(**):

$$F(s) = \int_{-\infty}^{\infty} e^{-st} f(t)u(t)dt = \int_{0_-}^{\infty} e^{-st} f(t)dt. \tag{3.2}$$

The latter integral defines the *one-sided or unilateral Laplace transform,* or simply the Laplace transform of $f(t)$. The lower limit $t = 0_-$ (as distinguished from $t = 0$ or $t = 0_+$) in a one-sided Laplace transform is taken in order to include the effect of any discontinuity at $t = 0$, such as an impulse function and independent initial conditions such as currents in inductances $i_L(0_-)$ and voltages across capacitances $v_C(0_-)$.

The direct Laplace transform (3.2) may also be indicated as $\mathbf{L}\{f(t)\} = F(s)$ so that **L** implies the Laplace transform and means that once the integral in equation 3.2 has been evaluated, $f(t)$, which is a time domain function, is transformed to $F(s)$, which is a frequency domain function.

(*)The terms "two-sided" or bilateral are used to emphasize the fact that both positive and negative times are included in the range of integration.

(**)In transient analysis of electric circuits $t = 0_-$ is denoted as the time just before the switching action, and $t = 0_+$ as the time just after the switching action, representing radically different states of the circuit. Mathematically, $f(0_-)$ is the limit of $f(t)$ as t approaches zero through negative values ($t < 0$), or the limit from the right, and $f(0_+)$ is the limit as t approaches zero through positive values ($t > 0$), or the limit from the left.

However, the Laplace transform of a function $f(t)$ exists only if the integral (3.2) converges, or

$$\int_0^\infty |f(t)|e^{-\sigma_1 t}dt < \infty, \quad \text{where} \quad \sigma_1 = \text{Re}(s).$$

This means that if the magnitude of $f(t)$ is restricted, or increases not faster than the exponential, i.e.,

$$|f(t)| < Me^{\alpha t} \tag{3.3}$$

for all positive t the integral will converge and the region of convergence is given by $\alpha < \sigma_1 < \infty$, as shown in Fig. 3.1(a). A function $f(t)$ which fits this condition is shown in Fig. 3.1(b). The physically possible functions of time, or functions which are common in practice, always have a Laplace transform. (An example of the function, which does not satisfy conditions of equation 3.3, is e^{t^2}, but not t^n or n^t.)

If we have a transformation $\mathbf{L}\{f(t)\}$, then we must have an inverse transformation $\mathbf{L}^{-1}\{F(s)\} = f(t)$, which is mathematically defined as

$$f(t) = \frac{1}{2\pi j}\int_{c-j\infty}^{c+j\infty} e^{st}F(s)ds. \tag{3.4}$$

3.3 LAPLACE TRANSFORM OF SOME SIMPLE TIME FUNCTIONS

For a better understanding of Laplace transformations, we shall begin by using this technique to determine the Laplace transforms for those time functions most frequently encountered in circuit analysis.

3.3.1 Unit-step function

As was mentioned already in Chapter 2 (see section 2.5), very often in circuit analysis a switching action takes place at an instant that is defined as $t = 0$ (or

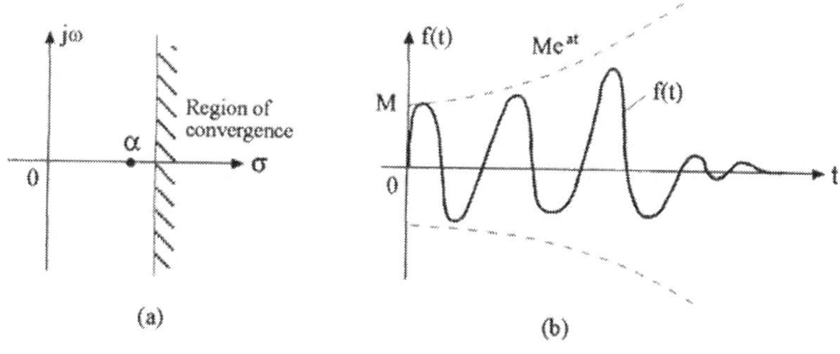

Figure 3.1 The illustration of the region of convergence in the Laplace transform definition (a); the function increasing (b).

$t_0 = 0$). We may indicate this action by using a *unit-step function*, which is

$$u(t) = \begin{cases} 0 & t < 0 \quad (t < t_0) \\ 1 & t > 0 \quad (t > t_0), \end{cases}$$

as shown in Fig. 3.2.

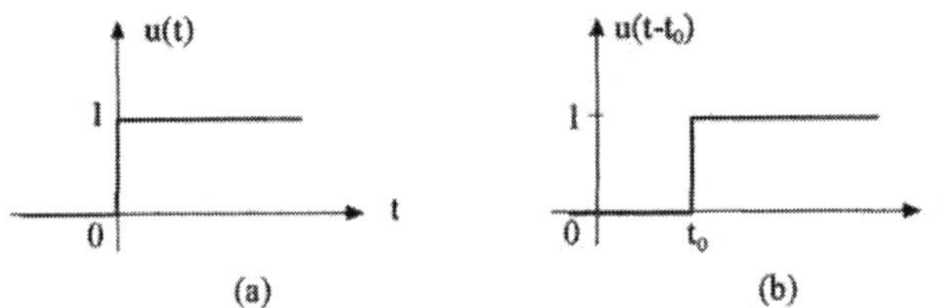

Figure 3.2 The unit-step function: $u(t)$ (a) and $u(t - t_0)$ (b).

Thus, the unit-step function is zero for all values of its argument (time), which are less than zero (or negative in the case of $(t - t_0)$) and which is unity for all positive values of its argument. By multiplying, for example, the voltage source value V_s by the unit-step function: $v(t) = V_s u(t)$, we indicate that this voltage source is connected to the network at the moment of time $t = 0$ (or if $v(t) = V_s u(t - t_0)$, at the time $t - t_0$).

In accordance with the Laplace transform definition (equation 3.2), we may write

$$\mathbf{L}\{u(t)\} = \int_{0_-}^{\infty} e^{-st} u(t) dt = \int_{0}^{\infty} e^{-st} dt = -\frac{1}{s} e^{-st} \Big|_0^{\infty} = \frac{1}{s}$$

for $\mathrm{Re}[s] = \sigma > 0$, i.e., that the region of convergence is the right half of the s-plane, except for the j-axis. Therefore,

$$u(t) \leftrightarrow \frac{1}{s}. \tag{3.5}$$

3.3.2 Unit-impulse function

Another singularity function, which is often used for circuit analysis, is the *unit-impulse function*. As was stated earlier, the impulse function is defined as

$$\delta(t) = 0 \quad \text{for} \quad t \neq 0 \quad \text{and} \quad \int_{-\infty}^{\infty} \delta(t) dt = 1.$$

Therefore, we have for any function $f(t)\delta(t) = f(0)\delta(t)$ since $\delta(t) = 0$ for $t \neq 0$. Now, by definition of Laplace transform

$$\mathbf{L}\{\delta(t)\} = \int_{0_-}^{\infty} e^{-st} \delta(t) dt = \int_{0_-}^{0_+} e^{0} \delta(t) dt = \int_{0_-}^{0_+} \delta(t) dt = 1.$$

Thus,

$$\delta(t) \leftrightarrow 1. \tag{3.6}$$

3.3.3 Exponential function

The next function of great interest is the *exponential function* $f(t) = e^{at}$ with a real, positive or negative, i.e.,

$$\mathbf{L}\{e^{at}u(t)\} = \int_{0_-}^{\infty} e^{-st}e^{at}\,dt = \int_{0_-}^{\infty} e^{-(s-a)t}\,dt = -\frac{1}{s-a}e^{-(s-a)}\Big|_0^{\infty} = \frac{1}{s-a}. \tag{3.7}$$

For both positive and negative a, the converge conditions are $\text{Re}[s] > a$, then $s - a > 0$, and $e^{-(s-a)t} \to 0$ as $t \to \infty$. Thus,

$$e^{\pm at}u(t) \leftrightarrow \frac{1}{s \mp a}, \tag{3.7a}$$

where a is always positive. Considering a imaginary quantity $a = \pm j\omega$ yields

$$e^{\pm j\omega t}u(t) \leftrightarrow \frac{1}{s \mp j\omega} \tag{3.7b}$$

for $\text{Re}[s] > 0$, since $|e^{\pm j\omega t}| = e^0 = 1$.

3.3.4 Ramp function

As an additional example, let us consider the *ramp function* $tu(t)$:

$$\mathbf{L}\{tu(t)\} = \int_{0_-}^{\infty} te^{st}\,dt.$$

By a straightforward integration by parts $[u = t,\ v = -(1/s)e^{-st}]$:

$$\mathbf{L}\{tu(t)\} = -t\frac{1}{s}e^{-st}\Big|_0^{\infty} - \int_0^{\infty} -\frac{1}{s}e^{-st}\,dt = 0 - \frac{1}{s^2}e^{-st}\Big|_0^{\infty} = \frac{1}{s^2}.$$

Therefore,

$$tu(t) \leftrightarrow \frac{1}{s^2}. \tag{3.8}$$

3.4 BASIC THEOREMS OF THE LAPLACE TRANSFORM

For further evaluation of Laplace transform techniques, several basic theorems will be introduced.

3.4.1 Linearity theorem

This theorem is based on *linearity properties* of integrals: if $f_1(t)$ and $f_2(t)$ have Laplace transforms $F_1(s)$ and $F_2(s)$ respectively, then

$$\mathbf{L}\{f_1(t) + f_2(t)\} = F_1(s) + F_2(s), \tag{3.9}$$

i.e., the Laplace transform of the sum of two (or more) time functions is equal to the sum of the transforms of the individual time functions, and conversely

$$\mathbf{L}^{-1}\{F_1(s) + F_2(s)\} = \mathbf{L}^{-1}\{F_1(s)\} + \mathbf{L}^{-1}\{F_2(s)\} = f_1(t) + f_2(t). \tag{3.10}$$

It is also obvious that for any constant K

$$Kf(t) \leftrightarrow KF(s). \tag{3.11}$$

From this it follows that the Laplace transform of a constant (for example, of a constant voltage/current source) for $t \geq 0$, is its value divided by s:

$$V_0 u(t) \leftrightarrow \frac{V_0}{s}. \tag{3.12}$$

As an example of the use of the linearity theorem, we will show the easiest way of obtaining the Laplace transform of the sinusoidal function $\sin \omega t$. Since

$$\sin \omega t = \frac{1}{2j}(e^{j\omega t} - e^{-j\omega t}),$$

in accordance with equation 3.7b, we have

$$\mathbf{L}\{\sin \omega t\} = \frac{1}{2j}\left(\frac{1}{s - j\omega} - \frac{1}{s + j\omega}\right) = \frac{(s + j\omega) - (s - j\omega)}{2j(s^2 + \omega^2)} = \frac{\omega}{s^2 + \omega^2}. \tag{3.13}$$

As a second example, let us consider the exponential of the form $(1 - e^{-at})$ which is often met in circuit analysis:

$$\mathbf{L}\{(1 - e^{-at})u(t)\} = \frac{1}{s} - \frac{1}{s + a} = \frac{a}{s(s + a)}. \tag{3.14}$$

As an example of using the opposite relationship (equation 3.10), let us determine the inverse Laplace transform of

$$F(s) = \frac{1}{(s + a)(s + b)}. \tag{3.15}$$

Using the partial-fraction expansion (see further on), we can split equation 3.15 into two parts:

$$F(s) = \frac{1}{(b - a)(s + a)} - \frac{1}{(b - a)(s + b)},$$

whose identity to equation 3.15 can be easily verified. In accordance with equation 3.10, we have

$$f(t) = \frac{1}{b-a} e^{-at}u(t) - \frac{1}{b-a} e^{-bt}u(t) = \frac{1}{b-a}(e^{-at} - e^{-bt})u(t).$$

Thus,

$$\frac{1}{b-a}(e^{-at} - e^{-bt})u(t) \leftrightarrow \frac{1}{(s+a)(s+b)}. \tag{3.16}$$

3.4.2 Time differentiation theorem

Time differentiation and integration (see further on) are the main theorems of Laplace transform techniques, which allow us to transform the derivatives and integrals appearing in the time-domain circuit equations.

Let $F(s)$ be the known transform of a time function $f(t)$, then

$$\mathbf{L}\left\{\frac{df}{dt}\right\} = \int_{0_-}^{\infty} e^{-st} \frac{df}{dt}\, dt,$$

and its integration by parts: $u = e^{-st}$ and $dv = (df/dt)dt$ gives

$$\mathbf{L}\left\{\frac{df}{dt}\right\} = f(t)e^{-st}\Big|_{0_-}^{\infty} - \int_{0_-}^{\infty} f(t)(-s)e^{-st}dt$$

$$= \lim_{t\to\infty} f(t)e^{-st} - f(0_-) + s\int_{0_-}^{\infty} f(t)e^{-st}dt.$$

The first limit must approach zero (since $F(s)$ exists) and the last integral is $F(s)$. Thus,

$$\mathbf{L}\left\{\frac{df}{dt}\right\} = sF(s) - f(0_-). \tag{3.17}$$

When the initial value of a function is zero, we simply have

$$\mathbf{L}\left\{\frac{df}{dt}\right\} = sF(s). \tag{3.17a}$$

By taking the derivative of a derivative, it may be shown that the differentiation properties for higher-order derivatives are

$$\mathbf{L}\left\{\frac{d^2f}{dt}\right\} = s^2F(s) - sf(0_-) - f'(0_-) \tag{3.18a}$$

$$\mathbf{L}\left\{\frac{d^3f}{dt}\right\} = s^3F(s) - s^2f(0_-) - sf'(0_-) - f''(0_-). \tag{3.18b}$$

In conclusion, when all initial conditions are zero, differentiating once with respect to t in the time domain corresponds to one multiplication by s in the frequency domain; differentiating twice in the time domain corresponds to multiplication by s^2 in the frequency domain and so on. Therefore, differentiation in the time domain is equivalent to multiplication by operands, which, of course, results in a substantial simplification. Note that when the initial conditions are not zero, by applying the differentiation theorem their presence is taken into account.

To demonstrate the use of the differential properties of the Laplace transform, let us consider the following example.

Example 3.1

Using Laplace transform techniques, find the current $i(t)$ in the series RL circuit driven by a constant voltage source, Fig. 3.3(a). Assume $L = 5$ H, $R = 4\,\Omega$, $v_s(t) = 6u(t)$ V and the initial value of the current is 4 A.

Solution

In accordance with KVL the loop equation is

$$5\frac{di}{dt} + 4i = 6u(t). \tag{3.19}$$

Assuming that the Laplace transform of the current is $I(s)$ and using the Laplace transform rules, with which we are already familiar, we transform the time domain equation 3.19 into the frequency domain.

$$5[sI(s) - 4] + 4I(s) = \frac{6}{s}. \tag{3.20}$$

Solving equation 3.20 for $I(s)$ yields

$$I(s) = 1.5\frac{0.8}{s(s+0.8)} + \frac{4}{s+0.8}, \tag{3.20a}$$

and with equations 3.7 and 3.14 we obtain

$$i(t) = 1.5(1 - e^{-0.8t})u(t) + 4e^{-0.8t}u(t) = (1.5 + 2.5e^{-0.8t})u(t)\text{ A}. \tag{3.20b}$$

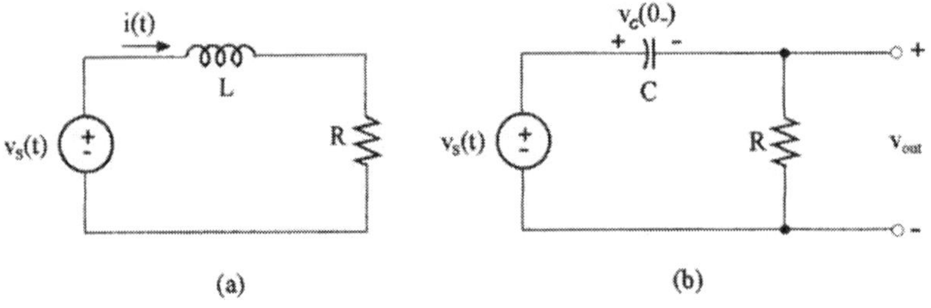

Figure 3.3 A circuit under study in Example 3.1 (a); circuit under study in Example 3.2 (b).

Note that instead of solving a differential equation 3.19, we actually solved the algebraic equation 3.20.

The time differentiation theorem also helps us to establish additional Laplace transform pairs. For example, consider $\mathbf{L}\{\cos \omega t\, u(t)\}$. Using equation 3.17a yields

$$\mathbf{L}\{\cos \omega t\, u(t)\} = \mathbf{L}\left\{\frac{1}{\omega}\frac{d}{dt}(\sin \omega t)u(t)\right\} = \frac{1}{\omega} s \frac{\omega}{s^2+\omega^2} = \frac{s}{s^2+\omega^2},$$

i.e.,

$$\cos \omega t\, u(t) \leftrightarrow \frac{s}{s^2+\omega^2}. \tag{3.21}$$

3.4.3 Time integration theorem

Let $F(s)$ be the known transform of a time function $f(t)$; then the Laplace transform of an integral as the time function can be determined in accordance with the definition (equation 3.2)

$$\mathbf{L}\left\{\int_{0_-}^{t} f(\tau)d\tau\right\} = \int_{0}^{\infty} e^{-st}\left[\int_{0}^{t} f(\tau)d\tau\right]dt.$$

Integrating by parts: $u = \int_{0_-}^{t} f(\tau)d\tau$ and $dv = e^{-st}$, yields

$$\begin{aligned}\mathbf{L}\left\{\int_{0_-}^{t} f(\tau)d\tau\right\} &= \left[\int_{0_-}^{t} f(\tau)d\tau\right]\left[-\frac{1}{s}e^{-st}\right]\Bigg|_{0_-}^{\infty} - \int_{0_-}^{\infty} -\frac{1}{s}e^{st}f(t)dt \\ &= \int_{0_-}^{0_-} f(t)dt\left(-\frac{1}{s}e^{0}\right) + \int_{0_-}^{\infty} f(\tau)d\tau - \left(-\frac{1}{s}e^{-\infty}\right) \\ &\quad + \frac{1}{s}\int_{0_-}^{\infty} e^{-st}f(t)dt.\end{aligned}$$

Since the first two terms on the right have vanished (note again that $\mathrm{Re}(s)$ is sufficiently large so $f(t)e^{-st} \to 0$ as $t \to \infty$) and the last integral is the Laplace transform of $f(t)$, we obtain

$$\int_{0_-}^{t} f(\tau)d\tau \leftrightarrow \frac{F(s)}{s} \tag{3.22}$$

which means that the integration in the time domain corresponds to the division by s in the frequency domain. In some cases, when the integral in equation 3.21 is taken for the low limit not zero but any positive or negative quantity a (for example when the capacitance in the electric circuit was precharged; thus the voltage across the capacitance is

$$v_C = \frac{1}{C}\int_{-\infty}^{t} i_C dt,$$

dividing the whole integral into two integrals, we obtain

$$\mathbf{L}\left\{\int_{-\infty}^{t} f(\tau)d\tau\right\} = \mathbf{L}\left\{\int_{-\infty}^{0_-} f(\tau)d\tau + \int_{0_-}^{t} f(\tau)d\tau\right\} = \mathbf{L}\{F_0\} + \mathbf{L}\left\{\int_{0_-}^{t} f(\tau)d\tau\right\}$$

$$= \frac{F(s)}{s} + \frac{F_0}{s}, \qquad (3.23)$$

where F_0 is the value of the first integral (initial capacitance voltage) and $F(s)$ is the Laplace transform of the considered function $f(t)$. To demonstrate how the integration theorem helps us in circuit analysis, we shall consider the following example.

Example 3.2

Using Laplace transform techniques, find the output voltage $v_{out}(t)$ in the series RC circuit shown in Fig. 3.3(b). Assume $R = 5\ \Omega$, $C = 0.5$ F, with an initial voltage $v_C = 3$ V and $v_s(t) = 12u(t)$ V.

Solution

The voltage loop equation in the time domain is

$$12u(t) = 2\int_{-\infty}^{t} i(t)dt + 5i(t). \qquad (3.24)$$

Taking the Laplace transform of both sides of equation 3.24 and since $v_C(0_-) =$ 3 V, we obtain

$$\frac{12}{s} = \frac{3}{s} + \frac{2}{s}I(s) + 5I(s). \qquad (3.25)$$

Solving equation 3.25 for $I(s)$ yields

$$I(s) = \frac{1.8}{s + 0.4}.$$

Since $v_{out} = 5i(t)$, its Laplace transform is

$$V_{out}(s) = 5\frac{1.8}{s + 0.4} = \frac{9}{s + 0.4},$$

which immediately gives $v_{out}(t)u(t) = 9e^{-0.4t}$ V.

It should be emphasized that if the time functions are zero at $t = 0$ (zero initial conditions) the linearity, differentiation and integration rules for phasor transform are identical to those for Laplace transform (only $j\omega$ has to be replaced by s). Consequently, the phasor impedance treatment of electric circuits and the Laplace transform impedance (see further on) analysis are identical. (Of course, we have to remember that these two techniques have different meanings:

the phasor analysis gives the *sinusoidal steady-state response*, while the Laplace transform relates to *zero-state response* to any Laplace transformable function.)

In conclusion, consider the complex exponential function $e^{(-\sigma+j\omega)t}$ and its transform equivalent

$$e^{(-\sigma+j\omega)t}u(t)\leftrightarrow\frac{1}{s+(\sigma+j\omega)}. \tag{3.26}$$

After separating real and imaginary parts of both sides of equation 3.26 and using linearity properties, we obtain two additional transform pairs

$$e^{-\sigma t}\cos\omega t\ u(t)\leftrightarrow\frac{s+\sigma}{(s+\sigma)^2+\omega^2}, \tag{3.26a}$$

$$e^{-\sigma t}\sin\omega t\ u(t)\leftrightarrow\frac{\omega}{(s+\sigma)^2+\omega^2}. \tag{3.26b}$$

Now let V and $\overset{*}{V}$ be a complex conjugate pair, then using linearity properties again, we obtain

$$\begin{aligned}\mathbf{L}^{-1}\left\{\frac{V}{s+(\sigma+j\omega)}+\frac{\overset{*}{V}}{s+(\sigma-j\omega)}\right\}&=Ve^{-(\sigma+j\omega)t}+\overset{*}{V}e^{-(\sigma-j\omega)t}\\&=2|V|e^{-\sigma t}\cos(\omega t+\psi)u(t),\end{aligned} \tag{3.27}$$

where $\psi=\angle V$.

Table 3.1 summarizes some of the more useful transform pairs (some of them were obtained above).

3.4.4 Time-shift theorem

Consider the transform of a time function shifted τ seconds in time as shown in Fig. 3.4. Using the definition of the Laplace transform, we obtain

$$\mathbf{L}\{f(t-\tau)u(t-\tau)\}=\int_{0_-}^{\infty}f(t-\tau)u(t-\tau)e^{-st}dt=\int_{\tau}^{\infty}f(t-\tau)e^{-st}dt.$$

Let $t-\tau=\theta$, then

$$\int_{\tau}^{\infty}f(t-\tau)e^{-st}dt=\int_{0_-}^{\infty}f(\theta)e^{-s(\tau+\theta)}d\theta=e^{-s\tau}\int_{0_-}^{\infty}f(\theta)e^{-s\theta}d\theta=e^{-s\tau}F(s).$$

Thus,

$$f(t-\tau)u(t-\tau)\leftrightarrow e^{-s\tau}F(s), \tag{3.28}$$

i.e., shifting by τ seconds in the time domain results in multiplication by $e^{-s\tau}$ in the frequency domain.

As an example of the application of this theorem, consider half a period of a sinusoidal function, as shown in Fig. 3.5(a). It can be represented as the sum of

Table 3.1 Laplace transform pairs

	$F(s)=\mathbf{L}\{f(t)\}$	$f(t)=\mathbf{L}^{-1}\{F(s)\}$
1	1	$\delta(t)$
2	$\frac{1}{s}$	$u(t)$
3	$\frac{1}{s^2}$	$tu(t)$
4	$\frac{1}{s^n}$	$\frac{t^{n-1}}{(n-1)!}u(t) \quad n=1, 2, \ldots$
5	$\frac{1}{s+a}$	$e^{-at}u(t)$
6	$\frac{1}{s(s+a)}$	$\frac{1}{a}(1-e^{-at})u(t)$
7	$\frac{1}{s^2(s+a)}$	$\frac{1}{a^2}[at-(1-e^{at})]u(t)$
8	$\frac{1}{(s+a)^2}$	$te^{-at}u(t)$
9	$\frac{s}{(s+a)^2}$	$(1-at)e^{-at}u(t)$
10	$\frac{1}{(s+a)^n}$	$\frac{t^{n-1}}{(n-1)!}e^{-at}u(t) \quad n=1, 2, \ldots$
11	$\frac{1}{(s+a)(s+b)}$	$\frac{1}{b-a}(e^{-at}-e^{-bt})u(t)$
12	$\frac{1}{s(s+a)(s+b)}$	$\frac{1}{ab}\left[1+\frac{1}{a-b}(be^{-at}-ae^{-bt})\right]u(t)$
13	$\frac{s}{(s+a)(s+b)}$	$\frac{1}{a-b}(ae^{-at}-be^{bt})u(t)$
14	$\frac{s}{s^2+\omega^2}$	$\frac{1}{\omega}\sin\omega t\, u(t)$
15	$\frac{s}{s^2+\omega^2}$	$\cos\omega t\, u(t)$
16	$\frac{1}{s(s^2+\omega^2)}$	$\frac{1}{a^2}(1-\cos\omega t)u(t)$
17	$\frac{s\sin\psi+\omega\cos\psi}{s^2+\omega^2}$	$\sin(\omega t+\psi)u(t)$
18	$\frac{s\cos\psi-\omega\sin\psi}{s^2+\omega^2}$	$\cos(\omega t+\psi)u(t)$
19	$\frac{1}{(s+a)^2+\omega^2}$	$\frac{1}{\omega}e^{-at}\sin\omega t\, u(t)$
20	$\frac{s+a}{(s+a)^2+\omega^2}$	$e^{-at}\cos\omega t\, u(t)$

Table 3.1 (Continued)

21	$\dfrac{1}{s[(s+a)^2+\omega^2]}$	$\dfrac{1}{a^2+\omega^2}\left[1-e^{-at}\left(\cos\omega t+\dfrac{a}{\omega}\sin\omega t\right)\right]u(t)$
22	$\dfrac{(s+a)\sin\psi+\omega\cos\psi}{(s+a)^2+\omega^2}$	$e^{-at}\sin(\omega t+\psi)u(t)$
23	$\dfrac{(s+a)\cos\psi-\omega\sin\psi}{(s+a)^2+\omega^2}$	$e^{-at}\cos(\omega t+\psi)u(t)$
24	$\dfrac{s}{(s^2+\omega^2)^2}$	$\dfrac{1}{2\omega}t\sin\omega t\,u(t)$
25	$\dfrac{s^2-\omega^2}{(s^2-\omega^2)^2}$	$t\cos\omega t\,u(t)$
26	$\dfrac{1}{(s^2+\omega^2)^2}$	$\left[\dfrac{1}{s\omega^3}\sin\omega t-\dfrac{1}{2\omega^2}t\cos\omega t\right]u(t)$
27	$\dfrac{1}{s^2-a^2}$	$\dfrac{1}{a}\sinh(at)u(t)$
28	$\dfrac{s}{s^2-a^2}$	$\cos(at)u(t)$
29	$\dfrac{s}{(s^2-a^2)^2}$	$\dfrac{1}{2a}t\sinh(at)u(t)$
30	$\dfrac{1}{(s+a)^2-b^2}$	$\dfrac{1}{b}e^{-at}\sinh(bt)u(t)$
31	$\dfrac{s+a}{(s+a)^2-b^2}$	$e^{-at}\cosh(bt)u(t)$
32	$\dfrac{1}{s[(s+a)^2-b^2]}$	$\dfrac{1}{a^2-b^2}\left[1-e^{-at}\left(\cosh bt+\dfrac{a}{b}\sinh bt\right)\right]u(t)$
33	$\dfrac{s}{(s+a)^2-b^2}$	$e^{-at}\left(\cosh bt-\dfrac{a}{b}\sinh bt\right)u(t)$
34	$\dfrac{1}{\sqrt{s}}$	$\dfrac{1}{\sqrt{\pi t}}$
35	$\dfrac{1}{s\sqrt{s}}$	$2\sqrt{\dfrac{t}{\pi}}$

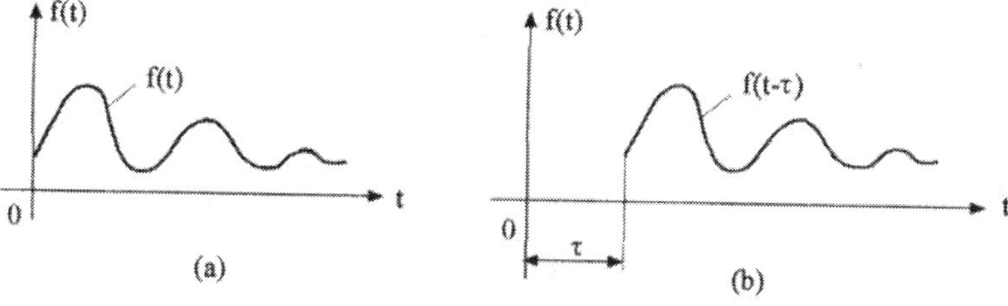

Figure 3.4 A function of time, $f(t)$, (a) and the same function delayed by τ (b).

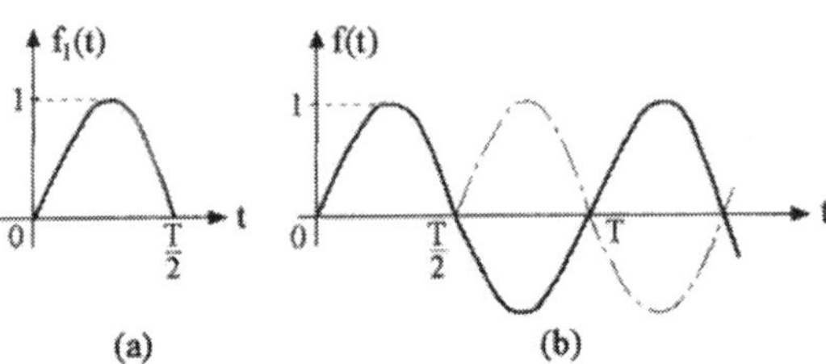

Figure 3.5 The positive half period of the sinusoidal function (a); shifted sinusoidal function (b).

two sinusoidal functions while the second one is delayed by half a period $T/2$, as shown in Fig. 3.5(b). Thus, $f_1(t)$ can be written as

$$f_1(t) = \sin \omega t + \sin \omega \left(t - \frac{T}{2}\right) u \left(t - \frac{T}{2}\right),$$

then in accordance with equations 3.13 and 3.28

$$\mathbf{L}\{f_1(t)\} = \frac{\omega(1 + e^{-sT/2})}{s^2 + \omega^2}. \tag{3.29}$$

The time-shift theorem is also useful in evaluating the Laplace transform of periodic time functions. Suppose that $f(t)$ is a periodic function (for $t \geq 0$) with period T, and $F_1(s)$ is the known transform of only the first period $f_1(t)$. Then the original $f(t)$ can be represented as the infinite sum of $f_1(t)$, delayed by an integer multiplied by T:

$$f(t) = \sum_{n=0}^{\infty} f_1(t - nT).$$

With the linearity and time-shift properties, the transform of $f(t)$ will be

$$F(s) = \sum_{n=0}^{\infty} e^{-nTs} F_1(s) = F_1(s) \sum_{n=0}^{\infty} e^{-nTs}. \tag{3.30}$$

The last sum in equation 3.30 is an infinitely decreasing geometric progression of the ratio e^{-Ts}, hence its sum is given by the formula $1/(1 - e^{-Ts})$. Therefore,

$$F(s) = \frac{F_1(s)}{1 - e^{-Ts}}, \tag{3.31}$$

where

$$F_1(s) = \mathbf{L}\{f_1(t)] = \int_{0_-}^{T} e^{-st} f(t) dt.$$

To illustrate the use of this transform theorem, let us apply it to the rectified sinus shown in Fig. 3.6. In accordance with equation 3.29 and using equation

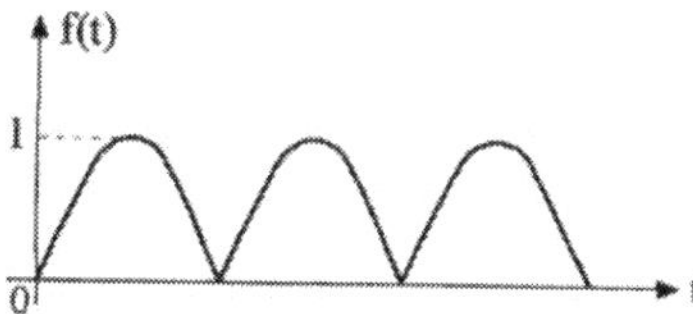

Figure 3.6 The sinusoidal shape function in full-wave rectification.

3.31, we obtain the transform of a periodic sinusoidal in full-wave rectification:

$$F(s) = \frac{\omega(1 + e^{-sT/2})}{(s^2 + \omega^2)(1 - e^{-Ts})}. \tag{3.32}$$

In another example of applying the time-shift theorem, let us find the Laplace transform of a triangular pulse train, Fig. 3.7(a). We first obtain the Laplace transform of the triangular pulse as the sum of ramp "1", shifted ramp "2" and shifted step functions "3" as shown in Fig. 3.7(b). Therefore,

$$F_1(s) = \frac{1}{Ts^2} - \frac{1}{Ts^2}e^{-sT} - \frac{1}{s}e^{-sT} = \frac{1}{Ts^2}(1 - e^{-sT}) - \frac{1}{s}e^{-sT}. \tag{3.33}$$

Now, to obtain the transform of a periodic pulse train, we divide equation 3.33 by $(1 - e^{-sT})$:

$$F(s) = \frac{1}{Ts^2} - \frac{e^{-sT}}{s(1 - e^{-sT})}. \tag{3.34}$$

3.4.5 Complex frequency-shift property

Shifting the origin of the transform in the frequency domain by s_0 has the same effect as multiplying the function $f(t)$ by e^{-s_0t} in the time domain. Indeed,

$$\mathbf{L}\{f(t)e^{-s_0t}\} = \int_{0_-}^{\infty} f(t)e^{-s_0t}e^{-st}dt = \int_{0_-}^{\infty} f(t)e^{-(s+s_0)t} = F(s + s_0). \tag{3.35}$$

This property of Laplace transform is especially useful in generating additional transforms.

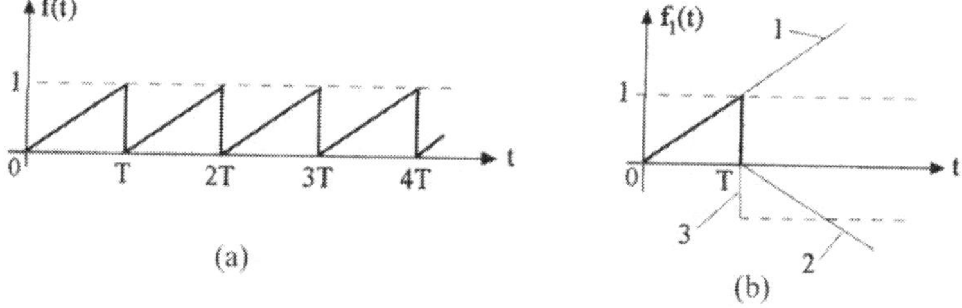

Figure 3.7 Triangular pulse train (a) and pulse representation (b).

For example, we can use the frequency-shift property to find the Laplace transform of $f(t) = e^{-\alpha t} \cos \omega_0 t\, u(t)$. Using the Laplace transform of $\cos \omega_0 t\, u(t)$ (equation 3.21), we have

$$e^{-\alpha t} \cos \omega_0 t\, u(t) \leftrightarrow \frac{s+\alpha}{(s+\alpha)^2 + \omega_0^2}, \tag{3.36}$$

which is the same as equation 3.26a.

As a second example, let us find the Laplace transform of $f(t) = te^{-s_0 t} u(t)$. With the help of (3.8), we have

$$te^{-s_0 t} u(t) \leftrightarrow \frac{1}{(s+s_0)^2}. \tag{3.37}$$

3.4.6 Scaling in the frequency domain

Scaling in the frequency domain, i.e. replacing s by s/a and dividing the transform by a, has the same effect as multiplying t by a in time domain. If

$$\mathbf{L}\{f(at)\} = \int_{0_-}^{\infty} f(at) e^{-st} dt,$$

then changing the variable $\lambda = at$, yields

$$\int_{0_-}^{\infty} f(\lambda) e^{-(s/a)\lambda} \left(\frac{1}{a}\right) d\lambda = \frac{1}{a} F\left(\frac{s}{a}\right).$$

Therefore,

$$f(at) \leftrightarrow \frac{1}{a} F\left(\frac{s}{a}\right), \tag{3.38a}$$

which is the same as

$$F(as) \leftrightarrow \frac{1}{a} f\left(\frac{t}{a}\right). \tag{3.38b}$$

This property also can be useful in obtaining additional transforms.

Example 3.3

Find the Laplace transform of the function $f_1(5t)$, if the Laplace transform of $f(t)$ is $F(s) = 1/(s^3 + 4)$.

Solution

In accordance with equation 3.38a

$$F_1(s) = \frac{1}{5} F\left(\frac{s}{5}\right) = \frac{1}{5} \frac{1}{(s/5)^3 + 4} = \frac{25}{s^3 + 500}.$$

3.4.7 Differentiation and integration in the frequency domain

Another property of interest will be obtained after examining the derivative of $F(s)$ with respect to s:

$$\frac{d}{ds} F(s) = \frac{d}{ds} \int_{0_-}^{\infty} e^{-st} f(t) dt.$$

Providing the differentiation of the integrand with respect to s gives the results:

$$\int_{0_-}^{\infty} -te^{-st} f(t) dt = \int_{0_-}^{\infty} [-tf(t)] e^{-st} dt,$$

which is simply the Laplace transform of $[-tf(t)]$. This means that *differentiation in the frequency domain* results in multiplication by $-t$ in the time domain:

$$-tf(t) \leftrightarrow \frac{d}{ds} F(s). \tag{3.39}$$

To illustrate the use of this rule, let us find the Laplace transform of higher powers of t. Noting that $tu(t) \leftrightarrow 1/s^2$, we apply the frequency domain differentiation theorem as follows:

$$\mathbf{L}\{-t^2 u(t)\} = \frac{d}{ds} \frac{1}{s^2} = -2\frac{1}{s^3},$$

or

$$\frac{t^2 u(t)}{2} \leftrightarrow \frac{1}{s^3}. \tag{3.40}$$

Continuing with the same procedure, we find

$$\frac{t^3}{3!} u(t) \leftrightarrow \frac{1}{s^4}, \tag{3.41}$$

and in general

$$\frac{t^{(n-1)}}{(n-1)!} u(t) \leftrightarrow \frac{1}{s^n}. \tag{3.42}$$

Next, let us examine the integration of $F(s)$ with respect to s and with the lower limit $s = \infty$:

$$\int_{\infty}^{s} F(s) ds = \int_{\infty}^{s} \left[\int_{0_-}^{\infty} f(t) e^{-st} dt\right] ds.$$

Interchanging the order of integration yields

$$\int_{\infty}^{s} F(s)ds = \int_{0_-}^{\infty} \left[\int_{\infty}^{s} e^{-st}ds\right] f(t)dt = \int_{0_-}^{\infty} \left[-\frac{1}{t}e^{-st}\right] f(t)dt$$

$$= \int_{0_-}^{\infty} -\frac{f(t)}{t} e^{-st}dt,$$

which is the Laplace transform of $f(t)/(-t)$. Thus, the *integration in the frequency domain* results in division by $-t$ in the time domain

$$\frac{f(t)}{-t} \leftrightarrow \int_{\infty}^{s} F(s)ds, \tag{3.43}$$

or by changing the limits in the integral

$$\frac{f(t)}{t} \leftrightarrow \int_{s}^{\infty} F(s)ds. \tag{3.43a}$$

For example, we have already obtained the pair (equation 3.14):

$$(1 - e^{-at}) \leftrightarrow \frac{a}{s(s+a)}, \quad \text{for} \quad t \geq 0.$$

With the frequency integration theorem

$$\mathbf{L}\left\{\frac{1-e^{-at}}{t}\right\} = \int_{s}^{\infty} \frac{a}{s(s+a)} ds.$$

In accordance with the integral tables, the last integral is

$$\int_{s}^{\infty} \frac{a}{s(s+a)} ds = -\ln \frac{s+a}{s}\bigg|_{s}^{\infty} = \ln \frac{s+a}{s}.$$

Therefore

$$\frac{1-e^{-at}}{t} \leftrightarrow \ln \frac{s+a}{s}, \quad \text{for} \quad t \geq 0. \tag{3.44}$$

The Laplace transform theorems and some properties which have been discussed here are summarized in Table 3.2.

3.5 THE INITIAL-VALUE AND FINAL-VALUE THEOREMS

These two fundamental theorems enable us to evaluate $f(0_+)$ and $f(\infty)$ by examining the limiting values of the transform $F(s)$.

Table 3.2 Laplace transform operations

Operation	$f(t),\ t\geq 0$	$F(s)$
Addition	$\sum_{i=1}^{n} f_i(t)$	$\sum_{i=1}^{n} F_i(t)$
Scalar multiplication	$af(t)$	$aF(s)$
Time differentiation, where $f(0_-)$, $f'(0_-)$ are the initial conditions	$\frac{df}{dt}$	$sF(s)-f(0_-)$
	$\frac{d^2f}{dt^2}$	$s^2F(s)-sf(0_-)-f'(0_-)$
Time integration, where $\int_{-\infty}^{0_-} f(t)dt$ is the initial condition	$\int_{0_-}^{t} f(t)dt$	$\frac{1}{s}F(s)$
	$\int_{-\infty}^{t} f(t)dt$	$\frac{1}{s}F(s)+\frac{1}{s}\int_{-\infty}^{0_-} f(t)dt$
Time shift	$f(t-a),\ a\geq 0$	$e^{-ad}F(s)$
Frequency shift	$f(t)e^{\mp at}$	$F(s\pm a)$
Frequency differentiation	$-tf(t)$	$\frac{dF(s)}{ds}$
Frequency integration	$\frac{f(t)}{t}$	$\int_{s}^{\infty} F(s)ds$
Scaling	$f(at),\ a\geq 0$	$\frac{1}{a}F\left(\frac{s}{a}\right)$
Initial value	$f(0_+)$	$\lim_{s\to\infty} sF(s)$
Final value, where all poles of $sF(s)$ lie in LHP	$f(\infty)$	$\lim_{s\to 0} sF(s)$
sin or cos multiplication in the time domain	$f(t)\sin(\omega t)$	$\frac{1}{2j}[F(s-j\omega)-F(s+j\omega)]$
	$f(t)\cos(\omega t)$	$\frac{1}{2}[F(s-j\omega)+F(s+j\omega)]$
Convolution	$f_1(t)*f_2(t)$	$F_1(s)F_2(s)$
Du Hamel integral	$\frac{d}{dt}[f_1(t)*f_2(t)]$ $=f_1(0)f_2(t)+\int_{0_-}^{t} f_1'(\tau)f_2(t-\tau)d\tau$ $=f_1(t)f_2(0)+\int_{0_-}^{t} f_1(\tau)f_2'(t-\tau)d\tau$	$sF_1(s)F_2(s)$
Time periodicity:		
(1) the transform of the first period	$\int_{0}^{T} f(t)e^{-st}dt$	$F_1(s)$
(2) the transform of periodical function	$f(t)=f(t+nT)$	$\frac{F_1(s)}{1-e^{-Ts}}$

The initial-value theorem: Consider the Laplace transform of the derivative (equation 3.17)

$$\mathbf{L}\left\{\frac{df}{dt}\right\} = sF(s) - f(0_-) = \int_{0_-}^{\infty} \frac{df}{dt} e^{-st} dt. \tag{3.45}$$

By breaking the integral into two parts and approaching s infinity, we obtain

$$\lim_{s\to\infty}\left(\int_{0_-}^{0_+} e^0 \frac{df}{dt} dt + \int_{0_+}^{\infty} e^{-st}\frac{df}{dt} dt\right) = \lim_{s\to\infty}\int_{0_-}^{0_+} df = f(0_+) - f(0_-), \tag{3.46}$$

since the second integral approaches zero with $s \to \infty$.

Now taking the limit of both sides of equation 3.45 and applying the results of equation 3.46 yields

$$\lim_{s\to\infty}[sF(s) - f(0_-)] = f(0_+) - f(0_-),$$

or, after removing $f(0_-)$ from the limit, we obtain

$$\lim_{s\to\infty}[sF(s)] = f(0_+).$$

Therefore, in general

$$\lim_{t\to 0_+} f(t) = \lim_{s\to\infty}[sF(s)], \tag{3.47}$$

i.e., the initial value of the time function $f(t)$ can be obtained from its Laplace transform by multiplying the transform by s and evaluating the limit of $sF(s)$ by letting s approach infinity. It should be noted that if $f(t)$ is discontinuous at $t = 0$, then the initial value is the limit as $t \to 0_+$, i.e., the limit from the right.

The initial value theorem is useful in checking the results of a transformation or an inverse transformation. Thus in Example 3.1 we obtained the transform of the current (equation 3.20a)

$$I(s) = \frac{1.2}{s(s+0.8)} + \frac{4}{s+0.8}. \tag{3.48}$$

Applying the initial-value theorem yields

$$i(0) = \lim_{s\to\infty}[sI(s)] = \lim_{s\to\infty}\left(\frac{1.2}{s+0.8} + \frac{s4}{s+0.8}\right) = 4\ \mathrm{A},$$

which is in agreement with the initial condition given.

The final-value theorem: To prove the final value theorem, let us again consider the Laplace transform of the derivative df/dt

$$\int_{0_-}^{\infty} \frac{df}{dt} e^{-st} dt = sF(s) - f(0_-), \tag{3.49}$$

and take the limit as $s \to 0$ for both sides of equation 3.49. Taking the limit for the left side of equation 3.49 yields

$$\lim_{s \to \infty} \int_{0_-}^{\infty} \frac{df}{dt} e^{-st} dt = \int_{0_-}^{\infty} \frac{df}{dt} dt = f(\infty) - f(0_-),$$

and for the right side

$$\lim_{s \to \infty} [sF(s) - f(0_-)] = \lim_{s \to \infty} [sF(s)] - f(0_-).$$

Equating these two results, we have

$$f(\infty) = \lim_{s \to \infty} [sF(s)],$$

or in general

$$\lim_{s \to \infty} f(t) = \lim_{s \to \infty} [sF(s)] \tag{3.50}$$

which is known as the final value theorem.

Of course, we can apply this theorem only if the limit of $f(t)$, as t becomes infinite, exists. In other words, this requires that all the poles of $F(s)^{(*)}$, except one simple pole at the origin (which gives the constant value of $f(t)$), lie within the left half of the s plane.

Considering again, for example, the transform for current (equation 3.48) from Example 3.1 and applying the final-value theorem yields

$$i(\infty) = \lim_{s \to 0} [sI(s)] = \lim_{s \to 0} \left(\frac{1.2}{s + 0.8} + \frac{s4}{s + 0.8} \right) = 1.5 \text{ A},$$

which is evident by inspection of the circuit in Fig. 3.3(a) in its steady-state behavior, i.e. at $t \to \infty$.

It is interesting to check the final value of the sinusoidal function. In accordance with (3.50), we obtain

$$\lim_{s \to 0} [sF(s)] = \lim_{s \to 0} \frac{s\omega}{s^2 + \omega^2} = 0.$$

However, it is evident that the sinusoidal function: $f(t) = \sin \omega t$ does not have a final value. Looking again at

$$F(s) = \frac{\omega}{s^2 + \omega^2} = \frac{\omega}{(s + j\omega)(s - j\omega)}$$

we can conclude that this transform fails the requirement that all the poles (except one) lie within the left half of the s plane, i.e. that $\text{Re}[s_k] < 0$ (here $\text{Re}[s_{1,2} = 0]$).

$^{(*)}$The roots of the denominator of $F(s)$ are considered as the poles of $F(s)$.

3.6 THE CONVOLUTION THEOREM

The convolution of two functions is defined as[*]

$$f_1(t)*f_2(t)=\int_{0_-}^{t} f(\tau)f(t-\tau)dt, \tag{3.51}$$

and its Laplace transform is given by

$$\mathbf{L}\{f_1(t)*f_2(t)\}=F_1(s)F_2(s). \tag{3.52}$$

Thus, the operation of convolution in the time domain is equivalent to multiplication in the frequency domain. Or, the inverse transform of the product of the transforms is the convolution of the individual inverse transforms. It is this property, among others, which makes the Laplace transform so useful in circuit analysis, especially since digital computers can be used for evaluating the convolution integral.

To prove the convolution theorem, let us calculate

$$J \triangleq \mathrm{L}\{\mathrm{f}_1(\mathrm{t})*\mathrm{f}_2(\mathrm{t})\}=\int_{0_-}^{\infty}\left[\int_{0_-}^{t}\mathrm{f}_1(\tau)\mathrm{f}_2(\mathrm{t}-\tau)d\tau\right]\mathrm{e}^{-st}dt,$$

and since $f(t-\tau)=0$ for all $\tau>t$, we may replace the upper limit "t" in the internal integral by "∞" and then interchange the order of integration:

$$J=\int_{0_-}^{\infty} f_1(\tau)\left[\int_{0_-}^{\infty} f_2(t-\tau)e^{-st}dt\right]d\tau.$$

Now in the inside integral we make the substitution $t'=t-\tau$ and $dt=dt'$ (note that the lower limit remained 0_-, since only for $t\geq 0_-$ does the function $f_2(t')\neq 0$). Thus,

$$J=\int_{0_-}^{\infty} f_1(\tau)\left[\int_{0_-}^{\infty} f_2(t')e^{-st'}dt'\right]e^{-s\tau}d\tau.$$

The bracketed term is $F_2(s)$, which is not a function of τ and can be pulled out of the integral, so we have

$$J=F_1(s)F_2(s).$$

Thus

$$f_1(t)*f_2(t)\leftrightarrow F_1(s)F_2(s). \tag{3.52a}$$

Since the right-hand side of equation 3.52a does not depend on the order of multiplication F_1 and F_2, consequently we can again conclude that the convolution is commutative.

As a simple example of the use of the convolution theorem, let us find the

[*]The lower limit in the convolution integral is taken here as $t=0_-$, like in a one-sided Laplace transform, in order to include the effect of any discontinuity at $t=0$.

convolution of $f_1(t) = t$ and $f_2(t) = e^{-at}$ for $t > 0$:

$$f_1(t) * f_2(t) = \mathbf{L}^{-1}\{F_1(s)F_2(s)\} = \mathbf{L}^{-1}\left\{\frac{1}{s^2}\frac{1}{s+a}\right\}. \tag{3.53}$$

The inverse transform of equation 3.53 can be obtained by the partial fraction expansion (see further on), so

$$\mathbf{L}^{-1}\left\{\frac{1}{s^2}\frac{1}{s+a}\right\} = \mathbf{L}^{-1}\left\{\frac{1}{as^2} - \frac{1}{a^2 s} + \frac{1}{a^2(s+a)}\right\} = \frac{1}{a}t - \frac{1}{a^2}(1 - e^{-at}), \quad t \geq 0.$$

Therefore,

$$t * e^{-at} = \left[\frac{1}{a}t - \frac{1}{a^2}(1 - e^{-at})\right] u(t). \tag{3.54}$$

The convolution theorem can be used for finding the Laplace transform of the functions which include square roots: $\sqrt{t}$. Indeed, if $F_1(s) \leftrightarrow f_1(t)$ then

$$F_1^2(s) \leftrightarrow f(t) = \int_0^t f_1(\tau) f_1(t - \tau) d\tau.$$

Changing the variable $\sigma = \tau - t/2$ and the integral limits respectively yields

$$f(t) = \int_{-t/2}^{t/2} f_1\left(\frac{t}{2} + \sigma\right) f_1\left(\frac{t}{2} - \sigma\right) d\sigma.$$

Now for $f_1(t) = 1/\sqrt{t}$ we obtain

$$f(t) = \int_{-t/2}^{t/2} \frac{d\sigma}{\sqrt{t/2 + \sigma}\sqrt{t/2 - \sigma}} = \int_{-t/2}^{t/2} \frac{d\sigma}{\sqrt{(t/2)^2 - \sigma^2}} = \sin^{-1}\frac{\sigma}{t/2}\bigg|_{\sigma=-t/2}^{\sigma=t/2} = \pi.$$

By taking the Laplace transform of both sides of $f(t) = \pi$ we have

$$F_1^2(s) = \frac{\pi}{s} \quad \text{or} \quad F_1 = \sqrt{\frac{\pi}{s}}.$$

Therefore,

$$\frac{1}{\sqrt{t}} \leftrightarrow \sqrt{\frac{\pi}{s}}.$$

Taking the integral of $f_1(t)$

$$\int_0^t \frac{1}{\sqrt{t}} dt = 2\sqrt{t},$$

and using the integration theorem, we finally have

$$\sqrt{t} \leftrightarrow \frac{\sqrt{\pi}}{2s\sqrt{s}}.$$

It is known, from basic circuit analysis, that the output voltage $v_{out}(t)$ at some point in a linear circuit driven by the input $v_{in}(t)$ can be obtained by convolving $v_{in}(t)$ with the impulse response $h(t)$ (response on a unit impulse at $t = 0$ with initial conditions zero)

$$v_{out}(t) = v_{in}(t) * h(t). \tag{3.55}$$

Taking the Laplace transform of both sides of equation 3.55 yields

$$V_{out}(s) = V_{in}(s)H(s),$$

where $H(s)$ is the transform of the impulse response, so

$$\mathbf{L}\{h(t)\} = H(s) = \frac{V_{out}(s)}{V_{in}(s)}. \tag{3.56}$$

The ratio (equation 3.56) was termed as the transfer function. Since the same rules are used by Laplace transform derivative and integral representations (with zero initial conditions) and by complex frequency analysis (see Table 3.3), there is considerable similarity between the transfer function and the Laplace transform of impulse response (equation 3.56). This is an important fact that will be used in Laplace transform techniques to analyze the transient behavior of some circuits.

Example 3.4

Find the transfer function $H(s) = V_{out}(s)/V_{in}(s)$ of the circuit shown in Fig. 3.8(a).

Solution

First we represent the circuit elements in the frequency domain as shown in Fig. 3.8(b). Then we find Z_{eq} of the parallel connection C and $(L + R_2)$:

$$Z_{eq}(s) = \frac{(2s+6)\frac{1}{s}}{2s+6+\frac{1}{s}} = \frac{s+3}{s^2+3s+0.5},$$

Table 3.3 Laplace transform impedances of R, L, C elements

Element	Time-domain relationship $i(t)$	s-domain relationship $I(s) = Ie^{st}$	Laplace transform with $i_L(0_-) = 0$, $v_C(0_-) = 0$	Impedance $Z(s) = \frac{V(s)}{I(s)}$
R	$v = Ri$	$V(s) = RIe^{st}$	$V(s) = RI(s)$	R
L	$v = L\frac{di}{dt}$	$V(s) = sLIe^{st}$	$V(s) = sLI(s)$	sL
C	$v = \frac{1}{C}\int idt$	$V(s) = \frac{1}{sC}I^{st}$	$V(s) = \frac{1}{sC}I(s)$	$\frac{1}{sC}$

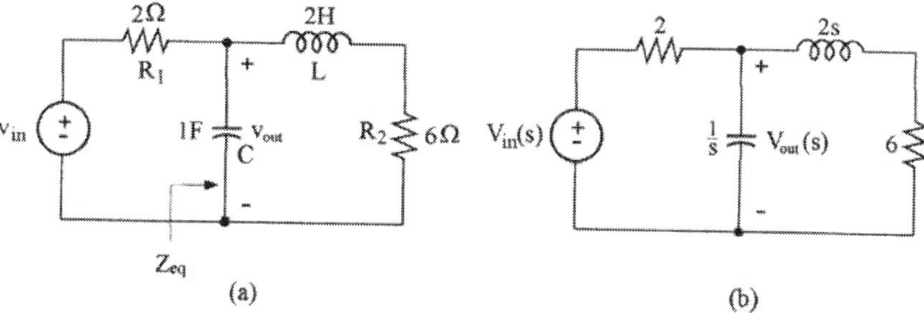

Figure 3.8 A circuit (a) and its Laplace transform model (b).

and the voltage

$$V_{out}(s) = V_{in}(s)\frac{Z_{eq}}{R_1 + Z_{eq}} = \frac{0.5s + 1.5}{s^2 + 3.5s + 2}.$$

Therefore, the transfer function is

$$H(s) = \frac{V_{out}}{V_{in}} = \frac{0.5s + 1.5}{s^2 + 3.5s + 2}. \tag{3.57}$$

3.6.1 Duhamel's integral

Suppose that the circuit response to a voltage unit-step function, $u(t) \leftrightarrow (1/s)$, is known and assigned as $g(t)$. Then, the Laplace transform of the input current might be found as

$$I_{in}(s) = \frac{V_{in}(s)}{Z_{in}(s)} = sV_{in}(s)\frac{1}{sZ_{in}(s)},$$

where

$$\frac{1}{sZ_{in}(s)} = G(s) \leftrightarrow g(t),$$

and

$$V_{in}(s) \leftrightarrow v_{in}(t).$$

Therefore, the Laplace transform of the current can be written as

$$sV(s)G(S).$$

Applying now the convolution theorem (equation 3.52a) and the differentiation properties (for the case when the initial values are zero) (equation 3.17a) we can obtain

$$[sV_{in}(s)][G(s)] \leftrightarrow v'_{in}(t) * g(t)$$

or

$$i_{in}(t) = \int_{0_-}^{t} v'_{in}(\tau)g(t-\tau)d\tau.$$

This integral is known as Duhamel's integral (one of its forms) or superposition integral, since the total response is obtained as superimposed responses to varying voltages delayed by $\Delta\tau$ (note that integration actually means summation).

When the initial values are none zero, we should subtract the initial voltage $v_{in}(0_-)$, in accordance with equation 3.17, from the first factor: $[sV_{in}(s)]$, which results in

$$I_{in}(s) = [sV_{in}(s) - v_{in}(0_-)][G(s)] + v_{in}(0_-)G(s).$$

This means that in the time domain the current is

$$i_{in}(t) = v_{in}(0_-)g(t) + \int_{0_-}^{t} v'_{in}(\tau)g(t-\tau)d\tau.$$

The other forms of Duhamel's integral in general notation are given in Table 3.2.

As a simple example of using Duhamel's integral, let us find the current for $t > T$ in the series RC circuit if the voltage forcing function is a triangular pulse, as shown in Fig. 3.7(b). The Laplace transform of the reaction to a unit-step function is

$$G(s) = \frac{1}{s\left(R + \frac{1}{sC}\right)} = \frac{1}{R\left(s + \frac{1}{RC}\right)},$$

which in time domain gives

$$g(t) = \frac{1}{R} e^{-\frac{t}{RC}}.$$

Using the first form of Duhamel's integral yields:

$$i(t) = \int_{0_-}^{t} v'_{in}(\tau)g(t-\tau)d\tau$$
$$= \int_{0}^{T} \frac{1}{TR} e^{-\frac{t-\tau}{RC}} d\tau = \frac{C}{T}\left(e^{\frac{T}{RC}} - 1\right) e^{-\frac{t}{RC}}.$$

(The reader may be convinced that the method using Duhamel's integral is much simpler than the straightforward solution.)

3.7 INVERSE TRANSFORM AND PARTIAL FRACTION EXPANSIONS

The analysis of a circuit by Laplace transforms yields the transform expression (like equation 3.57, for example) of the desired variable. The next step, therefore, is to go from the Laplace transform back to the time function, i.e. from the frequency domain to the time domain.

This section will represent methods more useful in engineering for finding $f(t)$ when $F(s)$ is known, avoiding the complex integration of equation 3.4. These methods convert $F(s)$ into a sum of terms, each of which can be found in Table 3.1 (or in more complete tables of Laplace transforms, in suitable handbooks). It is typically the case that $F(s)$ is the ratio of polynomials:

$$F(s) = \frac{N(s)}{D(s)} = \frac{a_n s^n + a_{n-1} s^{n-1} + \cdots + a_1 s + a_0}{b_m s^m + b_{m-1} s^{m-1} + \cdots + b_1 s + b_0}. \tag{3.58}$$

If the degree of $N(s)$ is larger or equal to the degree of $D(s)$, the numerator can be divided by the denominator to obtain the quotient $Q(s)$ and the remainder $R(s)$. Hence

$$F(s) = Q(s) + \frac{R(s)}{D(s)} = Q(s) + P(s), \tag{3.59}$$

where $R(s)/D(s)$ is a proper fraction. Let us consider the following example.

Example 3.5

Find the quotient and the remainder of the given $F(s)$

$$F(s) = \frac{s^3 + 5s^2 + 8s + 7}{s^2 + 4s + 3}.$$

Solution

Dividing the numerator by the denominator

$$\begin{array}{l|l} s^3 + 5s^2 + 8s + 7 & s^2 + 4s + 3 \\ \hline & s + 1 \end{array}$$

$$\underline{s^3 + 4s^2 + 3s}$$

$$s^2 + 5s + 7$$

$$\underline{s^2 + 4s + 3}$$

$$s + 4$$

yields

$$F(s) = s + 1 + \frac{s + 4}{s^2 + 4s + 3}. \tag{3.60}$$

Note that the time function, whose Laplace transform is the quotient polynomial, is obtained directly from

$$\mathbf{L}^{-1}\{q_{n-m}s^{n-m} + q_{n-m-1}s^{n-m-1} + \cdots + q_1 s + q_0\}$$
$$= q_{n-m}\delta^{(n-m)}(t) + \cdots + q_1\delta'(t) + q_0\delta, \quad (3.61)$$

where $\delta^{(n-m)}$, $\delta^{(n-m-1)}$, ... $\delta'(t)$ are derivatives of the unit impulse function.

For further treatment the proper fraction polynomials $R(s)$ and $D(s)$ have to be **coprime**, that is, any non-trivial common factor has to be cancelled out.

There are a few methods for expanding a proper fraction into partial fractions. We will discuss two of them: 1) equating coefficients and 2) Heaviside's expansion theorem. Each of them may be the best to use, depending on the situation.

3.7.1 Method of equating coefficients

(a) *Simple poles*

We first assume that the rational function $F(s)$ has simple poles, i.e. that the denominator of (3.58) has simple zeros. Then the proper fraction of equation 3.59 may be written as

$$P(s) = \frac{R(s)}{D(s)} = \frac{R(s)}{(s-p_1)(s-p_2)...(s-p_m)} = \frac{A_1}{s-p_1} + \frac{A_2}{s-p_2} + \cdots + \frac{A_m}{s-p_m}. \quad (3.62)$$

The constants A_i are known in mathematics as residues of the appropriate pole p_i. The equating coefficients method for determining A_i is illustrated in Example 3.6.

Example 3.6

Find the time function $f(t)$ if its Laplace transform is given by equation 3.60 (see Example 3.5).

Solution

First, we have to find the zeros of the equation $s^2 + 4s + 3 = 0$, which will be the poles

$$p_{1,2} = -2 \pm \sqrt{4-3} = -1, -3.$$

Therefore, in accordance with equation 3.62, the partial fraction expansion of the proper fraction (equation 3.60) is

$$\frac{s+4}{(s+1)(s+3)} = \frac{A_1}{s+1} + \frac{A_2}{s+3}. \quad (3.63)$$

Now combining two terms on the right side of this equation (by finding a common denominator) yields

$$\frac{s+4}{(s+1)(s+3)} = \frac{(A_1+A_2)s + (3A_1+A_2)}{(s+1)(s+3)}.$$

The constants A_1 and A_2 are found by equating like coefficients in the numerators. Thus,

$$A_1 + A_2 = 1, \quad 3A_1 + A_2 = 4,$$

or

$$A_1 = \frac{3}{2}, \quad A_2 = -\frac{1}{2}.$$

Therefore $F(s)$ is given by

$$F(s) = s + 1 + \frac{3/2}{s+1} - \frac{1/2}{s+3}.$$

Using the table of Laplace transform we obtain

$$f(t) = \delta'(t) + \delta(t) + \left(\frac{3}{2}e^{-t} - \frac{1}{2}e^{-3t}\right)u(t).$$

(b) *Multiple poles*

The following example illustrates the case when the denominator has repeated roots.

Example 3.7

Find the time domain function of the Laplace transform, which is given by

$$F(s) = \frac{s+2}{(s+3)^2}.$$

In this case, the expansion is given in the form

$$\frac{s+2}{(s+3)^2} = \frac{A_1}{s+3} + \frac{A_2}{(s+3)^2}.$$

Combining terms on the right gives

$$\frac{s+2}{(s+3)^2} = \frac{A_1(s+3) + A_2}{(s+3)^2}.$$

Equating like coefficients in the numerators yields

$$A_1 = 1, \quad 3A_1 + A_2 = 2 \quad \text{and} \quad A_2 = -1.$$

Therefore,

$$F(s) = \frac{1}{s+3} - \frac{1}{(s+3)^2},$$

and

$$f(t) = (e^{-3t} - te^{-3t})u(t).$$

In general, the method of equating the coefficients produces m simultaneous equations for determination A_i ($i = 1, 2 \ldots m$) constants. When m is large (usually more than three) and when the poles are complex, Heaviside's method is more appropriate.

3.7.2 Heaviside's expansion theorem

This method allows determining the unknown residue A_i by using one equation, which contains only one residue. To develop it we should again distinguish between different kinds of poles.

(a) *Simple poles*

Consider equation 3.62 and multiply both sides by $(s - p_1)$:

$$\frac{(s-p_1)R(s)}{(s-p_1)(s-p_2)\ldots(s-p_m)} = A_1 + \frac{(s-p_1)A_2}{(s-p_2)} + \cdots + \frac{(s-p_1)A_m}{(s-p_m)}.$$

Now we note that if $s = p_1$ then every term on the right side is zero, except A_1, while on the left side the $(s - p_1)$ terms in the numerator and denominator are cancelled. Therefore, A_1 can be evaluated as follows

$$A_1 = (s-p_1)\left.\frac{R(s)}{D(s)}\right|_{s=p_1} \tag{3.64}$$

or, in general,

$$A_k = (s-p_k)F(s)\Big|_{s=p_k} = \frac{N(p_k)}{\prod\limits_{\substack{i=1\\ i\neq k}}^{m}(p_k-p_i)}, \tag{3.64a}$$

where $F(s)$ is the Laplace transform function in proper fraction form and $N(s)$ is its numerator. Note that the substitution of the poles p_k for s in equation 3.64 and equation 3.64a has to be performed after canceling the term $(s - p_k)$ in the nominator and denominator.

Let us, for example, evaluate the residues of equation 3.63 from Example 3.6

$$F(s) = \frac{s+4}{(s+1)(s+3)},$$

using the general formula of equation 3.64. We have, since $p_1 = -1$ and $p_2 = -3$,

$$A_1 = (s+1)F(s) = \left.\frac{s+4}{s+3}\right|_{s=-1} = \frac{3}{2}$$

$$A_2 = (s+3)F(s) = \left.\frac{s+4}{s+1}\right|_{s=-3} = -\frac{1}{2},$$

which is, of course, identical to the results of Example 3.6.

Once the residues have been found, the Laplace transform, in accordance with partial fraction expansion, may be written as

$$F(s) = \frac{N(s)}{D(s)} = \sum_{k=1}^{m} \frac{A_k}{s - p_k}, \tag{3.65}$$

and the time domain function is

$$f(t) = \mathbf{L}^{-1}\{F(s)\} = \left(\sum_{k=1}^{m} A_k e^{p_k t} \right) u(t). \tag{3.66}$$

(b) *Multiple poles*

Suppose that the function $F(s)$ has a double pole at p_1 and the remaining poles are simple, which means that the denominator has a double zero at p_1 and hence

$$D(s) = (s-p_1)^2(s-p_2)\cdots(s-p_m).$$

Then the partial fraction expansion may be written as

$$F(s) = \frac{N(s)}{D(s)} = \frac{A_{11}}{s-p_1} + \frac{A_{12}}{(s-p_1)^2} + \sum_{k=2}^{m} \frac{A_k}{s-p_k}. \tag{3.67}$$

Multiplying both sides of equation 3.67 by $(s-p_1)^2$ and letting $s = p_1$, yields as before

$$A_{12} = (s-p_1)^2 F(s)\Big|_{s=p_1} = \frac{N(p_1)}{(p_1-p_2)(p_1-p_3)\ldots(p_1-p_m)}. \tag{3.68}$$

To find A_{11} we again multiply both sides of equation 3.67 by $(s-p_1)^2$

$$(s-p_1)^2 F(s)\Big|_{s=p_1} = A_{11}(s-p_1) + A_{12} + (s-p_1)^2 \sum_{k=2}^{m} \frac{A_k}{s-p_k}.$$

After differentiating the last expression with respect to s and letting $s = p_1$, we obtain

$$A_{11} = \left(\frac{d}{ds}[(s-p_1)^2 F(s)] \right)\Bigg|_{s=p_1}. \tag{3.69}$$

Note that the differentiation is performed after canceling the term $(s-p_1)^2$ in the bracketed expression, and substituting p_1 for s has to be done after differentiation.

In general, when $F(s)$ has a multiple pole p_q, which is repeated r times, i.e. the denominator of $F(s)$ contains a factor of $(s - p_q)^r$, the residues at the multiple pole are evaluated as

$$A_{q,i} = \frac{1}{(r-i)!}\left[\frac{d^{(r-i)}}{ds^{(r-i)}}(s-p_q)^r F(s)\right]_{s=p_q}. \tag{3.70}$$

The time domain expression corresponding to these terms can be obtained as

$$f(t) = \left[A_{q1} + A_{q2}t + A_{q3}\frac{t^2}{2!} + \cdots + A_{qr}\frac{t^{(t-r)}}{(r-1)!}\right]e^{p_q t}. \tag{3.71}$$

Example 3.8

Let $F(s) = \dfrac{s+2}{s^3(s+1)^2}$; find the time domain function.

Solution

The given Laplace transform has the poles $p_1 = 0$ repeated three times and $p_2 = -1$ repeated twice. Thus,

$$A_{11} = \left[\frac{1}{2!}\frac{d^2}{ds^2}\frac{s+2}{(s+1)^2}\right]_{s=0} = 4, \quad A_{12} = \left[\frac{1}{1!}\frac{d}{ds}\frac{s+2}{(s+1)^2}\right]_{s=0} = -3,$$

$$A_{13} = \left[\frac{s+2}{(s+1)^2}\right]_{s=0} = 2$$

and

$$A_{21} = \left[\frac{1}{1!}\frac{d}{ds}\frac{s+2}{s^3}\right]_{s=-1} = -4, \quad A_{22} = \left[\frac{s+2}{s^3}\right]_{s=-1} = -1.$$

Hence, in accordance with equation 3.71, the time domain function is

$$f(t) = [4 - 3t + t^2 - (4+t)e^{-t}]u(t).$$

(c) *Complex poles*

As we know from mathematics, polynomials with real coefficients may only have a pair of complex-conjugate poles, i.e. any complex pole in the denominator of $F(s)$ will be accompanied by its complex-conjugate pole. Then the corresponding residues are also a complex-conjugate pair, so only one of them must be found. Combining these two terms, yields to the appropriate time-domain fraction of the whole response.

Let the complex-conjugate pair be $p_1 = -\alpha \pm j\beta$ and the corresponding residues be A_1 and $\overset{*}{A}_1$, which are a complex-conjugate pair. Hence, using the exponential form

$$A_1 = |A_1|e^{j\psi_1} \quad \text{and} \quad \overset{*}{A}_1 = |A_1|e^{-j\psi_1}$$

we can write the expansion of $F(s)$, which is appropriate to complex poles, as

$$F_{c1} = \frac{|A_1|e^{j\psi_1}}{s+\alpha_1 - j\beta_1} + \frac{|A_1|e^{-j\psi_1}}{s+\alpha_1 + j\beta_1}, \tag{3.72}$$

and its time-domain inverse is

$$\begin{aligned}\mathbf{L}^{-1}\{F_{c1}(s)\} &= |A_1|e^{j\psi_1}e^{-(\alpha_1 - j\beta_1)t} + |A_1|e^{-j\psi_1}e^{-(\alpha_1 + j\beta_1)t} \\ &= |A_1|e^{-\alpha_1 t}[e^{j(\beta_1 t + \psi_1)} + e^{-j(\beta_1 t + \psi_1)}] \\ &= 2|A_1|e^{-a_1 t}\cos(\beta_1 t + \psi_1)u(t). \end{aligned} \tag{3.73}$$

This expression shows that the complex poles are associated with the time-domain response which is similar to the natural response of an underdamped second-order circuit.

Note that equation 3.73 can be simply obtained as a double real part of an inverse transform of only one of the fractions in equation 3.72. Indeed,

$$\begin{aligned} 2\,\mathrm{Re}\left[\mathbf{L}^{-1}\left\{\frac{|A_1|e^{j\psi_1}}{s+\alpha_1 - j\beta_1}\right\}\right] &= 2\,\mathrm{Re}[|A_1|e^{j\psi_1}e^{-(\alpha_1 - j\beta_1)t}] \\ &= 2|A_1|e^{-\alpha t}\cos(\beta_1 t + \psi_1)u(t). \end{aligned} \tag{3.74}$$

Example 3.9

Find $f(t)$ if $F(s)$ is given by

$$F(s) = \frac{s+4}{s^2 + 2s + 5}.$$

Solutions

First, we find the roots of the denominator, which are the poles of $F(s)$: $p_{1,2} = -1 \pm j2$. Hence,

$$F(s) = \frac{s+4}{(s+1-j2)(s+1+j2)}.$$

Now, in accordance with Heaviside's Expansion formula (equation 3.64), we have

$$A_1 = \left.\frac{s+4}{s+1+j2}\right|_{s=-1+j2} = \frac{3+j2}{j4} = \frac{1}{2} - j\frac{3}{4} = 0.9e^{-j32^\circ}.$$

Therefore,

$$f(t) = 2\,\mathrm{Re}[0.9e^{-j32^\circ}e^{-(1-j2)t}] = 1.8e^{-t}\cos(2t - 32^\circ)u(t).$$

The time-domain function of a complex-conjugate pair fraction can be obtained, using the complex residues in rectangular form $A_1 = a_{1r} + ja_{1i}$. Then the partial fraction, which is appropriate to the complex poles, will be

$$F_{c1}(s) = \frac{a_{1r}+ja_{1i}}{s+\alpha_1 - j\beta_1} + \frac{a_{1r}-ja_{1i}}{s+\alpha_1 + j\beta_1}, \tag{3.75}$$

and its inverse transform is

$$f_{c1}(t) = (a_{1r}+ja_{1i})e^{-(\alpha_1 - j\beta_1)t} + (a_{1r}-ja_{1i})e^{-(\alpha_1 + j\beta_1)t},$$

or, using Euler's formula,

$$f_{c1}(t) = e^{-\alpha_1 t}2(a_{1r}\cos\beta_1 t - a_{1i}\sin\beta_1 t)u(t). \tag{3.76}$$

In conclusion, it is worthwhile giving one other notation of a residue evaluation formula (see equations 3.64 and 3.64a).

If the pole factor $(s-p_1)$ cannot be easily canceled (for example when the denominator $D(s)$ is not given in factored form), then the residue A_1 must be treated as limit:

$$A_1 = \lim_{s\to p_1} R(s)\frac{(s-p_1)}{D(s)},$$

which in accordance with l'Hopital's Rule gives

$$A_1 = R(s)\lim_{s\to p_1}\frac{\dfrac{d}{ds}(s-p_1)}{\dfrac{d}{ds}(D(s))} = R(s)\frac{1}{D'(s)}\bigg|_{s=p_1}. \tag{3.77}$$

3.8 CIRCUIT ANALYSIS WITH THE LAPLACE TRANSFORM

There are two basic approaches to using Laplace transforms to find the circuit complete response. The first is to write differential equations describing the circuit and then to solve them using the Laplace transform of the variable and its derivatives. The advantage of this approach is that the Laplace transform provides an algebraic method for solving differential equations. Taking the inverse transform gives the time domain solution.

A second method of finding a circuit response is based on a model that directly describes relationships between the Laplace transforms of the circuit variables and its elements. This Laplace model is in some way similar to the frequency-domain circuits developed earlier.

First, we will discuss the differential equations approach. Consider the second-order circuit shown in Fig. 3.9. Then the KVL equation around the loop and the differential equation for v_C are

$$L\frac{di}{dt} + v_C + Ri = v_s, \quad C\frac{dv_C}{dt} = i.$$

These two equations are sufficient to solve the two unknown variables. The

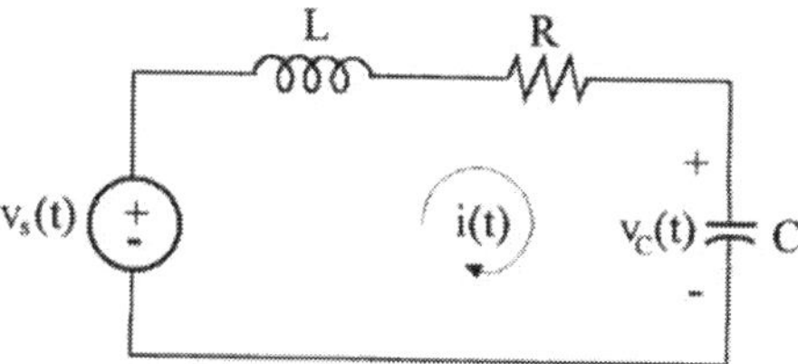

Figure 3.9 A second-order series connection *RLC* circuit.

Laplace transform of these two equations is

$$\begin{aligned} L\,[sI(s) - i(0\_)] + V_C(s) + RI(s) &= V_{in}(s) \\ C[sV_C(s) - v_C(0\_)] &= I(s). \end{aligned} \tag{3.78}$$

Let the initial condition be $i(0\_) = i(0) = I_0$ and $v_C(0\_) = v_C(0) = V_{CO}$ and after rearranging the equations 3.78 we have

$$\begin{aligned} (R + sL)I(s) + V_C(s) &= V_{in}(s) + LI_0 \\ -I(s) + sCV_C(s) &= CV_{CO}. \end{aligned} \tag{3.79}$$

Solving equation 3.79 for $I(s)$ using *Cramer's rule*, we have

$$I(s) = \frac{sCV_{in}(s)}{LCs^2 + RCs + 1} + \frac{sC[LI_0 - (1/s)V_{CO}]}{LCs^2 + RCs + 1} \tag{3.80}$$

or

$$I(s) = Y_{in}(s)V_{in}(s) + Y_{in}(s)W_0(s), \tag{3.81}$$

where

$$Y_{in}(s) = \frac{sC}{LCs^2 + RCs + 1}$$

is the Laplace transform of the input impedance and $W_0(s) = LI_0 - (1/s)V_{CO}$ is the initial condition representation. Equation 3.81 shows that $I(s)$ is the sum of two terms: one **due to the input source** and the second **due to all the initial conditions**. Going back to the time-domain, we can say that circuit variables contain two terms: a zero-state response (when all the initial conditions are zero) and a zero-input response (when all the inputs are zero).

Example 3.10

Let the elements of the circuit in Fig. 3.9 be normalized and have the values $L = 1$ H, $R = 3\,\Omega$ and $C = 1/2$ F. Let the voltage input $v_s(t) = (10 \sin t)u(t)$ and the initial condition $I_0 = 2$ A and $V_{CO} = 5$ V. Find $i(t)$.

Solution

Since the Laplace transform of the input voltage is $V_{in}(s) = 10/(s^2 + 1)$, expression

(3.80) after substituting the numerical values yields

$$I(s) = \frac{s}{s^2+3s+2}\frac{10}{s^2+1} + \frac{2s-5}{s^2+3s+2}.$$

The roots of the denominators are $s_1 = -1$, $s_2 = -2$ and $s_{3,4} = \pm j$. Therefore, using partial fractions, we obtain

$$I(s) = \frac{A_1}{s+1} + \frac{A_2}{s+2} + \frac{A_3}{s-j} + \frac{\overset{*}{A}_3}{s+j}.$$

Performing the computation, we obtain

$$A_1 = \left| \frac{10s}{(s+2)(s^2+1)} + \frac{2s-5}{s+2} \right|_{s=-1} = -12$$

$$A_2 = \left| \frac{10s}{(s+1)(s^2+1)} + \frac{2s-5}{s+1} \right|_{s=-2} = 13$$

$$A_3 = \left| \frac{10s}{(s^2+3s+2)(s+j)} \right|_{s=j} = \frac{j10}{(-1+3j+2)2j} = 1.58 \angle -71.6^{\circ}.$$

So,

$$i(t) = [-12e^{-t} + 13e^{2t} + 3.16\cos(t - 71.6^{\circ})]u(t).$$

The second approach which leads to more simplicity in Laplace transform circuit analysis uses the Laplace circuit model, which can be analyzed by frequency-domain methods. In these models, all the elements are expressed in terms of their impedances (admittances) at a complex frequency s (see Table 3.3 at the end of the chapter) and the voltage/current sources – by their Laplace transforms, i.e. as a function of s. Then one of the known methods (KVL, KCL, nodal/mesh analysis, Thévenin-Norton's theorem, etc.) can be used for identifying the desired variable transform. Finally, the time-domain response may be found with the help of the inverse transform (partial fraction expansion).

In the next few paragraphs we will illustrate how this technique may be used for circuit analysis with Laplace transform, starting with networks without initial energy stored (zero initial conditions).

3.8.1 Zero initial conditions

As an example, let us examine the circuit shown in Fig. 3.10(a). First, we convert the circuit to frequency-domain (or to its Laplace transform representation) as shown in Fig. 3.10(b). Let the voltage across the capacitance C_2, which is output voltage, be of interest. It may be found by the node equation. The output voltage is the node 1 voltage V_1. Therefore

$$\frac{V_1 - V_{in,1}}{R_1} + \frac{V_1}{R_2 + 1/sC_1} + \frac{V_1 - V_{in,2}}{R_3} + sC_2V_1 = 0.$$

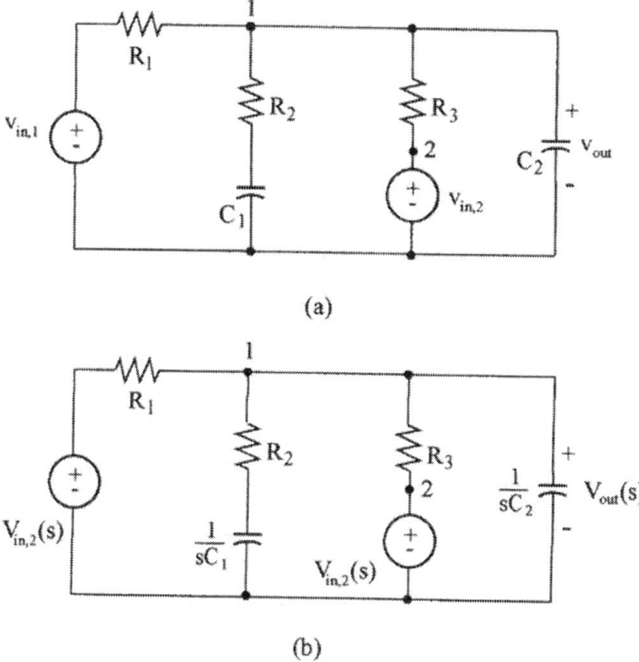

Figure 3.10 A two node circuit expressed in time-domain (a) and in the s-domain (b).

Solving for $V_1(V_{out})$ yields

$$V_{out}(s) = \frac{a_1 s + a_0}{b_2 s^2 + b_1 s + b_0}\left(\frac{V_{in,1}(s)}{R_1} + \frac{V_{in,2}(s)}{R_3}\right), \tag{3.82}$$

where

$$a_1 = R_1 R_2 R_3 C_1 \quad a_0 = R_1 R_3$$
$$b_1 = (R_1 R_2 + R_1 R_3 + R_2 R_3)C_1 \quad b_0 = R_1 + R_3$$
$$b_2 = R_1 R_2 R_3 C_1 C_2.$$

Now, in accordance with the transfer function concept

$$V_{out}(s) = H_1(s)V_{in,1}(s) + H_2(s)V_{in,2}(s), \tag{3.83}$$

and we can use the results in equation 3.83 for different inputs. It is obvious that

$$H_1(s) = (1/R_1)H_0(s), \quad H_2(s) = (1/R_3)H_0(s),$$

$$H_0(s) = \frac{a_1 s + a_0}{b_2 s^2 + b_1 s + b_0}.$$

To find $v_{out}(t)$ we need to evaluate the inverse transform of each term in equation 3.83

$$v_{out}(t) = \mathbf{L}^{-1}\{H_1(s)V_{in,1}(s)\} + \mathbf{L}^{-1}\{H_2(s)V_{in,2}(s)\}. \tag{3.84}$$

Example 3.11

Determine the voltage across the resistance R in the circuit shown in Fig. 3.11(a), which is already expressed in terms of the Laplace transform. The normalized elements are $L_1 = L_2 = 1$ H, $R = 1\ \Omega$, $v_1 = \cos tu(t)$, $v_2 = 1\delta(t)V$.

Solution

The first step is to convert the voltage sources to current sources and, after simplification, we obtain a simple circuit as shown in Fig. 3.11(b) and (c). Thus

$$I_0(s) = \frac{1}{L_1(s^2+1)} + \frac{1}{sL} = \frac{s^2+s+1}{s(s^2+1)}$$

$$Y(s) = \frac{1}{s} + \frac{1}{s} + \frac{1}{1} = \frac{s+2}{s},$$

and

$$v_R = I_0(s)\frac{1}{Y(s)} = \frac{s^2+s+1}{(s+2)(s^2+1)}.$$

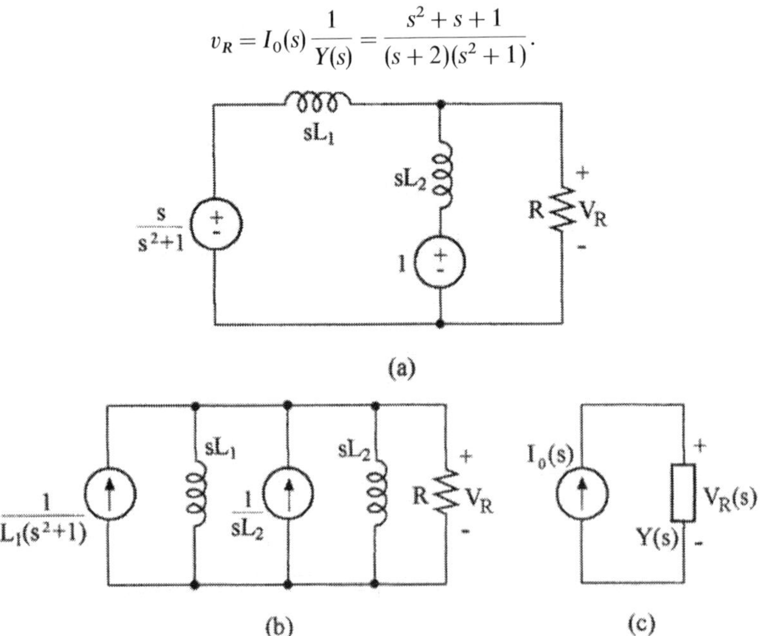

Figure 3.11 A circuit under study in Example 3.11: frequency-domain representation (a); circuit with current sources (b); final circuit (c).

Using the partial fraction expansion yields

$$V_R = \frac{A_1}{s+2} + \frac{A_2}{s-j} + \frac{\overset{*}{A}_2}{s+j}.$$

Therefore

$$A_1 = \left|\frac{s^2+s+1}{s^2+1}\right|_{s=-2} = 0.6$$

$$A_2 = \left|\frac{s^2+s+1}{(s+2)(s+j)}\right|_{s=j} = \frac{j}{(s+j)2j} = 0.2236\angle -26.6^\circ.$$

Then the desirable voltage in time-domain is

$$V_R(t) = 0.6e^{-2t} + 0.447\cos(t - 26.6^\circ)\ \text{V} \quad \text{for} \quad t \geq 0.$$

3.8.2 Non-zero initial conditions

As noted in the beginning of this chapter, the important advantage of the Laplace transform method is taking "automatically" into account the initial conditions. In the Laplace model approach, it is done by the appropriate frequency-domain equivalent of an inductor L with initial current and a capacitor C with initial voltage.

First, consider the initially charged inductor shown in Fig. 3.12(a). Since

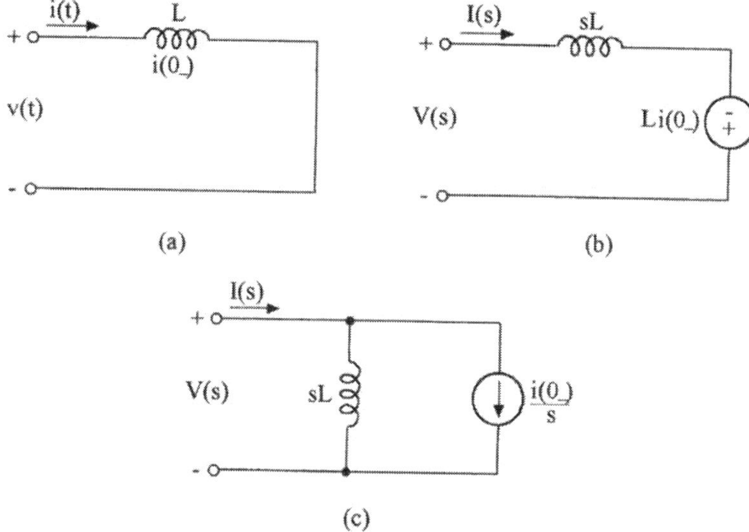

Figure 3.12 A Laplace model of an inductor with initial current: time-domain circuit (a); inductor representation in series with a voltage source (b); inductor representation in parallel with a current source (c).

$v(t) = L(di/dt)$, then

$$V(s) = sLI(s) - Li(0_-). \tag{3.85}$$

In accordance with this expression, the Laplace model for the inductor might be represented by a voltage source in series with an uncharged inductor, as shown in Fig. 3.12(b). An alternative Laplace model for the inductor can be obtained by converting the voltage source into the current source as shown in Fig. 3.12(c). Note that the voltage source in Fig. 3.12(b) is the transform of an impulse, while the current source in Fig. 3.12(c) is the transform of a step-function.

Considering the initially charged capacitor shown in Fig. 3.13(a) and in accordance with $i = C(dv/dt)$, yields

$$I(s) = sCV(s) - Cv(0_-). \tag{3.86}$$

The Laplace model of a capacitor with equation 3.86 is shown in Fig. 3.13(b). (It is the dual of the inductor model in Fig. 3.12(b).) By converting the current source in Fig. 3.13(b) into the voltage source, the second alternative of the capacitor model can be obtained as shown in Fig. 3.13(c). The voltage and current sources due to non-zero initial conditions, as represented above, are called **initial-condition generators**.

By using initial-condition generators, the Laplace transform circuit model is completed and can be analysed by frequency-domain methods when the initial conditions are not zero. The following examples illustrate these techniques.

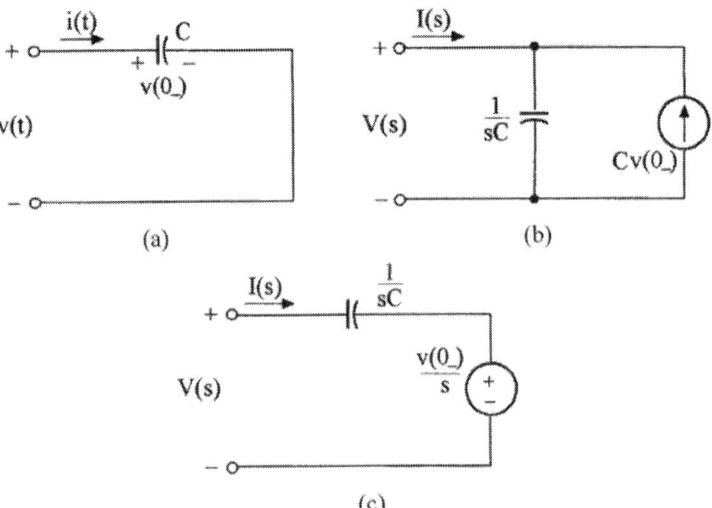

Figure 3.13 A Laplace model of initial charged capacitor: time-domain circuit (a); capacitor representation in parallel with a current source (b); capacitor representation in series with a voltage source (c).

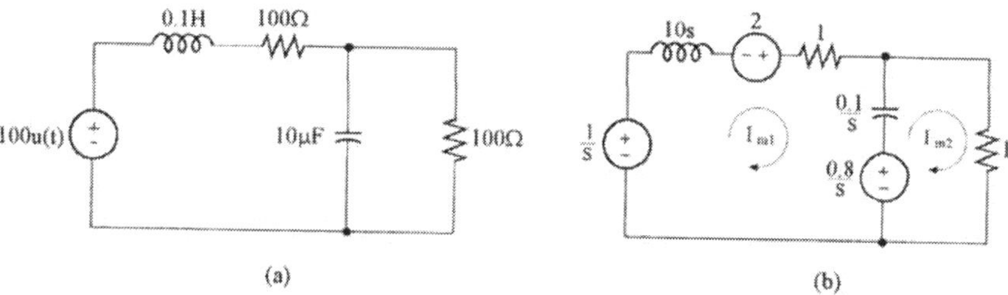

Figure 3.14 The given circuit of Example 3.12 (a); its normalized Laplace model (b).

Example 3.12

Find the complete response of the current $i(t)$ in the circuit shown in Fig. 3.14(a), if $i(0_-) = 0.2$ A and $v_C(0_-) = 80$ V.

Solution

To work with more convenient numbers, we first normalize them by choosing the impedance normalization factor K_m and frequency normalization factor K_f. Let $K_m = 10^{-2}$ and $K_f = 10^{-4}$, then $R_{new} = 10^{-2}R_{old} = 1\,\Omega$, $L_{new} = (10^{-2}/10^{-4})L_{old} = 10\,\text{H}$ and $C_{new} = (1/10^{-2}10^{-4})C_{old} = 0.1\,\text{F}$. The Laplace model circuit with normalized elements is shown in Fig. 3.14(b). Note that, to keep the same currents, voltage sources are also normalized in accordance to K_m. Using mesh analysis, we have

$$\left(1 + 10s + \frac{0.1}{s}\right) I_{m1} - \frac{0.1}{s} I_{m2} = \frac{1}{s} + 2 - \frac{0.8}{s}$$

$$-\frac{0.1}{s} I_{m1} + \left(1 + \frac{0.1}{s}\right) I_{m2} = \frac{0.8}{s}.$$

Solving for I_{m1} gives

$$I_{m1}(s) = \frac{0.2s^2 + 0.04s + 0.01}{s(s^2 + 0.2s + 0.02)},$$

with the poles $p_1 = 0$ and $p_{2,3} = -0.1 \pm j0.1$. Using the partial fraction expansion yields

$$I_{m1}(s) = \frac{A_1}{s} + \frac{A_2}{s + 0.1 - j0.1} + \frac{\overset{*}{A}_2}{s + 0.1 + j0.1},$$

where

$$A_1 = \left|\frac{0.2s^2 + 0.04s + 0.01}{s^2 + 0.2s + 0.02}\right|_{s=0} = 0.5$$

$$A_2 = \left|\frac{0.2s^2 + 0.04s + 0.01}{s(s + 0.1 + j0.1)}\right|_{s=-0.1+j0.1} = 0.212 \angle 135^\circ.$$

Then the current in time-domain is

$$i(t) = 0.5 + 0.424e^{-0.1t}\cos(0.1t + 135^\circ)\ \text{A} \quad \text{for} \quad t \geq 0.$$

Returning to the original circuit, i.e. that the actual natural frequency of the circuit is

$$s_{old} = \frac{s_{new}}{K_f} = 10^4(-0.1 \pm j0.1) = 10^3 \pm j10^3,$$

then

$$i(t) = 0.5 + 0.424e^{-10^3 t}\cos(10^3 t + 135^\circ). \tag{3.87}$$

Inspection of the circuit in Fig. 3.14(a) shows that the steady-state value of the current is 0.5 A, which is in agreement with the above results. Also checking the initial value of the current gives

$$i(t) = 0.5 + 0.424\cos 135^\circ = 0.2\ \text{A}.$$

The waveform of the current (equation 3.87) begins at a value of 0.2 A and approaches 0.5 A with decayed oscillation in approximately 5 ms.

Example 3.13

The circuit of Fig. 3.15(a) is in steady-state behavior. At $t = 0$ the second voltage source is applied in series with the capacitor. Find the transient response of the capacitor voltage $v_C(t)$.

Solution

Using the superposition approach, we construct the Laplace model circuit in which the second voltage source acts alone (Fig. 3.15(b)). Then the Laplace transform of a desirable voltage can be written as

$$V_{C2}(s) = -\frac{V_2}{s}\frac{1/sC}{Z_{in}(s)},$$

where

$$Z_{in}(s) = \frac{(10 + 0.1s)100}{110 + 0.1s} + \frac{10^4}{s} = \frac{10(s^2 + 200s + 11 \cdot 10^4)}{s(0.1s + 110)}.$$

Therefore,

$$V_{C2}(s) = -\frac{(0.1s + 110)10^4}{(s^2 + 200s + 11 \cdot 10^4)s} = \frac{10^4}{s}\frac{0.1s + 110}{(s + a)^2 + \omega^2}.$$

Using the method of equating the coefficients, this voltage can be obtained as

$$V_{C2}(s) = -\frac{10}{s} + 10\frac{s + a}{(s + a)^2 + \omega^2},$$

where $a = 100$ 1/s and $\omega = 316$ 1/s.

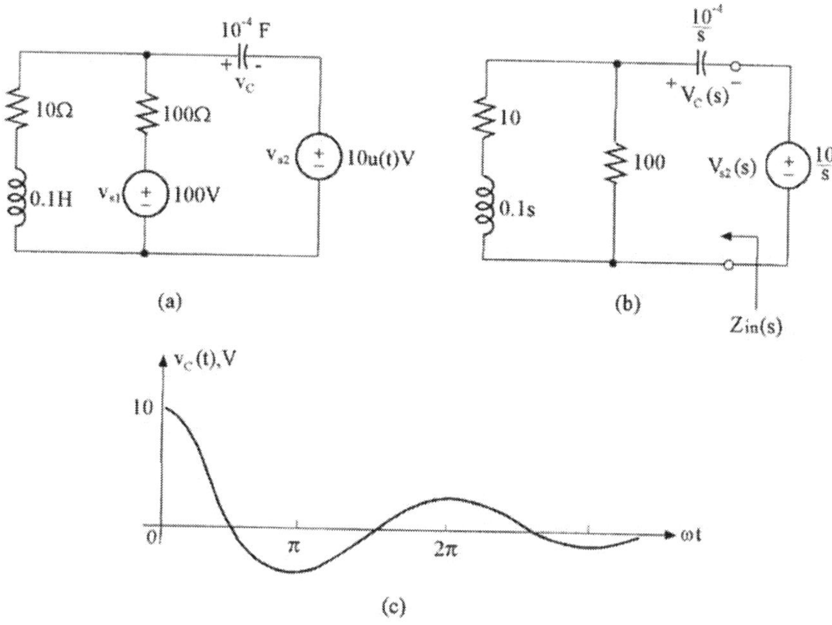

Figure 3.15 The given circuit of Example 3.13 (a); the Laplace model of the circuit driven by only the second source (b); the capacitor voltage waveform (c).

In accordance with the table of Laplace transform pairs, we obtain

$$v_{C2}(t) = 10(e^{-at}\cos\omega t - 1)u(t)\ \mathrm{V}.$$

Since the capacitor voltage v_{C1} caused by the first voltage source is 10 V, the entire capacitor voltage will be

$$v_C(t) = v_{C1} + v_{C2} = 10e^{-100t}\cos 316t\ \mathrm{V} \quad \text{for} \quad t \geq 0,$$

which is shown in Fig. 3.15(c).

3.8.3 Transient and steady-state responses

With the Laplace transform we can determine the transient and steady-state responses of the circuit variables. In order not to get involved in complicated notations, we will consider a simple example.

Let us say that the desired response of a circuit is described by a second-order differential equation

$$a_2\frac{d^2y}{dt^2} + a_1\frac{dy}{dt} + a_0y(t) = b_1\frac{dw}{dt} + b_0w(t), \tag{3.88}$$

and the initial conditions are $y(0_-) = y_0$, $y'(0_-) = y'_0$. Taking the Laplace transform of both sides of equation 3.88 gives (remember that $w(t)$ applies for $t \geq 0$)

$$a_2[s^2 Y(s) - sy_0 - y'_0] + a_1[sY(s) - y_0] + a_0 Y(s) = b_1 sW(s) + b_0 W(s),$$

or

$$(a_2 s^2 + a_1 s + a_0)Y(s) + (-a_2 y_0 s - a_2 y'_0 - a_1) = (b_1 s + b_0)W(s).$$

Solving for the Laplace transform $Y(s)$ yields

$$Y(s) = W(s)\frac{b_1 s + b_0}{a_2 s^2 + a_1 s + a_0} + \frac{W_0(s)}{a_2 s^2 + a_1 s + a_0}, \tag{3.89}$$

where $W_0(s) = a_2 y_0 s + a_2 y'_0 + a_1 y_0$, i.e., includes all of the terms that involve the initial conditions of y and its derivatives.

Noting that the first expression includes the transfer function $H(s) = Y(s)/W(s)$ (since it is separated from the initial conditions) we can finally write

$$Y(s) = H(s)W(s) + B(s), \tag{3.90}$$

where $B(s) = W_0(s)/(a_2 s^2 + a_1 s + a_0)$.

Now, taking the inverse Laplace transform of equation 3.90 via partial fraction expansion, we pay attention that the term $H(s)W(s)$ has two groups of poles: due to $W(s)$ and $H(s)$, while the term $B(s)$ only has poles due to $H(s)$. Therefore, the time response of $y(t)$ can be grouped into two kinds of terms

$$y(t) = \mathbf{L}^{-1}\{H(s)W(s)\}|_{p_w} + \mathbf{L}^{-1}\{H(s)W(s) + B(s)\}|_{p_h}, \tag{3.91}$$

where the first term includes all the partial fractions which correspond to the pole p_w of $W(s)$, while the second one includes all the partial fractions which correspond to the poles p_h of $H(s)$. Now, if all the poles of $H(s)$ are strictly in the left half of the s-plane (LHP), which is the most practical case, the steady-state value of $y(t)$ is entirely due to the first term:

$$y_{ss}(t) = \lim \mathbf{L}^{-1}\{H(s)W(s)\}|_{p_w}, \tag{3.92a}$$

i.e., y_{ss} will be non-zero if and only if $W(s)$ has at least one pole on the j-axis or in the right half of the s-plane (RHP). It means that only the input sources determine the *steady-state response*.

The *natural response* is determined in accordance with the second term of equation 3.91:

$$y_{nat}(t) = \mathbf{L}^{-1}\{H(s)W(s) + B(s)\}|_{p_h} \tag{3.92b}$$

which is entirely due to the poles of $H(s)$ and, if all of them are in the LHP, the natural response must eventually die out. However, the transient response or *complete response*, which is given by (3.91), is obviously determined by the poles of the circuit ($H(s)$) and the poles of the input sources ($W(s)$). Note again that the transfer function denominator roots determine all the natural frequencies, i.e. the roots of $a_2 s^2 + a_1 s + a_0 = 0$ in the above example.

Example 3.14

Determine the forced and natural responses of the output voltage in the circuit of Fig. 3.16(a) assuming that the capacitor was pre-charged with $v_{CO} = 6\,\mathrm{V}$ and $v_g = 4e^{-t}u(t)\,\mathrm{V}$.

Solution

First, we construct the Laplace transform model, shown in Fig. 3.16(b), of the given circuit. Next, we write the nodal equation for this circuit model:

$$V_1 - \frac{4}{s+1} + \frac{V_1}{s} + \frac{V_1 - 6/s}{2/s} - 0.58\frac{V_1}{s} = 0,$$

or

$$\frac{V_1(0.5s^2 + s + 0.42)}{s} = \frac{4}{s+1} + 3.$$

Solving for V_1 yields

$$V_{out}(s) = V_1(s) = \frac{8s}{(s+1)(s^2 + 2s + 0.84)} + \frac{6s}{s^2 + 2s + 0.4}.$$

The natural frequencies are $p_{1h} = -0.6$, $p_{2h} = -1.4$ and the forced frequency is

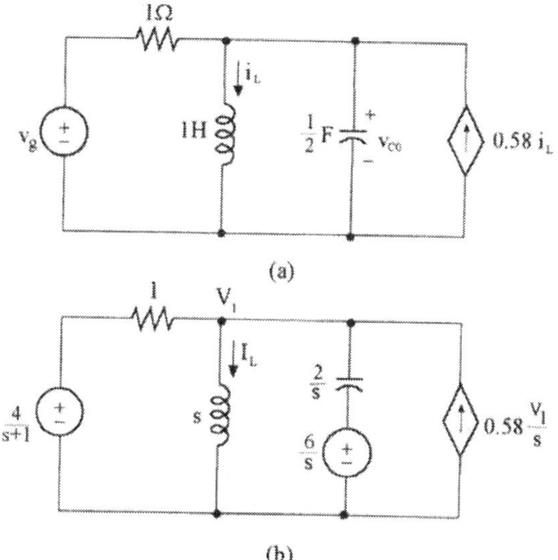

Figure 3.16 The circuit under study in Example 3.14 (a) and its Laplace model (b).

$p_w = -1$. Therefore, the residues of the first term are

$$A_{1h} = \left.\frac{8s}{(s+1)(s+1.4)}\right|_{s=-0.6} = -15, \quad A_{2h} = \left.\frac{8s}{(s+1)(s+0.6)}\right|_{s=-1.4} = -35,$$

$$A_w = \left.\frac{8s}{s^2+2s+0.4}\right|_{s=-1} = 50,$$

and the residues of the second term are

$$A'_{1h} = \left.\frac{6s}{s+1.4}\right|_{s=-0.6} = -4.5, \quad A'_{2h} = \left.\frac{6s}{s+0.6}\right|_{s=-1.4} = 10.5.$$

The time-domain responses are:

the forced response

$$v_{out,f} = 50e^{-t}u(t),$$

the natural response

$$\begin{aligned} v_{out,h} &= (-15e^{-0.6t} - 35e^{-1.4t} - 4.5e^{-0.6t} + 10.5e^{-1.4t})u(t) \\ &= (-19.5e^{-0.6t} - 24.5e^{-1.4t})u(t), \end{aligned}$$

and the complete transient response is

$$v_{out} = (50e^{-t} - 19.5e^{-0.6t} - 24.5e^{-1.4t})u(t)\ \mathrm{V},$$

which proves the initial voltage

$$v_{out}(0) = v_{CO} = 50 - 19.5 - 24.5 = 6\,\mathrm{V}.$$

3.8.4 Response to sinusoidal functions

Circuit responses to sinusoidal inputs are widely met in Power Systems Analysis. The transient analysis of such circuits by the Laplace transform might be simplified if the sinusoidal input function is taken as a complex function $\tilde{e}(t) = \tilde{E}e^{j\omega t}$ where $\tilde{E} = E_m e^{j\psi}$ is the phasor. The relation between the complex function and the actual input is given by

$$e(t) = \mathrm{Im}\{\tilde{e}(t)\} = \mathrm{Im}\{\tilde{E}e^{j\omega t}\} = E_m \sin(\omega t + \psi).$$

The Laplace transform of the complex input will be just $\tilde{E}(s) \leftrightarrow \tilde{E}/(s - j\omega)$. Then the Laplace transform of the output will be

$$\tilde{X}(s) = \tilde{E}(s)H(s) = \frac{E_m e^{j\psi}}{s - j\omega} H(s). \tag{3.93}$$

Taking the inverse transform yields

$$\tilde{x}(t) = E_m e^{j\psi} H(j\omega)e^{j\omega t} + E_m e^{j\psi} \sum_{k=1}^{n} \lim_{s \to p_k} \frac{(s - p_k)H(s)}{s - j\omega}. \tag{3.94a}$$

If the initial conditions are not zero, they have to be evaluated as imaginary quantities $I_L(0_-) = ji_L(0_-)$, $V_C(0_-) = jv_C(0_-)$ (since the imaginary part of the complex representation of phasors corresponds to the time-domain functions). Then, in accordance with equation 3.91, the complete complex response will be

$$\tilde{x}(t) = E_m e^{j\psi} H(j\omega) e^{j\omega t} + \sum_{k=1}^{n} \lim_{s \to p_k} \left[E_m e^{j\psi} \frac{(s - p_k)H(s)}{s - j\omega} + (s - p_k)B(s) \right] e^{p_k t}. \tag{3.94b}$$

With the complex response in equation 3.94a or 3.94b the actual response will be

$$x(t) = \text{Im}\{\tilde{x}(t)\}. \tag{3.95}$$

We will illustrate this method by the following example.

Example 3.15

In the circuit shown in Fig. 3.17(a), the switch closes at time $t = 0$ after having been opened for a long time. Find $i_2(t)$ assuming that the circuit is driven by the sinusoidal voltage source $v_g = 180\sin(314t + 30°)$ V.

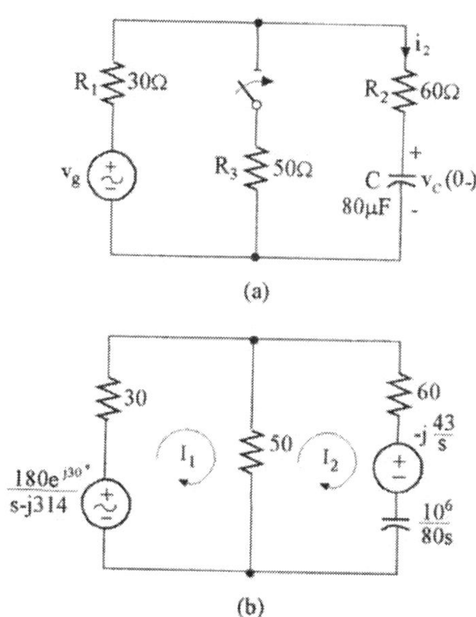

Figure 3.17 The circuit under study in Example 3.15 (a) and its Laplace model (b).

Solution

To determine the initial condition we must first calculate the capacitor steady-state voltage (before the switch is closed). The voltage source complex representation is $\tilde{v}_g = 180e^{j30^\circ}e^{j314t}$. So,

$$V_C(j\omega) = \frac{V_g(j\omega)}{R_1 + R_2 + 1/j\omega C}\frac{1}{j\omega C}$$

$$= \frac{180e^{j30^\circ}}{90 - j1/(314\cdot 80\cdot 10^{-6})}\frac{1}{j314\cdot 80\cdot 10^{-6}} = 72.3e^{-j36.2^\circ}\text{ V}.$$

Therefore, the voltage across the capacitor at $t = 0_-$

$$v_C(0_-) = 72.3\sin(-36.2) = -43.0\text{ V}.$$

Now we will construct the Laplace transform model circuit shown in Fig. 3.17(b). The Laplace transform of the voltage source, which is taken as a complex function, is $V_g(s) = 180e^{j30^\circ}/(s - j314)$. The initial capacitor voltage $v_C(0_-) = -43$ V is replaced by an initial-condition generator whose value is equal to the Laplace transform of this voltage multiplied by j:

$$V_{CO} = j\frac{(-43)}{s}.$$

In accordance with mesh analysis

$$80I_1(s) - 50I_2(s) = \frac{180e^{j30^\circ}}{s - j314}$$

$$-50I_1(s) + \left(110 + \frac{12.5\cdot 10^3}{s}\right)I_2(s) = j\frac{43}{s}.$$

Using Cramer's rule yields

$$I_2(s) = \frac{1.43e^{j30^\circ}s}{(s - j314)(s + 159)} + j\frac{0.546}{s + 159}.$$

Taking the inverse Laplace transform (with the help of the Laplace transform pairs, Table 3.1) we obtain

$$\tilde{i}_2(t) = 1.43e^{j30^\circ}\left[\frac{1}{-j314 - 159}(-j314e^{j314t} - 159e^{-159t})\right] + j0.546e^{-159t}$$

$$= 128e^{j(314t + 56.9^\circ)} + (0.646e^{-j33.2^\circ} + j0.546)e^{-159t}.$$

Finally, the imaginary part of the above expression gives the time-domain current

$$i_2(t) = 1.28\sin(314t + 56.9°) + [0.646\sin(-33.2°) + 0.548]e^{-159t}$$
$$= 1.28\sin(314t + 56.9°) + 0.159e^{-159t}\text{ A}.$$

3.8.5 Thévenin and Norton equivalent circuits

Thévenin/Norton's theorem can also be useful by circuit analysis via Laplace transform techniques. As we have already noted the Thévenin/Norton equivalent can be applied to the Laplace transform circuit model, using frequency-domain analysis. Here, the Thévenin/Norton equivalent can be especially helpful in reducing the initial state response to the zero-state response.

Consider any active one-port network shown in Fig. 3.18(a). The switch, after having been opened for a long time, closes at $t = 0$ (or any given time t_0) and the external branch ab is connected to the network. The Thévenin equivalent of the given network after closing the switch is shown in Fig. 3.18(b). It is obvious that this circuit is initially quiescent, so its response is ZSR (zero-state response). Therefore, the Laplace transform for the current I_{ab} is

$$I_{ab}(s) = \frac{V_{oc}(s)}{Z_{Th}(s) + Z_{ab}(s)}. \tag{3.96}$$

Anyway, it should be emphasized that the complete response of currents and voltages of the given network (except the branch ab) results from the superposition of two networks (a) and (b) in Fig. 3.18, i.e. the ZSR of currents and voltages has to be added to their previous steady-state values. The following example illustrates this technique.

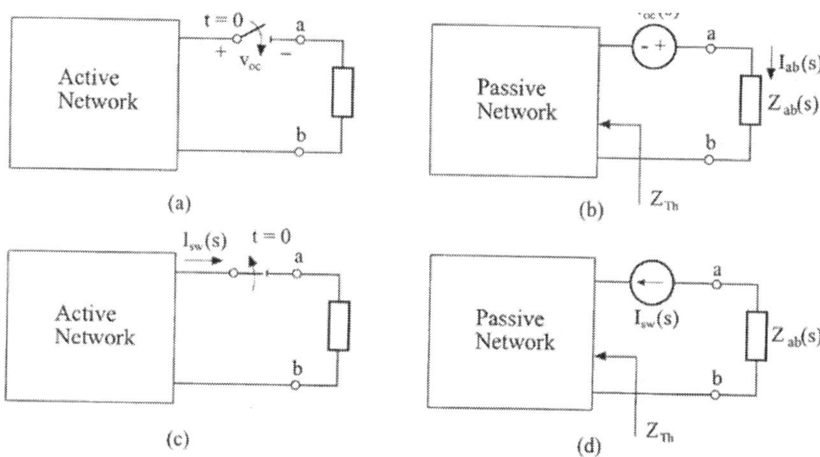

Figure 3.18 The illustration of Thévenin and Norton equivalents in Laplace transform representations: the active network with an open switch (a); the Thévenin equivalent with a voltage source (b); the active network with a closed switch (c); the Norton representation with a current source (d).

Example 3.16

The switch in the circuit shown in Fig. 3.19(a) closes after having been opened for a long time. Find the currents through the capacitor $i_C(t)$ and through the inductor $i_L(t)$.

Solution

The open circuit voltage across the switch is

$$V_{oc} = V_g \frac{R_2}{R_1 + R_2} = 200 \frac{10}{10 + 10} = 100 \text{ V}.$$

The Thévenin equivalent impedance of the circuit is

$$Z_{Th} = \frac{(R_1 + sL)R_2}{R_1 + R_2 + sL} = \frac{100 + s}{20 + 0.1s}.$$

The Thévenin equivalent of the Laplace transform circuit is shown in Fig. 3.19(b). Thus, the Laplace transform of the capacitor current is

$$I_C(s) = \frac{V_{oc}(s)}{Z_{Th} + Z_{ab}} = \frac{100}{s} \frac{1}{(100 + s)/(20 + 0.1s) + 10^3/s}$$

$$= \frac{100(0.1s + 20)}{s^2 + 200s + 20 \cdot 10^3} = \frac{100(0.1s + 20)}{(s + 100 - j100)(s + 100 + j100)},$$

where roots of the denominator are $p_{1,2} = -100 \pm j100$. Therefore,

$$A_{1C} = \left| \frac{100(0.1s + 20)}{s + 100 + j100} \right|_{s = -100 + j100} = 5 - j5 = 5\sqrt{2}e^{-j45^\circ},$$

and, in accordance with equation 3.73, the inverse Laplace transform will be

$$i_C(t) = 10 \cdot \sqrt{2} e^{-100t} \cos(100t - 45^\circ) \text{ A}.$$

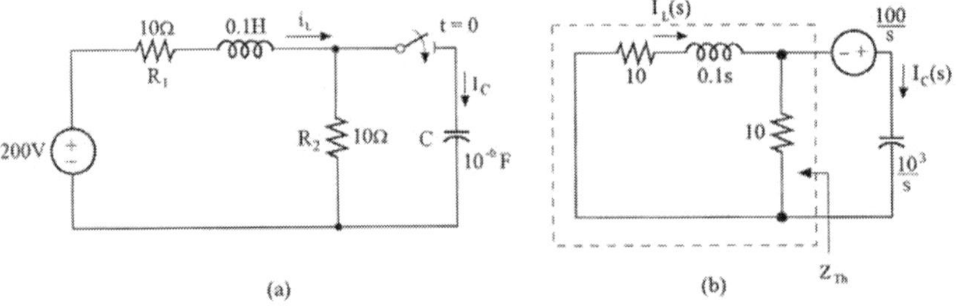

Figure 3.19 Circuit for Example 3.16 (a) and its Thévenin equivalent in *s*-domain (b).

To find the inductor current in circuit Fig. 3.19(b) we first use the current divider formula

$$I_L(s) = I_C(s)\frac{R_2}{R_1 + sL + R_2} = \frac{100(0.1s + 20)\cdot 10}{(s^2 + 200s + 20\cdot 10^3)(0.1s + 20)}$$

$$= \frac{1000}{(s + 100 - j100)(s + 100 + j100)},$$

which yields

$$A_{1L} = \left|\frac{1000}{s + 100 + j100}\right|_{s = -100 + j100} = -j5 = 5e^{-j90^\circ},$$

and

$$i_L(t) = 10e^{-100t}\cos(100t - 90^\circ) = 10e^{-100t}\sin 100t \text{ A}.$$

The steady-state value of the inductor current in Fig. 3.19(a), i.e. before the switch is closed:

$$I_L(0_-) = \frac{V_g}{R_1 + R_2} = \frac{200}{10 + 10} = 10 \text{ A}.$$

Therefore, the complete response of the current is

$$i_L(t) = 10 + 10e^{-100t}\sin 100t \text{ A}.$$

Note that initial capacitance current $i_C(0) = 10\sqrt{2}\cos(-45^\circ) = 10$ A is in agreement with its value, which can also be obtained by inspection of the circuit in Fig. 3.19(a):

$$i_C(0) = I_L(0_-) = 10 \text{ A}.$$

This result may also be obtained by straightforward calculation of $i_L(0)$ in accordance with the above formula: $i_L(0) = 10 + 10\cdot e^0 \sin 0 = 10$ A.

When the switch in any branch is opened after having been closed for a long time, as shown in Fig. 3.18(c), the equivalent circuit can be constructed by using a current source insert instead of the switch as shown in Fig. 3.18(d). The value of the current source is equal, and its direction is opposite, to the current flowing through the closed switch (short circuit current) just before its opening. Therefore, the rest of the network is passive, i.e. all the network sources are killed and it can be represented by its Thévenin impedance, as shown in Fig. 3.18(d). It is obvious again that this circuit is having zero initial conditions. For getting the complete response, the ZSR of the circuit in Fig. 3.18(d) has to be superimposed on the previous steady-state regime of the circuit in Fig. 3.18(c).

Example 3.17

In the circuit shown in Fig. 3.20(a), the switch is opened at time $t_1 = 0.2$ s, while the whole circuit has been driven by the voltage source $v_g = 10u(t)$ V since $t =$

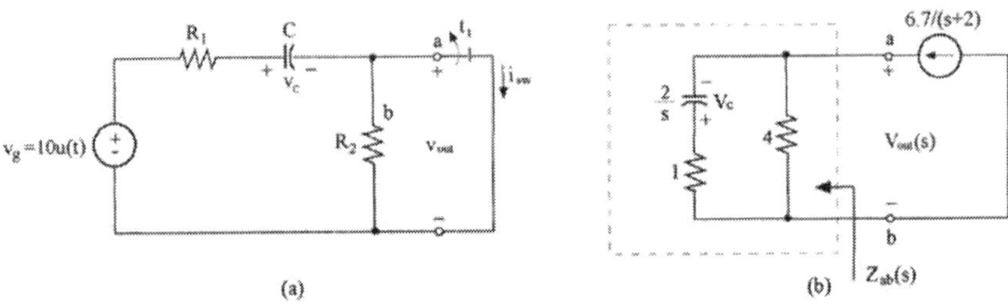

Figure 3.20 Circuit for Example 3.17 (a) and its Norton equivalent in *s*-domain (b).

0. Let $R_1 = 1\,\Omega$, $R_2 = 4\,\Omega$ and $C = 1/2\,\mathsf{F}$. Find the output voltage v_{out} and capacitance voltage v_C versus time.

Solution

First, we construct the Laplace transform circuit having zero initial conditions. For this purpose, we must find the current through the switch at the time $t = t_1$.

$$i_{sw}(t) = \frac{V_g}{R_1}e^{-at} = 10e^{-2t}, \quad t \geq 0,$$

since $a = 1/(R_1C) = 2\,\mathsf{s}^{-1}$ and $i_{sw}(0) = (10/1) = 10\,\mathsf{A}$.

Changing the variable $t = t_1 + t'$ yields

$$i_{sw}(t') = 10e^{-2t_1}e^{-2t'} = 6.7e^{-2t'}, \quad t' \geq 0,$$

and the transformed current is

$$I_{sw}(s) = 6.7\frac{1}{s+2}.$$

Next we calculate the Laplace transform internal impedance measured at the *ab* terminals (see Fig. 3.20(b)).

$$Z_{ab}(s) = \frac{4(1+2/s)}{4+1+2/s} = 0.8\frac{s+2}{s+0.4}.$$

The Laplace transform of the output voltage is

$$V_{out}(s) = Z_{ab}(s)I_{sw}(s) = 5.36\frac{1}{s+0.4},$$

and taking the inverse transform we obtain

$$v_{out}(t') = 5.36e^{-0.4t'}\,\mathsf{V}, \quad t' \geq 0,$$

since, because the voltage before opening the switch was zero, the complete response is the same. Next, we use voltage division to obtain the expression for the transformed capacitor voltage in Fig. 3.20(b):

$$V_C(s) = -V_{out}(s)\frac{2/s}{1+2/s} = -10.72\frac{1}{(s+0.4)(s+2)}.$$

In accordance with the Laplace transform pairs (see Table 3.1) we have

$$v_{C(ZSR)}(t') = \frac{-10.72}{0.4-2}(e^{-2t'} - e^{-0.4t'}) = 67(e^{-2t'} - e^{-0.4t'})\,\text{V}, \quad t \geq 0.$$

To get the complete response, we have to find the previous capacitor voltage, i.e., before the switch was opened (see circuit in Fig. 3.20(a))

$$v_{C(pr)}(t) = 10(1-e^{-2t}) = (10 - 6.7e^{-2t'})\,\text{V}.$$

Therefore, the complete response is

$$v_C(t') = v_{C(ZSR)} + v_{C(pr)} = 10 - 6.7e^{-0.4t'}\,\text{V}, \quad t' \geq 0.$$

Note that, according to this expression, the capacitor voltage at $t'=0$ is 3.3 V, which is equal to the capacitor voltage at the moment of the switch commutation in Fig. 3.20(a).

3.8.6 The transients in magnetically coupled circuits

The Laplace transform techniques are also very useful for the analysis of coupled circuits. Consider the magnetically coupled circuits shown in Fig. 3.21(a). The KVL mesh equations are

$$\begin{aligned} R_1 i_1 + L_1\frac{di_1}{dt} + M\frac{di_2}{dt} &= v_1(t) \\ R_2 i_2 + L_2\frac{di_2}{dt} + M\frac{di_1}{dt} &= v_2(t). \end{aligned} \tag{3.97}$$

Assuming non-zero initial conditions, the Laplace transform of equation 3.97

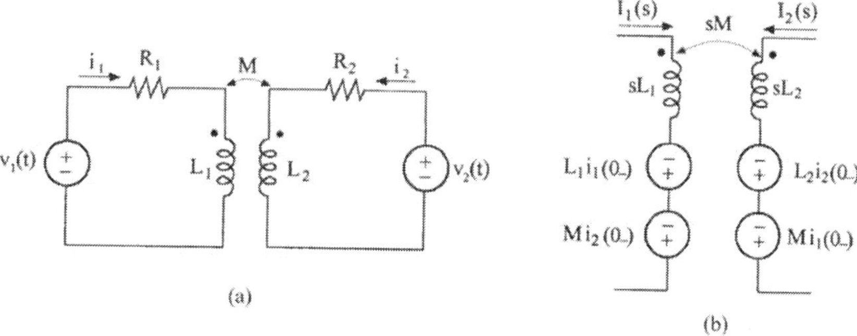

Figure 3.21 Magnetically coupled circuit (a) and its Laplace model (b).

gives

$$
\begin{aligned}
R_1 I_1(s) + sL_1 I_1(s) - L_1 i_1(0_-) + sMI_2(s) - Mi_2(0_-) &= V_1(s) \\
R_2 I_2(s) + sL_2 I_2(s) - L_2 i_2(0_-) + sMI_1(s) - Mi_1(0_-) &= V_2(s).
\end{aligned}
\tag{3.98}
$$

Combining terms yields

$$
\begin{aligned}
Z_1(s)I_1(s) + sMI_2(s) &= V_1(s) + L_1 i_1(0_-) + Mi_2(0_-) \\
sMI_1(s) + Z_2(s)I_2(s) &= V_2(s) + L_2 i_2(0_-) + Mi_1(0_-).
\end{aligned}
\tag{3.99}
$$

The Laplace transform circuit model in Fig. 3.21(b) represents equations equation 3.99 by the s-domain impedances and two voltage sources in each loop. The voltage sources $Mi_1(0_-)$ and $Mi_2(0_-)$ represent the time-domain effect of the initial stored energy in the mutual inductance due to the currents i_1 and i_2. Solving equation 3.99 for $I_1(s)$ and $I_2(s)$ gives

$$
I_1(s) = \frac{(V_1(s) + B_1)(s + a_2) - (V_2(s) + B_2)(M/L_2)s}{L_1(1 - k^2)(s^2 + as + b)} \tag{3.100a}
$$

$$
I_2(s) = \frac{(V_2(s) + B_2)(s + a_1) - (V_1(s) + B_1)(M/L_1)s}{L_2(1 - k^2)(s^2 + as + b)}, \tag{3.100b}
$$

where

$$
\begin{aligned}
a_1 = R_1/L_1, \quad a_2 = R_2/L_2, \quad k = \frac{M}{\sqrt{L_1 L_2}}, \\
a = \frac{a_1 + a_2}{1 - k^2}, \quad b = \frac{a_1 a_2}{1 - k^2},
\end{aligned}
\tag{3.101}
$$

and

$$
\begin{aligned}
B_1 &= L_1 i_1(0_-) + Mi_2(0_-) \\
B_2 &= L_2 i_2(0_-) + Mi_1(0_-),
\end{aligned}
\tag{3.102}
$$

are initial-condition equivalent generators of the first and the second loops respectively. It is worthwhile noting that in accordance with equation 3.100 the mutual coupled circuit has a second order characteristic equation: $s^2 + as + b = 0$, i.e., the mutual inductance does not increase the order of the circuit response.

Example 3.18

The mutually coupled circuit in Fig. 3.22(a) has $V_{in} = 120\,\text{V}$, $R = 60\,\Omega$, $L = 0.2\,\text{H}$, $M = 0.1\,\text{H}$. The switch is closed at $t = 0$ after having been opened for a long time. Find the currents $i_1(t)$ and $i_2(t)$ for $t \geq 0$.

Solution

First, we must find the initial conditions:

$$
i_2(0_-) = 0 \quad \text{and} \quad i_1(0_-) = \frac{V}{R} = 2\,\text{A}.
$$

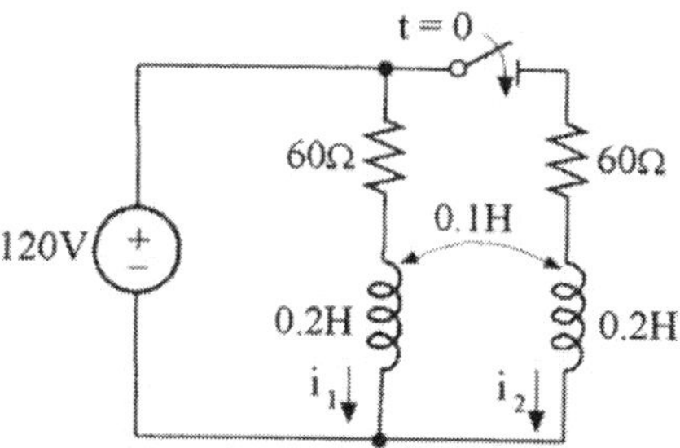

Figure 3.22 Circuit for Example 3.18.

In accordance with equations 3.101 and 3.102

$$a_1 = a_2 = \frac{R}{L} = 300\ \mathrm{s}^{-1}, \quad k = \frac{M}{L} = 0.5, \quad a = \frac{2a_1}{1-k^2} = \frac{600}{1-0.5^2} = 800\ \mathrm{s}^{-1},$$

$$b = \frac{a_1^2}{1-k^2} = \frac{300^2}{1-0.5^2} = 12\cdot 10^4\ \mathrm{s}^{-1}, \quad B_1 = Li_1(0_-) = 0.2\cdot 2 = 0.4,$$

$$B_2 = Mi_1(0_-) = 0.1\cdot 2 = 0.2.$$

With the help of equation 3.100, we obtain the transformed currents

$$I_1(s) = \frac{(120/s+0.4)(s+300) - (120/s+0.2)0.5s}{0.2\cdot 0.75(s^2+800s+12\cdot 10^4)} = \frac{2(s^2+600s+12\cdot 10^4)}{(s^2+800s+12\cdot 10^4)}$$

$$= \frac{2(s^2+600s+12\cdot 10^4)}{s(s+200)(s+600)}$$

$$I_2(s) = \frac{(120/s+0.2)(s+300) - (120/s+0.4)0.5s}{0.2\cdot 0.75(s^2+800s+12\cdot 10^4)} = \frac{800(s+300)}{(s^2+800s+12\cdot 10^4)}$$

$$= \frac{800(s+300)}{s(s+200)(s+600)},$$

i.e., the poles are $p_0 = 0$, $p_1 = -200$, $p_2 = -600$ 1/s. Therefore, the appropriate residues are:

$$A_{10} = \lim_{s\to 0} sI_1(s) = \frac{2\cdot 12\cdot 10^4}{12\cdot 10^4} = 2, \qquad A_{20} = 2,$$

$$A_{11} = \left.\frac{2(s^2+600s+12\cdot 10^4)}{s(s+600)}\right|_{s=-200} = -1, \quad A_{12} = -1,$$

$$A_{21} = \left.\frac{2(s^2+600s+12\cdot 10^4)}{s(s+200)}\right|_{s=-600} = -1, \quad A_{22} = -1,$$

which gives the time-domain currents

$$i_1(t) = (2 - e^{-200t} + e^{-600t})\ \text{A}$$

$$i_2(t) = (2 - e^{-200t} + e^{-600t})\ \text{A}.$$

In conclusion, it is worthwhile mentioning that the *Laplace transform technique is also widely used for solving electromechanical problems.* Consider, for example, the starting transients of a no-load shunt exciting d.c. motor (see Fig. 3.23). The torque equation is

$$T = mi = J\frac{d\omega}{dt},$$

where the motor torque T (Nm) is proportional to the current, J (kgm^2) is the moment of inertia and ω (rad/s) is the angular velocity.

The Kirchhoff's-law voltage equation for the motor is

$$V = Ri + L\frac{di}{dt} + k\omega,$$

where the term $k\omega$ is the generated, or back, voltage which is proportional to the angular velocity, and R, L are the resistance and the inductance of the armature winding. With zero-initial conditions the Laplace transform of these two equations will be

$$mI = Js\Omega$$

$$\frac{V}{s} = (R + sL)I + k\Omega,$$

where $\Omega(s)$ and $I(s)$ are the Laplace transform of the angular frequency and the current respectively. Solving the above equations for Ω and I yields

$$\Omega = V\frac{m}{JL}\frac{1}{s\left(s^2 + \frac{R}{L}s + \frac{km}{JL}\right)}$$

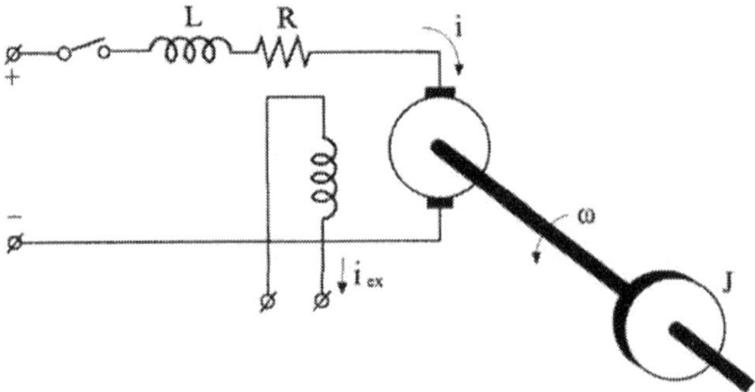

Figure 3.23 An electromechanical system with a shunt exciting d.c. motor.

and

$$I = V\frac{1}{L}\frac{1}{s^2 + \dfrac{R}{L}s + \dfrac{km}{JL}}.$$

The roots of the denominator are $s_{1,2} = -\alpha \pm \beta$, where $\alpha = R/2L$ and $\beta = \sqrt{\alpha^2 - (km/JL)}$. Thus, in accordance with the table of Laplace transform pairs, we obtain

$$\omega(t) = \frac{V}{k}\left[1 - \left(\cosh \beta t + \frac{\alpha}{\beta}\sinh \beta t\right)e^{-\alpha t}\right],$$

where $V/k = \omega_0$ is the no-load angular velocity, and

$$i(t) = \frac{V}{\beta L}e^{-\alpha t}\sinh \beta t,$$

where $i(0_-) = 0$ because of zero-initial conditions and $i(\infty) = 0$ since the motor is no-loaded and the losses in this example were neglected.

The condition of oscillations is $R^2 < 4k^2L/J$, and then $s_{1,2} = -\alpha \pm j\beta$.

Chapter #4

TRANSIENT ANALYSIS USING THE FOURIER TRANSFORM

4.1 INTRODUCTION

Like the Laplace transform, discussed in the previous chapter, the Fourier transform is very useful for transient analysis of electrical circuits. The Fourier transform, just as the Laplace transform, converts a function of time (time-domain function) into a function of frequency (frequency-domain function). However, in distinction to the Laplace transform, the Fourier transform transforms the time functions into a function of $j\omega$, a pure imaginary frequency, rather than a function of $s = c + j\omega$, which is a complex frequency. (More about the relation between Fourier and Laplace transforms further on.) From another point of view, the Fourier transform extends the Fourier series, which represents any **periodic** (but not **non-periodic**) function by an infinite sum of harmonics of different frequencies. The Fourier coefficients of such harmonics are functions of multiple $n\omega_0$ of a basic frequency ω_0, and are therefore discrete quantities corresponding to the integer n.

However, in circuit analysis there are many cases in which the forcing functions are non-periodic: such as different kinds of pulses and signals in communication engineering systems, or pulses resulting from lightning or some other strokes in power engineering systems. In these cases, as will be seen in this chapter, we may be able to find a Fourier transform $\mathbf{F}(j\omega) = |\mathbf{F}(j\omega)|e^{j\Psi(\omega)}$ of a **non-periodic** function, whose amplitude, $|\mathbf{F}(j\omega)|$ and phase $\Psi(\omega)$ spectra are continuous rather than discrete, i.e., they are functions of ω (but not of $n\omega$). The conversion of a non-periodic time function to a function of frequency ω allows us to analyze the transient behavior of any linear circuit by using their frequency characteristics, such as: impedance $Z(j\omega)$, admittance $Y(j\omega)$ and/or transfer coefficient $K(j\omega)$. This means that we will be able to use all the methods of steady-state analysis by applying them to the transient analysis, which again will reduce the integro-differential operating in the time domain to more simple algebraic operations in the frequency domain.

Thus, the Fourier transform extends the phasor concept, which has been developed for sinusoidal (periodic) functions to non-periodic functions, which are more general than just sinusoids.

4.2 THE INTER-RELATIONSHIP BETWEEN THE TRANSIENT BEHAVIOR OF ELECTRICAL CIRCUITS AND THEIR SPECTRAL PROPERTIES

The study of transient behavior of electrical circuits in the previous chapter shows that this behavior is largely related to their frequency characteristics. This was especially evident from applying the Laplace transform. Thus, if the voltage was applied to the input of an electric circuit, whose Laplace transform was given as

$$\mathbf{V}_1(s) \leftrightarrow v_1(t),$$

then the Laplace transform of its response, for example of the current, can be found as

$$\mathbf{I}_1(s) = \mathbf{V}_1(s)/\mathbf{Z}(s) = \mathbf{V}_1(s)\mathbf{Y}(s), \tag{4.1a}$$

where

$$\mathbf{Z}(s) = \mathbf{Z}(j\omega)|_{j\omega=s} \quad or \quad \mathbf{Y}(s) = \mathbf{Y}(j\omega)|_{j\omega=s}.$$

Here the complex impedance $\mathbf{Z}(j\omega)$ or the admittance $\mathbf{Y}(j\omega)$ are actually the frequency characteristics of the circuit. In exactly the same way, we can represent the transfer function

$$\mathbf{H}(s) = \mathbf{H}(j\omega)|_{j\omega=s},$$

and with this function the Laplace transform of the output voltage will be

$$\mathbf{V}_2(s) = \mathbf{H}(s)\mathbf{V}_1(s). \tag{4.1b}$$

In finding the response function of the circuit to any forcing function the properties of the circuit are completely determined by its frequency characteristic $\mathbf{Z}(j\omega)$, $\mathbf{Y}(j\omega)$ or $\mathbf{H}(j\omega)$. This relationship between the frequency characteristics of the system and its behavior in transients is obvious when taking into consideration the physical properties of the circuit elements. Thus the inductance, which prevents an abrupt change of current, is characterized by changing, in particular by increasing, its reactance (X) by increasing the frequency; and in the same way the capacitance, which prevents an abrupt change of the voltage, is characterized by changing, in particular by increasing, its susceptance (B) by increasing the frequency. The availability of resonant oscillations during the transients also depends on the characteristics of the system, i.e., its impedance/admittance.

Expressions like equation 4.1 are completely analogous to the phasor expressions for different harmonics in the steady-state analysis. Using the Fourier series in a non-sinusoidal analysis, we might write

$$\mathbf{I}_n = \mathbf{V}_n/\mathbf{Z}(j\omega) = \mathbf{Y}(j\omega)\mathbf{V}_n,$$

where $j\omega = jn\omega_0$ is the discrete frequency of different harmonics: $\mathbf{I}_n = \mathbf{I}(jn\omega_0)$ and $\mathbf{V}_n = \mathbf{V}(jn\omega_0)$. Thus, the discrete spectra of the current, $\mathbf{I}_n$ in accordance with the above expression, can be found if we are able to find the spectra of the forcing function, for instance, $\mathbf{V}_n$ by knowing its Laplace transform $\mathbf{V}(s)$ or straightforwardly by applying the Fourier series coefficient formulas.

However, if the forcing function is non-periodical, which happens in many cases where this function is exponential, rectangular or any kind of pulse, its spectra cannot be found by just replacing s by $j\omega$. (More precisely, as will be shown further on, in some special cases of the above functions, the frequency characteristics can be found anyway by replacing s by $j\omega$.) As is known, using the Fourier series this problem cannot be solved either, since the Fourier series is appropriate only for periodic functions.

Our goal in this chapter, therefore, is to develop a method which allows extending the phasor concept to non-periodic functions. The solution is the Fourier transform, which is, as we already mentioned, an extension of the Fourier series to non-periodic functions.

4.3 THE FOURIER TRANSFORM

4.3.1 The definition of the Fourier transform

Let us proceed to define the Fourier transform by first recalling the spectrum presentation of the periodic function. In the simplest case of one harmonic $A\sin(\omega t + \psi)$ (Fig. 4.1a) its amplitude and phase spectra will be as shown in Fig. 4.1b. Using the complex form of sinusoids,

$$\mathrm{A}\sin(\omega t + \psi) = \frac{A}{2j}\left[e^{j(\omega t + \psi)} - e^{-j(\omega t + \psi)}\right] = \frac{A}{2}\left[e^{j(\omega t + \psi - \pi/2)} + e^{-j(\omega t + \psi - \pi/2)}\right]^{(*)},$$

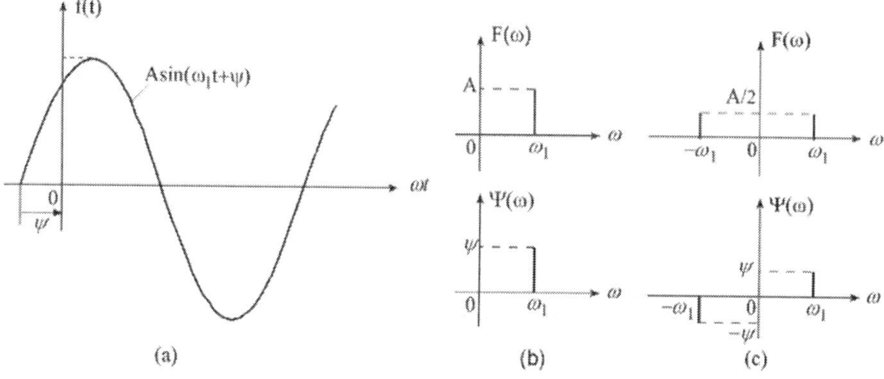

Figure 4.1 Sinusoidal faction (a) and its spectra: in real notation (b) and complex notation (c).

(*)Note that in this expression the additional angle $-\pi/2$ appears for a cosine presentation of the sinusoidal function. It should be noted, however, that in this book we are using sine presentation rather than cosine.

we may take into consideration also negative frequencies as in the second term, in the brackets, of this expression. (It has to be mentioned that the definition of a negative frequency has a purely mathematical meaning without any physical connection.) In this case, the amplitude and phase spectra will be as shown in Fig. 4.1(c). As can be seen, both spectra are presented by two ordinates correspondingly for positive and negative frequencies. The amplitude spectrum components are symmetrical about the vertical axis and the phase spectrum components are symmetrical about the origin.

As is known, if any current or voltage wave is not sinusoidal, but periodical, it may be represented by the *infinite Fourier series.* In trigonometrical form it will be:

$$f(t)=\frac{a_0}{2}+\sum_{n=1}^{\infty} A_n \sin(n\omega_0 t+\psi_n),$$

where amplitudes A_n and phases ψ_n can be expressed as

$$A_n=\sqrt{a_n^2+b_n^2},\quad \psi_n=\tan^{-1}\frac{a_n}{b_n},$$

and

$$\begin{aligned} a_0&=\frac{2}{T}\int_0^T f(t)dt,\\ a_n&=\frac{2}{T}\int_0^T f(t)\cos n\omega_0 t\, dt, \\ b_n&=\frac{2}{T}\int_0^T f(t)\sin n\omega_0 t\, dt,\quad n=0,1,2,\ldots. \end{aligned} \tag{4.2}$$

In accordance with the above expressions, the amplitude $A_n=f(n\omega_0)$ and phase $\psi_n=f(n\omega_0)$ spectra of a non-sinusoidal function may be sketched, as functions of $n\omega_0$. With a complex, or exponentional form of the Fourier series:

$$f(t)=\sum_{n=-\infty}^{\infty} \mathbf{C}_n e^{jn\omega_0 t}, \tag{4.3}$$

where

$$\mathbf{C}_n=\frac{1}{2}(a_n-jb_n),\quad \mathbf{C}_{-n}=\frac{1}{2}(a_n+jb_n),\quad \mathbf{C}_0=\frac{1}{2}a_0$$

and (4.3a)

$$\angle\mathbf{C}_n=-\tan^{-1}\frac{b_n}{a_n};\quad \angle\mathbf{C}_{-n}=-\angle\mathbf{C}_n=\tan^{-1}\frac{b_n}{a_n}.$$

The amplitude$|\mathbf{C}_n|=f(n\omega_0)$ and phase $\angle\mathbf{C}_n=f(n\omega_0)$ spectra will be functions of both positive and negative frequencies: the amplitude spectrum will be symmetrical about the vertical axis, and the phase spectrum will be symmetrical about the origin. Note that both the amplitude and phase spectra are discrete functions of harmonic frequencies.

Coefficients $\mathbf{C}_n$ can be determined by substituting the expressions for a_n and b_n (equation 4.2) into equation 4.3a by changing the limits of integration, i.e.,

$$\mathbf{C}_n = \frac{1}{T}\int_{-T/2}^{T/2} f(t)(\cos n\omega_0 t - j\sin n\omega_0 t)dt = \frac{1}{T}\int_{-T/2}^{T/2} f(t)e^{-jn\omega_0 t}dt.$$

An example of such *discrete spectra of a periodic function* is a train of rectangular pulses, having duration d, period T and amplitude V_0, which is shown in Fig. 4.2(a) and its spectra are shown in Fig. 4.2(b) and (c). Here the magnitudes are $|\mathbf{C}_n| = \varepsilon V_0 Sa(n\varepsilon\pi)$, where $Sa(x)$ is called a **sampling function** or **sinc function**

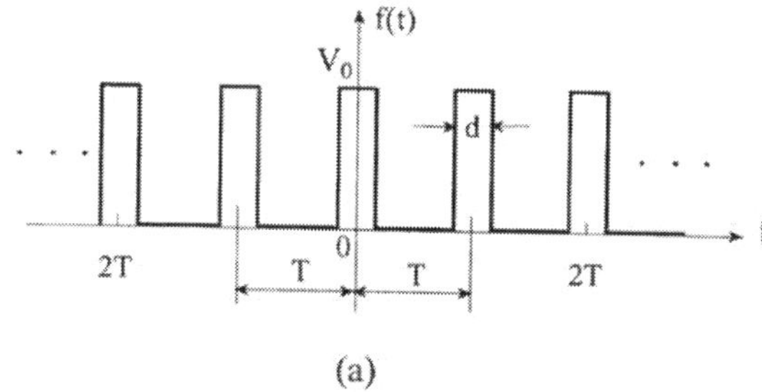

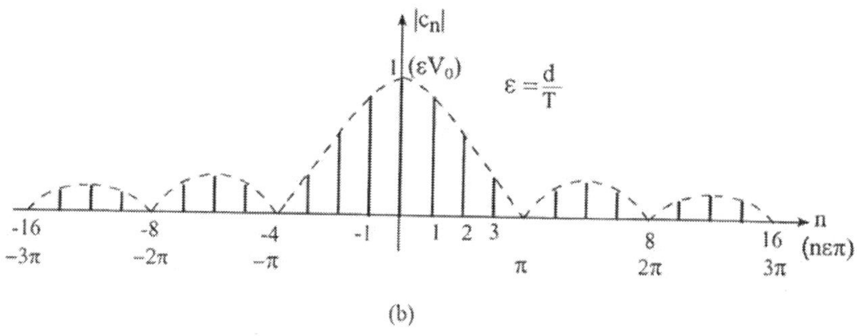

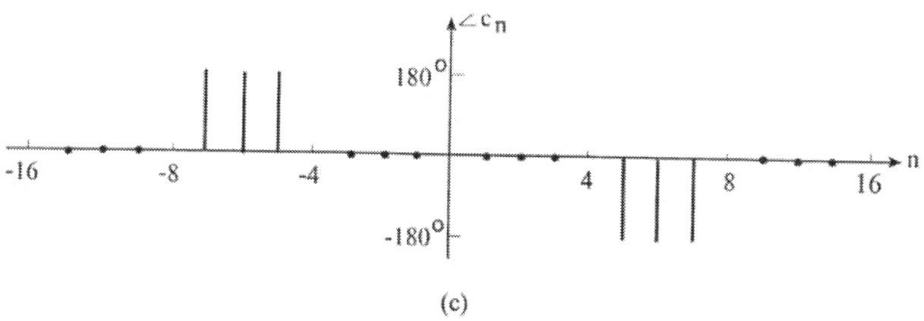

Figure 4.2 A train of rectangular pulses (a), its discrete amplitude (b) and phase (c) spectras.

(in mathematics: sinc $x = (\sin \pi x)/\pi x = \mathrm{Sa}(\pi x)$) and it might be calculated with most mathematical programs like MATHCAD, MATHLAB etc).

However, there are many important forcing functions that are not periodic functions, such as a single rectangular pulse, an impulse function, a step function and a rump function. Another example of a non-periodic function is an impulse voltage waveform, which appears in high-voltage transmission lines, when a stroke of lightning influences the line conductors[(*)].

Frequency spectra may also be obtained for such non-periodic functions; however, they will be continuous spectra, rather than discrete. These spectra can be obtained by using the **Fourier transform**, which is an extension of the **Fourier series** for non-periodic functions. With such spectra we will be able to extend the frequency analysis and the phasor concept to non-periodic functions.

Thus, the Fourier transform, in contrast to the Fourier series, is a function of the continuous frequency ω (but not of discrete frequency $n\omega$) and corresponds to the time-domain non-periodic function. To develop the Fourier transform technique we shall consider the non-periodic function $f(t)$, Fig. 4.3a, as defined on an infinite interval.

This function should satisfy the Dirichlet conditions: in any finite interval, $f(t)$ has at most a finite number of finite discontinuities, a finite number of maxima and minima and $\int_{-\infty}^{\infty} |f(t)|dt < \infty$, i.e., the integral converges. To be able to extend the use of the Fourier series to a non-periodic function we will define a new function $f_{per}(\mathrm{t})$ which is identical to $f(t)$ on the interval

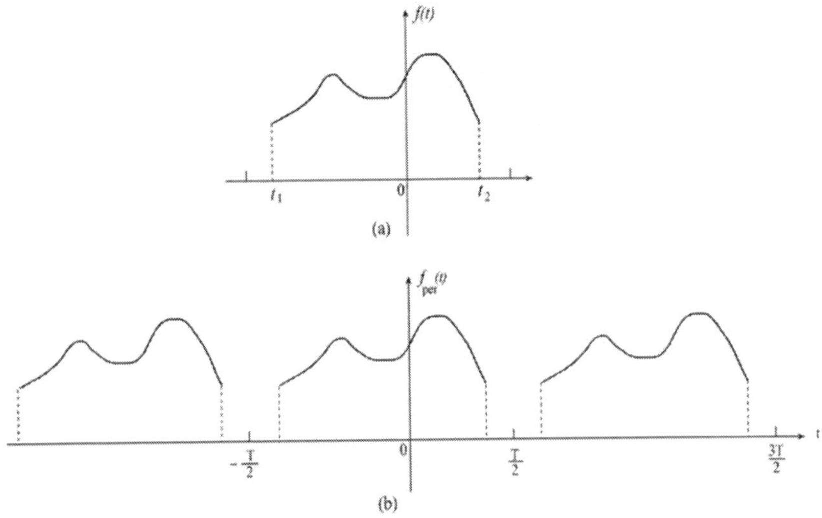

Figure 4.3 A non-periodic function (a) and its periodic extension (b).

(*) Gonen, T. (1988) *Electric Power Transmission System Engineering*. Wiley, New York, Chichester, Brisbane, Toronto, Singapore.

$-T/2 < t < T/2$ and is periodic of any period $T > t_2 - t_1$ as shown in Fig. 4.3b. Such a function $f_{per}(t)$ is said to be the *periodic extension of* $f(t)$ and might be represented by the Fourier series. The given non-periodic function $f(t)$, therefore, is also given by the same Fourier series, but only in the interval $(-T/2, T/2)$. Outside of this interval, this function cannot be represented by the Fourier series. Using the exponential form of the Fourier series for $f_{per}(\mathrm{t})$ we will have

$$f_{per}(t) = \frac{1}{2} \sum_{n=-\infty}^{\infty} \mathbf{C}_n e^{jn\omega_0 t}, \tag{4.5}$$

where

$$\mathbf{C}_n = \frac{1}{T} \int_{-T/2}^{T/2} f_{per}(t) e^{-jn\omega_0 t}\, dt. \tag{4.6}$$

Our intention is to let $T \to \infty$, in which case

$$f_{per}(t) \to f(t). \tag{4.7}$$

We will then have extended the Fourier series concept to the non-periodic function $f(t)$ by considering it to be periodic with an infinite period. Since now $T \to \infty$ and then $\omega_0 = 2\pi/T$ becomes vanishingly small, we may represent this limit by a differential, i.e., $\omega_0 \to d\omega$ so that:

$$\frac{1}{T} = \frac{\omega_0}{2\pi} \to \frac{d\omega}{2\pi} \quad (\text{when } T \to \infty). \tag{4.8}$$

Now, the harmonic discrete frequency $n\omega_0$ will approach the continuous frequency variable ω, since ω_0 becomes vanishingly small $(\omega_0 \to 0)$ and all the nearby frequencies approach a smoothly changing frequency ω_0. In other words, n tends to infinity as ω_0 approaches zero, such that the product is finite:

$$n\omega_0 \to \omega. \tag{4.9}$$

Substituting equation 4.6 into equation 4.5, and taking into consideration equations 4.7 and 4.9, we may obtain

$$f(t) = \frac{1}{2\pi} \sum \frac{2\pi}{T} \left[\int_{-\infty}^{\infty} f(t) e^{-j\omega t}\, dt \right] e^{j\omega t}. \tag{4.10}$$

The inner integral (in brackets) is a function of $j\omega$ (not of t) and we assign it to $\mathbf{F}(j\omega)$, so that

$$\mathbf{F}(j\omega) = \int_{-\infty}^{\infty} f(t) e^{-j\omega t}\, dt, \tag{4.11}$$

and it is the **Fourier transform** of $f(t)$. Then as

$$T \to \infty \left(\frac{1}{T} \to \frac{d\omega}{2\pi} \ (\text{equation 4.8}) \right) \quad \text{and} \quad \frac{2\pi}{T} \to d\omega,$$

the sum in equation 4.10 becomes an integral:

$$f(t) = \frac{1}{2\pi} \int_{-\infty}^{\infty} \mathbf{F}(j\omega) e^{j\omega t} d\omega, \tag{4.12}$$

which is called the **inverse Fourier transform**.

These two expressions above are known as the Fourier transform pair

$$\begin{cases} \mathbf{F}(j\omega) = \displaystyle\int_{-\infty}^{\infty} f(t) e^{-j\omega t} dt & \text{(4.13a)} \\ f(t) = \dfrac{1}{2\pi} \displaystyle\int_{-\infty}^{\infty} \mathbf{F}(j\omega) e^{j\omega t} d\omega, & \text{(4.13b)} \end{cases}$$

which are also often stated symbolically as

$$\begin{cases} \mathbf{F}(j\omega) = \mathscr{F}[f(t)] & \text{(a)} \\ f(t) = \mathscr{F}^{-1}[\mathbf{F}(j\omega)], & \text{(b)} \end{cases} \tag{4.14}$$

where $\mathscr{F}$ denotes the operation of taking the Fourier transform. These two expressions in equation 4.14 may also be indicated as

$$f(t) \leftrightarrow \mathbf{F}(j\omega). \tag{4.15}$$

The Fourier transform as seen in equation 4.13a is a transformation of the function $f(t)$ from the *time domain* to the *frequency domain* and corresponds to the Fourier coefficient expressions in equation 4.3a. Equation 4.13b, the inverse transform, is an opposite transformation of the complex function $\mathbf{F}(j\omega)$ from the *frequency domain* into the *time domain* and is a direct analogy to the Fourier series (equation 4.3). Another explanation of these two analogies is to say that the Fourier transform is a *continuous* representation (with ω being a continuous variable) of a non-periodic function, whereas the Fourier series is a *discrete* representation (with $n\omega_0$ being a discrete variable) of a periodic function. Finally, it must be indicated that the Fourier transform-pair relationship is unique: for a given function $f(t)$ there is one specific $\mathbf{F}(j\omega)$ and for a given $\mathbf{F}(j\omega)$ there is one specific $f(t)$.

The following examples show how we can use the above-developed expressions to find the Fourier transform of a non-periodic function and its spectra.

Example 4.1

Let us find the Fourier transform of the exponential function

$$f(t) = e^{-at} u(t) \quad (a > 0). \tag{4.16}$$

Using equation 4.13a we have

$$\mathscr{F}[e^{-at}u(t)] = \int_{-\infty}^{\infty} e^{-at} u(t) e^{-j\omega t} dt = \int_{0}^{\infty} e^{-(a+j\omega)t} dt = \frac{1}{-(a+j\omega)} e^{-(a+j\omega)} \Big|_{0}^{\infty}.$$

Because $a > 0$, the upper limit results in zero (since the imaginary part $e^{-j\omega t}$ represents the rotation features of the exponential amplitude e^{-at} and is therefore bound while the exponential approaches 0)[(*)]. Thus, we have

$$\mathscr{F}[e^{-at}u(t)] = \frac{1}{a + j\omega}, \tag{4.17a}$$

or

$$e^{-at}u(t) \leftrightarrow \frac{1}{a + j\omega} \quad (a > 0), \tag{4.17b}$$

and

$$\mathbf{F}(j\omega) = \frac{1}{a + j\omega}. \tag{4.17c}$$

In accordance with the obtained expressions (14.17) the amplitude spectra of an exponential function will be

$$[\mathbf{F}(j\omega)] = \frac{1}{\sqrt{a^2 + \omega^2}}, \tag{4.18}$$

and its phase spectrum

$$\Psi(\omega) = -\tan^{-1}\frac{\omega}{a}. \tag{4.19}$$

The exponential function (14.16) and its spectra (14.18) and (14.19) are shown in Fig. 4.4a–c.

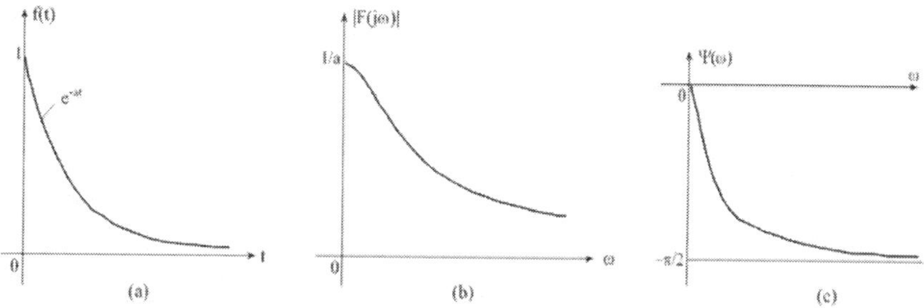

Figure 4.4 Exponential function (a) and its amplitude (b) and phase (c) spectra.

(*)Note that the function e^{-at} $(a > 0)$ does not have a Fourier transform, since the integral $\int_{-\infty}^{\infty} e^{-at}e^{-j\omega t}dt$ of its lower limit approaches infinity (infinite value).

Example 4.2

As another example, let us find the Fourier transform of the single *rectangular pulse*.

$$f(t) = \begin{cases} V_0 & -\frac{d}{2} < t < \frac{d}{2} \\ 0 & -\frac{d}{2} > t > \frac{d}{2}, \end{cases}$$

which is shown in Fig. 4.5(a). By the definition in equation 4.13a we have

$$\mathbf{F}(j\omega) = \int_{-\infty}^{\infty} f(t)e^{-j\omega t}dt = V_0 \int_{-d/2}^{d/2} e^{-j\omega t}dt = \frac{V_0}{-j\omega}(e^{-j(\omega d/2)} - e^{j(\omega d/2)}), \quad (4.20a)$$

$$\mathbf{F}(j\omega) = \frac{2V_0}{\omega}\left(\frac{e^{j(\omega d/2)} - e^{-j(\omega d/2)}}{j2}\right) = \frac{2V_0}{\omega}\sin\frac{\omega d}{2} = V_0 d\frac{\sin(\omega d/2)}{\omega d/2}, \quad (4.20b)$$

or shortly

$$f(t) = \left\{ \text{at time} \quad \begin{matrix} \frac{d}{2} < t < \frac{d}{2}: & V_0 \\ -\frac{d}{2} > t > \frac{d}{2}: & 0 \end{matrix} \right\} \leftrightarrow V_0 d \,\mathrm{Sa}\left(\frac{\omega d}{2}\right), \quad (4.20c)$$

where $Sa(\omega d/2)$ is the sampling function. This function yields the **continuous spectrum** of a rectangular pulse (Fig. 4.5) which is shown in Fig. 4.5(b). Whenever $\omega d/2 = k\pi$ $(k = 1, 2, ...)$ or the frequencies $\omega = 2\pi/d, 4\pi/d, ...$ the above spectrum curve crosses the *ω-axis*, i.e., is zero. For $\omega d/2 = (\pi/2)(2k+1)$ the spectrum curve reaches the maximum points which are $\mathbf{F}(j\omega)|_{\max} = V_0 d/(\pi/2)(2k+1)$. Note that in this case the phase spectrum equals zero.

We should again emphasize that this spectrum is a continuous function of ω as opposed to the *discrete spectrum* of a periodic sequence of rectangular pulses (as shown in Fig. 4.2). Note also that the value of the continuous spectrum in

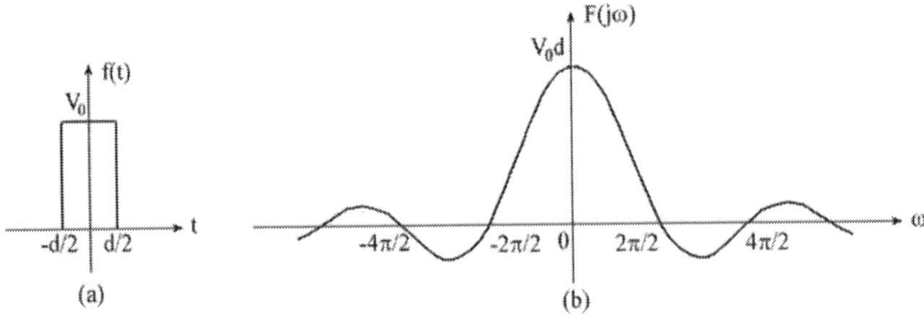

Figure 4.5 A rectangular pulse (a) and its spectrum (b)

equation 4.20 is as a product of the value of the pulse and the value of its duration, i.e., dimensionally it is indicated as "*volts times seconds*", or "*volts per unit frequency*". In the case of a discrete spectrum the value of its magnitude is given just in the same dimension as a periodic function (volts, or amperes). In order to better understand the above differences we should analyze more deeply the relationship between the Fourier transform and Fourier series and look into some of the properties of $\mathbf{F}(j\omega)$.

4.3.2 Relationship between a discrete and continuous spectra

To define the relationship between the discrete spectra of a non-sinusoidal periodic function and the continuous spectra of a non-periodic function we should use the complex form of the Fourier series. In Table 4.1 the basic formulas of the Fourier series and of the Fourier transform are given.

As was previously mentioned, any periodic function has discrete or line spectra for both its magnitude and phase. However, as the period increases, the lines of the discrete spectra become more dense with more lines. In Fig. 4.6 the changing of the magnitude spectrum of a train of rectangular pulses by increasing its period, which is the same as decreasing the scaled duration $\varepsilon = d/T$, is shown. The solid line, or the envelope of the magnitude spectrum, in this figure crosses the frequency axis at the $n\omega_0 d/2 = \pi k$, or $n\omega_0 = 2\pi k/d$ $(n = k/\varepsilon,\ k = \pm 1, \pm 2, \ldots)$, which means that the zero point of the envelope does not depend on T, but only on the duration d. However, when period increases, the number of lines, N, between the origin and the first zero, which is the same as between two adjoining zeros, increases directly proportional to T. Note also that the product $n\omega_0$ remains the same as can be seen from Fig. 4.6.

This number of lines might be calculated as

$$N = n - 1 = \frac{2\pi}{\omega_0 d} - 1 = \frac{1}{\varepsilon} - 1. \tag{4.21}$$

The line magnitudes of the discrete spectrum are inversely proportional to the period or directly proportional to the scaled duration ε, as can be seen from

Table 4.1 Basic formulas of Fourier series and Fourier transform

Fourier series (periodic function)	Fourier transform (non-periodic function)
$f(t) = \frac{1}{2}\sum_{n=-\infty}^{\infty} c_n e^{jn\omega_0 t}$	$f(t) = \frac{1}{2\pi}\int_{-\infty}^{\infty} \mathbf{F}(j\omega)e^{j\omega t}d\omega$
$\mathbf{C}_n = \frac{2}{T}\int_{-T/2}^{T/2} f(t)e^{-jn\omega_0 t}dt$	$\mathbf{F}(j\omega) = \int_{-\infty}^{\infty} f(t)e^{-j\omega t}dt$
$\mathbf{C}_n = c_n e^{j\psi_n}$	$\mathbf{F}(j\omega) = \lvert\mathbf{F}(j\omega)\rvert e^{j\psi(\omega)}$

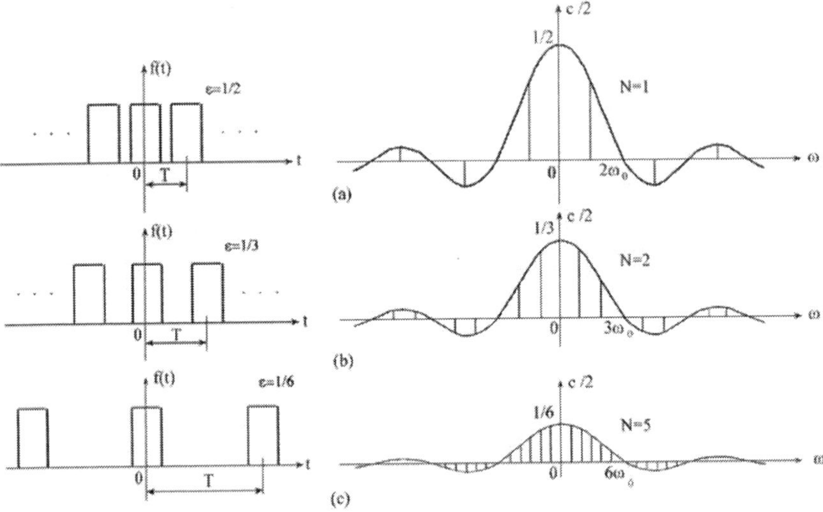

Figure 4.6 The changing of the spectrum as duration ε decreases.

expression (4.4) and Fig. 4.6, i.e. the broken-line curve is lower and approaches zero, as $T \to 0$, and finally coincides with the horizontal axis.

If we multiply the Fourier coefficients $1/2\mathbf{C}_n$ by period T and then let the period become infinite ($T \to \infty$), then the frequency interval $\omega_0 = 2\pi/T$ between the lines of the discrete spectrum approaches zero and the discrete spectrum will turn into a continuous spectrum of a non-periodic rectangular pulse of duration d. On the other hand the Fourier coefficients, being multiplied by period T, become non-dependent on the period and their magnitudes will not change, i.e. the envelope curve is not dependent on T and follows the expression

$$\mathbf{F}(j\omega) = \int_{-d/2}^{d/2} e^{-j\omega t} dt = d\,\frac{\sin(\omega d/2)}{\omega d/2}.$$

The above explanation is illustrated in Fig. 4.7.

It is obvious that the above example of a rectangular pulse can be generalized to any other kind of non-periodic function. In this case it is always possible to choose such a T, that

$$\mathbf{F}(j\omega) = \int_{-\infty}^{\infty} f(t)e^{-j\omega t} dt = \int_{-T/2}^{T/2} f(t)e^{-j\omega t} dt. \tag{4.22}$$

The periodic function, which coincides with the above non-periodic function $f(t)$ in the interval $-T/2$, $T/2$ and is of period T will have the line spectrum in accordance with the equation

$$\frac{1}{2}\mathbf{C}_n = \frac{1}{T}\int_{-T/2}^{T/2} f(t)e^{-jn\omega_0 t} dt. \tag{4.23}$$

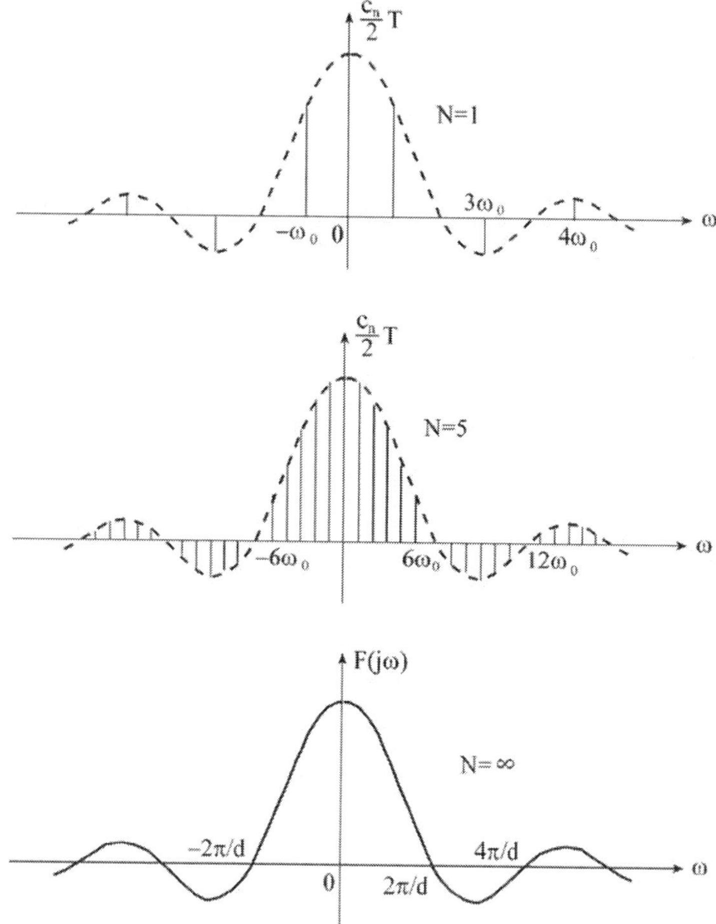

Figure 4.7 The transformation of a discrete spectrum into a continuous one.

By comparing these two equations 4.22 and 4.23 one can conclude that the continuous spectrum $\mathbf{F}(j\omega)$ (4.22) of the non-periodic function, being scaled by $1/T$, is identical to the envelope of the line spectrum (4.23) of the periodically repeated given function:

$$\frac{1}{2}\mathbf{C}_n\bigg/\frac{1}{T}=\frac{T}{2}\mathbf{C}_n=\int_{-T/2}^{T/2} f(t)e^{-jn\omega_0 t}\,dt=\mathbf{F}(j\omega)|_{\omega=n\omega_0}. \tag{4.24}$$

From the above it also follows that the phase spectrum:

$$\Psi_n=\Psi(\omega)|_{\omega=n\omega_0}. \tag{4.25}$$

Since the continuous spectrum is actually a line spectrum scaled by frequency

$1/T$, its measurement is in the unit of a function multiplied by a unit of time, as was previously mentioned.

4.3.3 Symmetry properties of the Fourier transform

Our objective in this section is to establish several of the symmetrical properties of the Fourier transform in order to use them in our further studies. Replacing $e^{-j\omega t}$ in equation 4.13a by trigonometric functions, using Euler's identity, we will get

$$\mathbf{F}(j\omega) = \int_{-\infty}^{\infty} f(t)\cos\omega t\, dt - j\int_{-\infty}^{\infty} f(t)\sin\omega t\, dt \tag{4.26}$$

All the functions, $\cos\omega t$ and $\sin\omega t$, are real functions of time, therefore both integrals in equation 4.26 are real functions of ω. Thus, we may write

$$\mathbf{F}(j\omega) = A(\omega) - jB(\omega) = |\mathbf{F}(j\omega)|e^{j\phi(\omega)}, \tag{4.27}$$

where

$$A(\omega) = \int_{-\infty}^{\infty} f(t)\cos\omega t\, dt \tag{4.27a}$$

$$B(\omega) = \int_{-\infty}^{\infty} f(t)\sin\omega t\, dt \tag{4.27b}$$

and

$$|\mathbf{F}(j\omega)| = \sqrt{A^2(\omega) + B^2(\omega)} \tag{4.27c}$$

$$\phi(\omega) = \tan^{-1}\frac{-B(\omega)}{A(\omega)}. \tag{4.27d}$$

Replacing ω by $(-\omega)$ shows that $A(\omega)$ and $|\mathbf{F}(j\omega)|$ are both even functions of ω, and $B(\omega)$ and $\phi(\omega)$ are both odd functions of ω. Let us now consider three cases.

(a) *Function* f(t) *is an even function of* t

As is known, an even function is symmetrical about the vertical axis, and an odd function is symmetrical about the origin. Since the *cosine* and *sine* are even and odd functions of t respectively, then $f(t)\cos\omega t$ is an even function and $f(t)\cos\omega t$ is an odd function of t. Therefore the integral of symmetrical limits in equation 4.27b is zero, i.e., $B(\omega) = 0$ and

$$\mathbf{F}(j\omega) = A(\omega) = \int_{-\infty}^{\infty} f(t)\cos\omega t\, dt = 2\int_{0}^{\infty} f(t)\cos\omega t\, dt. \tag{4.28}$$

With those results we may conclude that the Fourier transform of an even function is a real even function of ω and the phase function in (4.27d) is zero or π for all ω. Replacing $e^{-j\omega t}$ in (4.13b) by trigonometrical functions, yields

$$f(t) = \frac{1}{2\pi}\int_{-\infty}^{\infty} \mathbf{F}(j\omega)\cos\omega t\, d\omega + j\frac{1}{2\pi}\int_{-\infty}^{\infty} \mathbf{F}(j\omega)\sin\omega t\, d\omega. \qquad (4.29)$$

Since $\mathbf{F}(j\omega)$ is a real and even function of ω, the second integrand is an odd function of ω which results in a zero imaginary part of equation 4.29. Thus, in this case

$$f(t) = \frac{1}{\pi}\int_{0}^{\infty} A(\omega)\cos\omega t\, d\omega. \qquad (4.30)$$

Comparing the equations 4.28 and 4.30, we may see that the arguments ω and t might be interchanged, i.e., considering

$$F(jt) = F(-jt)$$

(since it is an even function) as a function of t, then its spectrum should be $f(\omega) = f(-\omega)$ as shown in Fig. 4.8 (a and b).

(b) *Function* f(t) *is an odd function of* t

In this case the function $f(t)\cos\omega t$ is an odd function of t and $f(t)\sin\omega t$ is an even function of t. Therefore, the integral in equation 4.27a is zero, i.e. $A(\omega) = 0$ and

$$\mathbf{F}(j\omega) = -jB(j\omega) = -j2\int_{0}^{\infty} f(t)\sin\omega t\, dt, \qquad (4.31)$$

i.e., $F(j\omega)$ is a pure imaginary and odd function of ω and therefore $\phi(\omega)$ is

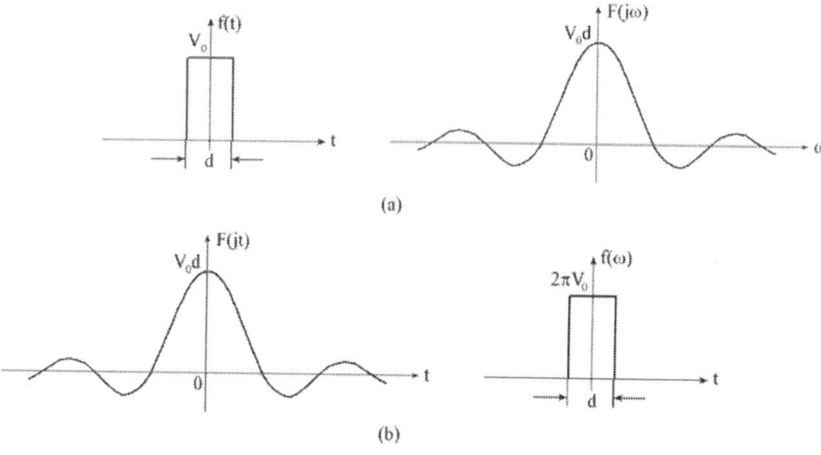

Figure 4.8 Interchange between the function and its spectrum: (a) a rectangular pulse $f(t)$ and its spectrum $F(j\omega)$, (b) a rectangular pulse spectrum $f(\omega)$ of the time sinc function.

$\pm\pi/2$. The function $F(j\omega)\cos\omega t$ is an odd function of ω, therefore the first integral in equation 4.29 turns into zero and

$$f(t)=\frac{1}{\pi}\int_0^{\infty} B(\omega)\sin\omega t\, dt. \tag{4.32}$$

The interchanging properties of the time function and its spectrum are applicable also in this case, i.e. consideration of the function

$$F(-jt)=-F(jt)$$

as a function of time yields its spectrum as $f(\omega)$.

(c) *Function* f(t) *is a non-symmetrical function, i.e., neither even nor odd*

Any non-symmetrical function can be presented as the sum of an even and odd function, i.e.,

$$f(t)=f_e(t)+f_o(t).$$

However,

$$f(-t)=f_e(t)-f_o(t),$$

which means that such a function does not obey either an even or odd function definition. Performing summation and subtraction of the above expression we obtain

$$f_e(t)=\frac{1}{2}[f(t)+f(-t)],\quad f_o(t)=\frac{1}{2}[f(t)-f(-t)].$$

With this result of splitting a non-symmetrical function into two subfunctions: even and odd, we may prove that in this case $\mathbf{F}(j\omega)$ is a complex function, whose real part is even while the imaginary part is an odd function of ω. Finally, we note that the replacement of ω by $-\omega$ in equation 4.27 gives the conjugate complex of $\mathbf{F}(j\omega)$, i.e.,

$$\mathbf{F}(-j\omega)=A(\omega)+jB(\omega)=\overset{*}{\mathbf{F}}(j\omega),$$

and we have

$$\mathbf{F}(j\omega)\mathbf{F}(-j\omega)=\mathbf{F}(j\omega)\overset{*}{\mathbf{F}}(j\omega)=|\mathbf{F}(j\omega)|^2=A^2(\omega)+B^2(\omega). \tag{4.33}$$

4.3.4 Energy characteristics of a continuous spectrum

If $f(t)$ is a periodic function of either the voltage across or the current through a circuit, then $(1/T)\int_0^T f^2(t)dt$ is proportional to the average power delivered to this circuit. With a complex form of the Fourier series, applied to a non-sinusoidal function, we obtain

$$\frac{1}{T}\int_0^T f^2(t)dt=\sum_{n=-\infty}^{\infty}\left(\frac{\mathbf{C}_n}{2}\right)^2, \tag{4.34}$$

which might be interpreted as the sum of the powers of all the amplitude spectrum components of a given function(*). In accordance with the previously explained relationship between the discrete and continuous spectra (para. 4.3.2), we can easily obtain an expression similar to equation 4.34, but for the non-periodic function $f(t)$. For this purpose we first multiply equation 4.34 by T and replace $(1/2)\mathbf{C}_n$ by $(1/T)|\mathbf{F}(j\omega)|_{\omega=n\omega_0}$:

$$\int_0^T f^2(t)dt = \frac{1}{T}\sum |\mathbf{F}(j\omega)|^2_{\omega=n\omega_0}. \tag{4.35}$$

Now, when $T\to\infty$ and $n\omega_0=\omega$, then

$$\frac{1}{T}=\frac{\omega_0}{2\pi}=\left.\frac{\omega/n}{2\pi}\right|_{n\to\infty}\to\frac{d\omega}{2\pi}$$

and replacing the sum in equation 4.35 by the integral, we obtain

$$\lim_{T\to\infty}\sum_{n=-\infty}^{\infty}\frac{1}{T}|\mathbf{F}(j\omega)|^2_{\omega=n\omega_0}=\frac{1}{2\pi}\int_{-\infty}^{\infty}|\mathbf{F}(j\omega)|^2 d\omega. \tag{4.36}$$

or, finally

$$\int_{-\infty}^{\infty} f^2(t)dt=\frac{1}{2\pi}\int_{-\infty}^{\infty}|\mathbf{F}(j\omega)|^2 d\omega, \tag{4.37}$$

This equation is a very useful expression known as *Parseval's theorem*, which confirms the connection between the energy associated with $f(t)$ and its spectrum. In other words, equation 4.37 shows that *the energy of the signal can be calculated either by an integration over all the time of applying the signal in the time domain or by an integration over all the frequencies in the frequency domain.*

In accordance with this theorem, we are able to calculate the energy associated with any bandwidth of a given function by integrating $|\mathbf{F}(j\omega)|^2$ over an appropriate frequency interval, i.e., that portion of the total energy lying within the chosen interval, or *energy density* (J/Hz). In other words, the shape of $|\mathbf{F}(j\omega)|^2$ gives the "picture" of energy distribution in the spectrum of a non-periodic function, as is shown, for example, in Fig. 4.9. For instance, 90% of the total energy of a rectangular pulse is concentrated into the frequency interval from $\omega=0$ to $\omega=2\pi/d$. The narrower the pulse the wider the bandwidth interval, where most of the energy is concentrated.

In the physical world, we may find examples of this phenomenon. For instance, a lightning stroke, which is of very short duration, produces observable signal frequencies over the entire communication spectrum from the relatively low frequencies used in radio reception to the considerably higher ones used in television reception.

(*)The above equation is also known by the statement that the power delivered to the circuit by a non-sinusoidal function is equal to the algebraic sum of the powers of all the harmonics, which represent this function.

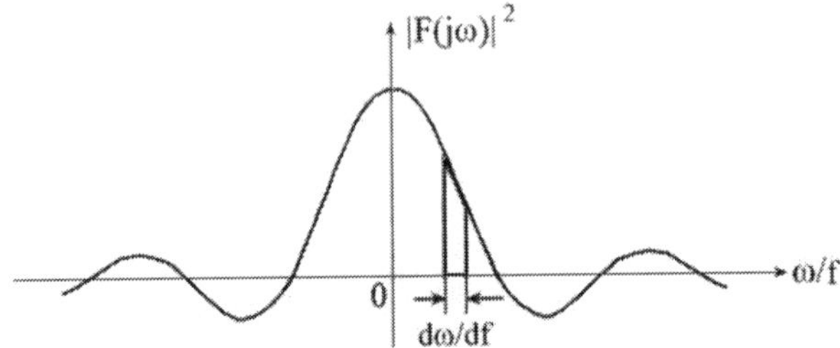

Figure 4.9 The amount of energy $|\mathbf{F}(j\omega t)|^2 d\omega$ associated with $f(t)$ lying in the bandwidth $d\omega$.

Example 4.3

As an example of using Parseval's theorem, let us assume that a 5 kV impulse of rectangular form, shown in Fig. 4.5(a), is applied to the input of an electrical circuit. Let us find the energy delivered to the circuit if $R_{in} = 1\ \Omega$ and the duration of the impulse $\tau = 2$ ms.

Solution

The Fourier transform of such an impulse in accordance with equation 4.20b is

$$\mathbf{F}(j\omega) = V_0\tau\ \mathrm{Sa}\left(\frac{\omega\tau}{2}\right) = V_0\tau\frac{\sin(\omega\tau/2)}{\omega\tau/2},$$

which in this case is pure a real function. Using Parseval's theorem, we have

$$W_{1\Omega} = \frac{1}{2\pi}\int_{-\infty}^{\infty}|F(j)|^2 d\omega = \frac{(V_0\tau)^2}{2\pi}\int_{-\infty}^{\infty}\left(\frac{\sin(\omega\tau/2)}{\omega\tau/2}\right)^2 d\omega.$$

By changing the variable $x = \omega\tau/2$ we have $d\omega = (2/\tau)dx$ and

$$W_{1\Omega} = \frac{V_0^2\tau}{\pi}\int_{-\infty}^{\infty}\left(\frac{\sin x}{x}\right)^2 dx = \frac{25\cdot 10^6\cdot 2\cdot 10^{-3}}{\pi}3.142 = 50\ \mathrm{kJ}^{(*)}.$$

The same might be calculated straightforwardly

$$W_{1\Omega} = \int_{-\infty}^{\infty} v^2(t)dt = 25\cdot 10^6\cdot t\Big|_{-\tau/2}^{\tau/2} = 25\cdot 10^6\cdot 2\cdot 10^{-3} = 50\ \mathrm{kJ}.$$

(*)The value of the integral in this expression can be calculated with computer programs like MATLAB, MATCAD or tables of integrals.

4.3.5 The comparison between Fourier and Laplace transforms (similarities and differences)

As was shown in section 3.2 the one-sided Laplace transform is a function of s:

$$\mathbf{F}(s) = \int_0^\infty f(t)e^{-st}dt, \tag{4.38}$$

where s is a complex argument with a real part c and an imaginary part ω, i.e., $s = c + j\omega$. It also was shown that the Laplace transform exists only if the integral in equation 4.38 converges, i.e., the function $f(t)$ is restricted:

$$|f(t)| < Me^{\alpha t} \quad \text{while} \quad \alpha < c < \infty. \tag{4.38a}$$

The Fourier transformation is defined over the entire time and not just for the positive values of time. However, in the circuit analysis, as was previously mentioned, the forcing functions and their responses are usually initiated at $t = 0$. Therefore, for such functions the Fourier transform (equation 4.13a) might be written as

$$\mathbf{F}(j\omega) = \int_0^\infty f(t)e^{-j\omega t}dt. \tag{4.39}$$

Comparing the above two equations 4.38 and 4.39, we may find that, by assuming in the Laplace transform (equation 4.38) that $c = 0$ and $s = j\omega$, both transforms are quite similar. However, the integral in equation 4.39 converges, if

$$|f(t)| < Me^{\alpha t} \quad \text{while} \quad \alpha < 0. \tag{4.39a}$$

This restriction is stronger than equation 4.38a, and means that the given function $f(t)$ does not exceed some exponentially decreasing functions. Some of the functions useful in circuit analysis do not meet this condition. For instance, functions such as unit functions, ramp functions, increasing exponential functions, and periodic functions belong to this category. For the function which does possess condition (equation 4.39a) we may find the Fourier transform by just replacing s by $j\omega$ in the Laplace transform, i.e.,

$$\mathbf{F}(j\omega) = \mathbf{F}(s)|_{s=j\omega}. \tag{4.40}$$

This way of finding the Fourier transform or function spectra for most of the non-periodic functions is the simplest and most convenient one, i.e., for this purpose we can simply use the Table of Laplace transform pairs (see Table 3.1).

The inverse Fourier transform (equation 4.13b) is also similar to the inverse Laplace transform

$$f(t) = \frac{1}{2\pi j}\int_{c-j\infty}^{c+j\infty} \mathbf{F}(s)e^{st}ds, \tag{4.41}$$

if we assume in equation 4.13b that $c = 0$ and $s = j\omega$, which means that the integration in equation 4.41 takes places on an imaginary axis. The restriction (equation 4.39a) also satisfies equation 4.38a, thus $\alpha < c$ in equation 4.38a also means $\alpha < 0$ in (4.39a), since $c = 0$. Therefore, for all the functions which meet condition of equation 4.39a, we may use all the rules for finding the inverse Fourier transform by applying those derived for the Laplace transform in Chap. 3. Finally we should note that, for functions that do not meet conditions of equation 4.39a, we still may calculate their Fourier transform, however not straightforwardly (see further on).

4.4 SOME PROPERTIES OF THE FOURIER TRANSFORM

Keeping in mind the similarity between Fourier and Laplace transformations, a brief account of the properties of the Fourier transform will be given here (the proof is similar to that given in section 3.4 for the Laplace transform).

(a) *Property of linearity*

If $f_1(t)$ and $f_2(t)$ have Fourier transforms $\mathbf{F}_1(j\omega)$ and $\mathbf{F}_2(j\omega)$ respectively, then

$$\mathscr{F}[f_1(t) \pm f_2(t)] = \mathbf{F}_1(j\omega) \pm \mathbf{F}_2(j\omega), \tag{4.42}$$

i.e., the Fourier transform of the sum (difference) of two (or more) time functions is equal to the sum (difference) of the transforms of the individual time functions and conversely:

$$\mathscr{F}^{-1}[\mathbf{F}_1(j\omega) \mp \mathbf{F}_2(j\omega)] = \mathscr{F}^{-1}[\mathbf{F}_1(j\omega)] \pm \mathscr{F}^{-1}[\mathbf{F}_2(j\omega)] = f_1(t) \pm f_2(t). \tag{4.43}$$

It is also obvious that for any constant K

$$Kf(t) \leftrightarrow K\mathbf{F}(j\omega). \tag{4.44}$$

The above properties are also known as *superposition* and *homogeneity* properties.

(b) *Differentiation properties*

Let us derive the transformation of the derivative of function $f(t)$. If $\mathbf{F}(j\omega)$ is the Fourier transform of $f(t)$, then

$$\mathscr{F}\left\{\frac{df}{dt}\right\} = \int_{-\infty}^{\infty} e^{-j\omega t}\frac{df}{dt}\,dt.$$

Its integration by parts, $u = e^{-j\omega t}$ and $dv = df$, gives

$$\mathscr{F}\left\{\frac{df}{dt}\right\} = f(t)e^{-j\omega t}\Big|_{-\infty}^{\infty} + j\omega\int_{-\infty}^{\infty} f(t)e^{-j\omega t}\,dt = j\omega\mathbf{F}(j\omega), \tag{4.45}$$

since the first term in this expression gives zero for both limits $t = \pm\infty$ (note that the given function, having a Fourier transform, must vanish to zero when

$|t| \to \infty$). Thus, differentiating a time-domain function corresponds to the multiplication of a frequency-domain function $\mathbf{F}(j\omega)$ by the factor $j\omega$. So we may write

$$\mathcal{F}\left\{\frac{df}{dt}\right\} = j\omega\mathbf{F}(j\omega). \tag{4.46}$$

This result may be readily extended to the general case for derivatives of order n

$$\mathcal{F}\left\{\frac{d^n f}{dt^n}\right\} = (j\omega)^n\mathbf{F}(j\omega). \tag{4.46a}$$

For the one-sided Fourier transform in which the first term in equation 4.45 turns into $f(0)$, so in this case, we will have

$$\mathcal{F}\left\{\frac{df}{dt}\right\} = j\omega\mathbf{F}(j\omega) - f(0), \tag{4.46b}$$

which is similar to the differentiation property of the Laplace transform (it is obvious due to the similarity between the one-sided Fourier transform and Laplace transform).

(c) *Integration properties*

Let $G(j\omega)$ be a spectrum of an integral $g(t) = \int_{-\infty}^{t} f(\tau)d\tau$. In accordance with the differentiation theorem, we may find that the Fourier transform of the function $f(t) = dg/dt$ will be

$$\mathbf{F}(j\omega) = j\omega G(j\omega).$$

Thus,

$$G(j\omega) = \frac{\mathbf{F}(j\omega)}{j\omega},$$

or

$$\mathcal{F}\left\{\int_{-\infty}^{t} f(\tau)d\tau\right\} = \frac{\mathbf{F}(j\omega)}{j\omega}, \tag{4.47}$$

i.e., the integration in the time domain corresponds to the division by $j\omega$ in the frequency domain. For the one-sided Fourier transform this result will not be changed, since the integral from $-\infty$ to 0 turns into zero and the lower limit in equation 4.47 will simply be zero. However, in order for the function $g(t)$ to be transformable, $g(\infty)$ must be equal to 0 (in other words, this requires that $g(\infty) = \int_{-\infty}^{\infty} f(\tau)d\tau = F(0) = 0$). If this condition is not satisfied, then the more general result is

$$\mathcal{F}\left\{\int_{-\infty}^{t} f(\tau)d\tau\right\} = \frac{\mathbf{F}(j\omega)}{j\omega} + \pi\mathbf{F}(0)\delta(\omega). \tag{4.47a}$$

(More explanations and examples of using this result can be seen further on in the following sections.)

(d) *Scaling properties*

Next let us consider one of the most interesting properties of the Fourier transformation – the effect of changing the time scale of a function, i.e. replacing argument t by a new one at, where a is some positive constant. If the given function is $f(t)$, the time-scaled function becomes $f(at)$. Taking the Fourier transformation of such a function, we have

$$\mathscr{F}\{f(at)\} = \int_{-\infty}^{\infty} f(at)e^{-j\omega t}\,dt. \tag{4.48}$$

By changing the variable $\lambda = at$ the differential dt becomes $d\lambda/a$ and substituting this in equation 4.48, we obtain

$$\mathscr{F}\{f(at)\} = \frac{1}{a}\int_{-\infty}^{\infty} f(\lambda)e^{-j(\omega/a)\lambda}\,d\lambda = \frac{1}{a}\mathbf{F}\left(j\frac{\omega}{a}\right). \tag{4.49}$$

From this relation we may conclude that the scaling of the variable t in the time domain results in a reciprocal scaling of the variable ω in the frequency domain. In addition, there is a scaling of the spectrum magnitude $\mathbf{F}(j\omega)$ by $1/a$.

Scaling properties of the Fourier transform provide a mathematical justification for the phenomenon described in the preceding sections that shortening the duration of a pulse, i.e., expressing it in a larger scale ($a > 1$) as $f(at)$, results in an a times wider spectrum $\mathbf{F}(j\omega/a)$ being expressed in a smaller scale ω/a. Thus, for instance, a pulse $f(t)$ which occurs from 0 to 1 s after scaling by $a = 5$, transforms to a pulse of the same form which will occur from 0 to 1/5 s (since $f(t_1) = f[a(t_1/a)]$ where t_1/a is a new time after scaling). The frequency spectrum $\mathbf{F}(j(\omega/5))$ will be five times wider because of the new frequency scale.

(e) *Shifting properties*

As another significant property of the Fourier transform, let us consider the effect of shifting, or delaying, in the time domain. That is, let us find the transform of $f(t-\tau)$ where τ is a shifting constant. By defining a new variable of integration $\lambda = t - \tau$ in equation 4.13b, we have

$$f(t-\tau) = \frac{1}{2\pi}\int_{-\infty}^{\infty} \mathbf{F}(j\omega)e^{j\omega(t-\tau)}\,d\omega = \frac{1}{2\pi}e^{-j\omega\tau}\int_{-\infty}^{\infty} \mathbf{F}(j\omega)e^{j\omega t}\,d\omega,$$

or

$$\mathscr{F}[f(t-\tau)] = e^{-j\omega t}\mathbf{F}(j\omega). \tag{4.50}$$

The physical meaning of this result is that a *delay* in the time domain (the function $f(t-\tau)$ is delayed τ seconds in respect to $f(t)$) corresponds to a phase shift by $-\omega\tau$ in the frequency domain.

(f) *Interchanging* t *and* ω *properties*

Finally, let us consider once again the property of interchanging t and ω in the Fourier transform pairs. In the discussion about the symmetrical properties (section 4.3.3), we have already considered such an interchanging. Now let us show that this property is general and can be applied to any function $f(t)$-symmetrical and non-symmetrical. To prove this statement, we first change the sign of ω in equation 4.13b and put the factor $1/2\pi$ inside the integral:

$$f(t)=\int_{-\infty}^{\infty}\frac{1}{2\pi}\mathbf{F}(-j\omega)e^{-j\omega t}d(-\omega)=\int_{-\infty}^{\infty}\frac{\mathbf{F}(-j\omega)}{2\pi}e^{-j\omega t}d\omega. \tag{4.51}$$

Secondly, we multiply both sides of equation 4.13a by $1/2\pi$ and change the sign of ω:

$$\frac{1}{2\pi}\mathbf{F}(-j\omega)=\frac{1}{2\pi}\int_{-\infty}^{\infty}f(t)e^{j\omega t}dt. \tag{4.52}$$

Now, by interchanging t and ω in equations 4.51 and 4.52, we have

$$f(\omega)=\int_{-\infty}^{\infty}\frac{\mathbf{F}(-jt)}{2\pi}e^{-j\omega t}dt, \tag{4.53a}$$

$$\frac{1}{2\pi}\mathbf{F}(-jt)=\frac{1}{2\pi}\int_{-\infty}^{\infty}f(\omega)e^{j\omega t}d\omega, \tag{4.53b}$$

or in short

$$\frac{1}{2\pi}\mathbf{F}(-jt)\leftrightarrow f(\omega). \tag{4.54}$$

Comparing this expression with equation 4.15 we may state that, if the time function $f(t)$ has as its spectrum the function $\mathbf{F}(j\omega)$, then the time function $\mathbf{F}(-jt)$ will have as its spectrum the function $f(\omega)$.

With the help of these properties, we can get a new set of transform pairs by simply using known ones. For instance by applying equation 4.54, with the results of Example 4.2 given in equation 4.20, we have

$$\frac{1}{2\pi}\frac{2V_0}{-jt}\sin\frac{-jt\tau}{2}\leftrightarrow\begin{cases}V_0 & |\omega|<\dfrac{\tau}{2}\\[2ex] 0 & |\omega|>\dfrac{\tau}{2},\end{cases}$$

or after taking $-j$ out of sine:

$$V_0\tau\, Sa\frac{t\tau}{2}\leftrightarrow\begin{cases}2\pi V_0 & |\omega|<\dfrac{\tau}{2}\\[2ex] 0 & |\omega|>\dfrac{\tau}{2},\end{cases}$$

which means that the rectangular pulse in the frequency domain represents a spectrum of a sinc function in the time domain as shown in Fig. 4.8.

Other properties of the Fourier transform may be readily derived in a way and manner used in connection with the Laplace transform due to the similarity between both transforms. The above discussed Fourier transform properties and some other important ones are summarized in Table 4.2.

Table 4.2 Fourier transform operations

	Operation	$f(t)$	$\mathbf{F}(j\omega)$
1	Addition	$\sum_{n=1}^{n} f_i(t)$	$\sum_{n=1}^{n} \mathbf{F}_i(j\omega)$
2	Scalar multiplication	$Kf(t)$	$K\mathbf{F}(j\omega)$
3	Time differentiation:		
	(a) two-sided transform	$\frac{d}{dt}f(t)$	$j\omega\mathbf{F}(j\omega)$
	(b) one-sided transform	$\frac{d}{dt}f(t)$	$j\omega\mathbf{F}(j\omega) - f(0)$
4	Time integration		
	(a) $\int_{-\infty}^{\infty} f(\tau)d\tau = 0$	$\int_{-\infty}^{t} f(\tau)d\tau$	$\frac{\mathbf{F}(j\omega)}{j\omega}$
	(b) $\int_{-\infty}^{\infty} f(\tau)d\tau \neq 0$	$\int_{-\infty}^{t} f(\tau)d\tau$	$\frac{\mathbf{F}(j\omega)}{j\omega} + \pi\mathbf{F}(0)\delta(\omega)$
5	Time-shift	$f(t \pm a)$	$e^{\pm j\omega a}\mathbf{F}(j\omega)$
6	Frequency-shift	$f(t)e^{\pm j\omega_0 t}$	$\mathbf{F}[f(\omega \pm \omega_0)]$
7	Time-scaling	$f(at)$	$\frac{1}{a}\mathbf{F}\left(j\frac{\omega}{a}\right)$
8	Frequency differentiation	$(-jt)f(t)$	$\frac{d}{dt}\mathbf{F}(j\omega)$
9	Frequency integration	$\frac{f(t)}{(-jt)}$	$\int_{-\infty}^{\infty} \mathbf{F}(j\omega)d\omega$
10	Convolution in time domain	$f_1(t) * f_2(t)$	$\mathbf{F}_1(j\omega)\mathbf{F}_2(j\omega)$
11	Multiplication in time domain		
	(a) by sine	$f(t)\sin\omega_1 t$	$\frac{1}{2j}(\mathbf{F}[j(\omega-\omega_1)] - \mathbf{F}[j(\omega+\omega_1)])$
	(b) by cosine	$f(t)\cos\omega_1 t$	$\frac{1}{2}(\mathbf{F}[j(\omega-\omega_1)] + \mathbf{F}[j(\omega+\omega_1)])$
12	Parseval's theorem	$\int_{-\infty}^{\infty} f^2(t)dt$	$\frac{1}{2\pi}\int_{-\infty}^{\infty} \lvert\mathbf{F}(j\omega)\rvert^2 d\omega$

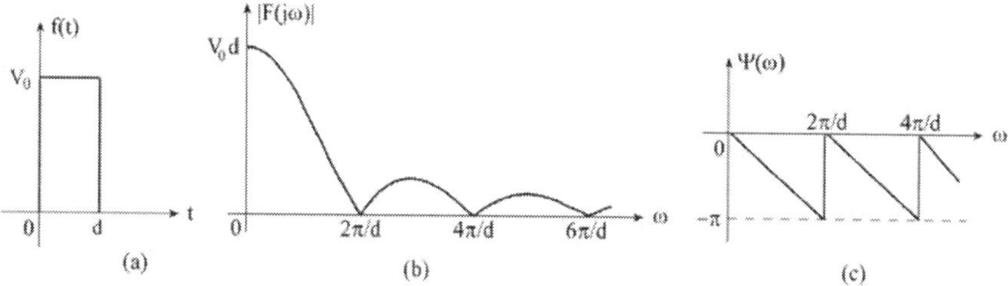

Figure 4.10 A pulse (a), its magnitude (b) and phase (c) spectra ($d = \tau$).

As an example of using Fourier transform properties let us derive the spectrum of the rectangular pulse shown in Fig. 4.10(a). This pulse is positioned at $0 < t < \tau$ and may be considered as shifting in respect to the pulse of Example 4.2 (note that $\tau = d$). Therefore, in order to obtain its spectrum we shall use the time shifting property. With the results of equation 4.20 in Example 4.2, and using equation 4.50, we have

$$\mathscr{F}\{f_{shift}(t)\} = \mathscr{F}\left\{F\left(t - \frac{\tau}{2}\right)\right\} = \frac{2V_0}{\omega}\sin\frac{\omega\tau}{2}e^{-j\frac{\omega\tau}{2}}$$

Thus, the magnitude spectrum (Fig. 4.10(b))

$$|\mathbf{F}(j\omega)| = V_0\tau\ \mathrm{Sa}\frac{\omega\tau}{2},$$

which is the same as in Example 4.2. The phase spectrum, however, will be

$$\Psi(\omega) = -\frac{\tau}{2}\omega,$$

which is declined lines changing from 0 to $-\pi$, as shown in Fig. 4.10(c), i.e., taking into consideration the sign of $\sin(\omega\tau/2)$, we have

$$\Psi(\omega) = -\frac{\tau}{2}\omega \qquad \text{for } 0 < \omega < \frac{2\pi}{\tau} \qquad \left(\sin\frac{\omega\tau}{2} > 0\right)$$

$$\Psi(\omega) = -\frac{\tau}{2}\omega + \pi \quad \text{for } \frac{2\pi}{\tau} < \omega < \frac{4\pi}{\tau} \quad \left(\sin\frac{\omega\tau}{2} < 0\right).$$

Our conclusion from this example is that time shifting does not influence the magnitude spectrum of the function, but changes its phase spectrum.

4.5 SOME IMPORTANT TRANSFORM PAIRS

For our future study of the Fourier transform technique, we shall develop the Fourier transform expression for those functions frequently used in circuit analysis. For this purpose we will do it either straightforwardly, using equations 4.13, or by applying the Fourier transform properties listed in Table 4.2.

4.5.1 Unit-impulse (delta) function

As we have already discussed in the previous chapter, the *unit-impulse* or *delta* function is defined as a time function which is zero when its argument is less or greater than zero and which is infinite when its argument is zero, while having a unit area, i.e.,

$$\delta(t-t_0)=0 \qquad t-t_0\neq 0\ (t\neq t_0) \tag{4.55a}$$

$$\int_{-\infty}^{\infty}\delta(t-t_0)=1 \quad t-t_0=0\ (t=t_0). \tag{4.55b}$$

If the switching operation occurs at $t=0$ (which always can be done by choosing $t_0=0$), we have

$$\delta(t)=0 \qquad t\neq 0 \tag{4.56a}$$

$$\int_{0_-}^{0_+}\delta(t-t_0)=1 \quad t=0. \tag{4.56b}$$

Multiplication of the delta function by a constant will not affect equation 4.55a and equation 4.56a, because the value of this function must still be zero when the argument is not zero and approaches infinity at $t=0$. However, this multiplication will change the integrals' value in equation 4.55b and equation 4.56b:

$$\int_{-\infty}^{\infty}A\delta(t)dt=\alpha. \tag{4.57}$$

This means that the area under the impulse is now equal to the multiplying factor, which is called the strength of the impulse. Following this rule we may interpret the multiplication of the delta function by any other function as follows:

$$\int_{-\infty}^{\infty}\mathrm{f(t)\delta(t)dt=f(0)},\quad \text{or} \tag{4.58a}$$

$$\int_{-\infty}^{\infty}f(t)\delta(t-t_0)dt=f(t_0). \tag{4.58b}$$

In this case, therefore, the strength of the impulse is the value of that function at the time for which the impulse argument is zero. For instance, the strength of the impulse multiplied by sine-function $f(t)\delta(t)=\sin(\omega t+60^\circ)\ \delta(0)$ is $\sqrt{3}/2$.

This property of a unit impulse function is sometimes called the **sampling property**. The graphical symbol for an impulse, used commonly, is an arrowhead line erected at the moment of the time when the impulse is applied (Fig. 4.11).

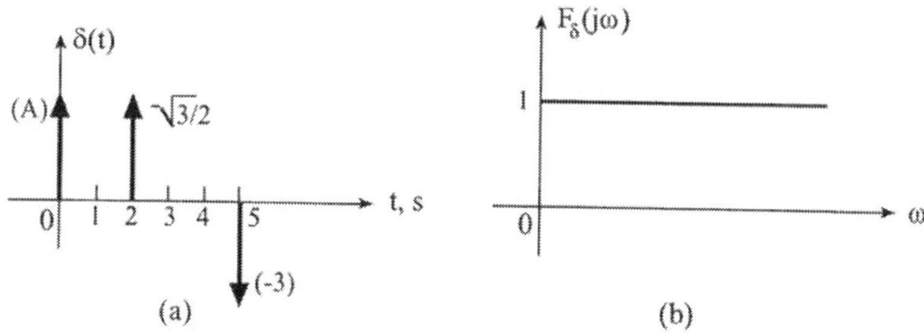

Figure 4.11 Positive and negative impulses of different strengths are plotted at the time of their appearances (a), a spectrum of impulse function (b).

The strength of the impulse is usually indicated by adjusting the arrow, as shown in Fig. 4.11(a).

Now bearing in mind the above properties of the impulse function and using the equation for finding the Fourier transform, we obtain

$$\mathscr{F}\{\delta(t-t_0)\} = \int_{-\infty}^{\infty} e^{-j\omega t}\delta(t-t_0) = e^{-j\omega t_0} \tag{4.59a}$$

and

$$\mathscr{F}\{\delta(t)\} = \int_{-\infty}^{\infty} e^{-j\omega t}\delta(t) = e^{-j\omega t}\Big|_{t=0} = 1, \tag{4.59b}$$

or

$$\mathbf{F}_{\delta}(j\omega) = 1. \tag{4.60}$$

This function is shown in Fig. 4.11(b) as the straight line of a unit magnitude. Note that the spectrum of the impulse function is infinite, since it goes to infinity. The result of equation 4.59a may also be written as

$$\mathscr{F}\{\delta(t-t_0)\} = e^{-j\omega t_0} = \cos\omega t_0 - j\sin\omega t_0. \tag{4.61}$$

Therefore, the energy density of a delta function is unity:

$$|\mathscr{F}\{\delta(t-t_0)\}|^2 = \cos^2\omega t_0 + \sin^2\omega t_0 = 1. \tag{4.62}$$

This result states that the energy (released in a unit input resistance) per unit bandwidth is unity at all frequencies. Since the impulse function has an infinite bandwidth, the total energy in the unit impulse is infinitely large (note that a unit impulse function is only a mathematical model of real pulse source functions which are, of course, bound).

In order to find the reverse Fourier transform of a unit impulse spectrum, we shall use the property of the Fourier transform which states that there is a unique one-to-one correspondence between a time function and its Fourier transform. Therefore, we can say that the inverse Fourier transform of $e^{-j\omega t_0}$

is $\delta(t-t_0)$, thus

$$\mathcal{F}\{e^{-j\omega t_0}\} = \frac{1}{2\pi}\int_{-\infty}^{\infty} e^{-j\omega t_0} e^{j\omega t} d\omega = \delta(t-t_0), \tag{4.63}$$

or in the symbolic way:

$$\delta(t-t_0) \leftrightarrow e^{-j\omega t_0}. \tag{4.64}$$

Next, by using the property of interchanging arguments t and ω in Fourier pairs, we may readily obtain from equation 4.64.

$$e^{j\omega_0 t} \leftrightarrow 2\pi\delta(\omega-\omega_0), \tag{4.65a}$$

which might be interpreted as a Fourier pair for a unit impulse in the frequency domain located at $\omega=\omega_0$. By changing the sign of the pulse location ω_0 to $-\omega_0$, we obtain

$$e^{-j\omega_0 t} \leftrightarrow 2\pi\delta(\omega+\omega_0). \tag{4.65b}$$

By letting $\omega_0 = 0$ we obtain

$$1 \leftrightarrow 2\pi\delta(\omega), \tag{4.66a}$$

from which it follows that

$$K \leftrightarrow 2\pi K\delta(\omega). \tag{4.66b}$$

Thus, the frequency spectrum of a constant K function in the time domain is a $2\pi K$ strength impulse in the frequency domain. An interpretation of this result is that a d.c. voltage or current forcing function, whose frequency is considered as zero, i.e., $\omega_0 = 0$, has its Fourier transform in accordance with equation 4.66.

Although the time functions in equation 4.65 are complex functions of time, which are not appropriate in the existing world of reality, with their help we can obtain in a very simple way the frequency spectra of such important functions as sine and cosine. Thus,

$$\mathcal{F}\{\cos\omega_0 t\} = \mathcal{F}\left\{\frac{1}{2}(e^{j\omega t}+e^{-j\omega t})\right\} = \pi\delta(\omega-\omega_0)+\pi\delta(\omega+\omega_0), \tag{4.67}$$

or

$$\cos\omega_0 t \leftrightarrow \pi[\delta(\omega-\omega_0)+\delta(\omega+\omega_0)], \tag{4.68a}$$

and similarly

$$\sin\omega_0 t \leftrightarrow j\pi[\delta(\omega+\omega_0)-\delta(\omega-\omega_0)]. \tag{4.68b}$$

The above expressions indicate that the *frequency spectra of sinusoidal functions* are given as a pair of impulses, located at $\omega = \pm\omega_0$.

This result actually corresponds to the representation of a sinusoidal function by imaginary frequencies $s \pm j\omega_0$, which was used in our previous study of circuit analysis for instance, in the symbolic or complex method.

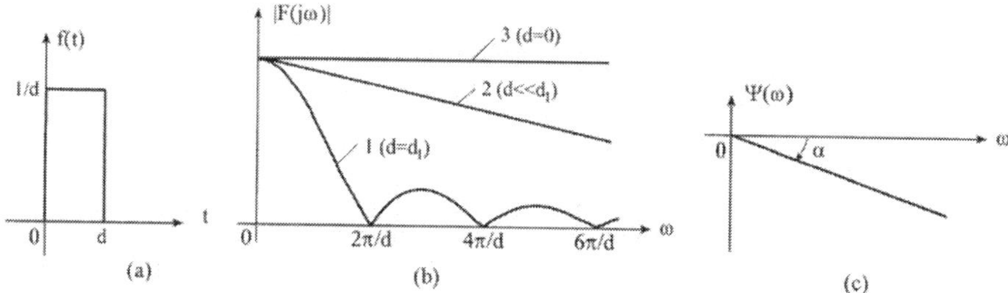

Figure 4.12 A short pulse (a) and its amplitude (b) and phase (c) spectra.

Example 4.4

Consider once again the rectangular pulse shown in Fig. 4.12(a). This pulse of a unit area (since $d(1/d) = 1$) approaches a unit impulse when $d \to 0$. In accordance with the result of Example 4.2 (see equation 4.20b) its spectrum is

$$\mathbf{F}(j\omega) = \frac{2}{\omega d} \sin \frac{\omega d}{2}. \tag{4.69}$$

By approaching $d \to 0$ in equation 4.69, we will obtain the Fourier transform of the unit impulse:

$$\mathbf{F}_\delta(j\omega) = \lim_{d \to 0} \frac{\sin(\omega d/2)}{\omega d/2} = 1. \tag{4.70}$$

Figure 4.12(b) shows the transformation of the spectrum (equation 4.69) into the spectrum (equation 4.70). The zero points of the spectrum, given for a sinc function (1) at $k(2\pi/d)$ $(k = 1, 2, \ldots)$, move to the right along the frequency axis to higher frequencies (2) so that for $d = 0$, the whole spectrum approaches a straight line (3). Note that the phase spectrum of the impulse function, applied at $t_0 = 0$, is zero. However, the phase spectrum of the impulse, applied at the time t_0, will be in accordance with equation 4.64, $\mathbf{F}(j\omega) = e^{-j\omega t_0}$, which gives

$$\Psi(\omega) = -t_0 \omega. \tag{4.71}$$

Graphically it is a straight line having an angle of declination $\alpha \propto \tan^{-1}(-t_0)$ as shown in Fig. 4.12(c).

4.5.2 Unit-step function

Our next consideration will be the unit-step function $u(t)$. In the previous chapter we introduced this function, which usually indicates a switching or failure action. It is defined (Fig. 4.13) as

$$u(t) = \begin{cases} 0 & t < 0 \\ 1 & t > 0, \end{cases} \tag{4.72a}$$

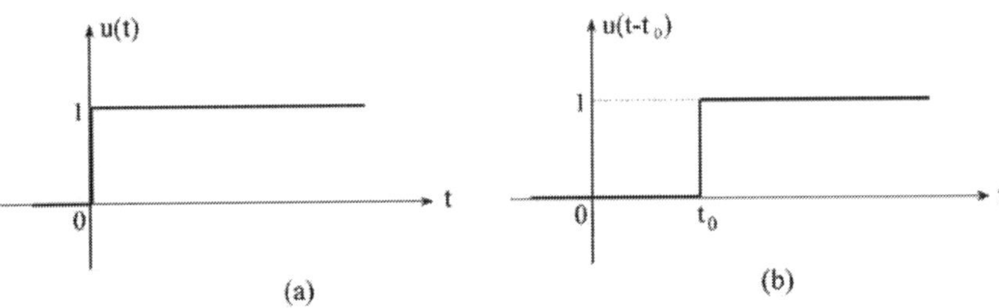

Figure 4.13 A unit-step function: at $t=0$ (a) and at $t=t_0$ (b).

or

$$u(t-t_0)=\begin{cases}0 & t<t_0\\ 1 & t>t_0.\end{cases} \tag{4.72b}$$

Thus, the unit-step function is zero for all values of its argument (time) which are less than zero ($t<0$) or less than $t_0(t>t_0)$ (note that in both cases the argument (time) is just negative), and is unity for all positive values of its argument ($t>0$) or ($t>t_0$). In order to find the Fourier transform of the unit-step function, we must indicate that this function is the kind of function whose transform cannot be obtained straightforwardly. This happens because the integral in equation 4.13 is unbound, which means that the unit-step function does not approach zero as t approaches infinity. One common way of achieving the Fourier transform of the unit-step function is by representing it as a sum of a constant and a *signum function* (Fig. 4.14):

$$u(t)=\frac{1}{2}[1+\mathrm{sgn}(t)]=\frac{1}{2}+\frac{1}{2}\mathrm{sgn}(t). \tag{4.73}$$

In accordance with equation 4.66 the transform of the first member in equation 4.73 will be $\pi\delta(\omega)$. As is known the second member in equation 4.73, a **signum**

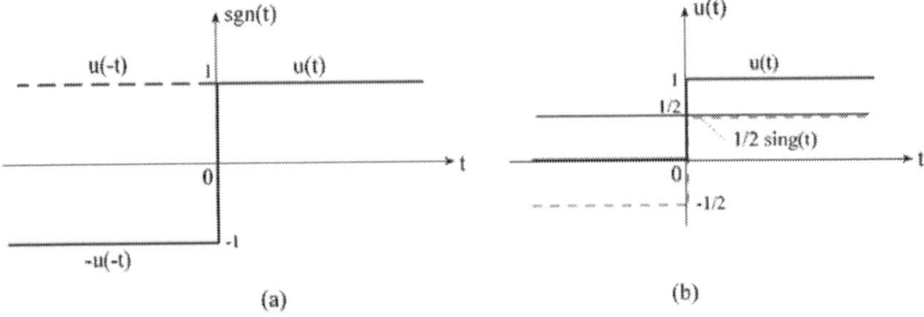

Figure 4.14 A signum function (a) and a representation of a unit function by the sum of a constant and a signum function (b).

function, is defined as

$$\mathrm{sgn}(t)\begin{cases} -1 & t<0 \\ 1 & t>0. \end{cases} \tag{4.74a}$$

or

$$\mathrm{sgn}(t) = u(t) - u(-t). \tag{4.74b}$$

The signum function can also be written as

$$\mathrm{sgn}(t) = \lim_{c\to 0}\,[e^{-ct}u(t) - e^{ct}u(-t)].$$

Factor $e^{\pm ct}$ is used here (as a convergence factor) to insure the approaching of unit step zero, as t gets very large (i.e., when $t\to\infty$). On the other hand, by approaching $c\to 0$, we are getting back to the originally given signum function. Using the definition of the Fourier transform, we obtain

$$\begin{aligned}\mathscr{F}\{\mathrm{sgn}(t)\} &= \lim_{c\to 0}\left[\int_0^{\infty} e^{-ct}e^{-j\omega t}\,dt - \int_{-\infty}^{0} e^{ct}e^{j\omega t}\,dt\right] \\ &= \lim_{c\to 0}\left(\frac{-e^{-ct}}{c+j\omega}\bigg|_0^{\infty} - \frac{-e^{ct}}{c-j\omega}\bigg|_{-\infty}^{0}\right) = \lim_{c\to 0}\frac{-j2\omega}{c^2+\omega^2} = \frac{2}{j\omega}.\end{aligned}$$

Thus,

$$\mathrm{sgn}(t)\leftrightarrow\frac{2}{j\omega}, \tag{4.75}$$

and

$$\mathscr{F}\{u(t)\} = \mathscr{F}\left\{\frac{1}{2}\right\} + \mathscr{F}\left\{\frac{1}{2}\,\mathrm{sgn}(t)\right\} = \pi\delta(\omega) + \frac{1}{j\omega},$$

or

$$u(t)\leftrightarrow\pi\delta(\omega) + \frac{1}{j\omega}. \tag{4.76}$$

The first term represents an impulse, in the frequency domain, of strength π occurring at $\omega = 0$. The second term is the same as the Laplace transform of a unit-step function in which s has been replaced by $j\omega$.

The magnitude and phase spectra of the unit-step function are shown in Fig. 4.15. Note that the magnitude spectrum of a unit-step contains the harmonics of all the frequencies, however the energy density at the low frequency harmonics is much higher. When $\omega\to 0$ the magnitude spectrum and its energy approach infinity. In general, any unbound signal is characterized by an infinite amount of energy.

Sudden spouts of d.c. or a.c. currents of industrial frequency (for instance by starting motors or short-circuiting) are similar to a unit-step function with a

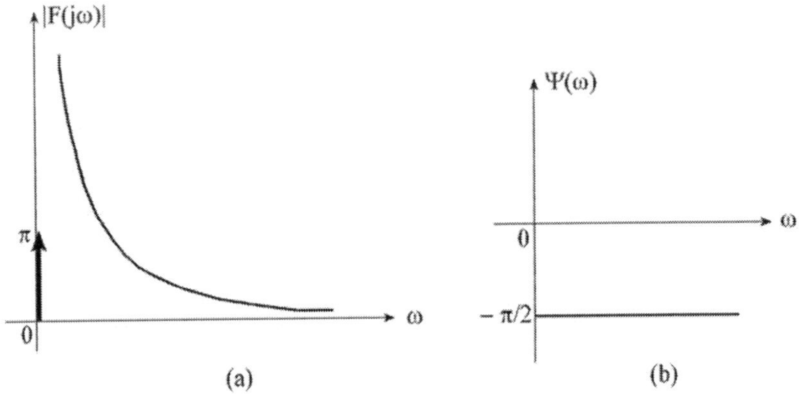

Figure 4.15 Magnitude (a) and phase (b) spectra of a unit-step function.

high energy density at low frequencies. This is the reason that most interference occurs on low-frequency radio broadcasts (long waves) and are almost invisible on high frequencies (short waves).

4.5.3 Decreasing sinusoid

Such a sine function is defined as

$$f(t) = e^{-at} \sin \omega_0 t \, u(t),$$

and its Fourier transform might be found as

$$\mathbf{F}(j\omega) = \int_0^{\infty} e^{-at} \sin \omega_0 t e^{-j\omega t} dt = \frac{\omega_0}{(a + j\omega)^2 + \omega_0^2}. \tag{4.77}$$

This results in magnitude spectrum

$$|\mathbf{F}(j\omega)| = \frac{\omega_0}{\sqrt{(a^2 + \omega_0^2 - \omega^2)^2 + 4a^2\omega^2}}, \tag{4.78}$$

and phase spectrum

$$\Psi(\omega) = -\tan^{-1} \frac{2a\omega}{a^2 + \omega_0^2 - \omega^2}. \tag{4.79}$$

The curves of $|\mathbf{F}(\mathbf{j}\omega)|$ and $\Psi_{(\omega)}$ are shown in Fig. 4.16, where $\omega_{(\max)} = \sqrt{\omega^2 + a^2}$.

4.5.4 Saw-tooth unit pulse

We can use the differentiation property to find the Fourier transform avoiding the straightforward integration of 4.13(b), which is in many cases extremely

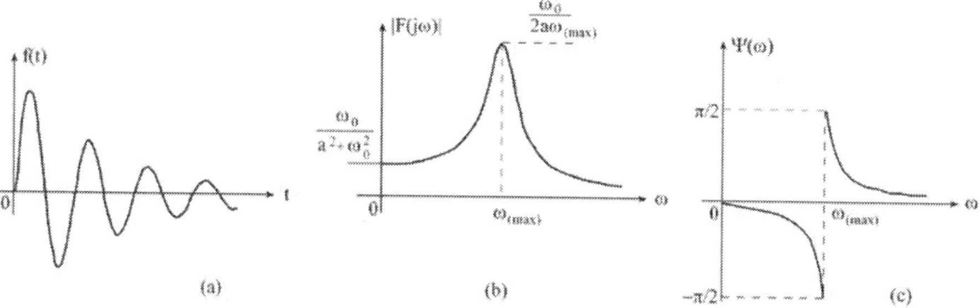

Figure 4.16 A decreasing sinusoidal function (a) and its magnitude (b) and phase (c) spectra.

difficult. Let us illustrate this method on the saw-tooth pulse, Fig. 4.17, where $\mathbf{F}(j\omega)$ represents the unknown spectrum of this pulse. After a single differentiation the saw-tooth pulse (a) takes the form (b). Now we add an equal and opposite impulse to the signal in (b) to cancel the appearing one. The result is the rectangular pulse remaining in (c). The second integration gives two impulses in (d), whose transform can be easily found, as $(1/a)(1-e^{-j\omega a})$. Hence, by

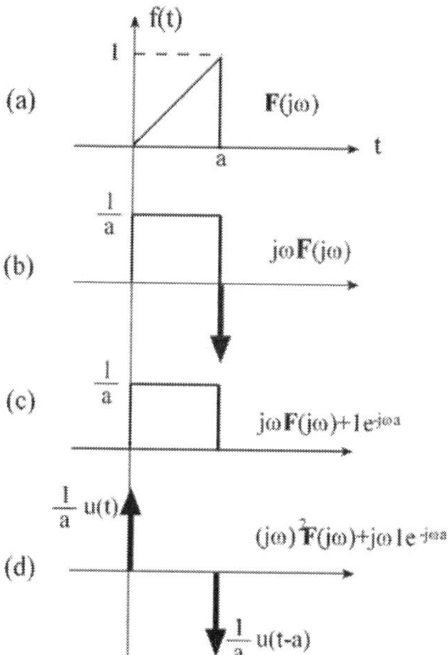

Figure 4.17 A unit saw-tooth pulse (a), its first differentiation (b), after adding a unit impulse (c) and after the second differentiation (d).

equaling:

$$\frac{1}{a}(1 - e^{-j\omega a}) = (j\omega)^2 \mathbf{F}(\omega) + j\omega e^{-j\omega a},$$

we obtain

$$\mathbf{F}(\omega) = \frac{-1 + (1 + j\omega)e^{-j\omega a}}{a\omega^2}.$$

This method, actually, is generalized because of the fact that any signal may be approximated as a piecewise-linear, in which case the signal reduces to impulses after two (or three) differentiations.

4.5.5 The Fourier transform of a periodic time function

Here we face the same problem, which we had in section 4.5.2 looking for the Fourier transform of a unit-step function. Any periodic function is, obviously, unbound, since it does not approach zero, as t approaches infinity. In order to obtain the Fourier transform of a periodic function we should distinguish between two cases: *two-sided* and *one-sided transforms*. The two-sided Fourier transform of a sinusoidal function, as shown in Fig. 4.17(a), has already been found in section 4.5.1 (see equation 4.68)].

However, in circuit analysis, the most frequently used forcing periodic functions are sinusoidal functions applied at $t = 0$, shown in Fig. 4.18(b). In this case, we can define such a function as

$$f(t) = \sin \omega_0 t \; u(t). \tag{4.80}$$

Using the Fourier transform property of multiplication by sine/cosine in the time domain (see entry 11 in Table 4.2) we have

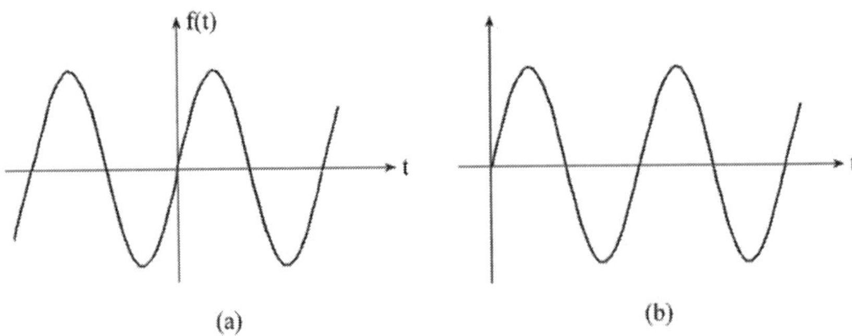

Figure 4.18 Sinusoidal function: (a) for both sides of the t-axis and (b) for only the positive side of the t-axis ($t > 0$).

$$F\{u(t)\sin\omega_0 t\} = \frac{1}{2j}(F_u[j(\omega-\omega_0)] - F_u[j(\omega+\omega_0)])$$

$$= \frac{1}{2j}\left(\pi\delta(\omega-\omega_0) + \frac{1}{j(\omega-\omega_0)} - \pi\delta(\omega+\omega_0) - \frac{1}{j(\omega+\omega_0)}\right)$$

$$= \frac{\pi}{2j}[\delta(\omega-\omega_0) - \delta(\omega+\omega_0)] + \frac{\omega_0}{\omega_0^2-\omega^2}.$$

Thus

$$\sin\omega_0 t u(t) \leftrightarrow \frac{j\pi}{2}[\delta(\omega+\omega_0) - \delta(\omega-\omega_0)] + \frac{\omega_0}{\omega_0^2-\omega^2}. \tag{4.81}$$

Note that the second member on the right side of equation 4.81 might be readily obtained from the Laplace transform of the sinusoid by replacing s by $j\omega$:

$$\mathscr{F}\{\sin\omega_0 t\} = \mathscr{L}\{\sin\omega_0 t\}_{s=j\omega} = \left.\frac{\omega_0}{s^2+\omega_0^2}\right|_{s=j\omega} = \frac{\omega_0}{\omega_0^2-\omega^2} \tag{4.82}$$

The first member on the right side of equation 4.81 represents the switching property of the unit-step function.

Table 4.3 gives the Fourier transform pairs for most of the familiar time functions encountered in circuit analysis. They may be used for finding the inverse transform of frequency domain functions, as was done for the Laplace transform method.

4.6 CONVOLUTION INTEGRAL IN THE TIME DOMAIN AND ITS FOURIER TRANSFORM

In the circuit analysis technique, applying Fourier transforms, the multiplication of two transforms (namely, the transform of the forcing function and the system function) is frequently used to obtain the transform of the response function. The inverse-transform operation must be performed to obtain the response function. In a similar way, as was shown in the previous chapter (with respect to the Laplace transform), we may state that the inverse transform of the product of two Fourier transforms is the *convolution integral*, i.e.,

$$f_{res}(t) = \mathscr{F}^{-1}\{\mathbf{F}_1(j\omega)\mathbf{F}_2(j\omega)\} = f_1(t)*f_2(t), \tag{4.85a}$$

where

$$f_1(t)*f_2(t) = \int_{-\infty}^{\infty} f_1(\tau)f_2(t-\tau)d\tau = \int_{-\infty}^{\infty} f_1(t-\tau)f_2(\tau)d\tau. \tag{4.85b}$$

Two integrals given by equation 4.85b are the two forms of the convolution integral in a very general form. By using equation 4.85 we shall take into consideration the physically realizable properties of electrical systems. Thus, the

Table 4.3 Fourier transform pairs

	$f(t)$	$\mathbf{F}(j\omega)$
1	$\delta(t)$	1
2	$\delta(t-t_0)$	$e^{-j\omega t_0}$
3	1	$2\pi\delta(\omega)$
4	$u(t)$	$\pi\delta(\omega)+\dfrac{1}{j\omega}$
5	$\text{sgn}(t)$	$\dfrac{2}{j\omega}$
6	$e^{j\omega_0 t}$	$2\pi\delta(\omega-\omega_0)$
7	$e^{-at}u(t)$	$\dfrac{1}{a+j\omega}$
8	$te^{-at}u(t)$	$\dfrac{1}{(a+j\omega)^2}$
9	$\sin\omega_0 t$	$j\pi[\delta(\omega+\omega_0)-\delta(\omega-\omega_0)]$
10	$\cos\omega_0 t$	$\pi[\delta(\omega+\omega_0)+\delta(\omega-\omega_0)]$
11	$\sin\omega_0 t\, u(t)$	$\dfrac{j\pi}{2}[\delta(\omega+\omega_0)-\delta(\omega-\omega_0)]+\dfrac{\omega_0}{\omega_0^2-\omega^2}$
12	$\cos\omega_0 t\, u(t)$	$\pi[\delta(\omega+\omega_0)+\delta(\omega-\omega_0)]+\dfrac{j\omega_0}{\omega_0^2-\omega^2}$
13	$e^{-at}\sin\omega_0 t\, u(t)$	$\dfrac{\omega_0}{(a+j\omega)^2+\omega_0^2}$
14	$e^{-at}\cos\omega_0 t\, u(t)$	$\dfrac{a+\omega_0}{(a+j\omega)^2+\omega_0^2}$
15	$u(t+\tau/2)-u(t-\tau/2)$	$\tau\dfrac{\sin\omega\tau/2}{\omega\tau/2}$

response of the system cannot begin before the forcing function is applied. Let us say that $f_2(t)\equiv h(t)$ is the response of the system, usually resulting from the application of a unit impulse at $t=0$; (see section 3.6). Therefore, $h(t)$ cannot exist for $t<0$ which means that in the second integral of equation 4.85b the integrand is zero when $\tau<0$ and the low limit of integration may be changed and the response function is

$$f_{res}=f_1(t)*f_2(t)=\int_0^{\infty} f_1(t-\tau)h(\tau)d\tau. \tag{4.86a}$$

For the same reason, in the first integral of equation 4.85b, $f_2(t-\tau)\equiv h(t-\tau)$ cannot exist for $t<\tau$, which means that the integrand is zero when $t-\tau$ is negative. The upper limit in this integral, therefore, may be changed and the response function is

$$f_{res} = f_1(t) * f_2(t) = \int_{-\infty}^{t} f_1(\tau)h(t-\tau)d\tau. \tag{4.86b}$$

Before continuing our discussion of applying the Fourier transformation method in circuit analysis, let us consider an example of using the convolution integral.

Example 4.5

Using the convolution integral, find the output voltage $v_o(t)$ in the series RL circuit, if the input $v_i(t)$ is a rectangular voltage pulse of 6 V in amplitude that starts at $t = 0$ and has a duration of 1 s (Fig. 4.19(a)). Assume that $L = 5$ H and $R = 4\ \Omega$.

Mathematically the input voltage may be written as $v_i(t) = u(t) - u(t-1)$. The impulse response $h(t)$ for the given circuit (Fig. 4.19(a)) might be evaluated as follows. Using the phasor method for analyzing circuits in the frequency domain or the so-called symbolic method (see further on), we obtain

$$H(j\omega) = V_{o,\delta}(j\omega) = V_{i,\delta}\frac{R}{R + j\omega L} = 1\frac{4}{4 + j\omega 5} = 0.8\frac{1}{0.8 + j\omega},$$

or, with entry 7 in Table 4.3,

$$h(t) = 0.8e^{-0.8t}u(t).$$

Now, applying the convolution integral yields

$$v_o(t) = v_i(t) * h(t) = \int_{-\infty}^{\infty} 6[u(t-\tau) - u(t-\tau-1)][0.8e^{-0.8\tau}u(\tau)]\,d\tau \tag{4.87}$$

or separating equation 4.87 into two integrals, we have

$$v_0(t) = 4.8\int_{-\infty}^{\infty} u(t-\tau)e^{-0.8\tau}u(\tau)d\tau - 4.8\int_{-\infty}^{\infty} u(t-1-\tau)e^{-0.8\tau}u(\tau)d\tau.$$

The first integral should be taken in the limits from 0 to t and the second in

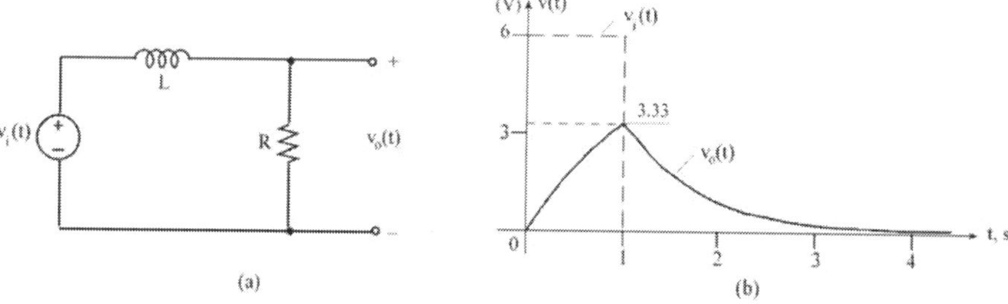

Figure 4.19 The given circuit (a) and the input $v_i(t)$ and output $v_o(t)$ voltages (b).

the limits from t to $t-1$. Thus,

$$v_0(t)=\begin{cases} 4.8\displaystyle\int_0^t e^{-0.8\tau}d\tau = 6(1-e^{-0.8t}) & 0<t<1 \\ -4.8\displaystyle\int_t^{t-1} e^{-0.8\tau}d\tau = 6(e^{0.8}-1)e^{-0.8t} = 7.35e^{-0.8t} & 1>t>\infty. \end{cases}$$

This function is shown in Fig. 4.19(b).

4.7 CIRCUIT ANALYSIS WITH THE FOURIER TRANSFORM

As we already know, the Fourier transform extends the Fourier series to a non-periodic function transforming the discrete spectra into continuous ones. Therefore, we can state that the Fourier transform represents the non-periodic function as an infinite sum of the harmonics, i.e. periodic functions possessing vanishingly small amplitudes. Therefore, we may apply the phasor concept and symbolic (complex) method used for steady-state analysis of the circuits driven by sinusoidal forcing functions.

Thus, considering the general circuit of Fig. 4.20 in the time domain we will obtain a differential equation, which describes the relation between the input (forcing) voltage $v_i(t)$ and the output (response) voltage $v_o(t)$:

$$a_0v_0(t)+a_1\frac{dv_o(t)}{dt}+a_2\frac{d^2v_o(t)}{dt^2}+\cdots=b_0v_i(t)+b_1\frac{dv_i(t)}{dt}+b_2\frac{d^2v_i(t)}{dt^2}+\cdots. \tag{4.88}$$

Taking the Fourier transform of both sides of equation 4.88 and using the differentiation and linearity properties, yields:

$$[a_0+a_1(j\omega)+a_2(j\omega)^2+\cdots]\mathbf{V}_o(j\omega)=[b_0+b_1(j\omega)+b_2(j\omega)^2+\cdots]\mathbf{V}_i(j\omega), \tag{4.89}$$

where $\mathbf{V}_i(j\omega)$ and $\mathbf{V}_o(j\omega)$ are the Fourier transforms of the input and output functions $v_i(t)$ and $v_o(t)$. From this result we may write

$$\mathbf{H}(j\omega)=\frac{\mathbf{V}_o(j\omega)}{\mathbf{V}_i(j\omega)}=\frac{b_0+b_1(j\omega)+b_2(j\omega)^2+\cdots}{a_0+a_1(j\omega)+a_2(j\omega)^2+\cdots} \tag{4.90}$$

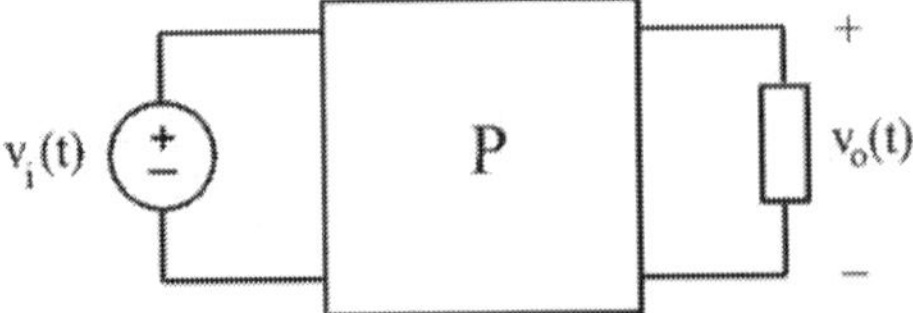

Figure 4.20 General passive circuit.

where $\mathbf{H}(j\omega)$ is identified as a network or system function (usually as an impulse response).

This is exactly the same result as would be obtained by the application of the phasor method and analyzing the circuit of Fig. 4.20 in the frequency domain. Note also that the same result could be achieved with the Laplace transform simply by replacing s by $j\omega$, in an expression like equation 3.58 in the previous chapter. Therefore, the above conclusion allows us to apply all the methods based on the phasor concept, using the impedances $\mathbf{Z}(j\omega)$ and admittances $\mathbf{Y}(j\omega)$ for finding the quantity $\mathbf{H}(j\omega)$ and solving other problems which relate to the Fourier transform. The only difference is that here the forcing functions, inputs, and the response functions, outputs, are Fourier transforms rather than phasors. This means that in the time domain the forcing functions and the responses are arbitrary (any) non-periodic functions rather than sinusoidal functions. In conclusion, just as the use of phasor (symbolic) transforms simplified the determination of the steady-state sinusoidal response, the use of Fourier transforms of various forcing functions can simplify the determination of the complete response of both the natural and forced components. The reason for this is quite simple: in both techniques, the differentiation in the time domain is represented in the frequency domain by multiplication by the factor $j\omega$; and similarly integration is related to division by the factor $j\omega$. By these means, relatively complicated differential and/or integral expressions are reduced to a relatively simple algebraic function of ω.

The next step in Fourier transform analysis is to find the time-domain description of the response transform for which we must evaluate an inverse Fourier transform technique. Some of the methods of this procedure will be developed in the following chapters. With the above remarks in mind, let us now consider some specific analysis problems.

Example 4.6

Let the decreasing exponential voltage $v_{in}(t) = e^{-5t}u(t)$ be applied to a given circuit and be related to the output voltage $v_o(t)$ by the equation

$$\frac{dv_o}{dt} + 3v_o = 3v_{in}.$$

Transforming the equation into the frequency domain using the Fourier technique, we have

$$(j\omega + 3)\mathbf{V}_o(j\omega) = 3\mathbf{V}_{in}(j\omega),$$

and the transfer function is

$$\mathbf{H}(j\omega) = \frac{\mathbf{V}_o(j\omega)}{\mathbf{V}_{in}(j\omega)} = \frac{3}{3 + j\omega}.$$

According to Table 4.3 the Fourier transform of the applied function will be

$$\mathbf{V}_{in}(j\omega) = \frac{1}{5 + j\omega}.$$

Therefore, the transform of the output voltage is

$$\mathbf{V}_o(j\omega) = \mathbf{H}(j\omega)\mathbf{V}_{in}(j\omega) = \frac{3}{(3+j\omega)(5+j\omega)}.$$

By partial fraction expansion (see section 3.7) we obtain

$$\mathbf{V}_o(j\omega) = 4.5\frac{1}{3+j\omega} - 4.5\frac{1}{5+j\omega}.$$

By using the linearity property of the Fourier transform and the table of Fourier transform pairs, we have

$$v_o(t) = 4.5(e^{-3t} - e^{-5t})u(t).$$

4.7.1 Ohm's and Kirchhoff's laws with the Fourier transform

Supposing that $\mathbf{V}_{in}(j\omega)$ is the Fourier transform of voltage $v_{in}(t)$, applied to the one-port circuit having the impedance $Z(j\omega)$, we may find the Fourier transform of the input current as

$$I_{in}(j\omega) = \frac{\mathbf{V}_{in}(j\omega)}{Z(j\omega)} = Y(j\omega)\mathbf{V}_{in}(j\omega). \tag{4.91}$$

This expression may be presented as Ohm's law in Fourier transform form.

For two-port circuits the input/output quantities might be found if the spectral characteristics of the transform coefficient $K(j\omega)$ or the transfer admittance/impedance $Y_{21}(j\omega)/Z_{21}(j\omega)$ are known.

Then the transform of the output voltage will be

$$\mathbf{V}_2(j\omega) = K_{21}(j\omega)\mathbf{V}_1(j\omega), \tag{4.92a}$$

or the output current will be

$$\mathbf{I}_2(j\omega) = Y_{21}(j\omega)\mathbf{V}_1(j\omega). \tag{4.92b}$$

Note that these two expressions are similar to Ohm's law in Fourier transform form.

In a similar way, using the phasor method, Kirchhoff's laws' equations can be written and analyzed.

4.7.2 Inversion of the Fourier transform using the residues of complex functions

The inverse Fourier transform for the above expressions can be found with the help of the residues of complex functions. Thus, if the Fourier transform of the given function is of the form (like in 4.91) we have

$$\frac{\mathbf{F}_1(j\omega)}{\mathbf{F}_2(j\omega)} \leftrightarrow \sum_{k=1}^{n} \frac{\mathbf{F}_1(j\omega_k)e^{j\omega_k t}}{\left[\frac{d}{d_j\omega}\mathbf{F}_2(j\omega)\right]_{\omega=\omega_k}} = j\sum_{k=1}^{n} \frac{\mathbf{F}_1(j\omega_k)e^{j\omega_k t}}{\left[\frac{d}{d\omega}\mathbf{F}_2(j\omega)\right]_{\omega=\omega_k}}, \tag{4.93a}$$

where ω_k are the roots of the equation $F_2(j\omega)=0$. If the expression in the denominator is of the form $\mathbf{F}_2(j\omega)=j\omega\mathbf{F}_3(j\omega)$, which means that $\mathbf{F}_2(j\omega)$ has a zero root $\omega_0=0$, then the inverse Fourier transform will be

$$\frac{\mathbf{F}_1(j\omega)}{j\omega\mathbf{F}_3(j\omega)} \leftrightarrow \frac{\mathbf{F}_1(0)}{\mathbf{F}_2(0)} + \sum_{k=1}^{n-1} \frac{\mathbf{F}_1(j\omega_k)e^{j\omega_k t}}{\omega_k\left[\frac{d}{d\omega}\mathbf{F}_3(j\omega)\right]_{\omega=\omega_k}}. \tag{4.93b}$$

These formulas (equation 4.93) are useful for cases in which a voltage/current source is applied (at $t=0$) to the circuit with zero initial conditions. Note that for such circuits all the voltages/currents for $t<0$ are zero, which means that a one-sided Fourier transform is used:

$$F(j\omega)=\int_0^\infty f(t)e^{-j\omega t}dt.$$

Formulas such as those in equation 4.93 are sometimes called “switching formulas” since they are used when the circuits are switched to different sources, i.e. for $t>0$.

Example 4.7

The *T*-circuit, shown in Fig. 4.21(a), is connected to the d.c. voltage source at $t=0$. Find the current $i_2(t)$ using a switching formula.

Solution

The transfer admittance of the circuit is

$$Y_{21}(j\omega)=\frac{1}{R+R//\frac{1}{j\omega C}}\cdot\frac{1/j\omega C}{R+\frac{1}{j\omega C}}=\frac{1}{2R+j\omega R^2C}.$$

Therefore, since the Fourier transform of the input voltage is $V/j\omega$, we have

$$I_2(j\omega)=Y_{21}(j\omega)V(j\omega)=\frac{V}{j\omega(2R+j\omega R^2C)}.$$

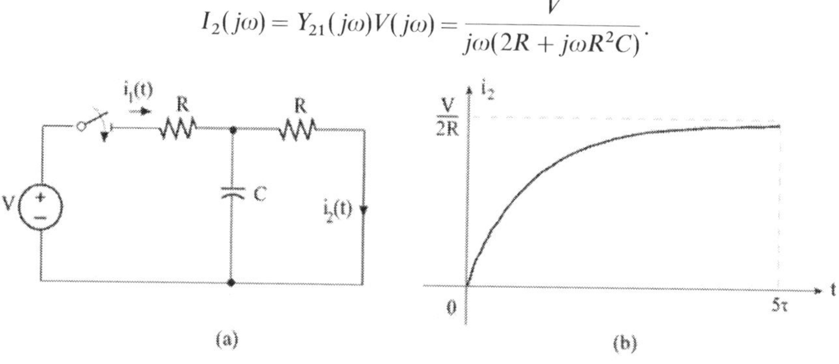

Figure 4.21 *T*-circuit (a) and the waveform of the current (b).

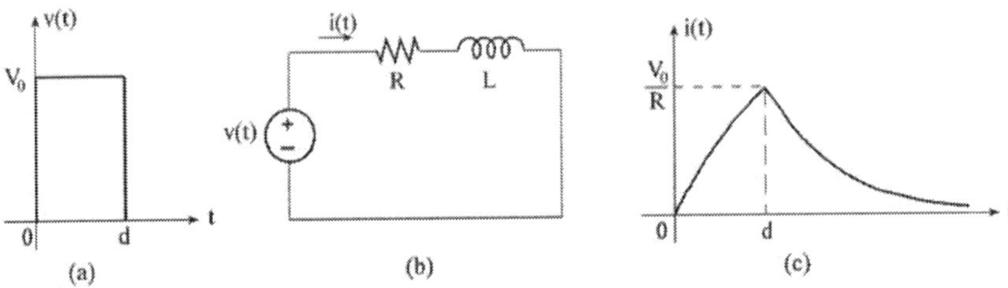

Figure 4.22 Input voltage waveform (a), a given circuit (b), and an input current waveform (c).

In this expression, in accordance to equation 4.93b, $F_3(j\omega) = 2R + j\omega R^2C$, $F_3'(j\omega) = jR^2C$ and the root $\omega_1 = j2/RC$. Thus,

$$i_2(t) = \frac{V}{2R} + \frac{Ve^{-\frac{2}{RC}t}}{j(2/RC)(jR^2C)} = \frac{V}{2R}\left(1 - e^{-\frac{2}{RC}t}\right).$$

This waveform of the current $i_2(t)$ is shown in Fig. 4.21(b).

Example 4.8

A rectangular pulse of voltage, Fig. 4.22(a), is applied to the series RL circuit shown in Fig. 4.22(b). Find the circuit current.

Solution

The transform of the given waveform of the applied voltage may be found by using equation 4.20a for the rectangular pulse in the interval 0–d, i.e.

$$V(j\omega) = \frac{V_0}{j\omega}(1 - e^{-j\omega d}).$$

The transform of the circuit impedance is simply $Z(j\omega) = R + j\omega L = jL(\omega - j\xi)$, where $\xi = R/L$ then, with Ohm's law, for the Fourier transforms in equation 4.91 we have

$$I(j\omega) = \frac{V_0(1 - e^{-j\omega d})}{j^2\omega L(\omega - j\xi)} = -\frac{V_0}{L}\frac{1 - e^{-j\omega d}}{\omega(\omega - j\xi)}.$$

or

$$I(j\omega) = -\frac{V_0}{L}\frac{1}{\omega(\omega - j\xi)} + \frac{V_0}{L}\frac{e^{-j\omega d}}{\omega(\omega - j\xi)} = I'(j\omega) + I''(j\omega).$$

The first part of the time-domain function $i(t)$ is found by using the partial fraction expansion (first multiplying the given fraction by $(j\xi)^2$):

$$\frac{-\xi^2}{(j\xi\omega)[j\xi(\omega - j\xi)]} = \frac{1}{j\xi\omega} + \frac{1}{j\xi(\omega - j\xi)} = -\frac{1}{\xi}\frac{1}{j\omega} + \frac{1}{\xi}\frac{1}{\xi + j\omega}.$$

In accordance with Table 4.3 (the 5th and 7th entries) and taking into consideration that $u(t)=\frac{1}{2}\operatorname{sgn}(t)+\frac{1}{2}$, we have:

$$\frac{1}{j\omega}\leftrightarrow-\frac{1}{2}\operatorname{sgn}(t)=-u(t)+\frac{1}{2},\quad \frac{1}{\xi+j\omega}\leftrightarrow e^{-\xi t}u(t).$$

Therefore,

$$i'(t)=-\frac{V_0}{L\xi}\left[-\left(u(t)-\frac{1}{2}\right)+e^{-\xi t}u(t)\right]=\frac{V_0}{R}(1-e^{-\xi t})-\frac{1}{2}\frac{V_0}{R}.$$

The second part of the current $i(t)$ differs from the first one by the sign and the shifting factor $e^{-j\omega\tau}$, therefore,

$$i''(t)=-\frac{V_0}{R}(1-e^{-\xi(t-d)})u(t-d)+\frac{1}{2}\frac{V_0}{R},$$

and finally

$$i(t)=i'(t)+i''(t)=\frac{V_0}{R}[(1-e^{-\xi t})u(t)-(1-e^{-\xi(t-d)})u(t-d)].$$

The same results might be obtained by using the switching formula of equation 4.93b. We may write the first part of $I(j\omega)$ as

$$I'(j\omega)=-\frac{V_0}{L}\frac{j}{j\omega(\omega-j\xi)},$$

where $F_1(j\omega)=j$, $F_3(j\omega)=\omega-j\xi$ and $\omega_1=j\xi$. Then,

$$i'(t)=-\frac{V_0}{L}\left(\frac{j}{-j\xi}+\frac{je^{-\xi t}}{j\xi}\right)=\frac{V_0}{R}(1-e^{-\xi t}),$$

and for the second part we have

$$i''(t)=-\frac{V_0}{R}(1-e^{-\xi(t-d)}).$$

Therefore,

$$i(t)=\frac{V_0}{R}[(1-e^{-\xi t})-(1-e^{-\xi(t-d)})],\quad \text{for}\quad t>0.$$

A plot of this waveform is shown in Fig. 4.22(c).

In general, the time domain current according to equation 4.91 may be found as an inverse Fourier formula

$$i(t)=\frac{1}{2\pi}\int_{-\infty}^{\infty}Y(j\omega)V_{in}(j\omega)e^{j\omega t}d\omega. \tag{4.94}$$

Let us consider, for instance, applying a constant voltage source V_0 to any one-port circuit having the admittance

$$Y(j\omega) = G(\omega) - jB(\omega) = \frac{\cos\varphi}{|Z(j\omega)|} - j\frac{\sin\varphi}{|Z(j\omega)|}. \tag{4.95}$$

Using for the input voltage the integral notation of a unit function(*)

$$u(t) = \frac{1}{2} + \frac{1}{\pi}\int_{-\infty}^{\infty} \frac{\sin\omega t}{\omega}\, d\omega \tag{4.96}$$

we have

$$\begin{aligned} i(t) &= \frac{V_0}{2Z(0)} + \frac{V_0}{\pi}\int_{-\infty}^{\infty} \frac{\sin(\omega t - \varphi)}{\omega|Z(j\omega)|}\, d\omega \\ &= \frac{V_0}{2Z(0)} + \frac{V_0}{\pi}\int_{0}^{\infty} \frac{\cos\varphi \sin\omega t}{\omega|Z(j\omega)|} - \frac{V_0}{\pi}\int_{0}^{\infty} \frac{\sin\varphi \cos\omega t}{\omega|Z(j\omega)|}\, d\omega, \end{aligned}$$

or with equation 4.95

$$i(t) = \frac{G(0)V_0}{2} + \frac{V_0}{\pi}\int_0^{\infty} G(\omega)\frac{\sin\omega t}{\omega}\, d\omega - \frac{V_0}{\pi}\int_0^{\infty} B(\omega)\frac{\cos\omega t}{\omega}\, d\omega. \tag{4.97}$$

This expression is valid for any instant of time; however, the current in the given circuit should be zero for $t < 0$. Then for $t > 0$ $i(-t)$ should be zero, which results in

$$\frac{G(0)V_0}{2} - \frac{V_0}{\pi}\int_0^{\infty} G(\omega)\frac{\sin\omega t}{\omega}\, d\omega - \frac{V_0}{\pi}\int_0^{\infty} B(\omega)\frac{\cos\omega t}{\omega}\, d\omega = 0. \tag{4.98}$$

By subtracting equation 4.98 from equation 4.97, we finally have

$$i_{tot} = i(t) - i(-t) = \frac{2V_0}{\pi}\int_0^{\infty} G(\omega)\frac{\sin\omega t}{\omega}\, d\omega \quad (\text{for } t > 0). \tag{4.99}$$

This formula can be used for finding the one-port current when only the resistive (active) spectrum of the one-port impedance is known. Thus, if the resistive spectrum of the circuit is given, the most efficient method of calculating the input current is the Fourier transform technique.

Example 4.9

As an example of using this method, let us examine a simple circuit shown in Fig. 4.23(a). (This circuit may be considered as a first moment simplified equivalent circuit of a power transformer, which is a capacitor, C, connected to a

(*)This presentation of a unit function is based on the known integral

$$\int_0^{\infty} \frac{\sin ax}{x}\, dx = \begin{cases} \pi/2 & \text{for } a > 0 \\ -\pi/2 & \text{for } a < 0 \end{cases}$$

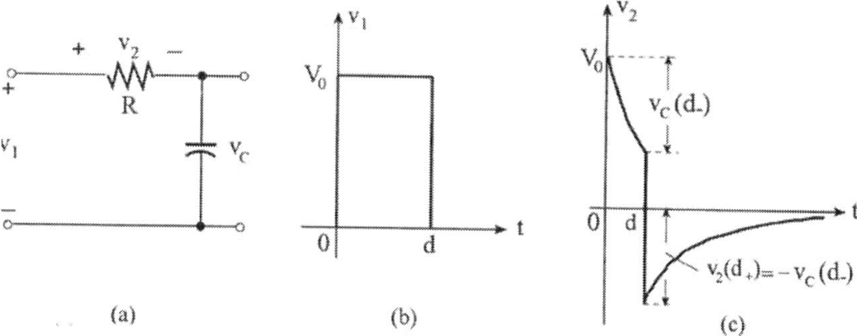

Figure 4.23 A given circuit (a), an applied voltage (b) and the waveform of the voltage across the resistance (c).

cable transmission line, represented by its characteristic resistance, R (see further on in Chapter 7).) Assume that a pulse voltage of rectangular form, 4.23(b), is applied to this circuit and find the voltage across the cable.

Solution

We first should find the real part of the transmission coefficient for $v_2(t)$

$$K_R(\omega) = \text{Re}\left[\frac{R}{R + 1/j\omega C}\right] = \text{Re}\left[\frac{j\omega CR}{1 + j\omega CR}\right] = \frac{(\omega CR)^2}{1 + (\omega CR)^2} = \frac{(\omega\tau)^2}{1 + (\omega\tau)^2},$$

where $\tau = RC$ (time constant). By treating the voltage pulse as two constant voltages shifted by time interval τ, we will have for the first voltage applied at $t = 0$

$$v_2'(t) = \frac{2V_0}{\pi}\int_0^\infty K_R(\omega)\frac{\sin\omega t}{\omega}\,d\omega = \frac{2V_0}{\pi}\int_0^\infty \frac{(\omega\tau)^2}{1 + (\omega\tau)^2}\frac{\sin\omega t}{\omega}\,d\omega.$$

By assigning $x = \omega\tau$ we have $\omega = x/\tau$, $d\omega = (1/\tau)dx$ and

$$v_2'(t) = \frac{2V_0}{\pi}\int_0^\infty \frac{x\sin(x/\tau)t}{1 + x^2}\,dx = \frac{2V_0}{\pi}\frac{\pi}{2}e^{-\frac{t}{\tau}} = V_0 e^{-\frac{t}{\tau}}.^{(*)}$$

The second part of the voltage differs from the first one by the sign and the shifting factor $e^{-j\omega d}$. Therefore,

$$v_2''(t) = -V_0 e^{-\frac{t-d}{\tau}}u(t - d).$$

(*)The integral in this expression is tabulated integral: $\int_0^\infty \frac{x\sin ax}{b^2 + x^2} = \frac{\pi}{2}e^{-(ab)}$.

Finally, we have

$$v_2(t) = v_2'(t) + v_2''(t) = V_0\left[e^{-\frac{t}{\tau}}u(t) - e^{-\frac{t-d}{\tau}}u(t-d)\right].$$

A plot of this waveform is shown in Fig.4.23(c).

4.7.3 Approximate transient analysis with the Fourier transform

In previous paragraphs, we have introduced how to use the Fourier transform for solving problems in circuit transient analysis; but as we have seen, only simple problems can be analyzed using the Fourier method straightforwardly. The main difficulty is in the evolution of the inverse transform integral.

However, the main significance of using the Fourier transform is in the fact that any impulse (such as signals in communication or lightning strokes in power systems) may be presented by its spectra and with the frequency characteristic of the circuit or system function (which usually is known) we can find the spectra of the system input or output response. Since there is a direct connection between Fourier transform techniques and sinusoidal steady-state analysis, the ratio of the phasor response to the phasor forcing function presents the transfer function or the system function

$$\frac{\mathbf{F}_o(j\omega)}{\mathbf{F}_{in}(j\omega)} = \mathbf{K}_{oi}(j\omega) = \frac{B}{A}e^{j(\beta-\alpha)},$$

where A and B are the magnitudes and α and β the phase angles of the input and output phasor for each value of ω. Moreover, we may conclude that the phasor analysis of linear circuits, which is presented in introductory courses, is but a special case of the more general techniques of Fourier transform analysis being studied here. As it was previously shown, the use of Fourier transforms and system functions enables us to handle non-sinusoidal, non-periodic forcing functions and responses. In many cases, when the analytical expression of a system function is not known, there is the possibility of achieving it experimentally. In both cases, the system function is given either analytically or experimentally. To find the time-domain response, we must apply the inverse Fourier transform

$$f(t) = \frac{1}{2\pi}\int_{-\infty}^{\infty}\mathbf{F}(j\omega)e^{j\omega t}d\omega, \tag{4.100}$$

where $\mathbf{F}(j\omega)$ may be presented, for instance, as a product of a forcing function $\mathbf{V}(j\omega)$ and a system function $\mathbf{K}(j\omega)$:

$$\mathbf{F}(j\omega) = \mathbf{V}_{in}(j\omega)\mathbf{K}_{oi}(j\omega),$$

However, in most practical cases, when the function is fairly complicated, the evaluation of an inverse Fourier transform can be extremely difficult. To find the time-domain description of the response function in such cases, we may apply approximate methods.

(a) *Method of trapezoids*

One of these methods is known as the method of **trapezoids**. To use this method only the real part of the integrand function in equation 4.100 is necessary. To show this, we must first simplify the inverse Fourier transform expression of equation 4.100. Let us assume

$$\mathbf{F}(j\omega) = G(\omega) - jB(\omega) \quad \text{and} \quad e^{j\omega t} = \cos \omega t + j \sin \omega t.$$

Then the integral in equation 4.100 will be

$$f(t) = \frac{1}{2\pi}\left\{\int_{-\infty}^{\infty} [G(\omega)\cos \omega t + B(\omega)\sin(\omega t)]\,d\omega + j\int_{-\infty}^{\infty} [G(\omega)\sin \omega t - B(\omega)\cos(\omega t)]\,d\omega\right\}.$$

The second integral in the above expression should be equal to zero, since the real-time function $f(t)$ cannot include an imaginary part. This decision also follows from the fact that the integrand is an odd function of ω ($G(\omega)$ is even and $B(\omega)$ is odd, therefore $G(\omega)\sin(\omega t)$ is odd and so is $B(\omega)\cos(\omega t)$). With the same consideration, we may conclude that the integrand of the first integral is an even function of ω. Therefore the first integral may be replaced by a double quantity of the same integral, but in limits of 0 and ∞:

$$f(t) = \frac{1}{\pi}\int_{0}^{\infty} [G(\omega)\cos \omega t + B(\omega)\sin(\omega t)]\,d\omega. \qquad (4.101a)$$

Furthermore, for the functions, which are zero, i.e. $f(t) = 0$, for $t < 0$ by changing the sign of t, we have

$$f(-t) = \frac{1}{\pi}\int_{0}^{\infty} [G(\omega)\cos \omega t + B(\omega)\sin \omega t]\,d\omega. \qquad (4.101b)$$

By adding equation 4.101b to equation 4.101a we obtain a simple expression for the inverse Fourier transform

$$f(t) = \frac{2}{\pi}\int_{0}^{\infty} G(\omega)\cos \omega t\, d\omega \quad \text{for} \quad f(t) = 0|_{t<0}. \qquad (4.102)$$

Usually function $G(\omega)$ is finite for $t = 0$ and $G(\omega) \to 0$ for $t \to \infty$, then we can provide the integration of equation 4.102 by parts:

$$f(t) = \frac{2}{\pi}\int_{0}^{\infty} \frac{1}{t} G(\omega) d\sin \omega t = \frac{2}{\pi t}\left\{G(\omega)\sin \omega t\Big|_{\omega=0}^{\omega=\infty} - \int_{0}^{\infty} \sin \omega t \frac{dG(\omega)}{d\omega}\,d\omega\right\},$$

which finally gives

$$f(t) = -\frac{2}{\pi t}\int_{0}^{\infty} \frac{dG(\omega)}{d\omega}\sin \omega t\, d\omega. \qquad (4.103)$$

With this expression, we may find the approximate time-domain response, if the frequency response $G(\omega)$ is known.

Suppose the analytical or experimental curve of $G(\omega)$ is known, as it is shown, for example, in Fig. 4.24(a). We then approximate the given curve $G(\omega)$ by the piecewise-linear curve $\overline{G}(\omega)$ so that a series of trapezoids can be built, whose bases are parallel to the ω axis, one side is perpendicular and the other is at an angle to the ω axis. In such a way, we have obtained in the above example three trapezoids g_1, g_2 and g_3 as shown in Fig. 4.24(b), which by their summation give the approximate curve $\overline{G}(\omega)$:

$$G(\omega) \cong \overline{G}(\omega) = \sum_{i=1}^{n} g_i(\omega).$$

Consider now a single trapezoid $g_{i(\omega)}$, which is shown in Fig. 4.25. For such a

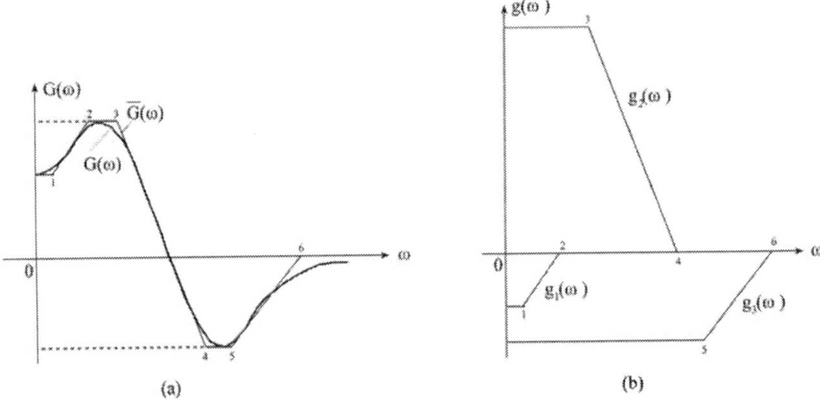

Figure 4.24 Given curve $G(\omega)$ (a) and its approximating trapezoids (b).

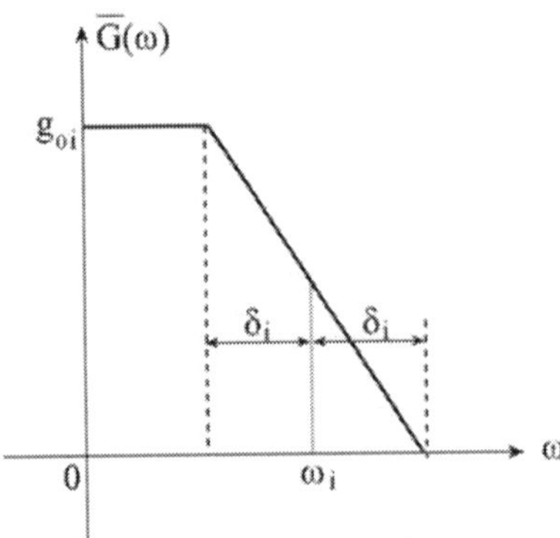

Figure 4.25 A single trapezoid of approximation curve $\overline{G}(\omega)$.

trapezoid its derivative will be:

$$\frac{dG(\omega)}{d\omega}=\begin{cases}0 & \text{for } 0<\omega<\omega_i-\delta_i\\ -g_{0i}/2\delta_i & \text{for } \omega_i-\delta_i<\omega<\omega_i+\delta_i.\end{cases}$$

Then the formula in equation 4.103 yields

$$f(t)=\frac{2}{\pi t}\frac{g_{0i}}{2\delta_i}\int_{\omega_i-\delta_i}^{\omega_i+\delta_i}\sin\omega t\,d\omega=-\frac{g_{0i}}{\pi\delta_i t^2}[\cos(\omega_i+\delta_i)t-\cos(\omega_i-\delta_i)t]$$

$$=2\frac{g_{0i}\omega_i}{\pi}\frac{\sin\omega_i t}{\omega_i t}\frac{\sin\delta_i t}{\delta_i t},$$

and

$$f(t)=\sum f_i(t)=\frac{2}{\pi}\sum g_{0i}\omega_i\mathrm{Sa}(\omega_i t)\,\mathrm{Sa}(\delta_i t). \tag{4.104}$$

The time response (equation 4.104) may be calculated using the tables of sinc function or with an appropriate computer program. It should be noted that the approximation of $G(\omega)$ by several trapezoids gives in many practical cases good results. The method of trapezoids, actually, is a generalized method because of the fact that any signal may by approximated as a piecewise-linear, in which case the signal reduces to impulses after two (or three) differentiations.

Example 4.10

As an example of using this method let us assume that, at the time $t=0$, an exponential pulse V_0e^{-at}, shown in Fig. 4.26(a), is applied to RL equivalent circuit, Fig. 4.26(b). Our goal is to find the voltage across the inductance, i.e., an output voltage. The Fourier transform of the output voltage may be found as

$$\mathbf{V}_o(j\omega)=\mathbf{V}_{in}(j\omega)\mathbf{K}_{oi}(j\omega).$$

Here:

$$\mathbf{V}_{in}(j\omega)=\frac{V_0}{a+j\omega}$$

(see 7th entry in Table 4.3) and

$$\mathbf{K}_{oi}(j\omega)=\frac{j\omega L}{R+j\omega L}=\frac{j\omega\tau}{1+j\omega\tau},$$

where $\tau=L/R$. The real part of $V_o(j\omega)$ then will be

$$G(\omega)=V_0\left|\frac{j\omega\tau}{(a+j\omega)(1+j\omega\tau)}\right|=V_0\frac{-\omega^2\tau(1+a\tau)}{(a-\omega^2\tau)^2+(\omega(1+a\tau))^2}.$$

The positive plot of this function, for $a=1$ ms and $\tau=5$ ms, is shown in

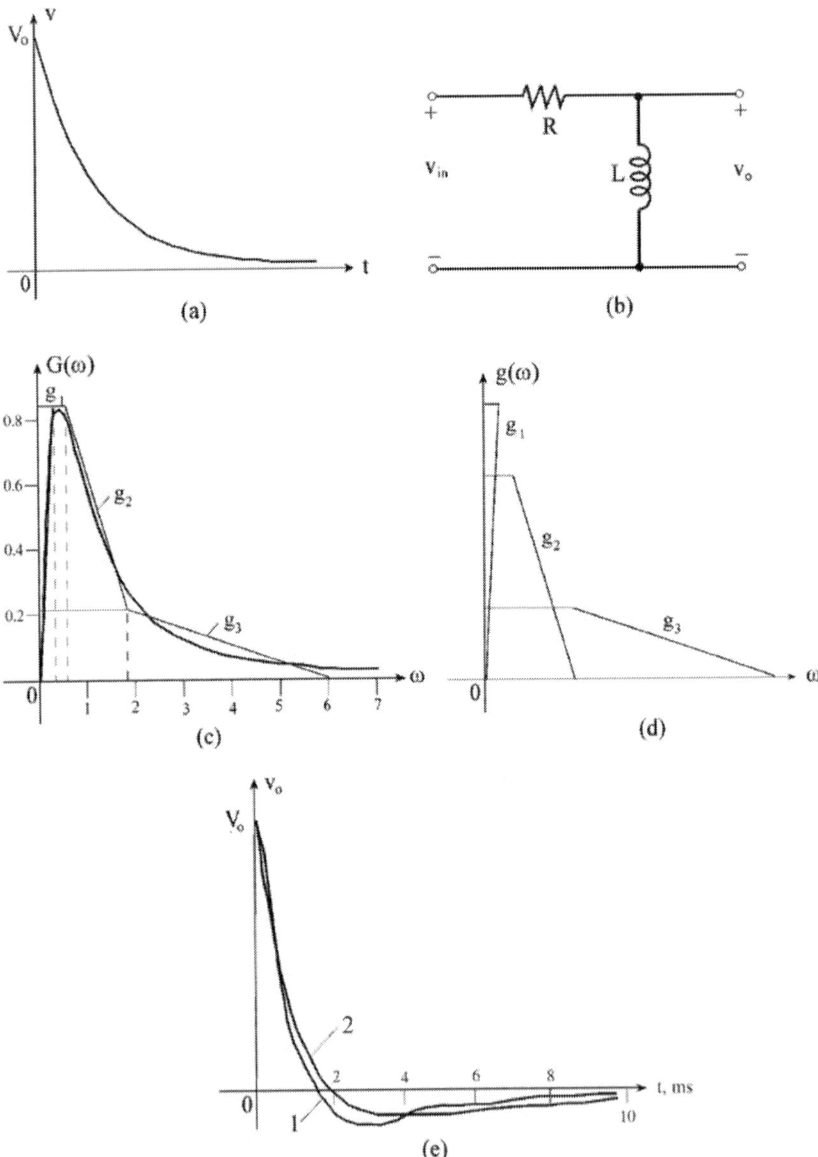

Figure 4.26 An exponential pulse (a), *RL* circuit (b), a positive plot of $|\mathbf{V}_o(j\omega)|$ (c), the obtained trapezoids (d) and the resulting curves of the output voltage $v_o(t)$.

Fig. 4.26(c). This plot might be divided into 4 trapezoids, as shown in Fig. 4.26(d). Then in accordance with equation 4.104 and the data obtained from Fig. 4.26(d) the time-domain response of the output voltage can be calculated, and the result is shown in Fig. 4.26(c), curve 1. Note that at the first moment the whole voltage applied to the circuit is transferred to the output: $v_o(0) = v_{in}(0)$, since the current, $i(0)$, is equal to zero.

This example is, of course, simple enough to use approximate methods and can be easily solved analytically, for instance with switching formula in equation 4.93. (The result is $0.25 \cdot (5e^{-t} - e^{-0.2t})$, which is also shown in Fig. 4.26(d), curve 2. However, we brought this example to illustrate the above method, which can be used for solving complicated problems using appropriate computer programs.

Chapter #5

TRANSIENT ANALYSIS USING STATE VARIABLES

5.1 INTRODUCTION

When the dynamic behavior of a circuit is under consideration, the equations representing the circuit, say in node or mesh analysis, are generally integro-differential. They can then be transformed into one scalar differential equation of the second or higher order. However, the differential equations of a circuit may also be written as a set of first-order differential equations, or when expressed in matrix form it results in a first-order vector differential equation of the form

$$\dot{\mathbf{x}} = f(\mathbf{x}, \mathbf{w}, t),$$

where $\mathbf{x}$ is a vector of unknown variables called *state variables*, $\mathbf{w}$ represents the set of inputs and t is the time.

The set of first-order differential equations written in such a form is called a *state equation* and the vector $\mathbf{x}$ represents the *state* of the network. State equations play an important role in the study of the dynamic behavior of a circuit. There are three basic advantages in using the state equations in this form. (1) There is an enormous amount of mathematical knowledge for solving such equations while the equations by themselves can be derived from formal topological properties of the circuit, using the matrix approach. (2) It can be easily and naturally extended to nonlinear and time-varying or switched networks and is, in fact, the approach most often used in characterizing such networks and (3) it is easily programmed for and solved by computers.

In this chapter, we shall formulate, derive and solve first-order vector differential equations, i.e. state equations. As before, we shall be limited here to linear, time-invariant circuits that may be reciprocal or nonreciprocal. On the other hand, this approach is applicable to circuits of any complicity, especially with computer-aided analysis. In this study, when using a computer is suggested, we are referring to the MATHCAD or MATHLAB programs which are also suitable for symbolic computation.

5.2 THE CONCEPT OF STATE VARIABLES

Two general methods of circuit analysis are usually studied in-depth in introductory courses in circuit analysis(*), namely nodal analysis and mesh analysis. Both of these methods are very useful for resistive d.c. and *RLC* a.c. circuits in their steady-state behavior. The basic variables in these two kinds of circuits, node voltages and mesh currents, were constant quantities, i.e. with no variation in time. Thus, the nodal and mesh equations in such circuits happen to be algebraic equations, without derivatives and integrals. However, node voltages or mesh currents when used as basic variables in *transient analysis* are expressed as a function of time. Therefore, the node and loop equations here are in general integro-differential equations of the second order.

Consider, as an example, the circuit in Fig. 5.1, in which the inductor current and two capacitor currents may be expressed as

$$i_{L2}=\frac{1}{L_2}\int_0^t (v_{n1}-v_{n2})d\tau + I_0, \tag{5.1a}$$

$$\begin{aligned} i_4 &= C_4\frac{dv_{C4}}{dt}=C_4\frac{dv_{n2}}{dt} \\ i_5 &= C_5\frac{dv_{C5}}{dt}=C_5\frac{dv_{n3}}{dt} \end{aligned} \tag{5.1b}$$

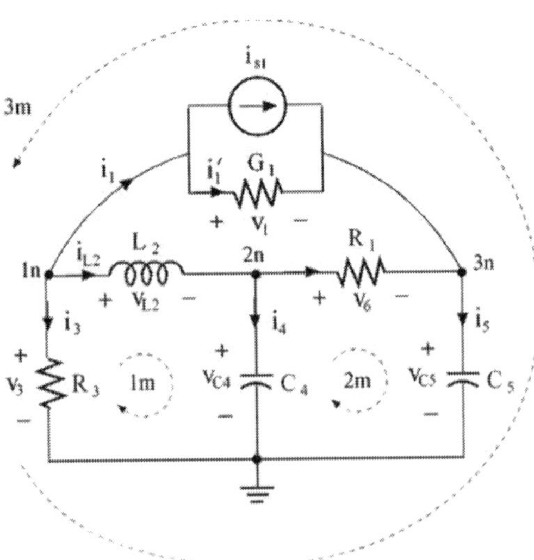

Figure 5.1 Circuit of the example for writing node and mesh equations.

(*) See for example W. H. Hayt and J. E. Kemmerly (1998) *Engineering Circuit Analysis*, McGraw-Hill.

Then the node equations may be written by inspection of the circuit as:

$$(G_1+G_2)v_{n1}+\frac{1}{L_2}\int_0^t v_{n1}d\tau-\frac{1}{L_2}\int_0^t v_{n2}d\tau-G_1v_{n3}=-i_{s1}-I_0$$

$$-\frac{1}{L_2}\int_0^t v_{n1}d\tau+G_6v_{n2}+C_4\frac{dv_{n2}}{dt}+\frac{1}{L_2}\int_0^t v_{n2}d\tau-G_6v_{n3}=I_0$$

$$-G_1v_{n1}-G_6v_{n3}+C_5\frac{dv_{n3}}{dt}=i_{s1}. \tag{5.2}$$

Once these equations are solved for the node voltages v_{n1}, v_{n2} and v_{n3}, the remaining variables are easily obtained.

However, the presence of the integrals of unknowns in node equations 5.2 causes some difficulties in the solution. The integrals can be eliminated by differentiating the equations in which they appear, but this will increase the order of the derivatives. An easier way of analyzing would be if we avoid the appearance of the integrals altogether. We note that an integral appears in the present example of node equations when the current of an inductor is eliminated by using equation 5.1a. In a similar way, the integrals appear in mesh equations when the voltages of the capacitors are eliminated by substituting their v–i relationship. Therefore these integrals will not appear if we leave both the capacitor voltages and inductor currents as variables using a mixed set of equations, i.e. based on Kirchhoff's laws.

Let us illustrate this idea of using capacitor voltages and inductor currents as unknown variables in the same example of the circuit in Fig. 5.1. We may write three independent KCL equations for the nodes $1n$, $2n$ and $3n$, and three KVL equations for loops (meshes) indicated by the dashed arrows:

$$\begin{aligned} &i_1'+i_{L2}+i_3=-i_{s1},\\ &-i_{L2}+i_4+i_6=0,\\ &-i_1'+i_5-i_6=i_{s1}, \end{aligned} \tag{5.3a}$$

$$\begin{aligned} &v_{L2}+v_{C4}-v_3=0,\\ &-v_{C4}+v_6+v_{C5}=0,\\ &v_3-v_{C5}-v_1=0. \end{aligned} \tag{5.5b}$$

Substituting equation 5.1b for i_4 and i_5, taking into consideration that $L_2(di_{L2}/dt)=v_{L2}$ and eliminating all branch voltages except for the capacitor voltages by using the v–i relationships, and after rearranging the terms, yields

$$\begin{aligned} &C_4\frac{dv_{C4}}{dt}=i_{L2}=i_6,\\ &C_5\frac{dv_{C5}}{dt}=i_1'+i_6+i_{s1},\\ &L_2\frac{di_{L2}}{dt}=-v_{C4}+R_3i_3 \end{aligned} \tag{5.4}$$

$$R_6 i_6 = v_{C5} - v_{C4} \tag{5.5a}$$

$$\begin{aligned} i_1' + i_3 &= i_{L2} - i_{s1} \\ R_1 i_1' - R_3 i_3 &= v_{C5}. \end{aligned} \tag{5.b}$$

These are six equations in six unknowns. However, we can reduce the number of equations that must be solved simultaneously. We note that equations 5.5a and 5.5b are algebraic, i.e., they contain no derivatives or integrals. They can be used to eliminate the rest of the unknown variables in (5.4) except v_{C4}, v_{C5} and i_{L2}, whose derivatives are involved in these equations. The algebraic equations 5.5a and 5.5b can be easily solved (the first one trivially) to yield

$$\begin{aligned} i_6 &= -\frac{1}{R_6} v_{C4} + \frac{1}{R_6} v_{C5} \\ i_1' &= \frac{1}{R_1 + R_3} v_{C5} + \frac{R_3}{R_1 + R_3} i_{L2} - \frac{R_3}{R_1 + R_2} i_{s1} \\ i_3 &= -\frac{1}{R_1 + R_3} v_{C5} + \frac{R_3}{R_1 + R_3} i_{L2} - \frac{R_1}{R_1 + R_3} i_{s1}. \end{aligned} \tag{5.6}$$

Finally, these equations can be substituted into equation 5.4 to yield, after rearrangement,

$$\begin{aligned} C_4 \frac{dv_{C4}}{dt} &= \frac{1}{R_6} v_{C4} - \frac{1}{R_6} v_{C5} + i_{L2} \\ C_5 \frac{dv_{C5}}{dt} &= -\frac{1}{R_6} v_{C4} - \frac{R_1 + R_3 + R_6}{R_6(R_1 + R_3)} v_{C5} + \frac{R_3}{R_1 + R_3} i_{L2} + i_{s1} \\ L_2 \frac{di_{L2}}{dt} &= -v_{C4} - \frac{R_3}{R_1 + R_3} v_{C5} + \frac{R_3 R_1}{R_1 + R_3} i_{L2} - \frac{R_1 R_3}{R_1 + R_3} i_{s1}, \end{aligned} \tag{5.7a}$$

or in matrix form, after dividing by the coefficients on the left,

$$\frac{d}{dt}\begin{bmatrix} v_{C4} \\ v_{C5} \\ i_{L2} \end{bmatrix} = \begin{bmatrix} \frac{1}{C_4 R_6} & -\frac{1}{C_4 R_6} & \frac{1}{C_4} \\ -\frac{1}{C_5 R_6} & \frac{R_1 + R_3 + R_6}{C_5 R_6 (R_1 + R_3)} & \frac{R_3}{C_5 (R_1 + R_3)} \\ -\frac{1}{L_2} & -\frac{R_3}{L_2 (R_1 + R_3)} & \frac{R_1 R_3}{L_2 (R_1 + R_3)} \end{bmatrix} \begin{bmatrix} v_{C4} \\ v_{C5} \\ i_{L2} \end{bmatrix} + \begin{bmatrix} 0 \\ \frac{1}{C_5} \\ -\frac{R_1 R_3}{L_2 (R_1 + R_3)} \end{bmatrix} i_{s1}. \tag{5.7b}$$

The resulting *matrix equation* 5.7b represents three first-order differential equations in three unknowns. It is called the **state equation** and the variables v_{C4}, v_{C5} and i_{L2} are called the **state variables**.

As can be seen, the advantage of this method is that no integrals appear, and subsequently no second derivatives occur as a result of the differentiation. The initial conditions, or **initial state** of the circuit, are the initial values of the capacitor voltages and inductor currents, which usually can be independently specified in the circuit, i.e. their values just after t_0 are determined by their values just before t_0. This is the second reason for choosing capacitor voltages and inductor currents as unknown variables.

Further advantages in describing the network by first-order differential equations are:

1) A simple systematic method for writing such equations can be formulated by using the graph theory.
2) A systematic matrix solution may be applied for solving these first-order differential equations. It may be easily programmed for a numerical and symbolic solution with appropriate computer software.
3) It is quite easy to extend the state-variable representation to time-varying and nonlinear networks.

The concept of *state variables*, or just **state**, satisfies two basic conditions of circuit analysis:

a) If at any time, say t_0, the state is known (which is the initial condition or initial state), then the state equations uniquely determine the state at any time $t > t_0$ for any given input. In other words, given the state of the circuit at time t_0 and all the inputs, the behavior of the circuit is completely determined for all $t > t_0$.

b) The state and the input uniquely determine the value of the remaining circuit variables.

Proof a) From the theory of differential equations we know that the initial values of the variables uniquely define, by differential equations, such as 5.7, the value of the variables for all $t \geq t_0$. In other words, the state $(v_C(t), i_L(t))$ can be expressed by the state equations in terms of the initial state.

Proof b) We may use the substitution (or compensation) principle, which states that in any linear circuit any voltage drop across a passive element, say the capacitance, may be substituted by an independent voltage source equal to this drop. In addition, any current through a passive element, say the inductance, may be substituted by an independent current source equal to this current. Hence, we will replace all the inductors by independent current sources whose values $i_L(t)$ are given by the found state variables and all the capacitors by independent voltage sources whose values are equal to the found state variables $v_C(t)$. As a result, we will obtain a pure resistive network in which any variable can be determined by any well-known method of resistive circuit analysis.

For example, let the desired output quantities be v_3 and v_6 in the circuit being considered in Fig. 5.1. Since $v_3 = R_3 i_3$ and $v_6 = R_6 i_6$, by multiplying the third and the first equations of 5.6 correspondingly by R_3 and R_6, we have

$$v_3 = -\frac{R_3}{R_1 + R_3} v_{C5} + \frac{R_1 R_3}{R_1 + R_3} i_{L2} - \frac{R_1 R_3}{R_1 + R_3} i_{s1}$$

$$v_6 = -v_{C4} + v_{C5},$$

where v_{C4}, v_{C5} and i_{L2} represent the voltage and current sources, which substitute the elements C_4, C_5 and L_2 subsequently. The above expressions in matrix form are

$$\begin{bmatrix} v_3 \\ v_6 \end{bmatrix} = \begin{bmatrix} 0 & -\dfrac{R_3}{R_1 + R_3} & \dfrac{R_1 R_3}{R_1 + R_3} \\ -1 & 1 & 0 \end{bmatrix} \begin{bmatrix} v_{C4} \\ v_{C5} \\ i_{L2} \end{bmatrix} + \begin{bmatrix} -\dfrac{R_1 R_3}{R_1 + R_3} \\ 0 \end{bmatrix} [i_{s1}]. \quad (5.8)$$

This matrix equation is called an output equation.

Both the state equation 5.7b and the output equation 5.8 equations may be written in compact matrix notation as

$$\dot{\mathbf{x}} = \mathbf{A}\mathbf{x} + \mathbf{b}\mathbf{w} \quad (5.9a)$$

$$\mathbf{y} = \mathbf{c}\mathbf{x} + \mathbf{d}\mathbf{w}, \quad (5.9b)$$

where $\mathbf{x}$ is the state vector, $\mathbf{w}$ is the input and $\mathbf{y}$ is the output vector. The meanings of matrixes, $\mathbf{A}$, $\mathbf{b}$, $\mathbf{c}$ and $\mathbf{d}$, which are dependent upon circuit elements, are obvious from equations 5.7b and 5.8.

Next, we shall consider the number of independent state variables that represent the transient behavior of a network.

5.3 ORDER OF COMPLEXITY OF A NETWORK

As is known, node-voltage, mesh-currents, and mixed variable equations (based on Kirchhoff's two laws) completely represent any electrical circuit. Recall that the number of independent node-voltage equations, i.e., number of independent Kirchhoff's current law (KCL) equations, is $B-(N-1)$, where B is the number of branches and N is the number of nodes. These numbers are determined only by the graph of the circuit and not by the types of the branches, i.e. they would not be influenced if the branches were all resistors, or if some were capacitors and/or inductors. However, in resistive circuits driven by d.c. sources the node or mesh equations are algebraic, with no variation in time. On the other hand, when capacitors or inductors are present, the equations will be integro-differential. Hence, the question is how many independent variables represent the circuit in its transient (dynamic) behavior. We know that each capacitor and each inductor introduces a variable in such behavior since the v-i characteristic of each contains a derivative or integral. We also know that, for a unique solution of differential equations, the arbitrary constants have to be determined.

The number of these constants is equal to the number of independent initial conditions that can be specified in a circuit. It is also known that the number of initial conditions is related to the energy-storing elements, capacitors and inductors, and in general is equal to the number of such elements in the circuit. The exceptions are the, so-called, **all-capacitor loops** and **all-inductor cut-sets**. Consider the circuit shown in Fig. 5.2. There are five energy-storing elements, but in this circuit there is an all-capacitor loop, consisting of two capacitors C_1 and C_2 and a voltage source, and an all-inductor cut-set (see dashed line in Fig. 5.2) consisting of three inductors L_3, L_4 and L_5. In this case, the capacitor voltages and inductor currents will be restricted by KVL and KCL, namely

$$v_{C1}+v_{C2}=v_{s8} \tag{5.10a}$$

$$i_{L4}+i_{L5}=i_{L3}, \tag{5.10b}$$

which means that one of the voltages and one of the currents can be determined if the other is known. This also means that the initial values of both v_{C1} and v_{C2} cannot be prescribed independently, nor can the initial values of all three currents i_{L3}, i_{L4} and i_{L5}. Therefore, each of the constraint relationships, such as equations 5.10a and 5.10b, reduce the number of independent variables.

In other words, the order of complexity of any network equals the total number of *energy-storing elements minus the number of all-capacitor loops and the number of all-inductor cut-sets*. Thus, the order of complexity of the circuit of Fig. 5.2 is $5-1-1=3$. Note that (1) all-capacitor loops may also consist of ideal voltage sources and all-inductor cut-sets may also include ideal current sources, and (2) only independent all-capacitor loops and all-inductor cut-sets are taken into account[(*)].

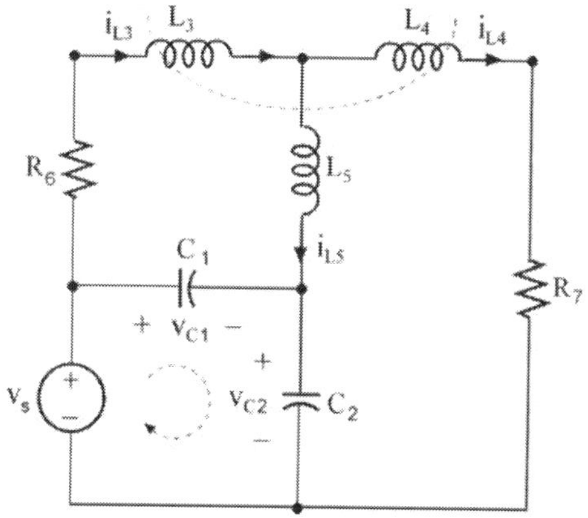

Figure 5.2 Circuit with an all-capacitor loop and an all-inductor cut-set.

[(*)]The opposite situation, when the circuit consists of all-inductor loops and all-capacitor cut-sets, does not influence the order of complexity, but it influences the values of the natural frequencies, namely $s=0$. For more about all-capacitor loops/cut-sets and all-inductor cut-sets/loops see in Balabanian, N. and Bickart T. A. (1969) *Electrical Network Theory*, John Wiley & Sons.

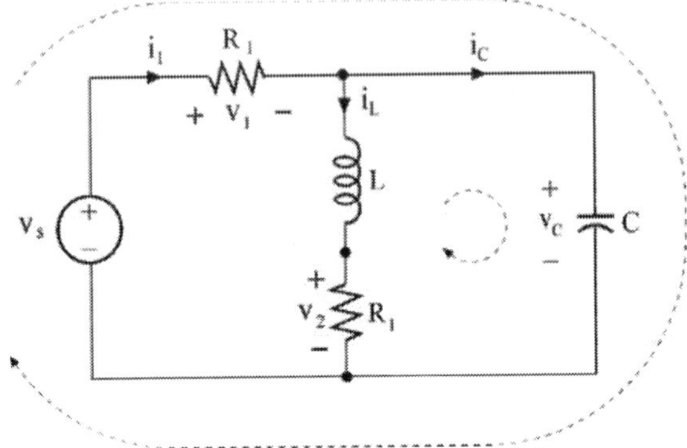

Figure 5.3 Second order circuit.

5.4 STATE EQUATIONS AND TRAJECTORY

Consider the circuit in Fig. 5.3. Let us use capacitor voltage v_C and inductor current i_L as state variables. Applying KCL to node $1n$ and KVL to the right loop and outer loop, we obtain

$$C\frac{dv_C}{dt} = -i_L + i_1, \quad L\frac{di_L}{dt} = v_C - R_2 i_L \tag{5.11}$$

$$R_1 i_1 + v_C = v_s, \tag{5.12}$$

Eliminating the non-desirable variable i_1 from equation 5.12 and substituting it into equation 5.11, after rearranging the terms, gives the state equations

$$\begin{aligned} \frac{dv_C}{dt} &= -\frac{1}{CR_1}v_C - \frac{1}{C}i_L + \frac{1}{CR_1}v_s, \\ \frac{di_L}{dt} &= \frac{1}{L}v_C - \frac{R_2}{L}i_L, \end{aligned} \tag{5.13}$$

or in *matrix form*

$$\frac{d\mathbf{x}(t)}{dt} = \mathbf{A}\mathbf{x}(t) + \mathbf{b}\mathbf{w}(t), \tag{5.14}$$

where:

$\mathbf{x}(t) = \begin{bmatrix} v_C(t) \\ i_L(t) \end{bmatrix}$ is a *vector of state variables*,

$\mathbf{A} = \begin{bmatrix} -\dfrac{1}{CR_1} & -\dfrac{1}{C} \\ \dfrac{1}{L} & -\dfrac{R_2}{L} \end{bmatrix}$ is a *constant* 2×2 *matrix*,

$$\mathbf{b} = \begin{bmatrix} -\dfrac{1}{R_1} \\ 0 \end{bmatrix} \quad \text{is a } \textit{constant vector},$$

$\mathbf{w}(t) = v_s(t)$ is the *scalar input*, or *input vector.*

For solving equation 5.14, the initial conditions of the inductor current and of the capacitor voltage have to be known. Thus, the pair $i_L(0) = I_0$ and $v_C(0) = V_0$ is called the initial state

$$\mathbf{x}_0 = \begin{bmatrix} I_0 \\ V_0 \end{bmatrix} \tag{5.15}$$

The zero input response, i.e., circuit response when $\mathbf{w}(t) = 0$,

$$\frac{d\mathbf{x}(t)}{dt} = \mathbf{A}\mathbf{x}(t) \tag{5.16}$$

is completely determined by the initial state equation 5.15. Thus, if we consider $[i_L(t), v_C(t)]$ as the coordinates of a point on the $i_L - v_C$ plane, then as t increases from 0 to ∞ the point $[i_L(t), v_C(t)]$ will trace a curve, which is called the *state-space trajectory* and the plane $i_L - v_C$ is called the *state-space* of the circuit. It is obvious that the trajectory curve starts at the initial point (I_0, V_0) and ends at the origin $(0, 0)$ when $t = \infty$. Since $v_C(t)$ and $i_L(t)$ are the components of the state vector $\mathbf{x}(t)$, the trajectory defines it in the state space. The velocity of the trajectory $(di_L/dt, dv_C/dt)$ can be obtained from the state equation 5.16. In other words, the trajectory of the state vector in a two-dimensional space characterizes the behavior of a second order circuit, i.e., for every t, the corresponding point of the trajectory specifies $i_L(t)$ and $v_C(t)$.

As an example, three different kinds of trajectory, for: a) overdamped, b) underdamped and 3) loss-less, are shown in Fig. 5.4(d). Note, that in the first case, the trajectory starts at (0.7, 0.9), when $t = 0$, and ends at the origin (0, 0), when $t = \infty$. In the second case, the trajectory is a shrinking spiral starting at the same point and terminating at the origin. Finally, when the circuit is loss-less (which of course is an ideal circuit) the trajectory is an ellipse centered at the origin whose semi-axes depend on the circuit parameters L and C and the initial state $[i_L(0), v_C(0)]$. The ellipse shape trajectory indicates that the response is oscillatory.

For suitably chosen different initial states (usually uniformly spaced points) in the $i_L - v_C$ plane we obtain a family of trajectories, called a *phase portrait*, as shown in Fig. 5.5(a).

As we have already mentioned, the state equations in matrix representation may be easily programmed to a numerical solution. Let us illustrate the approximate method for the calculation of the trajectory. We start at the initial point, determined by the initial state $\mathbf{x}_0[v_C(0), i_L(0)]^T$, and step forward a small interval of time to find an estimate of $\mathbf{x}$ at this new time. From this point we step

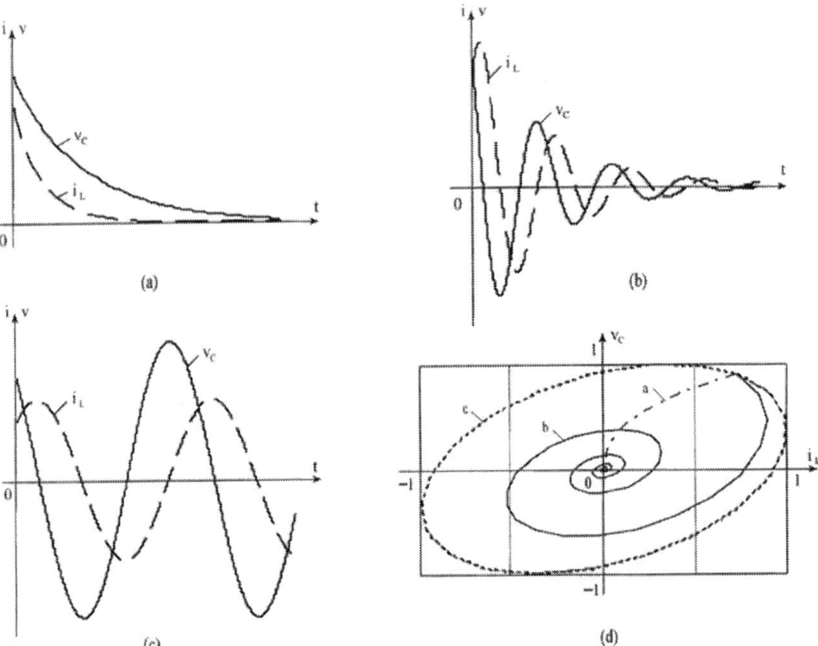

Figure 5.4 Waveforms for i_L and v_C in the second order circuits of an overdamped response (a), underdamped response (b), loss-less response (c) and state trajectories (d).

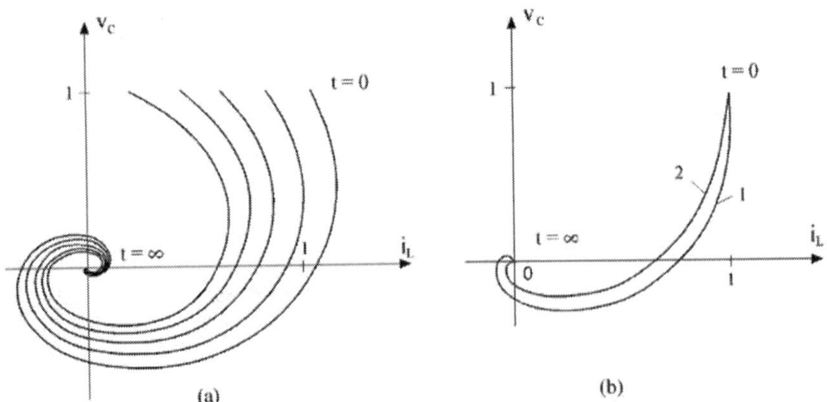

Figure 5.5 State trajectories: phase portrait (a) and for Example 5.1 (b): 1) an approximation with $\Delta t = 0.2$ s and 2) an exact trajectory.

forward again and estimate $\mathbf{x}$ after another short interval of time and so on. The estimate of $\mathbf{x}$ at the new time is found by evaluating $d\mathbf{x}/dt$ at the old time using the differential equation 5.16 and estimating the new value of $\mathbf{x}$ by the formula

$$\mathbf{x}_{new} = \mathbf{x}_{old} + \Delta t \left(\frac{d\mathbf{x}}{dt} \right)_{old}, \tag{5.17}$$

where Δt is the "step length". This step-by-step method is known as Euler's method.

Essentially, we are using a straight-line approximation to the function in each interval. In other words, this method is based on the assumption that if a sufficiently small interval of time Δt is chosen, then during that interval the trajectory velocity $d\mathbf{x}/dt$ is approximately constant. Thus, the straight-line segment, which approximates the trajectory on each step of calculation, is

$$\Delta \mathbf{x} = \left(\frac{d\mathbf{x}}{dt} \right)_{const} \Delta t.$$

It is obvious that the approximation calculated in this manner reaches the exact trajectory when Δt approaches zero. In practice, the value of Δt that should be selected depends primarily on the accuracy required and on the length of the time interval over which the trajectory is calculated. Once the trajectory is computed, the response of the circuit is easily obtained by plotting each of the state variables v_C, i_L versus time.

Example 5.1

Let us employ Euler's (first-order) method to calculate the state trajectory and capacitor voltage versus the time of the circuit shown in Fig. 5.3.

Solution

Let the values of the circuit elements be $R_1 = 1\ \Omega$, $R_2 = 1\ \Omega$, $L = 1$ H, $C = 1$ F and the initial state be $I_0 = 1$ A and $V_0 = 1$ V.

Then, substituting the above parameters in the matrix $\mathbf{A}$, we have the state equation 5.16 as

$$\frac{d\mathbf{x}}{dt} = \begin{bmatrix} -1 & -1 \\ 1 & -1 \end{bmatrix} \mathbf{x},$$

and the initial state is

$$\mathbf{x}(0) = \begin{bmatrix} 1 \\ 1 \end{bmatrix}$$

Let us pick $\Delta t = 0.1$ s. Using equation 5.17 yields the state at 0.1 s:

$$\mathbf{x}(0.1) = \begin{bmatrix} 1 \\ 1 \end{bmatrix} + 0.1 \begin{bmatrix} -1 & -1 \\ 1 & -1 \end{bmatrix} \begin{bmatrix} 1 \\ 1 \end{bmatrix} = \begin{bmatrix} 0.8 \\ 1 \end{bmatrix}.$$

Next, we can obtain the state at $t = 2\Delta t = 0.2$ s:

$$\mathbf{x}(0.2) = \begin{bmatrix} 0.8 \\ 1 \end{bmatrix} + 0.1 \begin{bmatrix} -1 & -1 \\ 1 & -1 \end{bmatrix} \begin{bmatrix} 0.8 \\ 1 \end{bmatrix} = \begin{bmatrix} 0.62 \\ 0.98 \end{bmatrix}.$$

From these two steps, we can write the state at $(k+1)\Delta t$ in terms of the state at $k\Delta t$

$$\mathbf{x}[(k+1)0.1] = \left(1 + 0.1 \begin{bmatrix} -1 & -1 \\ 1 & -1 \end{bmatrix}\right) \mathbf{x}(k\Delta t) = \begin{bmatrix} 0.9 & -0.1 \\ 0.1 & 0.9 \end{bmatrix} \mathbf{x}(k\Delta t).$$

In accordance with this formula the computer-aided calculation results are shown in Fig. 5.5(b). If we use $\Delta t = 0.01$, the resulting trajectory will coincide with the exact trajectory.

In conclusion, the general recurrence formula for approximating the trajectory may be written as(*)

$$\mathbf{x}[(k+1)\Delta t] = (1 + \Delta t\, \mathbf{A})\mathbf{x}(k\Delta t). \tag{5.18}$$

5.5 BASIC CONSIDERATIONS IN WRITING STATE EQUATIONS

In this section, we shall introduce a systematic method for writing state equations. This method is based on the topological properties of the network and is called the "proper tree" method. However, we must first consider KCL and KVL equations based on a cut-set and loop analysis.

5.5.1 Fundamental cut-set and loop matrixes

As is known from matrix analysis, the matrix formulation of independent KCL equations is given by using the reduced incident matrix **A**. Recall that for any connected graph, having N nodes and B branches, **A** has $N-1$ rows and B columns. Thus, the set of $N-1$ linearly independent KCL equations, written on the node basis, has the matrix form

$$\mathbf{Ai} = \mathbf{0}. \tag{5.19}$$

However, equation 5.19 is not the only way of writing KCL equations. It may also be done on the cut-set basis. A cut-set is defined as a set of k branches with the property that if all k branches are removed from the graph, it is separated into two parts. As an example, consider the graph shown in Fig. 5.6.

(*)For a more accurate approximation of the state-space trajectory, the Runge-Kutta fourth-order method can be used (see, for example in Bajpai, A. C., et al. (1974) *Engineering Mathematics*, John Wiley & Sons.

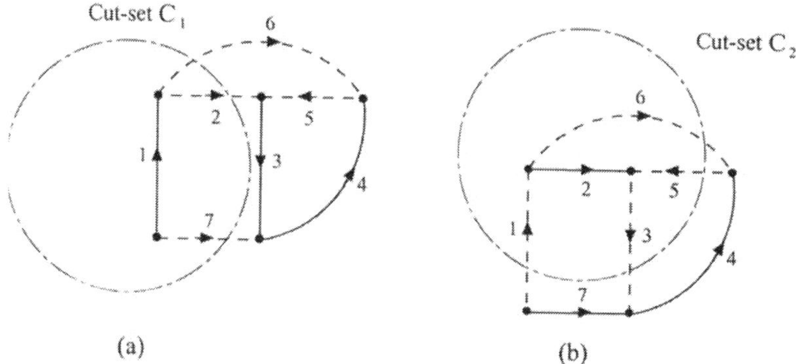

Figure 5.6 Two distinct cut-sets indicated by dashed lines.

Two distinct cut-sets are shown by dashed lines, namely $C_1 = (b_2, b_6, b_7)$ and $C_2 = (b_1, b_3, b_5, b_6)$. Recall now the generalized version of the KCL. By enclosing one of the cut parts of the circuit in the balloon-shaped surface, (see the dotted-dash line in Fig. 5.6(b)) we can write a KCL equation for this particular cut-set

$$-i_1 + i_3 - i_4 + i_5 = 0.$$

The number of such KCL equations is obviously equal to the number of distinct cut-sets. However, as we know, the number of independent KCL equations is $N-1$, where N is the number of nodes in the graph/circuit. Naturally, we are interested in writing linearly independent cut-set equations. For this purpose, we shall introduce the so-called **fundamental cut-set**. Choosing any tree in the graph, we define a fundamental cut-set as that associated with the tree branch, i.e. every tree branch together with some links constitutes a **unique cut-set** of the graph. Such a cut-set is shown, for example, in Fig. 5.7. As can be seen, removing the tree branch t_3 separates the tree into two parts T_1 and T_2. Then the links ℓ_a and ℓ_b together with twig t_3 constitute a unique cut-set. Indeed, removing any of the remaining links, even all of them (thin lines), cannot

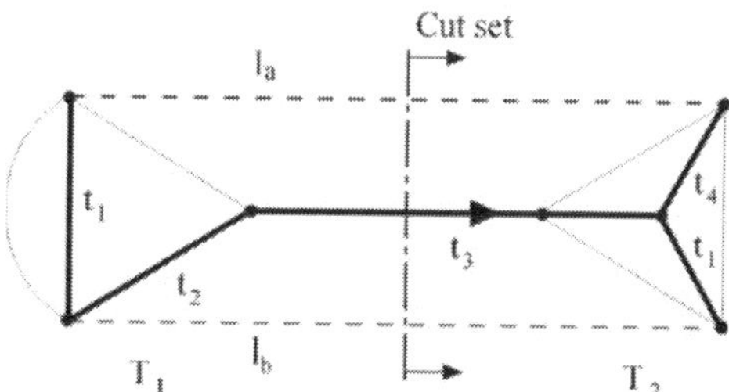

Figure 5.7 An example of a graph, tree and fundamental cut-set.

separate either T_1 or T_2 into two parts. Therefore, the above cut-set is unique. Obviously, each of the fundamental cut-sets is independent of any other, because each of them contains one and only one twig. Since the number of twigs in any tree is $N-1$, we can write $N-1$ linearly independent KCL equations following $N-1$ fundamental cut-sets. Note that the orientation of each fundamental cut-set is defined by the direction of the associated twig as shown in Fig. 5.7.

We will next consider the oriented graph of Fig. 5.8(a). A chosen tree is shown by heavy lines, and four fundamental cut-sets associated with four twigs (since a given graph has five nodes) are marked by dashed lines. For the sake of convenience, we first number the twigs from 1 to 4 and the links from 5 to 7, and adopt a reference direction for the cut-set, which agrees with the tree branch defining the cut-set. Applying KCL to the four cut-sets, we obtain

$$\begin{aligned}
&\text{cut-set 1:}\quad i_1 + i_7 = 0\\
&\text{cut-set 2:}\quad i_2 + i_6 + i_7 = 0\\
&\text{cut-set 3:}\quad i_3 - i_5 + i_6 - i_7 = 0\\
&\text{cut-set 4:}\quad i_4 - i_5 + i_6 = 0,
\end{aligned}$$

or in matrix form

$$\begin{array}{c} \text{cut sets} \\ \\ 1\\2\\3\\4 \end{array}
\begin{array}{c} \overbrace{\begin{array}{cccc}1&2&3&4\end{array}}^{\text{twigs}}\ \overbrace{\begin{array}{ccc}5&6&7\end{array}}^{\text{links}} \\
\begin{bmatrix} 1 & 0 & 0 & 0 & 0 & 0 & 1\\ 0 & 1 & 0 & 0 & 0 & 1 & 1\\ 0 & 0 & 1 & 0 & 1 & -1 & -1\\ 0 & 0 & 0 & 1 & -1 & 1 & 0 \end{bmatrix} \end{array}
\begin{bmatrix} i_1\\ i_2\\ i_3\\ i_4\\ i_5\\ i_6\\ i_7 \end{bmatrix} = \begin{bmatrix} 0\\0\\0\\0 \end{bmatrix} \tag{5.20}$$

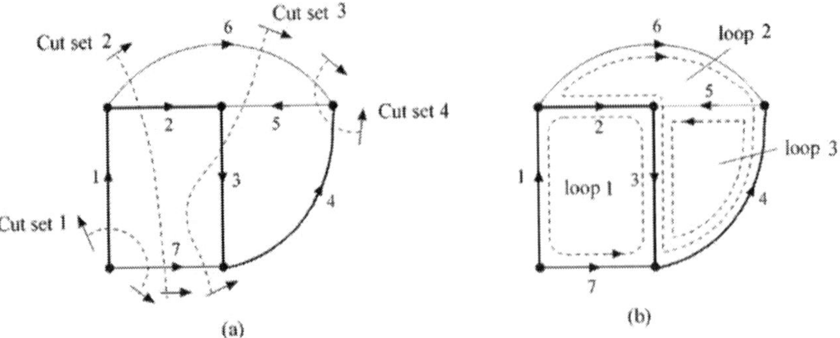

Figure 5.8 Fundamental cut-sets for the chosen tree (dashed lines) (a) and fundamental loops (dashed lines) (b).

In general, the KCL equations based on the fundamental cut-sets may be written in the short form:

$$\mathbf{Qi} = \mathbf{0}, \tag{5.21}$$

where $\mathbf{Q}$ is the *fundamental cut-set matrix* associated with the tree. The order of the $\mathbf{Q}$ matrix is $(N-1) \times B$, and its jk-th element is defined as follows:

$$q_{jk} \begin{cases} 1 & \text{if branch } k \text{ belongs to cut-set } j \text{ and has the same direction} \\ -1 & \text{if branch } k \text{ belongs to cut-set } j \text{ and has the opposite direction} \\ 0 & \text{if branch } k \text{ does not belong to cut-set } j. \end{cases}$$

Note that the fundamental cut-set matrix in equation 5.20 includes a unit sub-matrix of order $(N-1)$, which is the number of fundamental cut-sets and the number of twigs. Therefore,

$$\mathbf{Q} = [\mathbf{1}_t \quad \mathbf{Q}_\ell], \tag{5.22}$$

where $\mathbf{Q}_\ell$ is a sub-matrix of the order $(N-1) \times \ell$, i.e. it consists of $(N-1)$ rows and of ℓ (number of links) columns. The fundamental cut-set matrix $\mathbf{Q}$ will always have the form of equation 5.22 because each fundamental cut-set contains one and only one twig and its orientation agrees with the reference direction of the cut-set, by definition.

Next, we shall introduce the loop matrix. Mesh analysis, which is commonly studied in introductory courses in circuit analysis, is not the only method of writing a set of independent equations based on KVL. Another and actually more flexible method, which allows us to derive independent KVL equations, is based on the so-called **fundamental loop**. Every link of a co-tree (complement of the tree) together with some twigs, which are connected to the link, constitutes a unique loop associated with the link. Indeed, there cannot be any other path between two nodes of the tree, to which the link is connected. If there were two or more paths between two nodes of the tree, they will form a loop; this contradicts the main property of a tree. The set of fundamental loops is independent, since each of them contains one and only one link, i.e. every loop differs from another by at least one branch. Therefore, each link uniquely defines a fundamental loop. Hence, the number of fundamental loops is equal to the number of links, i.e. $B-(N-1)$. Each fundamental loop has a reference direction, which is defined by the direction of its associated link, as shown in Fig. 5.8(b).

So we use the fundamental loops to define $B-(N-1)$ linearly independent KVL equations. For the graph in Fig. 5.8(b), we may write the following three independent KVL equations:

$$\begin{aligned} &\text{Loop 1:} && v_3 + v_4 + v_5 = 0 \\ &\text{Loop 2:} && -v_2 - v_3 - v_4 + v_6 = 0 \\ &\text{Loop 3:} && -v_1 - v_2 - v_3 + v_7 = 0 \end{aligned}$$

or in matrix form

$$\begin{array}{c} \text{loops} \\ \downarrow \\ \\ 1 \\ 2 \\ 3 \end{array} \overset{\begin{array}{ccccccc} & & \text{twigs} & & & \text{links} & \\ 1 & 2 & 3 & 4 & 5 & 6 & 7 \end{array}}{\begin{bmatrix} 0 & 0 & 1 & 1 & 1 & 0 & 0 \\ 0 & -1 & -1 & -1 & 0 & 1 & 0 \\ -1 & -1 & 1 & 1 & 0 & 0 & 1 \end{bmatrix}} \begin{bmatrix} v_1 \\ v_2 \\ v_3 \\ v_4 \\ v_5 \\ v_6 \\ v_7 \end{bmatrix} = \begin{bmatrix} 0 \\ 0 \\ 0 \end{bmatrix} \tag{5.23}$$

In general, the KVL equations based on fundamental loops may be written in the short form:

$$\mathbf{Bv} = \mathbf{0}, \tag{5.24}$$

where $\mathbf{B}$ is the *fundamental loop matrix* associated with the tree. The order of the $\mathbf{B}$ matrix is $\ell \times B$, where ℓ is the number of loops, and its jk-th element is defined as follows:

$$b_{jk} \begin{cases} 1 & \text{if branch } k \text{ belongs to loop } j \text{ and has the same direction as the loop} \\ -1 & \text{if branch } k \text{ is in loop } j \text{ and has the opposite direction} \\ 0 & \text{if branch } k \text{ is not in loop } j. \end{cases}$$

Note that the fundamental loop matrix in equation 5.23 includes a unit sub-matrix of order ℓ, which is the number of fundamental loops and also the number of links. Therefore, we can express $\mathbf{B}$ in the form

$$\mathbf{B} = [\mathbf{B}_t \quad \mathbf{1}_\ell], \tag{5.25}$$

where $\mathbf{B}_t$ is a sub-matrix of $\ell \times (N-1)$, i.e. it consists of ℓ (number of links) rows and of $t = N - 1$ (number of twigs) columns. The unit matrix in $\mathbf{B}$ results from the fact that each fundamental loop contains one and only one link and by convention the reference directions of the fundamental loops are the same as that of the associated links.

Let us think that twig voltages are a set of the basic independent variables. Since each fundamental loop is formed from twigs and only one link, the link voltage can always be expressed in terms of twig voltages. Therefore, the branch voltages in any circuit can be determined by twig voltages, when the latter ones are used as independent variables. Indeed, in accordance with equations 5.24 and 5.25

$$[\mathbf{B}_t \quad \mathbf{1}_\ell] \begin{bmatrix} \mathbf{v}_t \\ \mathbf{v}_\ell \end{bmatrix} = \mathbf{0}, \tag{5.26}$$

where the branch voltage vector $\mathbf{v}$ is partitioned into two sub-vectors: $\mathbf{v}_t$ and

$\mathbf{v}_\ell$, which are, respectively, the twig-voltage sub-vector and link-voltage sub-vector. Performing the multiplication yields

$$\mathbf{B}_t\mathbf{v}_t + \mathbf{v}_\ell = \mathbf{0},$$

or

$$\mathbf{v}_\ell = -\mathbf{B}_t\mathbf{v}_t. \tag{5.27}$$

This means that link voltages are determined by twig voltages. Obviously, we can write the twig branch-voltage sub-vector as

$$\mathbf{v}_t = \mathbf{1}_t\mathbf{v}_t. \tag{5.28}$$

Combining equations 5.27 and 5.28, we have

$$\begin{bmatrix} \mathbf{v}_t \\ \mathbf{v}_\ell \end{bmatrix} = \begin{bmatrix} \mathbf{1}_t \\ -\mathbf{B}_t \end{bmatrix} \mathbf{v}_t, \tag{5.29}$$

or simply

$$\mathbf{v} = \begin{bmatrix} \mathbf{1}_t \\ -\mathbf{B}_t \end{bmatrix} \mathbf{v}_t, \tag{5.30}$$

which states that all the branch voltages in any circuit can be expressed in terms of twig voltages.

Now, let us again examine the fundamental cut-sets. Since each fundamental cut-set is formed from links and only one twig, we can express the twig-currents in terms of link-currents. Therefore, using the link-currents as basic independent variables, we can always determine the all branch currents by the independent variables. After partitioning the branch currents into twig-currents and link-currents, with equations 5.21 and 5.22, we have

$$[\mathbf{1}_t \quad \mathbf{Q}_\ell] \begin{bmatrix} \mathbf{i}_t \\ \mathbf{i}_\ell \end{bmatrix} = \mathbf{0}, \tag{5.31}$$

where $\mathbf{i}_t$ and $\mathbf{i}_\ell$ are, respectively, the twig-current and link-current sub-vectors. Then two matrixes in equation 5.31 can be multiplied to yield

$$\mathbf{i}_t + \mathbf{Q}_\ell\mathbf{i}_\ell = \mathbf{0},$$

or

$$\mathbf{i}_t = -\mathbf{Q}_\ell\mathbf{i}_\ell. \tag{5.32}$$

Combining equation 5.32 and the identity $\mathbf{i}_\ell = \mathbf{1}_\ell\mathbf{i}_\ell$, yields

$$\begin{bmatrix} \mathbf{i}_t \\ \mathbf{i}_\ell \end{bmatrix} = \begin{bmatrix} -\mathbf{Q}_\ell \\ \mathbf{1}_\ell \end{bmatrix} \mathbf{i}_\ell, \tag{5.33}$$

or

$$\mathbf{i} = \begin{bmatrix} -\mathbf{Q}_\ell \\ \mathbf{1}_\ell \end{bmatrix} \mathbf{i}_\ell, \tag{5.34}$$

which again states that all branch currents in any circuit can be expressed in terms of link currents.

A useful relation between two matrixes **Q** and **B** can now be determined. Recall *Tellegen's theorem* in the form

$$\mathbf{v}^T\mathbf{i} = \mathbf{0}. \tag{5.35}$$

By taking the transpose of $\boldsymbol{v}$ (equation 5.30), we obtain

$$\mathbf{v}^T = \left(\begin{bmatrix} \mathbf{1}_t \\ -\mathbf{B}_t \end{bmatrix} \mathbf{v}_t\right)^T = \mathbf{v}_t^T \begin{bmatrix} \mathbf{1}_t \\ -\mathbf{B}_t \end{bmatrix}^T = \mathbf{v}^T[\mathbf{1}_t - \mathbf{B}_t^T]. \tag{5.36}$$

After substituting equations 5.36 and 5.34 into equation 5.35 we have

$$\mathbf{v}_t^T[\mathbf{1}_t - \mathbf{B}_t^T]\begin{bmatrix} -\mathbf{Q}_t \\ \mathbf{1}_\ell \end{bmatrix}\mathbf{i}_\ell = \mathbf{0}, \quad \text{for all } \mathbf{v}_t \text{ and all } \mathbf{i}_\ell. \tag{5.37}$$

Since $\mathbf{v}_t^T \neq \mathbf{0}$ and $\mathbf{i}_\ell \neq \mathbf{0}$ then

$$[\mathbf{1}_t - \mathbf{B}_t^T]\begin{bmatrix} -\mathbf{Q}_t \\ \mathbf{1}_\ell \end{bmatrix} = \mathbf{0}. \tag{5.38}$$

Performing the multiplication, we obtain the identities

$$\mathbf{Q}_\ell = -\mathbf{B}_t^T \tag{5.39a}$$

and

$$\mathbf{B}_t = -\mathbf{Q}_\ell^T. \tag{5.39b}$$

This relationship between two sub-matrixes $\mathbf{Q}_\ell$ and $\mathbf{B}_t$ results from the fact that both fundamental cut-set matrix $\mathbf{Q}_\ell$ and fundamental loop matrix $\mathbf{B}_t$ give the topological relation between graph branches and fundamental cut-sets and fundamental loops respectively. Also, note that both matrixes $\mathbf{Q}_\ell$ and $\mathbf{B}_t$ come from the same tree.

Replacing $-\mathbf{B}_t$ by $\mathbf{Q}_\ell^T$ in equation 5.30, we obtain

$$\mathbf{v} = \begin{bmatrix} \mathbf{1}_t \\ \mathbf{Q}_\ell^T \end{bmatrix}\mathbf{v}_t = \mathbf{Q}^T\mathbf{v}_t, \tag{5.40}$$

which can be interpreted as a matrix transformation of twig-voltages into branch voltages. Similarly, replacing $-\mathbf{Q}_\ell$ by $\mathbf{B}_t^T$ in equation 5.34, we obtain

$$\mathbf{i} = \begin{bmatrix} \mathbf{B}_t^T \\ \mathbf{1}_\ell \end{bmatrix}\mathbf{i}_\ell = \mathbf{B}^T\mathbf{i}_\ell, \tag{5.41}$$

which is a matrix transformation of link-currents into branch currents.

Finally, substituting equations 5.40 and 5.41 into Tellengen's theorem (equation 5.35), we have

$$\mathbf{v}_t^T\mathbf{Q}\mathbf{B}^T\mathbf{i}_\ell = \mathbf{0}, \quad \text{for all } \mathbf{v}_t \text{ and } \mathbf{i}_\ell, \tag{5.42}$$

which can be reduced to the following relation between the matrixes

$$\mathbf{QB}^T = \mathbf{0}. \tag{5.43}$$

In conclusion, the following comments on loop and cut-set matrixes have to be made. The methods of circuit analysis based on loop and cut-set matrixes are more flexible, allowing more possible applications than the node and mesh analyses. So, as we remember, the mesh analysis based on mesh matrix **M** is restricted to the planar graph only, whereas the fundamental loop matrix **B**, based on tree, is applicable to any graph including non-planar, by means of allowing us to write a maximal number of linearly independent KVL equations.

The concept of duality is usually applied (in introductory courses) to planar graphs and planar circuits by means of node and mesh terms. By now, we may extend this concept to fundamental matrixes **B** and **Q**, pertaining to non-planar graphs and circuits. So, the listing of dual terms can be extended as follows:

Twig – Link,
Fundamental cut-set – Fundamental loop,
Twig voltage, v_t – Link current, i_ℓ,
Fundamental cut-set matrix, **Q** – Fundamental loop matrix, **B**.

Thus, two graphs, G_1 and G_2 having the same number of branches, are dual if the number of fundamental cut-sets of one of them is equal to the number of fundamental loops of the second and their **Q** and **B** matrixes are identical, namely

$$\mathbf{Q}_1 = \mathbf{B}_2.$$

5.5.2 "Proper tree" method for writing state equations

Our aim now is to write the state and output equations in the form of equation 5.9

$$\dot{\mathbf{x}}(t) = \mathbf{A}\mathbf{x}(t) + \mathbf{b}\mathbf{w}(t) \tag{5.44a}$$

$$\mathbf{y}(t) = \mathbf{c}\mathbf{x}(t) + \mathbf{d}\mathbf{w}(t), \tag{5.44b}$$

where **x** is the state vector containing all the capacitor voltages and all the inductor currents, **w** is the input vector containing all the independent voltage and current sources, driving the circuit and **y** is the desired output vector. **A**, **b**, **c** and **d** are constant matrixes whose elements depend on circuit parameters. Equation 5.44a is a first order matrix differential equation with constant matrix coefficients. $\dot{\mathbf{x}}$ is the first derivative of the state vector **x**, i.e. it consists of the derivatives of the state variables dv_C/dt and di_L/dt. We note that these quantities are given by currents in the capacitors $C(dv_C/dt)$ and voltages across inductors $L(di_L/dt)$. To evaluate capacitor currents in terms of other currents, we must write cut-set equations and to evaluate inductor voltages in terms of other voltages we must write loop equations. Therefore, it turns out that we could do this if, using the concept of cut-set and loop analysis, we chose a tree which

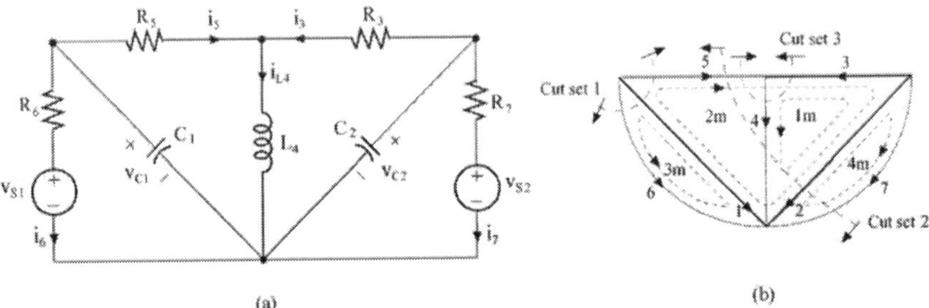

Figure 5.9 A circuit of the example for writing state equations (a), the oriented graph and proper tree (b).

includes all the capacitors but no inductors. Such a tree is called a *proper tree*(*) We can complete the proper tree if the number of twigs is larger than the number of capacitors by adding resistors and voltage sources. Thus, the inductors, the remaining resistors and possibly the current sources will constitute the co-tree links.

Following this method, we may write a fundamental cut-set equation for each capacitor-twig, in which the capacitor current $C(dv_C/dt)$ is expressed in terms of other currents. We may write a fundamental loop equation as well for each link inductor in which the inductor voltage $L(di_L/dt)$ is expressed in terms of other voltages. We shall also take into consideration that the basic variables in cut-set/loop analysis are twig voltages and link currents. Hence, we shall use the appropriate v–i relationships for resistive and active elements. Thus for twig resistors we use the form $v_t = Ri$ and for the link resistors $i_\ell = Gv$. For the same reason we put the voltage sources into the twigs and the current sources into the links. (To fulfill these requirements, we can use a source transformation and shifting techniques.) At this point, let us illustrate the above description by the following example. For the sake of generality, we will divide the solution procedure into five steps. Consider the circuit shown in Fig. 5.9(a).

Step 1 Choosing the state variables

The circuit contains two capacitors and one inductor. Therefore, the state variables are v_{C1}, v_{C2} and i_{L4}, and the state vector is

$$\mathbf{x} = \begin{bmatrix} v_{C1} \\ v_{C2} \\ i_{L4} \end{bmatrix}. \tag{5.45}$$

Step 2 Choosing the proper tree

(*)If a circuit contains an all-capacitor loop or an all-inductor cut-set, a proper tree does not exist. For such cases see in Balabanian, N. and Bickart, T. A. (1969) *Electrical Network Theory*, John Wiley & Sons.

The proper tree picked for the circuit, shown in Fig. 5.9(b), includes two capacitors and resistor R_3.

Step 3 Writing the fundamental cut-set equations

These equations are written in such a way that the capacitor currents are defined by other link currents and/or current sources (if such are present), and the remaining currents are written in terms of inductor currents and/or current sources.

$$\begin{aligned} &\text{cut-set 1:} \quad C_1 \frac{dv_{C1}}{dt} = -i_5 - i_6 \\ &\text{cut-set 2:} \quad C_2 \frac{dv_{C2}}{dt} = -i_{L4} + i_5 - i_7 \end{aligned} \tag{5.46}$$

$$\text{cut-set 3:} \quad G_3 v_3 + i_5 = i_{L4}. \tag{5.47}$$

Step 4 Writing the fundamental loop equations

The loop equations are written in such a way that the inductor voltages are defined by other twig voltages and/or voltage sources (if such are present), and the remaining voltages are written in terms of capacitor voltages and/or voltage sources

$$\text{Loop 1:} \quad L_4 \frac{di_{L4}}{dt} = v_{C2} - v_3 \tag{5.48}$$

$$\text{Loop 2:} \quad -v_3 + R_5 i_5 = v_{C1} - v_{C2} \tag{5.49}$$

$$\left.\begin{aligned} &\text{Loop 3:} \quad R_6 i_6 = v_{C1} - v_{s1} \\ &\text{Loop 4:} \quad R_7 i_7 = v_{C2} - v_{s2} \end{aligned}\right\}. \tag{5.50}$$

The last two steps lead to state equations

$$\begin{aligned} C_1 \frac{dv_{C1}}{dt} &= -i_5 - i_6 \\ C_2 \frac{dv_{C2}}{dt} &= -i_{L4} + i_5 - i_7 \\ L_4 \frac{di_{L4}}{dt} &= v_{C2} - v_3. \end{aligned} \tag{5.51}$$

Step 5 Expressing the right-hand side of the state equations in terms of state variables and/or inputs. In this example, currents i_5, i_6, i_7 and voltage v_3 have to be expressed in terms of the capacitor voltages v_{C1}, v_{C2} and the inductor current i_{L4}. By solving equations 5.50, we have

$$i_6 = \frac{1}{R_6} v_{C1} - \frac{1}{R_6} v_{s1}, \quad i_7 = \frac{1}{R_7} v_{C2} - \frac{1}{R_7} v_{s2}, \tag{5.52}$$

equations 5.47 and 5.49 form a set of two algebraic equations of two unknowns:

$$\begin{bmatrix} -1 & R_5 \\ G_3 & 1 \end{bmatrix}\begin{bmatrix} v_3 \\ i_5 \end{bmatrix} = \begin{bmatrix} v_{C1}-v_{C2} \\ i_{L4} \end{bmatrix} \tag{5.53}$$

Solving equation 5.53 yields

$$\begin{aligned} v_3 &= -\frac{1}{1+R_5G_3}v_{C1} + \frac{1}{1+R_5G_3}v_{C2} + \frac{R_5}{1+R_5G_3}i_{L4} \\ i_5 &= \frac{G_3}{1+R_5G_3}v_{C1} - \frac{G_3}{1+R_5G_3}v_{C2} + \frac{1}{1+R_5G_3}i_{L4}. \end{aligned} \tag{5.54}$$

Finally, equations 5.52 and 5.54 can be substituted into equation 5.51 to yield, after rearrangement and dividing through the equations by appropriate C_1, C_2, L_4,

$$\frac{d}{dt}\begin{bmatrix} v_{C1} \\ v_{C2} \\ i_{L4} \end{bmatrix} = \begin{bmatrix} -\dfrac{1+aR_6G_3}{R_6C_1} & \dfrac{aG_3}{C_1} & -\dfrac{a}{C_1} \\ \dfrac{aG_3}{C_2} & -\dfrac{1+aR_7G_3}{R_7C_2} & -\dfrac{1-a}{C_2} \\ \dfrac{a}{L_4} & \dfrac{1-a}{L_4} & -\dfrac{aR_5}{L_4} \end{bmatrix}\begin{bmatrix} v_{C1} \\ v_{C2} \\ i_{L4} \end{bmatrix} + \begin{bmatrix} \dfrac{1}{R_6C_1} & 0 \\ 0 & \dfrac{1}{R_7C_2} \\ 0 & 0 \end{bmatrix}\begin{bmatrix} v_{s1} \\ v_{s2} \end{bmatrix}, \tag{5.55}$$

where $a = 1/(1+R_5G_3)$.

Note that state equations here are written in the matrix form of equation 5.44a where the input vector (in this example) is $\mathbf{w} = [v_{s1}\ v_{s2}]^T$ and the meanings of matrixes **A** and **b** are obvious.

Suppose now that the remaining branch variables, i.e. v_3, i_5, i_6 and i_7 are a desired output. Then, using equations 5.54 and 5.52, we can express the output in terms of the state variables and the input as

$$\begin{bmatrix} v_3 \\ i_5 \\ i_6 \\ i_7 \end{bmatrix} = \begin{bmatrix} -a & a & aR_5 \\ aG_3 & -aG_3 & a \\ 1/R_6 & 0 & 0 \\ 0 & 1/R_7 & 0 \end{bmatrix}\begin{bmatrix} v_{C1} \\ v_{C2} \\ i_{L4} \end{bmatrix} + \begin{bmatrix} 0 & 0 \\ 0 & 0 \\ -1/R_6 & 0 \\ 0 & -1/R_7 \end{bmatrix}\begin{bmatrix} v_{s1} \\ v_{s2} \end{bmatrix}. \tag{5.56}$$

This is an output equation in the form of equation 5.44b, where the output vector is $\mathbf{y} = [v_3\ i_5\ i_6\ i_7]^T$ and the meanings of the constant matrixes are obvious.

Remark. The capacitor charges and the inductor fluxes can also be used as state variables. Then in the above example the state vector will be

$$\mathbf{x} = [q_1 \quad q_2 \quad \lambda_4]^T,$$

where $q_1 = C_1 v_{C1}$, $q_2 = C_2 v_{C2}$ and $\lambda_4 = L_4 i_{L4}$.

Substituting $v_{C1} = q_1/C_1$, $v_{C2} = q_2/C_2$ and $i_4 = \lambda_4/L_4$ in equation 5.55, and after simplification, we obtain

$$\frac{d}{dt}\begin{bmatrix} q_1 \\ q_2 \\ \lambda_4 \end{bmatrix} = \begin{bmatrix} -\dfrac{1 + aR_6G_3}{R_6C_1} & \dfrac{aG_3}{C_2} & -\dfrac{a}{L_4} \\ \dfrac{aG_3}{C_1} & -\dfrac{1 + aR_7G_3}{R_7C_2} & -\dfrac{1 + a}{L_4} \\ \dfrac{a}{C_1} & \dfrac{1 - a}{C_2} & -\dfrac{aR_5}{L_4} \end{bmatrix} \begin{bmatrix} q_1 \\ q_{C2} \\ \lambda_4 \end{bmatrix} + \begin{bmatrix} \dfrac{1}{R_6} & 0 \\ 0 & \dfrac{1}{R_7} \\ 0 & 0 \end{bmatrix} \begin{bmatrix} v_{s1} \\ v_{s2} \end{bmatrix} \tag{5.57}$$

which is the state equation using the charges and fluxes as state variables.

It is worthwhile mentioning that some other variables in the circuit may be used as state variables. For example, a current through a resistor in parallel with a capacitor or voltage across a resistor in series with an inductor can be treated as state variables. Also any linear combination of capacitor voltages or inductor currents may be used as state variables. This can be helpful in writing state equations when the circuit consists of all-capacitor loops or all-inductor cut-sets. The next step would be to solve the state equations. However, before doing so, we shall consider the general approach for deriving state equations in matrix form.

5.6 A SYSTEMATIC METHOD FOR WRITING A STATE EQUATION BASED ON CIRCUIT MATRIX REPRESENTATION

Consider a network whose elements are inductors, capacitors, resistors and independent sources. As stated, we assume that capacitors do not form a loop and inductors do not form a cut-set. We also assume that the network graph is connected and as a first step we will pick a *proper tree*. We can obviously include all capacitors into the tree branches, since they do not form any loop. Usually, it might be necessary to add some resistors and/or voltage sources in order to complete the tree. Then all the inductors will be assigned to the links. In the next step we shall partition the circuit branches into four sub-sets: the capacitive twigs, the resistive twigs, the inductive links and the resistive links. For the sake of specifics, we shall use an example to illustrate this procedure.

Consider again the circuit shown in Fig. 5.9(a). The circuit graph and the proper tree are shown in Fig. 5.9(b). The KCL equations for the fundamental cut-sets, in accordance with equation 5.31, are

$$[\mathbf{1}_t \quad \mathbf{Q}_\ell]\begin{bmatrix} \mathbf{i}_C \\ \mathbf{i}_G \\ -- \\ \mathbf{i}_L \\ \mathbf{i}_R \end{bmatrix} = \mathbf{0}, \tag{5.58}$$

where subvectors of twig and link currents are

$$\mathbf{i}_t = \begin{bmatrix} \mathbf{i}_C \\ \mathbf{i}_G \end{bmatrix}, \quad \mathbf{i}_\ell = \begin{bmatrix} \mathbf{i}_L \\ \mathbf{i}_R \end{bmatrix}$$

and $\mathbf{i}_C$, $\mathbf{i}_G$, $\mathbf{i}_L$ and $\mathbf{i}_R$ are in turn subvectors representing currents in capacitive and resistive (conductive) twigs and inductive and resistive links, respectively. In our example, these four subvectors are

$$\mathbf{i}_C = \begin{bmatrix} i_{C1} \\ i_{C2} \end{bmatrix}, \quad \mathbf{i}_G = [i_{G3}], \quad \mathbf{i}_L = [i_{L4}], \quad \mathbf{i}_R = \begin{bmatrix} i_{R5} \\ i_{R6} \\ i_{R7} \end{bmatrix} \tag{5.59}$$

and the equation 5.58 becomes

$$\begin{bmatrix} 1 & 0 & 0 & \overbrace{0}^{\mathbf{Q}_{CL}} & \overbrace{1 \quad 1 \quad 0}^{\mathbf{Q}_{CR}} \\ 0 & 1 & 0 & 1 & -1 \quad 0 \quad 1 \\ 0 & 0 & 1 & \underbrace{-1}_{\mathbf{Q}_{GL}} & \underbrace{1 \quad 0 \quad 0}_{\mathbf{Q}_{GR}} \end{bmatrix}_{\mathbf{Q}_\ell} \begin{bmatrix} i_{C1} \\ i_{C2} \\ i_{C3} \\ -- \\ i_{L4} \\ i_{R5} \\ i_{R6} \\ i_{R7} \end{bmatrix} = \mathbf{0} \tag{5.60}$$

The KVL equations may be written in the form (see equation 5.26)

$$[\mathbf{B}_t \quad \mathbf{1}_\ell]\begin{bmatrix} \mathbf{v}_C \\ \mathbf{v}_G \\ -- \\ \mathbf{v}_L \\ \mathbf{v}_R \end{bmatrix} = \mathbf{0}, \tag{5.61}$$

where

$$\mathbf{v}_t = \begin{bmatrix} \mathbf{v}_C \\ \mathbf{v}_G \end{bmatrix}, \quad \mathbf{v}_\ell = \begin{bmatrix} \mathbf{v}_L \\ \mathbf{v}_R \end{bmatrix}$$

are subvectors of twig and link voltages and $\mathbf{v}_C$, $\mathbf{v}_G$, $\mathbf{v}_L$, $\mathbf{v}_R$ are in turn subvectors representing voltages across the capacitive and resistive (conductive) twigs and inductive and resistive links, respectively. For the circuit in Fig. 5.9(a) the voltage subvectors are

$$\mathbf{v}_C = \begin{bmatrix} v_{C1} \\ v_{C2} \end{bmatrix}, \quad \mathbf{v}_G = [v_{G3}], \quad \mathbf{v}_L = [v_{L4}], \quad \mathbf{v}_R = \begin{bmatrix} v_{R5} \\ v_{R6} - v_{sR6} \\ v_{R7} - v_{sR7} \end{bmatrix} = \begin{bmatrix} v_\ell 5 \\ v_6 \\ v_7 \end{bmatrix} \tag{5.62}$$

where v_{sR6} represents v_{s1} and v_{sR7} represents v_{s2}. The KVL equation 5.61 becomes

$$\left[\begin{array}{cc:c:cccc} \multicolumn{2}{c}{\mathbf{B}_{LC}} & \mathbf{B}_{LG} & & & & \\ 0 & -1 & 1 & 1 & 0 & 0 & 0 \\ \hdashline -1 & 1 & -1 & 0 & 1 & 0 & 0 \\ -1 & 0 & 0 & 0 & 0 & 1 & 0 \\ 0 & -1 & 0 & 0 & 0 & 0 & 1 \\ \multicolumn{2}{c}{\mathbf{B}_{RC}} & \mathbf{B}_{RG} & & & & \\ \multicolumn{3}{c}{\mathbf{B}_t} & & & & \end{array}\right] \begin{bmatrix} v_{C1} \\ v_{C2} \\ v_{G3} \\ \hdashline v_{L4} \\ v_{R5} \\ v_6 \\ v_7 \end{bmatrix} = \mathbf{0}. \tag{5.63}$$

Note that $\mathbf{B}_t = -\mathbf{Q}_\ell^T$.

Next we shall use the *v-i*, or *i-v* characteristics to introduce branch equations. We will employ the concept of a generalized branch, i.e. combining passive and active elements together. However, we must now take into consideration four different branches: two for twigs and two for links, as shown in Fig. 5.10. As was mentioned earlier, we shall assume that the voltage sources are located in the link branches and the current sources are located in the twig branches. Therefore, in matrix form we have:

$$\begin{aligned} &\textit{capacitor twigs} \quad \mathbf{i}_C = \mathbf{C}\frac{d}{dt}\mathbf{v}_C + \mathbf{i}_{sC} \\ &\textit{inductor links} \quad \mathbf{v}_L = \mathbf{L}\frac{d}{dt}\mathbf{i}_L + \mathbf{v}_{sL} \end{aligned} \tag{5.64}$$

$$\begin{aligned} &\textit{resistor twigs} \quad \mathbf{i}_G = \mathbf{G}\mathbf{v}_G + \mathbf{i}_{sG} \\ &\textit{resistor links} \quad \mathbf{v}_R = \mathbf{R}\mathbf{i}_R + \mathbf{v}_{sR} \end{aligned} \tag{5.65}$$

where the matrixes **C**, **L**, **G** and **R** are the branch parameter matrixes; namely,

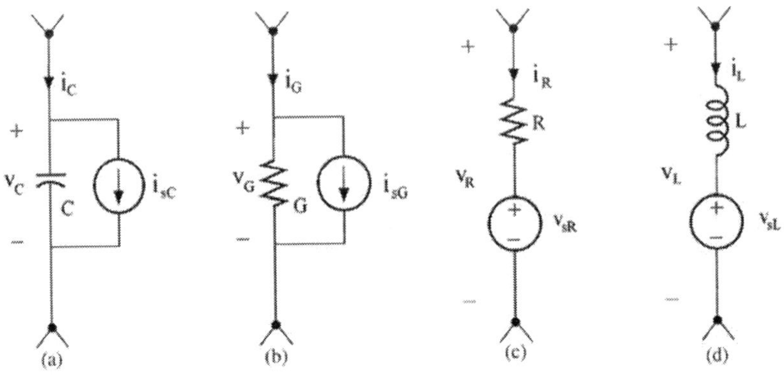

Figure 5.10 Generalized branches with independent sources: twig capacitor (a), twig resistor (b), link resistor (c) and link inductor (d).

the twig capacitance matrix, the link inductance matrix, the twig conductance matrix and the link resistance matrix, respectively. Note that **C**, **L**, **G** and **R** are square diagonal matrixes, but if the circuit consists of coupled elements (mutual inductances and/or dependent sources), **L**, **G** and **R** might not be diagonal any more. For the example in Fig. 5.9

$$\mathbf{C} = \begin{bmatrix} C_1 & 0 \\ 0 & C_2 \end{bmatrix}, \quad \mathbf{L} = [L_4] \tag{5.66}$$

$$\mathbf{G} = [G_3], \quad \mathbf{R} = \begin{bmatrix} R_5 & & 0 \\ & R_6 & \\ 0 & & R_7 \end{bmatrix}. \tag{5.67}$$

The vectors $\mathbf{v}_{sR}$, $\mathbf{v}_{sL}$ and $\mathbf{i}_{sG}$, $\mathbf{i}_{sC}$ represent the independent voltage and current sources, which in the present example are

$$\mathbf{v}_{sR} = \begin{bmatrix} 0 \\ v_{s1} \\ v_{s2} \end{bmatrix}, \quad \mathbf{v}_{sL} = \mathbf{0}, \quad \mathbf{i}_{sG} = \mathbf{0}, \quad \mathbf{i}_{sC} = \mathbf{0}. \tag{5.68}$$

Equation 5.64 can be rewritten to yield

$$\mathbf{C}\frac{d}{dt}\mathbf{v}_C = \mathbf{i}_C - \mathbf{i}_{sC}, \quad \mathbf{L}\frac{d}{dt}\mathbf{i}_L = \mathbf{v}_L - \mathbf{v}_{sL}. \tag{5.69}$$

To bring these equations to the form of state equations, we must eliminate the variables. For this purpose, we shall solve the KCL equation 5.58 and KVL equation 5.61 equations together with the branch equations 5.64 and 5.65.

Equations 5.58 and 5.61 can be rewritten as

$$\begin{bmatrix} \mathbf{i}_C \\ \mathbf{i}_G \end{bmatrix} = -\mathbf{Q}_\ell \begin{bmatrix} \mathbf{i}_L \\ \mathbf{i}_R \end{bmatrix} = -\begin{bmatrix} \mathbf{Q}_{CL} & \mathbf{Q}_{CR} \\ \mathbf{Q}_{GL} & \mathbf{Q}_{GR} \end{bmatrix} \begin{bmatrix} \mathbf{i}_L \\ \mathbf{i}_R \end{bmatrix} \tag{5.70a}$$

and

$$\begin{bmatrix} \mathbf{v}_L \\ \mathbf{v}_R \end{bmatrix} = -\mathbf{B}_t \begin{bmatrix} \mathbf{v}_C \\ \mathbf{v}_G \end{bmatrix} = -\begin{bmatrix} \mathbf{B}_{LC} & \mathbf{B}_{LG} \\ \mathbf{B}_{RC} & \mathbf{B}_{RG} \end{bmatrix} \begin{bmatrix} \mathbf{v}_C \\ \mathbf{v}_G \end{bmatrix} \tag{5.70b}$$

where in the following solution matrixes $\mathbf{Q}_\ell$ and $\mathbf{B}_t$ are partitioned into submatrixes. The order of each of the submatrixes in equations 5.70 is determined by the number of twigs (which is the number of rows) and by the number of corresponding links (which is the number of columns) in equation 5.70a and vice versa in equation 5.70b. For example, the number of rows in $\mathbf{Q}_{CL}$ (equation 5.70a) is equal to the number of capacitor currents in $\mathbf{i}_C$ (capacitor twigs) and the number of its columns is equal to the number of inductor currents in $\mathbf{i}_L$ (inductor links). It can also be shown that there are simple relations between $\mathbf{Q}_\ell$ and $\mathbf{B}_t$ submatrixes, namely

$$\mathbf{B}_{LC} = -\mathbf{Q}^T_{CL}, \quad \mathbf{B}_{LG} = -\mathbf{Q}^T_{GL}, \quad \mathbf{B}_{RC} = -\mathbf{Q}^T_{CR}, \quad \mathbf{B}_{RG} = -\mathbf{Q}^T_{GR}. \tag{5.71}$$

The undesirable variables $\mathbf{i}_C$ and $\mathbf{v}_L$ in equation 5.69 can now be expressed from equation 5.70 to yield

$$\mathbf{i}_C = -\mathbf{Q}_{CL}\mathbf{i}_L - \mathbf{Q}_{CR}\mathbf{i}_R \tag{5.72a}$$

$$\mathbf{v}_L = -\mathbf{B}_{LC}\mathbf{v}_C - \mathbf{B}_{LG}\mathbf{v}_G, \tag{5.72b}$$

and after substituting these two expressions into equation 5.69, we obtain

$$\begin{aligned} \mathbf{C}\frac{d}{dt}\mathbf{v}_C &= -\mathbf{Q}_{CL}\mathbf{i}_L - \mathbf{Q}_{CR}\mathbf{i}_R - \mathbf{i}_{sC} \\ \mathbf{L}\frac{d}{dt}\mathbf{i}_L &= -\mathbf{B}_{LC}\mathbf{v}_C - \mathbf{B}_{LG}\mathbf{v}_G - \mathbf{v}_{sL}. \end{aligned} \tag{5.73}$$

However, we still need to eliminate $\mathbf{i}_R$ and $\mathbf{v}_G$. Substituting $\mathbf{i}_G$ and $\mathbf{v}_R$ from equation 5.70 into equation 5.65, and after rearrangement, results in two simultaneous matrix equations in two unknowns $\mathbf{i}_R$ and $\mathbf{v}_G$,

$$\mathbf{R}\mathbf{i}_R + \mathbf{B}_{RG}\mathbf{v}_G = \mathbf{M} \tag{5.73a}$$

$$\mathbf{Q}_{GR}\mathbf{i}_R + \mathbf{G}\mathbf{v}_G = \mathbf{N}, \tag{5.73b}$$

where

$$\mathbf{M} = -\mathbf{B}_{RC}\mathbf{v}_C - \mathbf{v}_{sR} \quad \text{and} \quad \mathbf{N} = -\mathbf{Q}_{GL}\mathbf{i}_L - \mathbf{i}_{sG} \tag{5.74}$$

Solving these two equations by the substitution method yields

$$\mathbf{i}_R = \mathbf{R}_{eq}^{-1}(-\mathbf{B}_{RG}\mathbf{G}^{-1}\mathbf{N} + \mathbf{M}) \tag{5.75a}$$

$$\mathbf{v}_G = \mathbf{G}_{eq}^{-1}(-\mathbf{Q}_{GR}\mathbf{R}^{-1}\mathbf{M} + \mathbf{N}), \tag{5.75b}$$

where

$$\mathbf{R}_{eq} = \mathbf{R} - \mathbf{B}_{RG}\mathbf{G}^{-1}\mathbf{Q}_{GR} \tag{5.76a}$$

$$\mathbf{G}_{eq} = \mathbf{G} - \mathbf{Q}_{GR}\mathbf{R}^{-1}\mathbf{B}_{RG}. \tag{5.76b}$$

Finally, we substitute equation 5.75 with equation 5.74 in equation 5.73 to obtain, after rearrangement, the state representation is follows

$$\frac{d}{dt}\begin{bmatrix}\mathbf{v}_C \\ \mathbf{i}_L\end{bmatrix} = \underbrace{\begin{bmatrix}\mathbf{C} & \mathbf{0} \\ \mathbf{0} & \mathbf{L}\end{bmatrix}^{-1}}_{\mathbf{A}} \underbrace{\begin{bmatrix}\mathbf{A}^1_{11} & \mathbf{A}^1_{12} \\ \mathbf{A}^1_{21} & \mathbf{A}^1_{22}\end{bmatrix}}_{\mathbf{A}^1} \begin{bmatrix}\mathbf{v}_C \\ \mathbf{i}_L\end{bmatrix}$$

$$+ \underbrace{\begin{bmatrix}\mathbf{C} & \mathbf{0} \\ \mathbf{0} & \mathbf{L}\end{bmatrix}^{-1}}_{\mathbf{b}} \underbrace{\begin{bmatrix}\mathbf{b}^1_{11} & \mathbf{b}^1_{12} & \mathbf{b}^1_{13} & \mathbf{b}^1_{14} \\ \mathbf{b}^1_{21} & \mathbf{b}^1_{22} & \mathbf{b}^1_{23} & \mathbf{b}^1_{24}\end{bmatrix}}_{\mathbf{b}^1} \begin{bmatrix}\mathbf{i}_{sC} \\ \mathbf{i}_{sG} \\ \mathbf{v}_{sL} \\ \mathbf{v}_{sR}\end{bmatrix} \tag{5.77}$$

where the matrix terms are

$$\begin{aligned}\mathbf{A}^1_{11} &= \mathbf{Q}_{CR}\mathbf{R}_{eq}^{-1}\mathbf{B}_{RC} \quad & \mathbf{A}^1_{12} &= -\mathbf{Q}_{CL} - \mathbf{Q}_{CR}\mathbf{R}_{eq}^{-1}\mathbf{B}_{RG}\mathbf{G}^{-1}\mathbf{Q}_{GL} \\ \mathbf{A}^1_{22} &= \mathbf{B}_{LG}\mathbf{G}_{eq}^{-1}\mathbf{Q}_{GL} \quad & \mathbf{A}^1_{21} &= -\mathbf{B}_{LC} - \mathbf{B}_{LG}\mathbf{G}_{eq}^{-1}\mathbf{Q}_{GR}\mathbf{R}^{-1}\mathbf{B}_{RC}\end{aligned} \tag{5.78}$$

$$\begin{aligned}&\mathbf{b}^1_{11} = -\mathbf{1} \quad \mathbf{b}^1_{12} = -\mathbf{Q}_{CR}\mathbf{R}_{eq}^{-1}\mathbf{B}_{RG}\mathbf{G}^{-1} \quad \mathbf{b}^1_{13} = \mathbf{0} \quad \mathbf{b}^1_{14} = \mathbf{Q}_{CR}\mathbf{R}_{eq}^{-1} \\ &\mathbf{b}^1_{21} = \mathbf{0} \quad \mathbf{b}^1_{22} = \mathbf{B}_{LG}\mathbf{G}_{eq}^{-1} \quad \mathbf{b}^1_{23} = -\mathbf{1} \quad \mathbf{b}^1_{24} = -\mathbf{B}_{LG}\mathbf{G}_{eq}^{-1}\mathbf{Q}_{GR}\mathbf{R}^{-1}.\end{aligned} \tag{5.79}$$

Let us now use the above expressions to calculate the **A** and **b** matrixes in our example.

First we determine the submatrixes of the $\mathbf{Q}_\ell$ matrix

$$\mathbf{Q}_\ell \begin{bmatrix}\mathbf{Q}_{CL} & \mathbf{Q}_{CR} \\ \mathbf{Q}_{GL} & \mathbf{Q}_{GR}\end{bmatrix} = \left[\begin{array}{c|ccc} 0 & 1 & 1 & 0 \\ 1 & -1 & 0 & 1 \\ \hline -1 & 1 & 0 & 0\end{array}\right].$$

Then with equation 5.76 and equation 5.71 we have

$$\mathbf{R}_{eq}\begin{bmatrix}R_5 & & \mathbf{0} \\ & R_6 & \\ \mathbf{0} & & R_7\end{bmatrix} - \begin{bmatrix}1 \\ 0 \\ 0\end{bmatrix}\begin{bmatrix}\dfrac{1}{G_3}\end{bmatrix}[1 \quad 0 \quad 0] = \begin{bmatrix}\dfrac{1+R_5G_3}{G_3} & 0 & 0 \\ 0 & R_6 & 0 \\ 0 & 0 & R_7\end{bmatrix}$$

$$\mathbf{R}_{eq}^{-1} = \begin{bmatrix}aG_3 & 0 & 0 \\ 0 & 1/R_6 & 0 \\ 0 & 0 & 1/R_7\end{bmatrix}, \quad \text{where again } a = \frac{1}{1+R_5G_3}$$

$$\mathbf{G}_{eq}=[G_3]+[1 \quad 0 \quad 0]\begin{bmatrix}1/R_5 & 0 & 0\\ 0 & 1/R_6 & 0\\ 0 & 0 & 1/R_7\end{bmatrix}\begin{bmatrix}-1\\ 0\\ 0\end{bmatrix}=\left[\frac{1+R_5G_3}{R_5}\right]=\left[\frac{1}{aR_5}\right]$$

$$\mathbf{G}_{eq}^{-1}=[aR_5]$$

$$\mathbf{A}_{11}^{1}=\begin{bmatrix}1 & 1 & 0\\ -1 & 0 & 1\end{bmatrix}\begin{bmatrix}aG_3 & & 0\\ & 1/R_6 & \\ 0 & & 1/R_7\end{bmatrix}\begin{bmatrix}-1 & +1\\ -1 & 0\\ 0 & -1\end{bmatrix}$$

$$=-\begin{bmatrix}(aG_3+1/R_6) & -aG_3\\ -aG_3 & (aG_3+1/R_7)\end{bmatrix}$$

$$\mathbf{A}_{22}^{1}=[1][aR_5][-1]=-[aR_5]$$

$$\mathbf{A}_{12}^{1}=-\begin{bmatrix}0\\ 1\end{bmatrix}-\begin{bmatrix}1 & 1 & 0\\ -1 & 0 & 1\end{bmatrix}\begin{bmatrix}aG_3 & 0 & 0\\ 0 & 1/R_6 & 0\\ 0 & 0 & 1/R_7\end{bmatrix}\begin{bmatrix}-1\\ 0\\ 0\end{bmatrix}[1/G_3][-1]$$

$$=-\begin{bmatrix}1\\ 1-a\end{bmatrix}$$

$$\mathbf{A}_{21}^{1}=-[0 \quad -1]-[1][aR_5][1 \quad 0 \quad 0]\begin{bmatrix}1/R_5 & 0 & 0\\ 0 & 1/R_6 & 0\\ 0 & 0 & 1/R_7\end{bmatrix}\begin{bmatrix}-1 & +1\\ -1 & 0\\ 0 & -1\end{bmatrix}$$

$$=[a \quad (1-a)]$$

$$\begin{bmatrix}\mathbf{C} & \mathbf{0}\\ \mathbf{0} & \mathbf{L}\end{bmatrix}^{-1}=\begin{bmatrix}C_1 & 0 & 0\\ 0 & C_2 & 0\\ 0 & 0 & L_2\end{bmatrix}^{-1}=\begin{bmatrix}1/C_1 & 0 & 0\\ 0 & 1/C_2 & 0\\ 0 & 0 & 1/L_4\end{bmatrix}$$

Therefore the **A** matrix is

$$\mathbf{A}=\begin{bmatrix}\mathbf{C} & \mathbf{0}\\ \mathbf{0} & \mathbf{L}\end{bmatrix}^{-1}\begin{bmatrix}\mathbf{A}_{11}^{1} & \mathbf{A}_{12}^{1}\\ \mathbf{A}_{21}^{1} & \mathbf{A}_{22}^{1}\end{bmatrix}=\left[\begin{array}{cc|c}-\dfrac{1+R_6aG_3}{R_6C_1} & \dfrac{aG_3}{C_1} & -\dfrac{a}{C_1}\\ \dfrac{aG_3}{C_2} & -\dfrac{1+R_7aG_3}{R_7C_2} & -\dfrac{1-a}{C_2}\\ \hline \dfrac{a}{L_4} & \dfrac{1-a}{L_4} & -\dfrac{aR_5}{L_4}\end{array}\right],$$

which agrees with the results previously obtained (see equation 5.55).

To find the **b** matrix we will calculate equation 5.79. Since only the $\mathbf{v}_{sR}$ vector is present we need only two elements of **b**:

$$\mathbf{b}_{14}^{1} = \begin{bmatrix} 1 & 1 & 0 \\ -1 & 0 & 1 \end{bmatrix} \begin{bmatrix} aG_3 & 0 & 0 \\ 0 & 1/R_6 & 0 \\ 0 & 0 & 1/R_7 \end{bmatrix} = \begin{bmatrix} aG_3 & 1/R_6 & 0 \\ -aG_3 & 0 & 1/R_7 \end{bmatrix}$$

$$\mathbf{b}_{24}^{1} = -[1][aR_5][1 \quad 0 \quad 0] \begin{bmatrix} 1/R_5 & 0 & 0 \\ 0 & 1/R_6 & 0 \\ 0 & 0 & 1/R_7 \end{bmatrix} = -[a \quad 0 \quad 0]$$

Therefore, the reduced **b** matrix is

$$\mathbf{b} = \begin{bmatrix} \mathbf{C} & \mathbf{0} \\ \mathbf{0} & \mathbf{L} \end{bmatrix}^{-1} \begin{bmatrix} \mathbf{b}_{14}^{1} \\ \mathbf{b}_{24}^{1} \end{bmatrix} = \begin{bmatrix} aG_3/C_1 & 1/R_6C_1 & 0 \\ -aG_3/C_2 & 0 & 1/R_7C_2 \\ -a/L_4 & 0 & 0 \end{bmatrix}$$

which also agrees with the results in equation 5.55. Note that a voltage source in link 5 is absent ($v_{sR5} = 0$), therefore the above matrix can be reduced even more, namely

$$\mathbf{b} = \begin{bmatrix} 1/R_6C_1 & 0 \\ 0 & 1/R_7C_2 \\ 0 & 0 \end{bmatrix}$$

which is exactly the same as in equation 5.55.

Comparing the systematic method for writing state equations with the intuitive approach, which we first presented in the previous sections, we may conclude that it is rather complicated. In many practical instances, the final results can be arrived at much easier and faster by following the intuitive approach. However, the systematic method has an appreciable advantage for computer-aided analysis, since it can be easily programmed.

5.7 COMPLETE SOLUTION OF THE STATE MATRIX EQUATION

We will now turn to the solution of the state equation of the form of equation 5.44a, repeated here for convenience:

$$\dot{\mathbf{x}}(t) = \mathbf{A}\mathbf{x}(t) + \mathbf{b}\mathbf{w}(t). \tag{5.80}$$

5.7.1 The natural solution

We will begin by considering the natural or zero-input (non-forced) solution; that is $\mathbf{w}(t) = 0$. Equation 5.80 then simplifies to

$$\dot{\mathbf{x}}(t) = \mathbf{A}\mathbf{x}(t) \quad \text{or} \quad \dot{\mathbf{x}}(t) - \mathbf{A}\mathbf{x}(t) = \mathbf{0}. \tag{5.81}$$

It is customary to compare a vector problem with its scalar version. In this case, the scalar version of equation 5.81 is

$$\frac{dx(t)}{dt} = ax(t). \tag{5.82}$$

The solution of equation 5.82, that satisfies the initial condition $x(0)$, is

$$x(t) = e^{at}x(0).$$

Suppose we try the same form for the solution of equation 5.81, that is

$$\mathbf{x}(t) = \mathbf{e}^{\mathbf{A}t}\mathbf{x}(0). \tag{5.83}$$

where $\mathbf{e}^{\mathbf{A}t}$ is called the *matrix exponential* and is an example of a function of matrix $\mathbf{A}$.

5.7.2 Matrix exponential

In mathematics the matrix exponential is defined similarly to a scalar exponential (or complex exponential), i.e. in terms of the power series expansion:

$$\mathbf{e}^{\mathbf{A}t} = \mathbf{1} + \frac{t}{1!}\mathbf{A} + \frac{t^2}{2!}\mathbf{A}^2 + \cdots + \frac{t^k}{k!}\mathbf{A}^k + \cdots = \sum_{k=0}^{\infty}\frac{t^k}{k!}\mathbf{A}^k. \tag{5.84}$$

Since $\mathbf{A}$ is a square matrix of order n, the matrix exponential $\mathbf{e}^{\mathbf{A}t}$ is also a square matrix of order n.

Example 5.2

As an example, let us take the matrix of Example 5.1, namely

$$\mathbf{A} = \begin{bmatrix} -1 & -1 \\ 1 & -1 \end{bmatrix}$$

then

$$\mathbf{A}^2 = \begin{bmatrix} -1 & -1 \\ 1 & -1 \end{bmatrix}\begin{bmatrix} -1 & -1 \\ 1 & -1 \end{bmatrix} = \begin{bmatrix} 0 & 2 \\ -2 & 0 \end{bmatrix}, \quad \mathbf{A}^3 = \begin{bmatrix} 2 & -2 \\ 2 & 2 \end{bmatrix}$$

and

$$\begin{aligned} \mathbf{e}^{\mathbf{A}t} &= \begin{bmatrix} 1 & 0 \\ 0 & 1 \end{bmatrix} + t\begin{bmatrix} -1 & -1 \\ 1 & -1 \end{bmatrix} + \frac{t^2}{2}\begin{bmatrix} 0 & 2 \\ -2 & 0 \end{bmatrix} + \frac{t^3}{6}\begin{bmatrix} 2 & -2 \\ 2 & 2 \end{bmatrix} + \cdots \\ &= \begin{bmatrix} 1 - t + \dfrac{t^3}{3} + \cdots & -t + t^2 - \dfrac{t^3}{3} + \cdots \\ t - t^2 + \dfrac{t^3}{3} + \cdots & 1 - t + \dfrac{t^3}{3} + \cdots \end{bmatrix}. \end{aligned} \tag{5.85}$$

As can be seen from equation 5.85, each of the elements of the matrix $e^{\mathbf{A}t}$ is a continuous function of t. Term-by-term differentiation of the matrix exponential (equation 5.84) results in

$$\frac{d}{dt}(\mathbf{e}^{\mathbf{A}t}) = \mathbf{A} + t\mathbf{A}^2 + \frac{t^2}{2!}\mathbf{A}^3 + \frac{t^3}{3!}\mathbf{A}^4 + \cdots$$

$$= \mathbf{A}\left(\mathbf{1} + t\mathbf{A} + \frac{t^2}{2!}\mathbf{A}^2 + \frac{t^3}{3!}\mathbf{A}^3 + \cdots\right) = \mathbf{A}\mathbf{e}^{\mathbf{A}t}, \tag{5.86}$$

i.e., the formula for the derivative of a matrix exponential is the same as it is for a scalar exponential. Substituting equation 5.83 into the matrix differential equation 5.81, results in identity:

$$\mathbf{A}\mathbf{e}^{\mathbf{A}t}\mathbf{x}(0) = \mathbf{A}\mathbf{e}^{\mathbf{A}t}\mathbf{x}(0).$$

Thus, we have established that equation 5.83 is indeed the solution to equation 5.81.

We must now show that the inverse of a matrix exponential exists and equals $(\mathbf{e}^{\mathbf{A}t})^{-1} = \mathbf{e}^{-\mathbf{A}t}$. For the latter we can write

$$\mathbf{e}^{-\mathbf{A}t} = \mathbf{1} - \mathbf{A}t + \mathbf{A}^2\frac{t^2}{2!} - \mathbf{A}^3\frac{t^3}{3!} + \cdots + (-1)^k\mathbf{A}^k\frac{t^k}{k!} + \cdots.$$

Now let this series be multiplied by the series for the positive exponential in equation 5.84. This term-by-term multiplication results in **1** since all other terms are cancelled. Thus,

$$\mathbf{e}^{\mathbf{A}t}\mathbf{e}^{-\mathbf{A}t} = \mathbf{1}.$$

This result tells us that the matrix $\mathbf{e}^{-\mathbf{A}t}$ is the inverse of $\mathbf{e}^{\mathbf{A}t}$, since by definition the product of the matrix by its inverse gives a unit matrix. This result can be used, first of all, to show that in general if the initial vector $\mathbf{x}(0)$ is known for some time, for instance t_0, namely $\mathbf{x}_{nat}(t_0)$ then the solution will be

$$\mathbf{x}_n(t) = \mathbf{e}^{\mathbf{A}(t-t_0)}\mathbf{x}(t_0). \tag{5.87}$$

Indeed, substituting $t = t_0$, results in identity:

$$\mathbf{x}_n(t_0) = \mathbf{e}^{\mathbf{A}t_0}\mathbf{e}^{-\mathbf{A}t_0}\mathbf{x}(t_0) = \mathbf{1}\mathbf{x}(t_0),$$

where we have used

$$\mathbf{e}^{\mathbf{A}+\mathbf{B}} = \mathbf{e}^{\mathbf{A}}\cdot\mathbf{e}^{\mathbf{B}}.$$

(This can be verified by using equation 5.84 for both sides of equality.)

5.7.3 The particular solution

To find the complete solution to equation 5.80, we must now find the particular solution to the differential equation, i.e. the forced response. For this purpose, assume a solution of the form

$$\mathbf{x}_p(t) = \mathbf{e}^{\mathbf{A}t}\mathbf{q}(t), \tag{5.88}$$

where $\mathbf{q}(t)$ is an unknown function to be determined. In order to be a solution, equation 5.88 has to satisfy the differential equation. Substituting equation 5.88 in equation 5.80 gives

$$\frac{d}{dt}[\mathbf{e}^{\mathbf{A}t}\mathbf{q}(t)] = \mathbf{A}\mathbf{e}^{\mathbf{A}t}\mathbf{q}(t) + \mathbf{b}\mathbf{w}(t),$$

or

$$\mathbf{A}\mathbf{e}^{\mathbf{A}t}\mathbf{q}(t) + \mathbf{e}^{\mathbf{A}t}\frac{d\mathbf{q}(t)}{dt} = \mathbf{A}\mathbf{e}^{\mathbf{A}t}\mathbf{q}(t) + \mathbf{b}\mathbf{w}(t).$$

Thus

$$\frac{d\mathbf{q}(t)}{dt} = \mathbf{e}^{-\mathbf{A}t}\mathbf{b}\mathbf{w}(t). \tag{5.89}$$

Integrating, we obtain

$$\mathbf{q}(t) = \mathbf{q}(t_0) + \int_{t_0}^{t} \mathbf{e}^{-\mathbf{A}\tau}\mathbf{b}\mathbf{w}(\tau)d\tau.$$

Thus, the particular solution is

$$\mathbf{x}_p(t) = \mathbf{e}^{\mathbf{A}t}\mathbf{q}(t) = \mathbf{e}^{\mathbf{A}t}\mathbf{q}(t_0) + \int_{t_0}^{t} \mathbf{e}^{\mathbf{A}(t-\tau)}\mathbf{b}\mathbf{w}(\tau)d\tau.$$

To evaluate $\mathbf{q}(t_0)$, we use the complete solution being evaluated at t_0

$$\mathbf{x}(t)|_{t=t_0} = \mathbf{x}_n(t) + \mathbf{x}_p(t) = \mathbf{e}^{\mathbf{A}(t-t_0)}\mathbf{x}(t_0) + \mathbf{e}^{\mathbf{A}t}\mathbf{q}(t_0) + \left.\int_{t_0}^{t} \mathbf{e}^{\mathbf{A}(t-\tau)}\mathbf{b}\mathbf{w}(\tau)d\tau\right|_{t=t_0},$$

or

$$\mathbf{x}(t_0) = \mathbf{x}(t_0) + \mathbf{e}^{\mathbf{A}t_0}\mathbf{q}(t_0) + \mathbf{0},$$

which implies that $\mathbf{q}(t_0) = \mathbf{0}$.

Hence, finally the complete solution of the state equation 5.80 is

$$\mathbf{x}(t) = \mathbf{e}^{\mathbf{A}(t-t_0)}\mathbf{x}(t_0) + \int_{t_0}^{t} \mathbf{e}^{\mathbf{A}(t-\tau)}\mathbf{b}\mathbf{w}(\tau)d\tau. \tag{5.90}$$

To evaluate this solution the basic calculation is a determination of the matrix exponential $\mathbf{e}^{\mathbf{A}t}$. This will be discussed in the next subsection.

5.8 BASIC CONSIDERATIONS IN DETERMINING FUNCTIONS OF A MATRIX

In this section, we shall examine two methods of computing $\mathbf{e}^{\mathbf{A}t}$ in closed form. This matrix exponential is a particular function of a matrix. The simplest functions of a matrix are powers of a matrix and polynomials. As we have seen, the matrix exponential can be represented by an infinite series of such functions. The matrix polynomial has the form

$$f(\mathbf{A}) = \mathbf{A}^n + a_{n-1}\mathbf{A}^{n-1} + \cdots + a_1\mathbf{A} + a_0\mathbf{1}. \tag{5.91}$$

The generalization of polynomials is an infinite series:

$$f(\mathbf{A}) = a_0\mathbf{1} + a_0\mathbf{A} + a_2\mathbf{A}^2 + \cdots + a_k\mathbf{A}^k + \cdots = \sum_{k=0}^{\infty} a_k\mathbf{A}^k. \tag{5.92}$$

The function $f(\mathbf{A})$ is itself a matrix, and in the last case each of the matrix elements is an infinite series. This matrix series is said to converge if each of the element series converges.

We will begin with a brief description of some of the properties of matrixes that will be useful in our studies.

5.8.1 Characteristic equation and eigenvalues

An algebraic equation that often appears in network transient analysis is

$$\lambda\mathbf{x} = \mathbf{A}\mathbf{x}, \tag{5.93}$$

where $\mathbf{A}$ is a square matrix of order n. The problem is to find scalars λ and vectors $\mathbf{x}$ that satisfy this equation. A value of λ for which a nontrivial solution of $\mathbf{x}$ exists, is called an eigenvalue, or characteristic value of $\mathbf{A}$. The corresponding vector $\mathbf{x}$ is called an eigenvector, or characteristic vector, of $\mathbf{A}$. After collecting the terms on the left-hand side, we have

$$[\lambda\mathbf{1} - \mathbf{A}]\mathbf{x} = \mathbf{0}. \tag{5.94}$$

This equation will have a nontrivial solution for $\mathbf{x}$ only if the matrix $[\lambda\mathbf{1} - \mathbf{A}]$ is singular, i.e.,

$$\det[\lambda\mathbf{1} - \mathbf{A}] = \mathbf{0}. \tag{5.95}$$

This equation is known as the characteristic equation associated with $\mathbf{A}$. It is also closely related to the auxiliary (characteristic) equation of the corresponding differential equation of order n for the system. The determinant on the left-hand side of equation 5.95 is actually a polynomial of degree n in λ and is called the characteristic polynomial of $\mathbf{A}$. For each value of λ that satisfies the characteristic equation, a nontrivial solution of equation 5.94 can be found. To illustrate this procedure, consider the following example.

Example 5.3

Let us find the eigenvalues and eigenvectors of a matrix of the second order

$$\mathbf{A} = \begin{bmatrix} 2 & 1 \\ 3 & 4 \end{bmatrix}.$$

The characteristic polynomial is also of order two:

$$\det\left\{\lambda\begin{bmatrix}1 & 0\\ 0 & 1\end{bmatrix} - \begin{bmatrix}2 & 1\\ 3 & 4\end{bmatrix}\right\} = \det\begin{bmatrix}\lambda-2 & -1\\ -3 & \lambda-4\end{bmatrix} = \lambda^2 - 6\lambda + 5$$

$$= (\lambda-5)(\lambda-1) = g(\lambda).$$

Thus, $\lambda^2 - 6\lambda + 5 = 0$ is the characteristic equation of the matrix. The roots of the characteristic equation, or the eigenvalues, are

$$\lambda_1 = 5 \quad \text{and} \quad \lambda_2 = 1.$$

To obtain the eigenvector corresponding to the eigenvalue $\lambda_1 = 5$, we solve equation 5.94 by using the given matrix **A**. Thus

$$\left\{\begin{bmatrix}5 & 0\\ 0 & 5\end{bmatrix} - \begin{bmatrix}2 & 1\\ 3 & 4\end{bmatrix}\right\}\begin{bmatrix}x_1\\ x_2\end{bmatrix} = \begin{bmatrix}0\\ 0\end{bmatrix}$$

or

$$\begin{bmatrix}3 & -1\\ -3 & 1\end{bmatrix}\begin{bmatrix}x_1\\ x_2\end{bmatrix} = \begin{bmatrix}0\\ 0\end{bmatrix} \quad \text{and} \quad x_2 = 3x_1.$$

Therefore

$$\begin{bmatrix}x_1\\ x_2\end{bmatrix} = \begin{bmatrix}x_1\\ 3x_1\end{bmatrix} = \begin{bmatrix}1\\ 3\end{bmatrix}[x_1] \quad \text{for any value of } x_1.$$

The eigenvector corresponding to the eigenvalue $\lambda_2 = 1$ is obtained similarly.

$$\begin{bmatrix}-1 & -1\\ -3 & -3\end{bmatrix}\begin{bmatrix}x_1\\ x_2\end{bmatrix} = \begin{bmatrix}0\\ 0\end{bmatrix}$$

from which

$$\begin{bmatrix}x_1\\ x_2\end{bmatrix} = \begin{bmatrix}x_1\\ -x_1\end{bmatrix} = \begin{bmatrix}1\\ -1\end{bmatrix}[x_1] \quad \text{for any value of } x_1.$$

The first method to be discussed for finding functions of a matrix is based on the Caley-Hamilton theorem.

5.8.2 The Caley-Hamilton theorem

This theorem states that *every square matrix satisfies its own characteristic equation.* For example, if we substitute **A** for λ in the characteristic equation of Example 5, we obtain the matrix equation

$$g(\mathbf{A}) = \mathbf{A}^2 - 6\mathbf{A} + 5\cdot\mathbf{1} = \mathbf{0},$$

where, again, **1** is an identity matrix and **0** is a matrix whose elements are all zero. Thus,

$$\begin{bmatrix}2&1\\3&4\end{bmatrix}\begin{bmatrix}2&1\\3&4\end{bmatrix}-6\begin{bmatrix}2&1\\3&4\end{bmatrix}+5\begin{bmatrix}1&0\\0&1\end{bmatrix}=\begin{bmatrix}7&6\\18&19\end{bmatrix}-\begin{bmatrix}12&6\\18&24\end{bmatrix}+\begin{bmatrix}5&0\\0&5\end{bmatrix}$$

$$=\begin{bmatrix}0&0\\0&0\end{bmatrix}.$$

The equation is certainly satisfied in this example.

The Caley-Hamilton theorem permits us to reduce the order of a matrix polynomial of any higher order to be of an order no greater than $n-1$, where n is the order of the matrix. For example, if $\mathbf{A}$ is a square matrix of order 3, then its characteristic equation is

$$g(\lambda)=\lambda^3+a_2\lambda^2+a_1\lambda+a_0=0, \tag{5.96}$$

and by the Caley-Hamilton theorem we have

$$\mathbf{A}^3+a_2\mathbf{A}^2+a_1\mathbf{A}+a_0\mathbf{1}=\mathbf{0}.$$

Then

$$\mathbf{A}^3=-a_2\mathbf{A}^2-a_1\mathbf{A}-a_0\mathbf{1}. \tag{5.97}$$

Thus, $\mathbf{A}^3$ may be expressed in terms of the matrixes of an order not higher than 2 and identity matrix. Hence, the given polynomial of order 3 is reduced to a polynomial of order 2. To extend these results to polynomials of an even higher order, we multiply equation 5.97 throughout by $\mathbf{A}$ to obtain

$$\mathbf{A}^4=-a_2\mathbf{A}^3-a_1\mathbf{A}^2-a_0\mathbf{A}. \tag{5.98}$$

Substituting equation 5.97 for $\mathbf{A}^4$, we obtain

$$\mathbf{A}^4=(a_2^2-a_1)\mathbf{A}^2+(a_2a_1-a_0)\mathbf{A}+a_2a_0\mathbf{1}. \tag{5.99a}$$

To generalize these results, let us develop an iterative formula for expressing higher powers of $\mathbf{A}$. We assign the obtained coefficients in equation 5.99 by upper script, as follows

$$\mathbf{A}^4=a_2^{(1)}\mathbf{A}^2+a_1^{(1)}\mathbf{A}+a_0^{(1)}\mathbf{1}. \tag{5.99b}$$

Multiplying this expression throughout by $\mathbf{A}$, and collecting like terms, yields

$$\mathbf{A}^5=(-a_2a_2^{(1)}+a_1^{(1)})\mathbf{A}^2+(-a_1a_2^{(1)}+a_0^{(1)})\mathbf{A}+(-a_0a_2^{(1)})\mathbf{1}=a_2^{(2)}\mathbf{A}^2+a_1^{(2)}\mathbf{A}+a_0^{(2)}\mathbf{1},$$

where again $a_2^{(2)}, a_1^{(2)}, a_0^{(2)}$ are the new coefficients and a_2, a_1, a_0 are as before the coefficients of the characteristic equation 5.96. Now the iterative formula for this case, $n=3$, can be written as

$$\begin{aligned}\mathbf{A}^{3+k}&=(-a_2a_2^{(k-1)}+a_1^{(k-1)})\mathbf{A}^2+(-a_1a_2^{(k-1)}+a_0^{(k-1)})\mathbf{A}+(-a_0a_2^{(k-1)})\mathbf{1}\\&=a_2^{(k)}\mathbf{A}^2+a_1^{(k)}\mathbf{A}+a_0^{(k)}1.\end{aligned} \tag{5.100}$$

Note that this formula also works fine for the first calculation of $\mathbf{A}^4$ (equation

5.99) if the coefficients in equation 5.97 are assigned as $a_2^{(0)} = -a_2$, $a_1^{(0)} = -a_1$ and $a_0^{(0)} = -a_0$. Generalizing this result (equation 5.100) for any matrix of order n, we can write

$$\mathbf{A}^{n+k} = (-a_{n-1}a_{n-1}^{(k-1)} + a_{n-2}^{(k-1)})\mathbf{A}^{n-1} + (-a_{n-2}a_{n-1}^{(k-1)} + a_{n-3}^{(k-1)})\mathbf{A}^{n-2} + \cdots + (-a_0 a_{n-1}^{(k-1)})\mathbf{1}. \quad (5.101)$$

This gives us an expression for $\mathbf{A}^{n+k}$, $k = 0, 1, 2, \ldots$, in terms of $\mathbf{A}^{n-1}$, $\mathbf{A}^{n-2}, \ldots, \mathbf{A}$ and $\mathbf{1}$.

Continuing this process, we see that any power of $\mathbf{A}$ can be represented as a weighted polynomial in $\mathbf{A}$ of an order, at most $n-1$. Hence, functions of matrixes, including $\mathbf{e}^{\mathbf{A}t}$, that can be expressed as a polynomial(*)

$$f(\mathbf{A}) = \alpha_0\mathbf{1} + \alpha_1\mathbf{A} + \cdots + \alpha_k\mathbf{A}^k + \cdots = \sum_{k=0}^{\infty} \alpha_k\mathbf{A}^k, \quad (5.102)$$

may be reduced to the expression

$$f(\mathbf{A}) = \beta_0\mathbf{1} + \beta_1\mathbf{A} + \cdots + \beta_{n-1}\mathbf{A}^{n-1} = \sum_{k=0}^{n-1} \beta_k\mathbf{A}^k. \quad (5.103)$$

Here, the coefficients $\beta_0, \beta_1, \ldots, \beta_{n-1}$ are functions of $a_0, a_1, \ldots, a_{n-1}$ and $\alpha_0, \alpha_1, \ldots$. Their approximate calculation can be carried out by the iterative method used in the calculation of higher powers of $\mathbf{A}$ in equation 5.101 and by using equation 5.102. However this straightforward method can be lengthy.

Example 5.4

(a) Let us first calculate a simple matrix function $f(\mathbf{A}) = \mathbf{A}^4$, where $\mathbf{A}$ is the matrix of the previous example. Since the characteristic equation of $\mathbf{A}$ is $\lambda^2 - 6\lambda + 5 = 0$, we have

$$\mathbf{A}^2 = 6\mathbf{A} - 5\cdot\mathbf{1},$$

where $a_1 = -6$ and $a_0 = 5$. Using an iterative formula, and noting that in the first calculation $a_1^{(0)} = -a_1$ and $a_0^{(0)} = -a_0$, yields

$$\begin{aligned}\mathbf{A}^3 &= [-a_1 a_1^{(0)} + a_0^{(0)}]\mathbf{A} + (-a_0 a_1^{(0)})\mathbf{1} \\ &= [6\cdot 6 - 5]\mathbf{A} + (-5\cdot 6)\mathbf{1} = 31\mathbf{A} - 30\,\mathbf{1},\end{aligned}$$

where $a_1^{(1)} = 31$ and $a_0^{(1)} = -30$. Hence,

$$\mathbf{A}^4 = [(6)(31) - 30]\mathbf{A} - 5\cdot 31\,\mathbf{1} = 156\mathbf{A} - 155\,\mathbf{1},$$

and finally

$$\mathbf{A}^4 = 156\begin{bmatrix} 2 & 1 \\ 3 & 4 \end{bmatrix} - \begin{bmatrix} 155 & 0 \\ 0 & 155 \end{bmatrix} = \begin{bmatrix} 157 & 156 \\ 468 & 469 \end{bmatrix}$$

(*)In general, any analytic function of matrix $\mathbf{A}$ can be expressed as a polynomial in $\mathbf{A}$ of an order no greater than one less than the order of $\mathbf{A}$. For proof see N. Balabanian and T. A. Bickart (1969) *Electrical Network Theory*, John Wiley & Sons.

(b) As a second example, let us calculate a matrix potential $f(\mathbf{A})=\mathbf{e}^{\mathbf{A}t}$ for $t=$ 1 s, using the approximation up to fifth term:

$$\mathbf{e}^{\mathbf{A}} \cong \mathbf{1}+\mathbf{A}+\frac{1}{2!}\mathbf{A}^2+\frac{1}{3!}\mathbf{A}^3+\frac{1}{4!}\mathbf{A}^4$$

$$=\mathbf{1}+\mathbf{A}+\frac{1}{2}(-5\cdot\mathbf{1}+6\mathbf{A})+\frac{1}{6}(-30\cdot\mathbf{1}+31\mathbf{A})+\frac{1}{24}(155\cdot\mathbf{1}+156\mathbf{A})$$

$$=-12.96\cdot\mathbf{1}+15.67\mathbf{A}$$

and finally

$$\mathbf{e}^{\mathbf{A}} \cong \begin{bmatrix} -13 & 0 \\ 0 & -13 \end{bmatrix}+15.7\begin{bmatrix} 2 & 1 \\ 3 & 4 \end{bmatrix}=\begin{bmatrix} 18.4 & 15.7 \\ 47.1 & 49.8 \end{bmatrix}.$$

We shall next develop an easier, one-step method for finding β-coefficients in the function of matrix expression (equation 5.103). Let us return to the characteristic equation of matrix $\mathbf{A}$

$$g(\lambda)=|\lambda\mathbf{1}-\mathbf{A}|=\lambda^n+a_{n-1}\lambda^{n-1}+\cdots+a_1\lambda+a_0=0. \tag{5.104}$$

The eigenvalues $\lambda_1, \lambda_2, \ldots, \lambda_n$, which are the roots of the characteristic equation 5.104, obviously satisfy the equation 5.104 as well as matrix $\mathbf{A}$ (in accordance with the Caleg-Hamilton theorem). Therefore, using the same procedure as before, we can derive an expression similar to equation 5.103 for the eigenvalues instead of the matrix by itself, namely:

$$f(\lambda)=\beta_0+\beta_1\lambda+\beta_2\lambda^2+\cdots+\beta_{n-1}\lambda^{n-1}=\sum_{k=0}^{n-1}\beta_k\lambda^k. \tag{5.105}$$

It is understandable that this expression holds for any λ that is a solution of the characteristic equation 5.104, that is for any eigenvalue of the matrix $\mathbf{A}$.

(a) *Distinct eigenvalues*

Assume first that the eigenvalues are *distinct*; that is, that none is repeated. Substituting $\lambda_1, \lambda_2, \ldots, \lambda_n$ in equation 5.105 gives n equations in n unknown β's:

$$\begin{aligned} &\beta_0+\beta_1\lambda_1+\beta_1\lambda_1^2+\cdots+\beta_{n-1}\lambda_1^{n-1}=f(\lambda_1) \\ &\beta_0+\beta_1\lambda_1+\beta_2\lambda_2^2+\cdots+\beta_{n-1}\lambda_2^{n-1}=f(\lambda_2) \\ &\ldots\ldots\ldots\ldots\ldots\ldots\ldots\ldots\ldots\ldots\ldots\ldots \\ &\beta_0+\beta_1\lambda_n+\beta_2\lambda_n^2+\cdots+\beta_{n-1}\lambda_1^{n-1}=f(\lambda_n). \end{aligned} \tag{5.106}$$

The coefficients $\beta_0, \beta_1, \ldots, \beta_{n-1}$ can then be obtained as the solution to this linear system of scalar equations, i.e. the inversion of the set of equations 5.106 gives the solution. With the known β-coefficients, the function of the matrix representation problem is solved:

$$f(\mathbf{A}) = \sum_{k=0}^{n-1} \beta_k \mathbf{A}^k. \tag{5.107}$$

Example 5.5

Let us illustrate this process with the same simple example (as in Example 5.4):

(a) Find $f(\mathbf{A}) = \mathbf{A}^4$, if $\mathbf{A} = \begin{bmatrix} 2 & 1 \\ 3 & 4 \end{bmatrix}$

The characteristic equation is (see Example 5.3)

$$g(\lambda) = \lambda^2 - 6\lambda + 5 = 0.$$

Thus, the eigenvalues are

$$\lambda_1 = 5, \quad \lambda_2 = 1.$$

In accordance with equation 5.106, we have

$$\beta_0 + \beta_1 5 = 5^4,$$
$$\beta_0 + \beta_1 1 = 1^4.$$

Solving these simple equations for unknowns β_0 and β_1, gives

$$\beta_1 = 156, \quad \beta_0 = -155.$$

The solution for $\mathbf{A}^4$ is found by using equation 5.107

$$f(\mathbf{A}) = \mathbf{A}^4 = -155 \cdot \mathbf{1} + 156 \cdot \mathbf{A}$$

which is the same as the results obtained in the previous example.

(b) Find $f(\mathbf{A}) = \mathbf{e}^{\mathbf{A}t}$ for the same matrix $\mathbf{A}$

The equations for unknowns β_0 and β_1 in this case will be

$$\beta_0 + 5\beta_1 = e^{5t},$$
$$\beta_0 + \beta_1 = e^{t}.$$

Solving this equation gives

$$\beta_1 = \tfrac{1}{4}e^{5t} - \tfrac{1}{4}e^{t}, \quad \beta_0 = -\tfrac{1}{4}e^{5t} + \tfrac{5}{4}e^{t}.$$

Thus, the matrix exponential is

$$\mathbf{e}^{\mathbf{A}t} = (-\tfrac{1}{4}e^{5t} + \tfrac{5}{4}e^{t})\mathbf{1} + (\tfrac{1}{4}e^{5t} - \tfrac{1}{4}e^{t})\mathbf{A}$$
$$= (-\tfrac{1}{4}e^{5t} + \tfrac{5}{4}e^{t})\begin{bmatrix} 1 & 0 \\ 0 & 1 \end{bmatrix} + (\tfrac{1}{4}e^{5t} - \tfrac{1}{4}e^{t})\begin{bmatrix} 2 & 1 \\ 3 & 4 \end{bmatrix}.$$

By an obvious rearrangement, this becomes

$$\mathbf{e}^{\mathbf{A}t} = \begin{bmatrix} \frac{1}{4}e^{5t} + \frac{3}{4}e^{t} & \frac{1}{4}e^{5t} - \frac{1}{4}e^{t} \\ \frac{3}{4}e^{5t} - \frac{3}{4}e^{t} & \frac{3}{4}e^{5t} + \frac{1}{4}e^{t} \end{bmatrix}. \tag{5.108}$$

It is interesting to compare these results with those obtained in the previous example. The approximate, up to fifth term, evaluation of the exponents e^5 and e^1 $(t = 1\text{ s})$ gives

$$e^5 \cong 1 + 5 + \frac{1}{2!}5^2 + \frac{1}{3!}5^3 + \frac{1}{4!}5^4 = 65.4$$

$$e^1 \cong 1 + 1 + \frac{1}{2!} + \frac{1}{3!} + \frac{1}{4!} = 2.71.$$

Substituting these results in equation 5.108 yields

$$\mathbf{e}^{\mathbf{A}} \cong \begin{bmatrix} 18.4 & 15.6 \\ 47.0 & 49.7 \end{bmatrix}$$

which agrees with the previous results.

Therefore, the series form of the exponential may permit some approximate numerical results; it does not lead to a closed form. However, with the help of the Caley-Hamilton theorem, we obtained the closed-form equivalent for the exponential $\mathbf{e}^{\mathbf{A}t}$ (equation 5.107). We shall now return our consideration to the complete solution of the state equation in the form of equation 5.90, repeated here for convenience:

$$\mathbf{x}(t) = \mathbf{e}^{\mathbf{A}(t-t_0)}\mathbf{x}(t_0) + \int_{t_0}^{t} \mathbf{e}^{\mathbf{A}(t-\tau)}\mathbf{b}\mathbf{w}(\tau)d\tau. \tag{5.109}$$

The following example illustrates this computation.

Example 5.6

Find the complete solution of the state equation describing the circuit in Fig. 5.9, considered before. For the sake of convenience, it is redrawn here again in Fig. 5.11(a). Let the circuit element values be $C_1 = 1$ F, $C_2 = 2$ F, $L_4 = 1$ H, $G_3 = 1$ S, $R_5 = 1\ \Omega$, $R_6 = 2/7\ \Omega$, $R_7 = 1/3\ \Omega$.

Solution

Substituting these parameters into equation 5.55, we obtain the following **A** matrix

$$\mathbf{A} = \begin{bmatrix} -4 & \frac{1}{2} & -\frac{1}{2} \\ \frac{1}{4} & -\frac{7}{4} & -\frac{1}{4} \\ \frac{1}{2} & \frac{1}{2} & -\frac{1}{2} \end{bmatrix}. \tag{5.110}$$

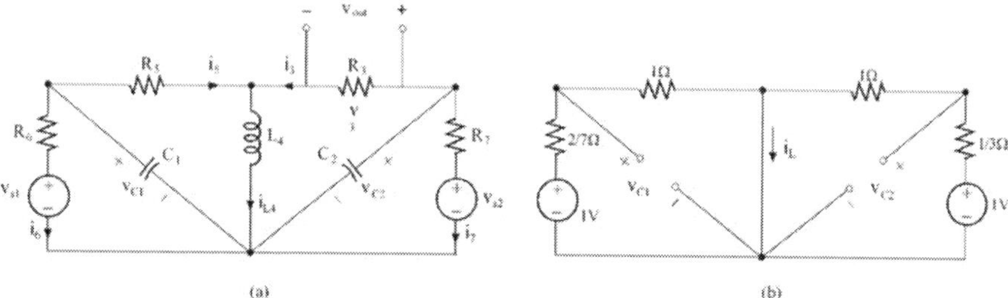

Figure 5.11 A circuit of Example 5.6 (a) and its steady-state equivalent (b).

The characteristic equation is

$$g(\lambda) = |\lambda \cdot \mathbf{1} - \mathbf{A}| = \begin{vmatrix} \lambda + 4 & -\frac{1}{2} & \frac{1}{2} \\ -\frac{1}{4} & \lambda + \frac{7}{4} & \frac{1}{4} \\ -\frac{1}{2} & -\frac{1}{2} & \lambda + \frac{1}{2} \end{vmatrix} = 0.$$

Thus,

$$g(\lambda) = (\lambda + 4)[(\lambda + \tfrac{7}{4})(\lambda + \tfrac{1}{2}) + \tfrac{1}{4}] = 0.$$

Simplifying yields

$$(\lambda + 4)(\lambda^2 + \tfrac{9}{4}\lambda + \tfrac{9}{8}) = 0. \tag{5.111}$$

Thus, the eigenvalues of **A** are

$$\lambda_{1,2} = -\frac{9}{8} \pm \sqrt{\left(\frac{9^2}{8^2} - \frac{9}{8}\right)} = -1.125 \pm 0.375$$

or

$$\lambda_1 = -0.75, \ \lambda_2 = -1.5, \ \lambda_3 = -4.$$

Using the results of equation 5.106, we can evaluate β_0, β_1, and β_2 from the equations

$$\beta_0 - 0.75\beta_1 + (-0.75)^2\beta_2 = e^{-0.75t}$$
$$\beta_0 - 1.5\beta_1 + (-1.5)^2\beta_2 = e^{-1.5t}$$
$$\beta_0 - 4\beta_1 + (-4)^2\beta_2 = e^{-4t},$$

which in the matrix form are

$$\begin{bmatrix} 1 & -0.75 & 0.5625 \\ 1 & -1.5 & 2.25 \\ 1 & -4 & 16 \end{bmatrix} \begin{bmatrix} \beta_0 \\ \beta_1 \\ \beta_2 \end{bmatrix} = \begin{bmatrix} e^{-0.75t} \\ e^{-1.5t} \\ e^{-4t} \end{bmatrix}. \tag{5.113}$$

The solution for β's is found by inversion, as

$$\begin{bmatrix}\beta_0\\ \beta_1\\ \beta_2\end{bmatrix}=\begin{bmatrix}1 & -0.75 & 0.5625\\ 1 & -1.5 & 2.25\\ 1 & -4 & 16\end{bmatrix}^{-1}\begin{bmatrix}e^{-0.75t}\\ e^{-1.5t}\\ e^{-4t}\end{bmatrix}$$

$$=\begin{bmatrix}2.462 & -1.6 & 0.1385\\ 2.256 & -2.533 & 0.2769\\ 0.4103 & -0.5333 & 0.1231\end{bmatrix}\begin{bmatrix}e^{-0.75t}\\ e^{-1.5t}\\ e^{-4t}\end{bmatrix}$$

$$=\begin{bmatrix}2.462e^{-0.75t} & -1.6e^{-1.5t} & 0.1385e^{-4t}\\ 2.256e^{-0.75t} & -2.533e^{-1.5t} & 0.2769e^{-4t}\\ 0.4103e^{-0.75t} & -0.5333e^{-1.5t} & 0.1231e^{-4t}\end{bmatrix}. \tag{5.114}$$

With β's now known, matrix $\mathbf{e}^{\mathbf{A}t}$ will be

$$\mathbf{e}^{\mathbf{A}t}=\begin{bmatrix}1 & 0 & 0\\ 0 & 1 & 0\\ 0 & 0 & 1\end{bmatrix}\beta_0+\begin{bmatrix}-4 & 0.5 & -0.5\\ 0.25 & -1.75 & -0.25\\ 0.5 & 0.5 & -0.5\end{bmatrix}\beta_1$$

$$+\begin{bmatrix}15.87 & -3.125 & 2.125\\ -1.563 & 3.063 & 0.438\\ -2.125 & -0.875 & -0.125\end{bmatrix}\beta_2.$$

Substituting equation 5.114 for β's and collecting like terms yields the final results

$$\mathbf{e}^{\mathbf{A}t}=\begin{bmatrix}-0.048 & -0.154 & -0.256\\ -0.077 & -0.229 & -0.384\\ 0.256 & 0.769 & 1.283\end{bmatrix}e^{-0.75t}+\begin{bmatrix}0.066 & 0.4 & 0.133\\ 0.2 & 1.2 & 0.4\\ -0.133 & -0.8 & -0.267\end{bmatrix}e^{-1.5t}$$

$$+\begin{bmatrix}0.985 & -0.246 & 0.123\\ -0.123 & 0.031 & -0.015\\ -0.123 & 0.031 & -0.015\end{bmatrix}e^{-4t}. \tag{5.115}$$

Now suppose that the initial state vector at $t_0=0$ is $\mathbf{x}(0)=[0.5\ \ 1.5\ \ 1]^T$, then the natural solution (for $\mathbf{w}(t)=\mathbf{0}$) in equation 5.109 is

$$\mathbf{x}_{nat}(t)=\mathbf{e}^{\mathbf{A}t}\mathbf{x}(0)=\begin{bmatrix}-0.511e^{-0.75t} & +0.767e^{-1.5t} & +0.246e^{-4t}\\ -0.766e^{-0.75t} & +2.30e^{-1.5t} & -0.031e^{-4t}\\ 2.564e^{-0.75t} & -1.534e^{-1.5t} & -0.031e^{-4t}\end{bmatrix}.$$

(5.116)

The next step is to find the particular or forced solution of the state equation. Let the input vector $\mathbf{w}(t) = [1 \;\; 1]^T$. Substituting the circuit parameters into matrix **b** in equation 5.55, we obtain

$$\mathbf{b} = \begin{bmatrix} 3.5 & 0 \\ 0 & 1.5 \\ 0 & 0 \end{bmatrix}. \tag{5.117}$$

Since the input is a constant (d.c.), evaluating the integral in equation 5.55 results, for $t_0 = 0$, in

$$\int_0^t \mathbf{e}^{\mathbf{A}(t-\tau)}\mathbf{bw}\, d\tau = -\mathbf{A}^{-1}\mathbf{e}^{\mathbf{A}(t-\tau)}\mathbf{bw}|_0^t = \mathbf{A}^{-1}[\mathbf{e}^{\mathbf{A}t} - \mathbf{1}]\mathbf{bw}, \tag{5.118}$$

where the inverse of the **A** matrix is found as follows

$$\mathbf{A}^{-1} = \begin{bmatrix} -4 & \frac{1}{2} & -\frac{1}{2} \\ \frac{1}{4} & -\frac{7}{4} & -\frac{1}{4} \\ \frac{1}{2} & \frac{1}{2} & -\frac{1}{2} \end{bmatrix}^{-1} = \begin{bmatrix} -0.222 & 0 & 0.222 \\ 0 & -0.5 & 0.25 \\ -0.222 & -0.5 & -1.528 \end{bmatrix}. \tag{5.119}$$

Performing now, all the calculations in equation 5.118, with equations 5.119, 5.115, 5.117 and $\mathbf{w} = [1 \;\; 1]^T$, we obtain the particular solution

$$\mathbf{x}_{par}(t) = \begin{bmatrix} 0.547e^{-0.75t} - 0.556e^{-1.5t} - 0.769e^{-4t} + 0.778 \\ 0.821e^{-0.75t} - 1.667e^{-1.5t} + 0.096e^{-4t} + 0.750 \\ -2.735e^{-0.75t} + 1.111e^{-1.5t} + 0.096e^{-1.5t} + 1.528 \end{bmatrix}. \tag{5.120}$$

The final result of the complete solution is simply obtained by combining the above two solutions: the natural (equation 5.116) and the particular (equation 5.120), which leads to

$$\mathbf{x}(t) = \mathbf{x}_{nat} + \mathbf{x}_{par} = \begin{bmatrix} 0.034e^{-0.75t} + 0.211e^{-1.5t} - 0.523e^{-4t} + 0.778 \\ 0.052e^{-0.75t} + 0.633e^{-1.5t} + 0.065e^{-4t} + 0.750 \\ -0.171e^{-0.75t} - 0.423e^{-1.5t} + 0.065e^{-4t} + 1.528 \end{bmatrix} \begin{bmatrix} v_{c1} \\ v_{c2} \\ i_{L4} \end{bmatrix}. \tag{5.121}$$

Figure 5.12 shows the state variables v_{c1}, v_{c2}, i_{L4} behavior versus time.

The computer calculation of the state variables in the above example, using the MATHCAD program is shown in Appendix I. (Note that the computing results are slightly different from those obtained above.)

To complete this example, suppose that voltage v_3 is of interest. Then the

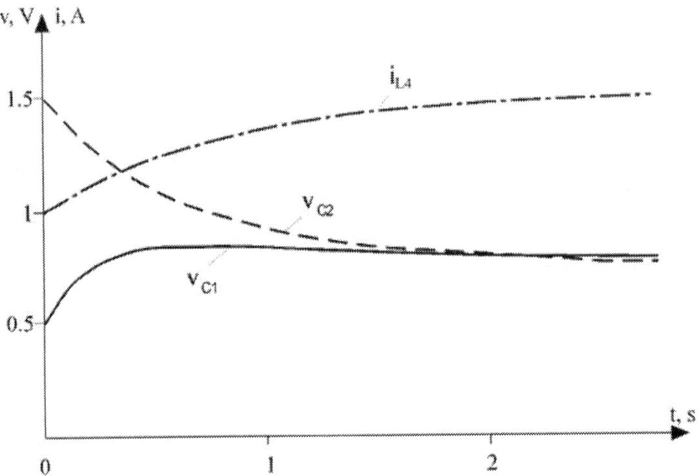

Figure 5.12 Two capacitor voltages and inductor current curves versus time of Example 5.6.

output equation 5.56 simplifies to

$$v_3(t) = [-a \quad a \quad aR_5]\mathbf{x}(t) = [-\tfrac{1}{2} \quad \tfrac{1}{2} \quad \tfrac{1}{2}]\begin{bmatrix} v_{c1} \\ v_{c2} \\ i_{L4} \end{bmatrix}.$$

Thus, the output voltage is

$$\begin{aligned} v_{out}(t) &= v_3 = \tfrac{1}{2}(-v_{c1} + v_{c2} + i_{L4}) \\ &= -0.077e^{-0.75t} - 0.0005e^{-1.5t} + 0.327e^{-4t} + 0.750\ \text{V}. \quad (5.122) \end{aligned}$$

Note that by inspection of the given circuit in its d.c. steady-state behavior, i.e. the capacitors are open-circuited and the inductor is short-circuited as shown in Fig. 5.11(b), we may find

$$v_{c1}(\infty) = \frac{v_{s1}}{R_5 + R_6} R_5 = \frac{1}{1 + 2/7} \cdot 1 = 0.778\ \text{V}$$

$$v_{c2}(\infty) = \frac{v_{s2}}{R_3 + R_7} R_3 = \frac{1}{1 + 1/3} \cdot 1 = 0.75\ \text{V}$$

$$i_L(\infty) = v_{c1}/R_5 + v_{c2}/R_3 = 0.778 + 0.75 = 1.528\ \text{A},$$

which is in agreement with the final results in equation 5.121.

(b) *Multiple eigenvalues*

If some of the eigenvalues of **A** (roots of the characteristic equation $g(\lambda) \neq 0$)

are not distinct and there are repeated values (for example $\lambda_1 = \lambda_2$), then in this case, the number of independent equations in 5.106 would be fewer than n unknown coefficients β. The following theorem allows us to extend the solution for finding all β's to the case of repeated eigenvalues.

Theorem:(*) Let $\mathbf{A}$ be the $n \times n$ matrix with n_0 distinct eigenvalues $\lambda_1, \lambda_2, \ldots, \lambda_{n0}$ and m multiple eigenvalues ($n_0 < n$, if no eigenvalue is repeated, then $n_0 = n$). Let the eigenvalue λ_i occur with multiplicity r_i, and define the polynomials

$$\mathbf{P}(\mathbf{A}) = \sum_{k=0}^{n-1} \beta_k \mathbf{A}, \tag{5.123}$$

and

$$P(\lambda) = \sum_{k=0}^{n-1} \beta_k \lambda^k. \tag{5.124}$$

Then the matrix function $f(\mathbf{A})$ is identical to the matrix polynomial $\mathbf{P}(\mathbf{A})$ (see 5.107) if the following conditions are obeyed:

for each distinct eigenvalue

$$f(\lambda_i) = P(\lambda_i) \quad i = 1, 2, \ldots, n_0 \tag{5.125a}$$

for each multiple eigenvalue

$$\frac{d^q}{d\lambda^q} f(\lambda)|_{\lambda=\lambda_i} = \frac{d^q}{d\lambda^q} P(\lambda)|_{\lambda=\lambda_i},$$

$$i = n_{0+1}, n_{0+2}, \ldots, n_{0+m}, \quad q = 0, 1, 2, \ldots, r_i - 1 \tag{5.125b}$$

that the first condition (equation 5.125a) gives us only n_0 ($n_0 < n$) independent equations for finding n unknown β-coefficients. However, the second condition (equation 5.125b) yields the remaining equations needed to solve for $\beta_0, \beta_1, \ldots, \beta_{n-1}$. For this purpose equation 5.125b shall be rewritten in terms of the unknown β's

$$\frac{d^q}{d\lambda^q} f(\lambda)|_{\lambda=\lambda_i} = \frac{d^q}{d\lambda^q} \sum_{k=0}^{n-1} \beta_k \lambda^k|_{\lambda=\lambda_i} = \sum_{k=q}^{k-1} k(k-1)\cdots(k-q+1)\beta_k \lambda_i^{k-q},$$

$$i = n_{0+1}, n_{0+2}, \ldots, n_{0+m}, \quad q = 0, 1, 2, \ldots, r-1 \tag{5.126}$$

The total number of independent equations, therefore, will be

$$n_0 + \sum_{1}^{m} r_i = n.$$

Example 5.7

As an example of the determination of a matrix function when $\mathbf{A}$ has multiple

---

(*)The proof can be found in the book by Balabanian N. and Bickart T. A. (1969) *Electrical Network Theory*, John Wiley & Sons.

eigenvalues, let us consider the same circuit in Fig. 5.11 of the previous example with slightly different parameters, namely: $R_6 = 1/3\,\Omega$, $R_7 = 2/5\,\Omega$ (the rest of the parameters are the same). Suppose we wish to find $\mathbf{e}^{\mathbf{A}t}$.

Solution

The **A** matrix in this case will be

$$\mathbf{A} = \begin{bmatrix} -\frac{7}{2} & \frac{1}{2} & -\frac{1}{2} \\ \frac{1}{4} & -\frac{3}{2} & -\frac{1}{4} \\ \frac{1}{2} & \frac{1}{2} & -\frac{1}{2} \end{bmatrix}$$

which yields the characteristic equation

$$g(\lambda) = \begin{bmatrix} \lambda + \frac{7}{2} & -\frac{1}{2} & \frac{1}{2} \\ -\frac{1}{4} & \lambda + \frac{3}{2} & \frac{1}{4} \\ -\frac{1}{2} & -\frac{1}{2} & \lambda + \frac{1}{2} \end{bmatrix}$$

$$= (\lambda + \tfrac{7}{2})(\lambda + \tfrac{3}{2})(\lambda + \tfrac{1}{2}) + \tfrac{1}{4}\lambda + \tfrac{3}{8} = (\lambda + \tfrac{7}{2})(\lambda^2 + 2\lambda + 1) = 0.$$

Thus, the eigenvalues are $\lambda_1 = -\frac{7}{2}$ and double $\lambda_2 = -1$, i.e. the multiplicity $r = 2$. Therefore, for the first distinct eigenvalue, in accordance with equation 5.125a, we have

$$\beta_0 + \beta_1(-\tfrac{7}{2}) + \beta_2(-\tfrac{7}{2})^2 = e^{-(7/2)t},$$

and for the double eigenvalue, in accordance with equation 5.125b we have

$$\beta_0 + \beta_1(-1) + \beta_2(-1)^2 = e^{-t}, \quad q = 0$$

$$\beta_1 + 2\beta_2(-1) = te^{-t}, \quad q = 1.$$

Since

$$\left.\frac{df(\lambda_2)}{d\lambda}\right|_{\lambda_2 = -1} = \left.\frac{d}{d\lambda_2}(e^{\lambda_2 t})\right|_{\lambda_2 = -1} = te^{-t},$$

the above equations in the matrix form are

$$\begin{bmatrix} 1 & -7/2 & 49/4 \\ 1 & -1 & 1 \\ 0 & 1 & -2 \end{bmatrix} \begin{bmatrix} \beta_0 \\ \beta_1 \\ \beta_2 \end{bmatrix} = \begin{bmatrix} e^{-3.5t} \\ e^{-t} \\ te^{-t} \end{bmatrix}.$$

The solution for β's gives

$$\begin{bmatrix} \beta_0 \\ \beta_1 \\ \beta_2 \end{bmatrix} = \begin{bmatrix} 0.16e^{-3.5t} + 0.84e^{-t} + 1.4te^{-t} \\ 0.32e^{-3.5t} - 0.32e^{-t} + 1.8te^{-t} \\ 0.16e^{-3.5t} - 0.16e^{-t} + 0.4te^{-t} \end{bmatrix}.$$

With β's known, the desired matrix is

$$\mathbf{e}^{\mathbf{A}t} = \begin{bmatrix} 1 & & 0 \\ & 1 & \\ 0 & & 1 \end{bmatrix} \beta_0 + \begin{bmatrix} -3.5 & 0.5 & -0.5 \\ 0.25 & -1.5 & -0.25 \\ 0.5 & 0.5 & -0.5 \end{bmatrix} \beta_1$$

$$+ \begin{bmatrix} 12.125 & -2.75 & 1.875 \\ -1.375 & 2.25 & 0.375 \\ -1.875 & -0.75 & -0.125 \end{bmatrix} \beta_2.$$

Substituting the β's from the previous solution, and after simplifying, we obtain

$$\mathbf{e}^{\mathbf{A}t} = \begin{bmatrix} 0.98e^{-3.5t}+0.02e^{-t}-0.05te^{-t} & -0.28e^{-3.5t}+0.28e^{-t}-0.2te^{-t} & 0.14e^{-3.5t}-0.14e^{-t}-0.15te^{-t} \\ -0.14e^{-3.5t}+0.14e^{-t}-0.1te^{-t} & 0.04e^{-3.5t}+0.96e^{-t}-0.4te^{-t} & -0.02e^{-3.5t}+0.02e^{-t}-0.3te^{-t} \\ -0.14e^{-3.5t}+0.14e^{-t}+0.15te^{-t} & 0.04e^{-3.5t}-0.04e^{-t}+0.6te^{-t} & 1.02e^{-3.5t}-0.02e^{-t}+0.45te^{-t} \end{bmatrix}.$$

(c) *Complex eigenvalues*

We shall illustrate the computation of a matrix exponential when some of the roots of the characteristic equation are complex quantities, considering the following example.

Example 5.8

Let the circuit in Fig. 5.11 (of the previous example) have the same parameters, excluding $R_6 = 2/5\ \Omega$ and $R_7 = 1/2\ \Omega$. Our purpose is again to compute $\mathbf{e}^{\mathbf{A}t}$.

Solution

We substitute the above parameters into the **A** matrix of equation 5.55 to yield

$$\mathbf{A} = \begin{bmatrix} -3 & \frac{1}{2} & -\frac{1}{2} \\ \frac{1}{4} & -\frac{5}{4} & -\frac{1}{4} \\ \frac{1}{2} & \frac{1}{2} & -\frac{1}{2} \end{bmatrix}.$$

Thus, the characteristic equation of **A** is

$$g(\lambda) = (\lambda + 3)(\lambda + \tfrac{5}{4})(\lambda + \tfrac{1}{2}) + \tfrac{1}{4}\lambda + \tfrac{3}{4} = 0,$$

or after a rearrangement of terms

$$(\lambda + 3)(\lambda^2 + \tfrac{7}{4}\lambda + \tfrac{7}{8}) = 0,$$

Therefore, the eigenvalues are

$$\lambda_1 = -3, \quad \lambda_{2,3} = -\tfrac{7}{8} \pm \sqrt{\tfrac{49-56}{64}} = -0.875 \pm j0.331.$$

Note that two complex eigenvalues are a conjugate pair. Thus, in accordance with equation 5.106, we have

$$\beta_0 + \beta_1(-3) + \beta_2(-3)^2 = e^{-3t}$$

$$\beta_0 + \beta_1(-0.875 + j0.331) + \beta_2(-0.875 + j0.331)^2 = e^{-0.875t}\, e^{j0.331t}$$

$$\beta_0 + \beta_1(-0.875 - j0.331) + \beta_2(-0.875 - j0.331)^2 = e^{-0.875t}\, e^{-j0.331t}.$$

Next, we solve these equations to yield for β's:

$$\beta_0 = 0.819e^{-3t} + e^{-0.875t}(3.86 \sin 0.331t + 0.811 \cos 0.331t)$$

$$\beta_1 = 0.378e^{-3t} + e^{-0.875t}(5.46 \sin 0.331t - 0.378 \cos 0.331t)$$

$$\beta_2 = 0.216e^{-3t} + e^{-0.875t}(1.39 \sin 0.331t - 0.216 \cos 0.331t).$$

Hence, matrix $\mathbf{e}^{\mathbf{A}t}$ will be

$$\mathbf{e}^{\mathbf{A}t} = \begin{bmatrix} 1 & & 0 \\ & 1 & \\ 0 & & 1 \end{bmatrix} \beta_0 + \begin{bmatrix} -3 & 0.5 & -0.5 \\ 0.25 & -1.25 & -0.25 \\ 0.5 & 0.5 & -0.5 \end{bmatrix} \beta_1$$

$$+ \begin{bmatrix} 8.875 & -2.375 & 1.625 \\ -1.187 & 1.563 & 0.313 \\ -1.625 & -0.625 & -1.125 \end{bmatrix} \beta_2.$$

Finally, substituting the above results for β's, after simplifying, we obtain

$$\mathbf{e}^{\mathbf{A}t} = \begin{bmatrix} 0.973e^{-3t} - 0.174\zeta_1 + 0.027\zeta_2 & -0324e^{-3t} - 0.572\zeta_1 + 0.324\zeta_2 & 0.162e^{-3t} - 0.470\zeta_1 - 0.162\zeta_2 \\ -0.162e^{-3t} - 0.280\zeta_1 + 0.162\zeta_2 & 0.054e^{-3t} - 0.787\zeta_1 + 0.946\zeta_2 & -0.027e^{-3t} - 0.930\zeta_1 + 0.027\zeta_2 \\ -0.162e^{-3t} + 0.470\zeta_1 + 0.162\zeta_2 & -0.054e^{-3t} + 1.86\zeta_1 - 0.054\zeta_2 & -0.027e^{-3t} + 0.960\zeta_1 + 1.027\zeta_2 \end{bmatrix}$$

where $\zeta_1 = e^{-0.875t} \sin 0.331t$, $\zeta_2 = e^{-0.875t} \cos 0.331t$.

Suppose we now wish to know the zero input response of the circuit to the initial vector, $\mathbf{x}(0) = [1 \;\; 1 \;\; 0]^T$, i.e. the capacitors are initially charged to 1 V each. Then,

$$\mathbf{x}_{nat}(t) = \mathbf{e}^{\mathbf{A}t}[1 \quad 1 \quad 0]^T = \begin{bmatrix} v_{c1} \\ v_{c2} \\ i_{L4} \end{bmatrix}$$

$$= \begin{bmatrix} 0.649e^{-3t} + e^{-0.875t}(-0.746 \sin 0.331t + 0.351 \cos 0.331t) \\ -0.108e^{-3t} + e^{-0.875t}(-1.073 \sin 0.331t + 1.108 \cos 0.331t) \\ -0.108e^{-3t} + e^{-0.875t}(2.329 \sin 0.331t + 0.108 \cos 0.331t) \end{bmatrix}.$$

These two voltage curves and one current curve versus time are shown in Fig. 5.13.

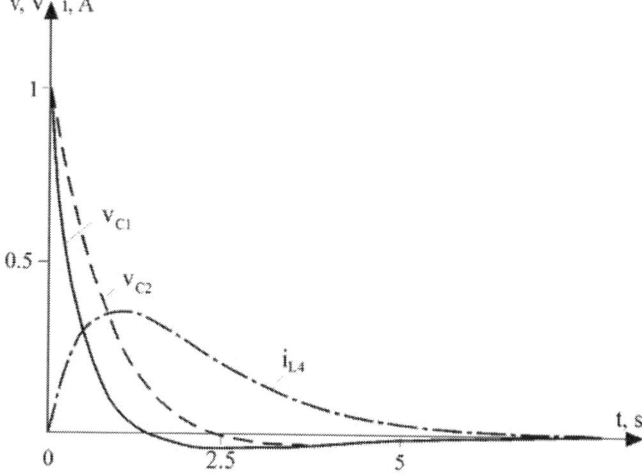

Figure 5.13 Two capacitor voltages and inductor current curves versus time of Example 5.8 in the case of complex-conjugate eigenvalues.

5.8.3 Lagrange interpolation formula

One other method of computing functions of a matrix is based on the Lagrange interpolation formula (this formula is also known as the Silvestre formula). Thus, knowing the eigenvalues λ's of matrix **A**, any function of **A** may be determined as:

$$f(\mathbf{A}) = \sum_{i=1}^{n} \left(\prod_{\substack{k=1 \\ k \neq 1}}^{n} \frac{\mathbf{A} - \lambda_k \mathbf{1}}{\lambda_i - \lambda_k} \right) f(\lambda_i), \tag{5.127}$$

where $\prod_{\substack{k=1 \\ k \neq 1}}^{n}$ means the product of terms $\frac{\mathbf{A} - \lambda_k \mathbf{1}}{\lambda_i - \lambda_k}$ where k takes the values $1, 2, \ldots, n$ but excluding $k = i$. For example, using the data of Example 5.6, equation 5.127 implies that

$$\mathbf{e}^{\mathbf{A}t} = \frac{(\mathbf{A} + 1.5\cdot\mathbf{1})(\mathbf{A} + 4\cdot\mathbf{1})}{(-0.75 + 1.5)(-0.75 + 4)} e^{-0.75t} + \frac{(\mathbf{A} + 0.75\cdot\mathbf{1})(\mathbf{A} + 4\cdot\mathbf{1})}{(-1.5 + 0.75)(-1.5 + 4)} e^{-1.5t}$$
$$+ \frac{(\mathbf{A} + 0.75\cdot\mathbf{1})(\mathbf{A} + 1.5\cdot\mathbf{1})}{(-4 + 0.75)(-4 + 1.5)} e^{-4t}.$$

Substituting matrix **A** (equation 5.110) and performing all the arithmetic, leads to

$$\mathbf{e}^{\mathbf{A}t} = \begin{bmatrix} -0.050 & -0.154 & -0.256 \\ -0.077 & -0.230 & -0.385 \\ 0.256 & 0.769 & 1.282 \end{bmatrix} e^{-0.75t} + \begin{bmatrix} 0.067 & 0.4 & 0.133 \\ 0.2 & 1.2 & 0.4 \\ -0.133 & -0.8 & -0.267 \end{bmatrix} e^{-1.5t}$$

$$+ \begin{bmatrix} 0.985 & -0.246 & 0.123 \\ -0.123 & 0.031 & -0.015 \\ -0.123 & 0.031 & -0.015 \end{bmatrix} e^{-4t}$$

which agrees with the previous results obtained in equation 5.115.

The Lagrange interpolation formula can be easily programmed, which is an advantage in computer-aided calculations.

5.9 EVALUATING THE MATRIX EXPONENTIAL BY LAPLACE TRANSFORM

In conclusion, let us introduce the Laplace transform application for solving the matrix differential equation. To simplify the procedure, we first apply the Laplace transform to the homogeneous equation (see equation 5.81):

$$\frac{d}{dt}\mathbf{x}(t) - \mathbf{A}\mathbf{x}(t) = \mathbf{0}. \tag{5.128}$$

Applying the Laplace transform to equation 5.128, we get

$$s\mathbf{X}(s) - \mathbf{X}(0) - \mathbf{A}\mathbf{X}(s) = \mathbf{0}, \tag{5.129}$$

where $\mathbf{X}(s)$ is the Laplace transform of $\mathbf{x}(t)$. Supposing that $\mathbf{X}(0) = 1$ (equation 5.129) can be written as follows:

$$(s\cdot\mathbf{1} - \mathbf{A})\mathbf{X}(s) = \mathbf{1}, \tag{5.130}$$

or

$$\mathbf{X}(s) = (s\cdot\mathbf{1} - \mathbf{A})^{-1}. \tag{5.131}$$

Now, we take the inverse transform to get $\mathbf{x}(t)$

$$\mathbf{x}(t) = L^{-1}\{(s\cdot\mathbf{1} - \mathbf{A})^{-1}\} = \mathbf{e}^{\mathbf{A}t}. \tag{5.132}$$

As can be seen, since we have taken $\mathbf{X}(0) = \mathbf{1}$, this expression is also equal to the matrix exponential $\mathbf{e}^{\mathbf{A}t}$.

Example 5.9

Let us apply this result to the simple circuit shown in Fig. 5.14, where the proper tree branches are emphasized.

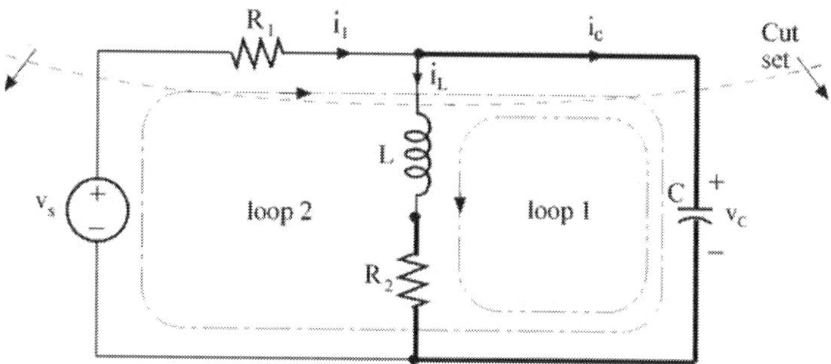

Figure 5.14 A circuit of Example 5.9.

Solution

The capacitor voltage v_C and the inductor current i_L are the state variables in this case. The fundamental cut-set equation and two fundamental loop equations yield

$$C\frac{dv_C}{dt} = -i_L + i_1$$

$$L\frac{di_L}{dt} = v_C - R_2 i_L$$

$$R_1 i_1 = -v_C + v_s \quad \text{or} \quad i_1 = -\frac{1}{R_1}v_C + \frac{1}{R_1}v_s.$$

To eliminate a non-desirable variable, i_1, in the first equation, in this simple case, the third equation shall be inserted into the first one for i_1. Thus, the state equations are

$$\frac{dv_C}{dt} = -\frac{1}{R_1 C}v_C - i_L + \frac{1}{R_1}v_s$$

$$\frac{di_L}{dt} = \frac{1}{L}v_C - \frac{R_2}{L}i_L,$$

or in the matrix form

$$\frac{d}{dt}\begin{bmatrix} v_C \\ i_L \end{bmatrix} = \begin{bmatrix} -1/R_1 C & -1 \\ 1/L & -R_2/L \end{bmatrix}\begin{bmatrix} v_C \\ i_L \end{bmatrix} + \begin{bmatrix} 1/R_1 \\ 0 \end{bmatrix}[v_s]. \tag{5.133}$$

Let the element values be $C = 1.0$ F, $L = 4/3$ H, $R_1 = 2/5\ \Omega$, $R_2 = 2/3\ \Omega$ and $v_s = 1$ V. This yields the coefficient matrixes **A** and **b**

$$\mathbf{A} = \begin{bmatrix} -5/2 & -1 \\ 3/4 & -1/2 \end{bmatrix}, \quad \mathbf{b} = \begin{bmatrix} 5/2 \\ 0 \end{bmatrix} \tag{5.134}$$

and the input matrix $\mathbf{w} = [v_s] = [1]$. Next, we find the matrix $[s\mathbf{1} - \mathbf{A}]$ and its determinant

$$s\mathbf{1} - \mathbf{A} = \begin{bmatrix} s + \frac{5}{2} & 1 \\ -\frac{3}{4} & s + \frac{1}{2} \end{bmatrix}$$

$$\det(s\mathbf{1} - \mathbf{A}) = (s + \tfrac{5}{2})(s + \tfrac{1}{2}) + \tfrac{3}{4} = s^2 + 3s + 2 = (s + 1)(s + 2).$$

The inverse matrix $[s\cdot\mathbf{1} - \mathbf{A}]^{-1}$ is now easily obtained as

$$[s\cdot\mathbf{1} - \mathbf{A}] = \begin{bmatrix} \dfrac{s + \frac{1}{2}}{(s+1)(s+2)} & \dfrac{1}{(s+1)(s+2)} \\ -\dfrac{\frac{3}{4}}{(s+1)(s+2)} & \dfrac{s + \frac{5}{2}}{(s+1)(s+2)} \end{bmatrix}$$

$$= \begin{bmatrix} -\dfrac{\frac{1}{2}}{s+1} + \dfrac{\frac{3}{2}}{s+2} & \dfrac{1}{s+1} - \dfrac{1}{s+2} \\ -\dfrac{\frac{3}{4}}{s+1} + \dfrac{\frac{3}{4}}{s+2} & \dfrac{\frac{3}{2}}{s+1} - \dfrac{\frac{1}{2}}{s+2} \end{bmatrix}.$$

A partial-fraction expansion was performed in the last step. The inverse Laplace transform of this expression is

$$L^{-1}[s\cdot\mathbf{1} - \mathbf{A}]^{-1} = \begin{bmatrix} -\frac{1}{2}e^{-t} + \frac{3}{2}e^{-2t} & e^{-t} - e^{-2t} \\ -\frac{3}{4}e^{-t} + \frac{3}{4}e^{-2t} & \frac{3}{2}e^{-t} - \frac{1}{2}e^{-t} \end{bmatrix} = \mathbf{e}^{\mathbf{A}t}. \qquad (5.135)$$

(It is left as an exercise for the reader to verify this result using one of the above given methods for determining a matrix exponential.)

Suppose that the initial conditions are $v_C = 1$ V and $i_L(0) = 0$, and then the natural response will be

$$\mathbf{x}_n(t) = \begin{bmatrix} v_{C,n} \\ i_{L,n} \end{bmatrix} = \mathbf{e}^{\mathbf{A}t}\begin{bmatrix} 1 \\ 0 \end{bmatrix} = \begin{bmatrix} -\frac{1}{2}e^{-t} + \frac{3}{2}e^{-2t} \\ -\frac{3}{4}e^{-t} + \frac{3}{4}e^{-2t} \end{bmatrix}. \qquad (5.136)$$

Note that the verification of equation 5.136 at $t = 0$ yields the initial values of $v_C(0)$ and $i_L(0)$. The particular solution of equation 5.133 may also be obtained with equation 5.135 using, for example, equation 5.118. Thus,

$$\mathbf{x}_p(t) = \mathbf{A}^{-1}[\mathbf{e}^{\mathbf{A}t} - \mathbf{1}]\mathbf{bw} = \begin{bmatrix} -\frac{1}{4} & \frac{1}{2} \\ -\frac{3}{8} & -\frac{5}{4} \end{bmatrix}\begin{bmatrix} -\frac{1}{2}e^{-t} + \frac{3}{2}e^{-2t} & e^{-t} - e^{-2t} \\ -\frac{3}{4}e^{-t} + \frac{3}{4}e^{-2t} & \frac{3}{2}e^{-t} - \frac{1}{2}e^{-2t} \end{bmatrix}\begin{bmatrix} \frac{5}{2} \\ 0 \end{bmatrix}$$

or after performing all the calculations

$$\mathbf{x}_{part}(t)\begin{bmatrix} v_{C,p} \\ i_{L,p} \end{bmatrix} = \begin{bmatrix} \frac{5}{4}e^{-t} - \frac{15}{8}e^{-2t} + \frac{5}{8} \\ -\frac{15}{8}e^{-t} + \frac{15}{16}e^{-2t} + \frac{15}{16} \end{bmatrix}.$$

By inspection (see the circuit in Fig. 5.13) it can be easily verified that the steady-state values of the capacitor voltage and the inductor current agree with those found below:

$$v_{C,p(\infty)} = \tfrac{5}{8}\,\mathrm{V} \quad \text{and} \quad i_{L,p(\infty)} = \tfrac{15}{16}\,\mathrm{A}.$$

The Laplace transform is one of the ways of evaluating the matrix exponential. However, if we are going to use the Laplace transform for circuit analysis, we may do it straightforwardly using the methods described in Chapter 3. The methods of matrix function evaluation, considered in this chapter, are the most general and suitable for computer-aided computation.

Chapter #6

TRANSIENTS IN THREE-PHASE SYSTEMS

6.1 INTRODUCTION

In the previous chapters we have discussed transients in single-phase circuits. However, practically all-electric power is generated, transmitted, distributed and utilized in three-phase systems. Three-phase networks are generally more complicated than single-phase circuits. The complication arises from the interconnection and displacement angle between phases, the triplicate number of components and the branches introduced by the three phases and, also, because of the need to sometimes consider mutual coupling between phases. Naturally, we started our study of transient analysis with single-phase circuits, while establishing the principles and different methods, and gaining experience in techniques of solving problems. Our continued analysis of transients in three-phase networks, therefore, will be based on our previous study.

There are two basic methods for the analysis and calculation of transients in three-phase circuits: 1) to extend the single-phase approach and 2) to use symmetrical components. The first approach is based on the use of the generalized current/voltage phasor of the three-phase system and the two axes representation of a synchronous machine. The single-phase approach, hence, considers the three-phase system as one entity and that a disturbance occurring at one point affects the whole system, and that the transient components excited are not symmetrical and do not obey the three-phase relationships like in steady-state behavior. The method of symmetrical components has been used for many years to calculate the steady-state behavior of three-phase networks when some part of the network happens to run under unbalanced conditions (primarily with an unbalanced load). The method may also be used to analyze unsymmetrical faults, such as: the single-phase to earth fault, the phase-to-phase short circuit, etc. The method of symmetrical components, actually, removes the unsymmetrical conditions and allows the computation to proceed much the same as for symmetrical three-phase short-circuit conditions, with, of course, some extra complications of the whole procedure.

In this chapter we will discuss the short-circuit faults (symmetrical as well as unsymmetrical) at different points of a three-phase system and the transient overvoltages. The emphasis will be placed on the analysis of the terminal short circuits of power transformers and generators.

6.2 SHORT-CIRCUIT TRANSIENTS IN POWER SYSTEMS

The dominant causes of disturbance of the normal operation of power systems are short-circuits. Short-circuit currents are generally of a magnitude many times that of their rated values. In consequence, high dynamic and thermal stresses are generated, which affect the electrical equipment. In the case of short circuit to earth, unacceptable contact potentials arise, which can lead to damage to the equipment and personal danger. Hence, in planning and designing electric power networks the highest consideration must be given to short-circuit analysis and short-circuit current estimations. Knowing the value of short-circuit currents and their flow is also necessary for the specification of protective devices. The following sections are dedicated to short-circuit transient analysis and different methods of calculating short-circuit currents.

In three-phase systems a distinction is made between the following kinds of short-circuits:

a) **Three-pole short-circuit**, in which the three voltages at the short-circuit point are all zero, and the three conductors are symmetrically loaded by the short-circuit currents, as shown in Fig. 6.1(a). Hence this kind of short-circuit fault is called symmetrical and the analysis of this kind of short circuit is performed on a single-phase representation. It should be noted that this kind of short circuit is relatively rare, but it is usually the most dangerous since the short-circuit currents developed in this fault are of the highest magnitude. They are

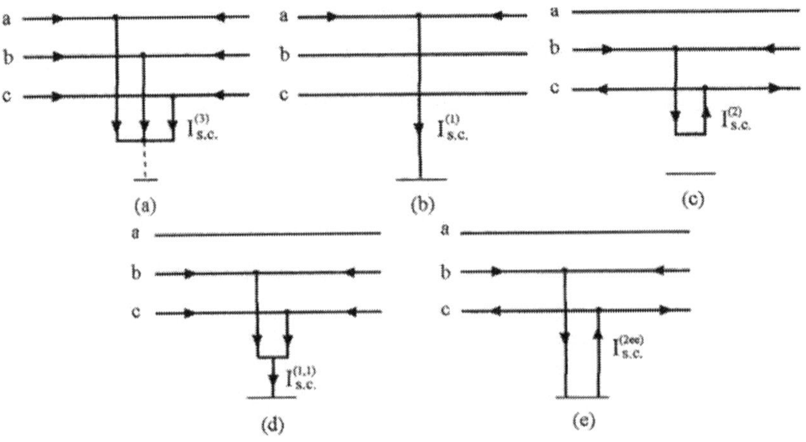

Figure 6.1 Designation of short-circuit faults: three-pole short circuit (a), a single-pole-ground short circuit (b), two-pole-ground fault (d) and double-earth fault (e).

important for specifying the equipment under the short-circuit fault. Since the three-phase voltage at this kind of fault drops to zero, stability problems arise and non-static loads, such as induction motors, run down to stand still (see further on in Chapter 8).

The other four kinds of short-circuits are entirely unsymmetrical conditions. In particular, the voltages at the short-circuit point are not all zero. As a result of the unsymmetrical conditions, mutual couplings are introduced between the phase conductor and the neutral conductor, if present.

b) **The single-pole short-circuit between one of the phases and earth**, Fig. 6.1(b). This kind of fault is the most frequently encountered. Sometimes when the network possesses a low neutral earth impedance, the fault current can even exceed the largest currents produced by a three-pole short-circuit.

c) **The two-pole short-circuit without an earth fault**, Fig. 6.1(c), in which only two phase voltages at the short-circuit point are zero. In this kind of short-circuiting the short-circuit currents are usually less than those produced by a three-pole short circuit. However, if the short-circuit location is close to the generator, the subsequent short-circuit current can become greater than in the three-pole case.

d) **The two-pole short-circuit with an earth fault**, Fig. 6.1(d). This kind of fault may occur in a system with a grounded neutral and has similar characteristics to the previous one.

e) **The double earth fault**, which occurs in a system with an isolated neutral, Fig. 6.1(e). The short-circuit currents in this case may not exceed the rated values, but are significant with regard to the determination of the contact potential and dimension of the earthing systems.

6.2.1 Base quantities and per-unit conversion in three-phase circuits

In the analysis of power networks it is common to use a so-called **"per unit"** system (denoted p.u.) for expressing network quantities rather than a system of actual units (Ω, A, V, etc.). According to this system all the quantities are expressed as fractions of reference quantities, or **base values**, such as base apparent power S_b (VA), base voltage, V_b, and/or base current, I_b. It is obvious that it is enough to choose only two of these quantities since they are related by the expression

$$S_b = \sqrt{3} V_b I_b. \tag{6.1}$$

Usually the base voltage is chosen, in addition to the base power, and the base current is calculated as

$$I_b = \frac{S_b}{\sqrt{3} V_b}, \tag{6.1a}$$

where V_b and I_b are the line quantities.

Hence, the p.u. quantities will be:

$$V_{pu} = \frac{V}{V_b} \quad \text{(p.u. voltage)} \tag{6.2a}$$

$$I_{pu} = \frac{I}{I_b} \quad \text{(p.u. current)} \tag{6.2b}$$

$$S_{pu} = \frac{S}{S_b}, \quad P_{pu} = \frac{P}{S_b}, \quad Q_{pu} = \frac{Q}{S_b} \quad \text{(p.u. power)} \tag{6.2c}$$

and the most important p.u. quantity, the p.u. impedance and its components:

$$Z_{pu} = \frac{Z_{\Omega}}{Z_b}, \quad R_{pu} = \frac{R_{\Omega}}{Z_b}, \quad X_{pu} = \frac{X_{\Omega}}{Z_b}. \tag{6.3}$$

Here the base impedance, Z_b, is established with Ohm's Law as

$$Z_b = \frac{V_b}{\sqrt{3}I_b} = \frac{V_b^2}{S_b}. \tag{6.4}$$

With equation 6.4 we can write

$$Z_{pu} = Z_{\Omega}\frac{S_b}{V_b^2}, \quad R_{pu} = R_{\Omega}\frac{S_b}{V_b^2}, \quad X_{pu} = X_{\Omega}\frac{S_b}{V_b^2}. \tag{6.5}$$

Note that in expressions (equations 6.3–6.5) the impedances and their components are per-phase quantities. It should also be denoted that all the expressions (equations 6.1–6.5) are proper for a one-phase network. In such a case the $\sqrt{3}$ must be omitted, and all the quantities are phase or just circuit values. With the known p.u. value, the actual value can be obtained as

$$Z_{\Omega} = Z_{pu}\frac{V_b^2}{S_b} = Z_{pu}\frac{V_b}{\sqrt{3}I_b}. \tag{6.6}$$

The p.u. system is widely used in "Electric machine and transformer" courses, where the parameters of electric machines and transformers and their characteristics are usually expressed in per-unit quantities. It stands to reason, therefore, that the p.u. system is used in "Power system" courses, since power systems consist, primarily, of synchronous generators, transformers and motors. All such equipment varies widely in size, power, voltages etc. However, for equipment of the same type the p.u. impedances, voltage drops and losses are in the same order, regardless of size.

For example, if the primary winding reactance of a 50 kVA, 6.6 kV single-phase transformer is $X_1 = 38.5\ \Omega$, then this reactance measured in p.u. will be

$$X_{1,pu} = \frac{X_{1,\Omega}}{Z_b} = \frac{38.5}{871} = 0.044,$$

where the base impedance is

$$Z_b = \frac{V_r^2}{S_r} = \frac{6.6^2 \cdot 10^3}{50} = 871\ \Omega.$$

Per-unit quantities are often expressed as a percentage. Percent quantities differ from per-unit by a factor of 100. Hence, the above p.u. reactance, in percent, will be $X_{1,\%} = 4.4\%$. All the transformers of the same series as the above transformer will have about the same percent reactance regardless of their power.

The p.u. values of different items of apparatus by themselves, such as transformers, synchronous generators, motors etc. are given in terms of their own kVA/MVA power and voltage ratings. Hence, for any power system in which several pieces of equipment are involved, it is necessary to refer all the given p.u. values to the system base values: base MVA power and base voltage. Thus, if $Z_{pu}^{(r)}$ is the per-unit impedance (reactance) for rated values, the same impedance (reactance) referred to the base values, will be

$$Z_{pu}^{(b)} = Z_{pu}^{(r)} \frac{S_b V_r^2}{S_r V_b^2}, \tag{6.7}$$

which shows that the "new" p.u. value is directly proportional to the ratio of powers and inversely proportional to the ratio of the squared voltages. If $V_r = V_b$, then

$$Z_{pu}^{(b)} = Z_{pu}^{(r)} \frac{S_b}{S_r}. \tag{6.7a}$$

As already mentioned, in a three-phase system X_{pu} is a per-phase reactance, $S_b(S_r)$ is a three-phase power and $V_b(V_r)$ is a line voltage.

The single base power chosen is to be relatively large, at least equal to, or larger than, the highest power source in the network. All the system impedances will then be related to this base power. The base voltages, however, differ in the dependence on the level of transformation. As we know, these voltages are intended for supplying the transmission and distribution lines over a range from a few thousand volts to a million volts. Hence, the entire power network may have many different voltage levels. By analyzing such a network, all the impedances must be referred to one voltage level. Since all voltages and currents are related directly or inversely as the turn ratio of transformers in any part of power systems, all voltages, currents, volt-amperes and impedances will have the same per-unit values regardless of where they appear in the system. Applying the per-unit values allows the elimination of different voltage levels and represents the entire network on a single voltage level. This is another reason for using a per-unit system of representing the power system quantities.

Let us discuss this topic in more detail. If some particular device is located on the voltage level, which differs from the base voltage level, which is chosen as a main or system base voltage (sb), its base quantities should be calculated as

$$V^{(b)} = \frac{1}{n_1 n_2 \ldots n_k} V^{(sb)}, \quad I^{(b)} = (n_1 n_2 \ldots n_k) I^{(sb)}, \tag{6.8}$$

where $n_1, n_2, \ldots, n_k$ are the turn ratios of the transformers, which are connected in series between the location of the device and the main base level (the turn ratios must be taken in the direction of the main voltage level towards the level of the device location).

The device's actual impedance, which referred (reflected) to the main voltage level, will be

$$Z_{\Omega}^{(sb)} = n_1^2 n_2^2 \cdots n_k^2 Z_{\Omega}^{(b)} = n_{eq}^{(2)} Z_{\Omega}^{(b)},$$

where

$$n_{eq} \cong V_{sb}/V_b \tag{6.9}$$

With n_{eq} the p.u. value of the impedance is

$$Z_{pu}^{(sb)} = Z_{\Omega}^{(sb)} \frac{S_b}{V_{sb}^2} = n_{eq}^2 Z_{\Omega}^{(b)} \frac{S_b}{V_{sb}^2} = Z_{\Omega}^{(b)} \frac{S_b}{V_b^2} = Z_{pu}^{(b)}. \tag{6.10}$$

This important result shows that the p.u. impedance value referred to the system (main) base voltage can be calculated with the same expression (equation 6.5) as has been referred to the base voltage of the equipment location, regardless of which system (main) base voltage is chosen.

It is important to note that for the same reason the p.u. impedance of a transformer is the same whether it is referred to the primary or secondary side. Indeed, let us assume that the p.u. impedance, which referred to the primary (step-down transformer), is Z_1, and that which referred to the secondary is $Z_2 = Z_1/n^2$, where n is the turn ratio ($n = N_1/N_2 = V_{1r}/V_{2r}$). then

$$Z_{1pu} = Z_1 \frac{S_r}{V_{1r}^2},$$

and

$$Z_{2pu} = Z_2 \frac{S_r}{V_{2r}^2} = Z_1 \frac{S_r}{n^2 V_{2r}^2} = Z_1 \frac{S_r}{V_{1r}^2} = Z_{1pu}.$$

Hence, the result is the same as the p.u. impedance, which is referred to the primary.

It shall be noted that, since the voltages at the sending V_1 and receiving V_2 ends of a transmission line are different (because of the voltage drop), the line rated voltage is usually taken as an average value

$$V_{\ell,r} = \frac{V_1 + V_2}{2} \tag{6.11}$$

The average values of the voltages are taken as base voltages for each of the voltage levels in the network(*).

The turn ratio, i.e., the ratio of rated voltages, of the power network transformers may not be the same as the ratio of the average voltages of different levels, so that the impedance referring can be done in two ways: approximate or exact. The referring in accordance to the base-average voltages is approximate. In this case the turns ratio of the transformers (or the ratio of their rated voltages) is taken equal to the ratio of the level voltages. If the base voltages are related by the turn ratios of the transformers, the referring is accounted as an exact one. Let us illustrate these two approaches of expressing p.u. impedances in the following example.

Example 6.1

Consider the three-phase network whose one-line diagram is shown in Fig. 6.2. The rating values and p.u. reactances of the generator and the transformers as well as the parameters of the transmission line and current-limiting reactor are indicated in this diagram. For the calculation of a short-circuit current draw the equivalent circuit and find all the p.u. reactances, which are referred to the generator voltage level in two ways: 1) approximately and 2) exactly.

Solution

We first have to specify the base volt-ampere power, which for a given network it is reasonable to choose a value of 100 MVA.

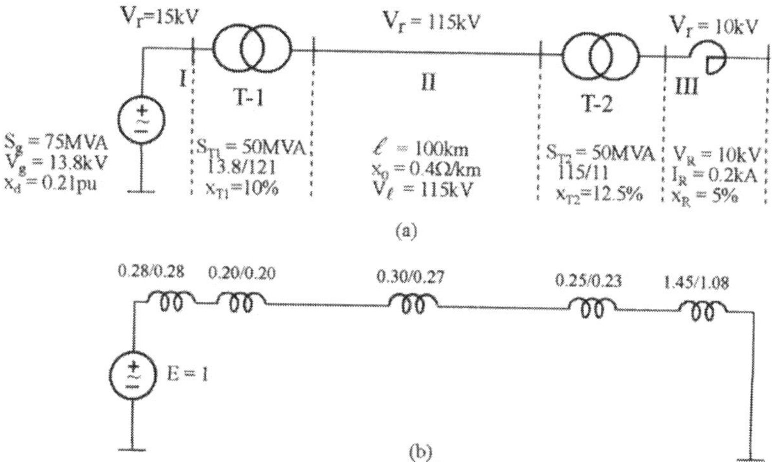

Figure 6.2 A one-line diagram of a given network (a) and its equivalent circuit in terms of p.u. (b).

(*)The average voltages are usually in accordance with those recommended by electric companies or general standards.

1) *Approximate evaluation.* The rated voltages on each level will be as base values, i.e., on the generator level: $V_{bI} = 13.8$ kV, on the line level: $V_{bII} = 115$ kV and on the distribution level: $V_{bIII} = 10$ kV.

The base currents in accordance with equation 6.1a are

$$I_{bI} = \frac{100}{\sqrt{3} \cdot 13.8} = 4.18 \text{ kA}, \quad I_{bII} = \frac{100}{\sqrt{3} \cdot 115} = 0.5 \text{ kA}, \quad I_{bIII} = \frac{100}{\sqrt{3} \cdot 10} = 5.77 \text{ kA}.$$

Then the p.u. reactances are obtained in accordance with equation 6.7a as:

for the generator $$X_g = 0.21 \frac{100}{75} = 0.28 \text{ pu},$$

for the sending end transformer

$$X_{T1} = 0.1 \frac{100}{50} = 0.2 \text{ pu},$$

for the receiving and transformer

$$X_{T2} = 0.125 \frac{100}{50} = 0.25 \text{ pu},$$

for the transmission line in accordance with equation 6.5

$$X_\ell = 0.4 \cdot 100 \frac{100}{115^2} = 0.30 \text{ pu},$$

for the current-limiting reactor

$$X_{rct} = 0.05 \frac{100}{3.46} = 1.45 \text{ pu},$$

where $S_{re} = \sqrt{3} \cdot 10 \cdot 0.2 = 3.46$ MVA is the reactor rating apparent power.

2) *Exact evaluation.* The base voltage on the generator level, as in the previous calculation, will be $V_{bI} = 13.8$ V. The base voltages on the line level and on the distribution level, in accordance to the turn ratio of the transformers, will be (equation 6.8),

$$V_{bII} = \frac{1}{13.8/121} = 121 \text{ kV} \quad \text{and} \quad V_{bIII} = \frac{1}{115/11} = 11.6 \text{ kV}$$

The base currents are (equation 6.1a)

$$I_{bI} = \frac{100}{\sqrt{3} \cdot 13.8} = 4.18 \text{ kA}, \quad I_{bII} = \frac{100}{\sqrt{3} \cdot 121} = 0.48 \text{ kA}, \quad I_{bIII} = \frac{100}{\sqrt{3} \cdot 11.6} = 4.98 \text{ kA}.$$

The per-unit reactances are obtained as:

for the generator (equation 6.7a)

$$X_g = 0.21 \frac{100}{75} = 0.28 \text{ pu},$$

i.e., the same as in the previous calculation (since the base voltage on the generator level did not change),

for the sending and transformer

$$X_{T1} = 0.1\,\frac{100}{50} = 0.2\ \text{pu},$$

i.e., it did not change either for the same reason,

for the receiving end transformer (equation 6.7)

$$X_{T2} = 0.125\,\frac{100}{50}\left(\frac{115}{121}\right)^2 = 0.23\ \text{pu}$$

(since the base voltage and the rated voltage are not equal),

for the transmission line (equation 6.5)

$$X_\ell = 0.4\cdot 100\,\frac{100}{121^2} = 0.27\ \text{pu}$$

for the current-limiting reactor (equation 6.7)

$$X_{re} = 0.05\,\frac{100}{3.46}\left(\frac{10}{11.6}\right)^2 = 1.08\ \text{pu}.$$

Finally, it might be good to point out that using per-unit quantities in short-circuit fault analysis simplifies to a great extent the numerical calculations manually and/or by using computers.

6.2.2 Equivalent circuits and their simplification

As the equivalent circuit of the power system network in per-unit quantities is established, the next step in short-circuit calculation is to simplify the network. Using the known methods of circuit analysis we may, in most cases, simplify the network so that only a single equivalent generator will feed the short-circuit fault through an equivalent impedance. The following will remind the reader of the most useful of these methods.

(a) *Series and parallel connections*

We start with the series and parallel connections, simplifying them by well-known formulas. Thus, if we have a few generators operating in parallel (usually at the same power station), as shown in Fig. 6.3, we may integrate them into a single one by using this formula (sometimes called Millman's formula).

$$E_{eq} = \frac{E_1 Y_1 + E_2 Y_2 + \cdots + E_n Y_n}{Y_1 + Y_2 + \cdots + Y_n} = \frac{\sum^n EY}{\sum^n Y}, \quad X_{eq} = \frac{1}{\sum^n Y}, \tag{6.12}$$

where

$$Y_1 = \frac{1}{X_1},\ Y_2 = \frac{1}{X_2},\ \cdots\ Y_n = \frac{1}{X_n}.$$

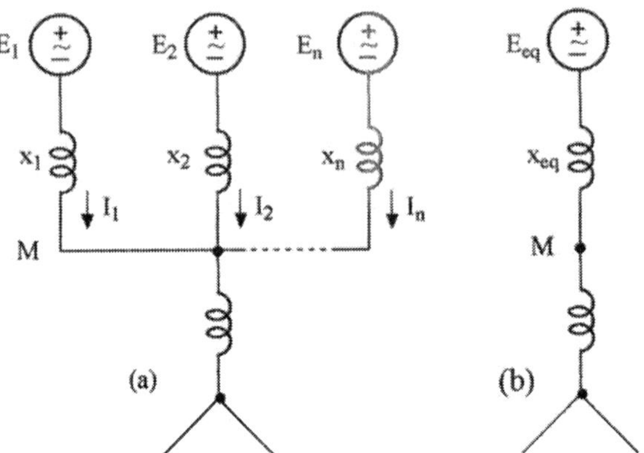

Figure 6.3 Three generators in parallel (a) and the equivalent circuit (b).

For only two generators the above formula will be

$$E_{eq} = \frac{E_1 X_1 + E_2 X_2}{X_1 + X_2}, \quad X_{eq} = \frac{X_1 X_2}{X_1 + X_2}. \tag{6.13}$$

These formulas are valid for any value of E's (EMF's) including zero. In particular, the load may be treated as a main generator having zero EMF ($E = 0$). Then such a generator can be combined with others, instead of connecting the zero potential point of the load with the point of the short-circuit fault, as shown in Fig. 6.4. This consideration of the load is approximate; however, it allows us to easily simplify the network. As can be seen in Fig. 6.4(b) the generators can be gradually integrated all together into one single generator, as shown in Fig. 6.4(c). With two more steps, as shown in Figs. 6.4(c) and (d) the given network is simplified to a single generator and a single reactance. In contrast to the above procedure, the connection of zero potential points, as shown in Fig. 6.4(a) (see the dashed line) gives rise to a more complicated circuit, which includes two loops.

(b) *Delta-star (and vice-versa) transformation*

The delta-star transformation can also be useful for the simplification of networks having a short-circuit fault. For introducing this technique, let us consider the network shown in Fig. 6.5(a). In the first step the star $X_3 - X_4 - X_5$ is replaced by delta (shown by dash lines) whose reactances are calculated by the following formulas

$$X_8 = X_3 + X_4 + \frac{X_3 X_4}{X_5}, \quad X_9 = X_4 + X_5 + \frac{X_4 X_5}{X_3}, \quad X_{10} = X_3 + X_5 + \frac{X_3 X_5}{X_4}.$$

Replacing the parallel connecting reactances with their equivalents, we obtain the circuit in Fig. 6.5(b). In the next step we transform the delta $X_8 - X_{11} - X_{12}$

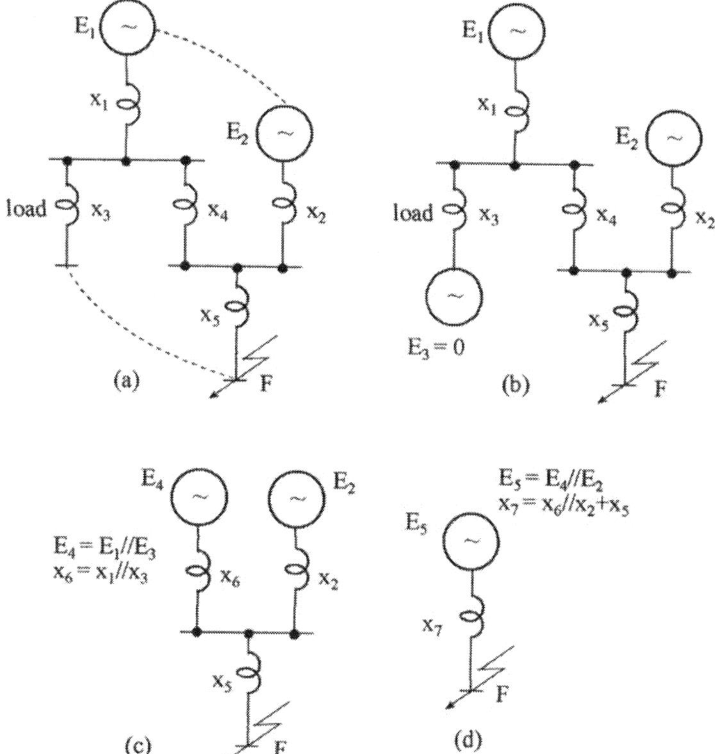

Figure 6.4 A network containing a load (a), the load has been replaced by a generator having zero EMF (b), two steps of simplifying the circuit (c) and (d).

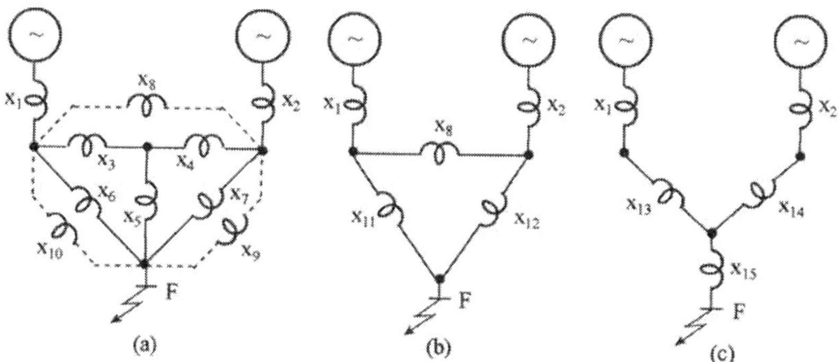

Figure 6.5 A given network (a), a network after a star-delta transformation (b) and a network after a delta-star transformation (c).

into a star by using the formulas

$$X_{13} = \frac{X_8 X_{11}}{X_8 + X_{11} + X_{12}}, \quad X_{14} = \frac{X_8 X_{12}}{X_8 + X_{11} + X_{12}}, \quad X_{15} = \frac{X_{11} X_{12}}{X_8 + X_{11} + X_{12}}.$$

The obtained circuit, Fig. 6.5(c), can now be simplified, as was previously done, into one having a single generator and a single impedance.

(c) *Using symmetrical properties of a network*

We may use symmetrical properties to simplify a given network. Consider the network shown in Fig. 6.6(a). If the rating values of transformers, reactors and cables are identical, the entire network is symmetrical relative to the fault point and can be simplified as shown in Fig. 6.6(b). The rest of the elements may not be included in this circuit, since the fault current will not flow through them. The obtained scheme has two parallel branches and can be easily simplified to a single reactance.

6.2.3 The superposition principle in transient analysis

By neglecting the magnetic saturation in synchronous machines and transformers (which is common practice in the transient analysis of power systems), the power network may be treated as a linear system. Hence, the principle of superposition can be applied to its analysis. As was shown in section 2.6, to find the short-circuit current at the fault point, we may superimpose two regimes:

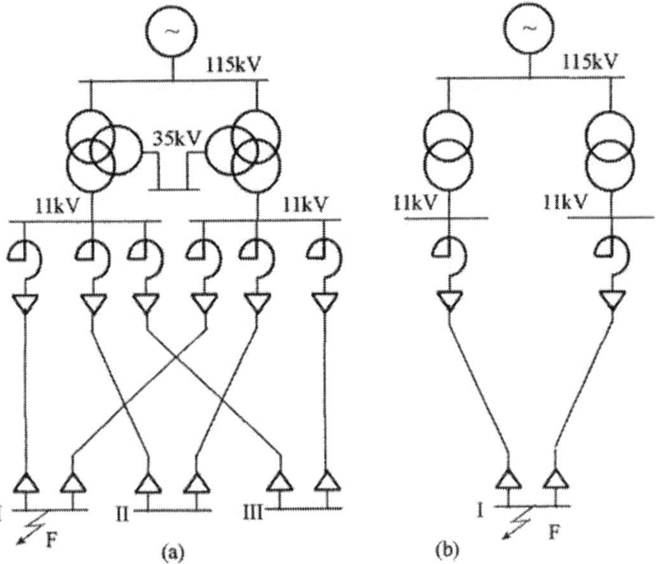

Figure 6.6 A symmetrical network (a) and its simplification (b).

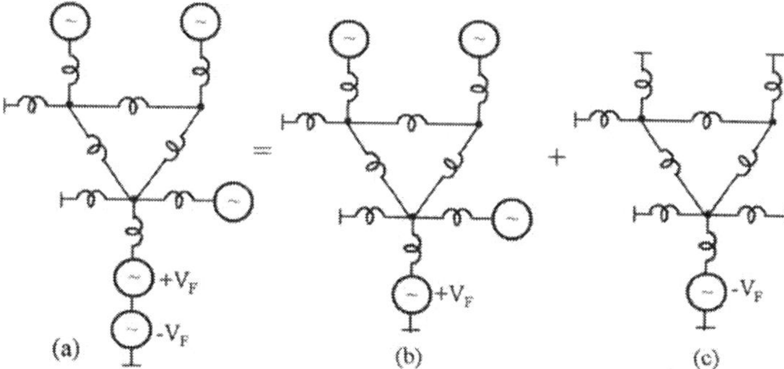

Figure 6.7 A given network (a), the network of a previous regime (b) and the network due to the fault.

1) a previous one, i.e. prior to fault and 2) the additional one, which arises due to the fault. To illustrate this technique, consider the network shown in Fig. 6.7. It is obvious that the fault conditions will not change, if we insert in the fault point two voltage sources equal in magnitude, but opposite in sign, as shown in this figure. The magnitude of these sources should be chosen equal to the voltage value at the fault prior to the fault (if this voltage is not known, the rated value can be used). Following the superposition principle the network in Fig. 6.7(a) can now be represented as two separate networks.

The first one, shown in Fig. 6.7(b), is actually the network of a normal operation, prior to the fault occurring. The second one, shown in Fig. 6.7(c), is the network of the fault regime. Usually the operational conditions (the voltages at the nodes and the branch currents) are known, so that only the network in figure (c) must be analyzed. This network is simpler than the given one, since it has only one source, and therefore might be easier to simplify to a single reactance. The total currents will be found by the summation of the normal condition currents and the fault currents found in the circuit of Fig. 6.7(c).

Example 6.2

The equivalent circuit of part of a power system is shown in Fig. 6.8. The p.u. impedances of the generators, transformers and transmission lines, as well as the generators' EMF's, are indicated on the scheme. (The one-line diagram of the network and the calculations of the p.u. values are given in Appendix II.) Simplify this network up to a single source and single impedance.

Solution

As a first step we replace two parallel EMF's, E_1 and E_5, by their equivalent one (since all the values are in per unit quantities, the indication p.u. is omitted)

$$E_{\mathrm{eq1}} = \frac{1.25/(0.64 + 0.19)}{1/0.83 + 1/4.55} = 1.06,$$

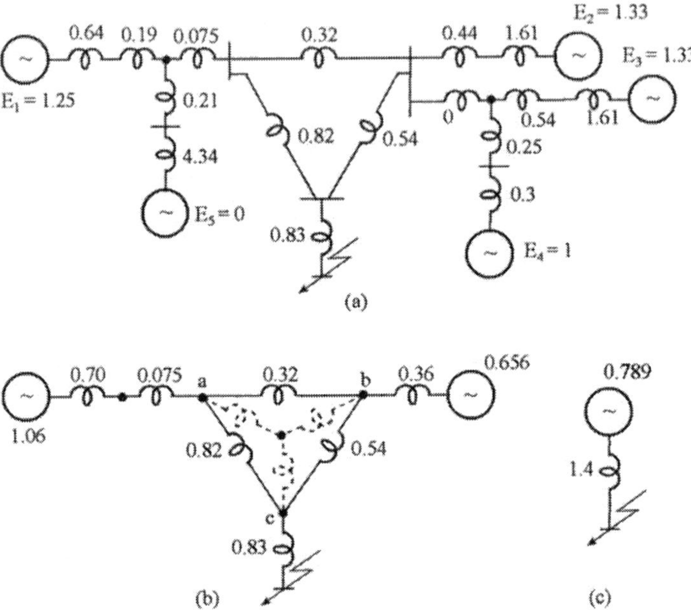

Figure 6.8 A given network (a), after the first step of simplification (b) and the resulting circuit (c).

and

$$X_{\text{eq1}} = \frac{1}{1/0.83 + 1/4.55} = 0.70.$$

In the same way we replace three parallel EMF's E_1, E_3 and E_4 by E_{eq2}

$$E_{\text{eq2}} = \frac{1.33/2.05 + 1.33/2.15 + 1/0.55}{1/2.05 + 1/2.15 + 1/0.55} = \frac{1.818}{2.771} = 0.656,$$

and

$$X_{\text{eq2}} = 1/2.771 = 0.360.$$

As a result we obtain the circuit shown in Fig. 6.8(b).

The next step is the delta-star transformation and replacing two EMF's by a total one

$$X_a = \frac{0.32 \cdot 0.82}{0.32 + 0.82 + 0.84} = \frac{0.262}{1.68} = 0.160,$$

$$X_b = \frac{0.32 \cdot 0.54}{1.68} = 0.100,$$

$$X_c = \frac{0.54 \cdot 0.82}{1.68} = 0.260.$$

Now, the total EMF is obtained as

$$E_{tot} = \frac{1.06 \cdot 0.46 + 0.656 \cdot 0.94}{0.46 + 0.94} = \frac{1.104}{1.4} = 0.789,$$

$$X_{tot} = \frac{0.46 \cdot 0.94}{1.4} + 0.26 + 0.83 = 1.40.$$

The resulting circuit is shown in Fig. 6.8(c).

6.3 SHORT-CIRCUITING IN A SIMPLE CIRCUIT

As we have already mentioned, in the majority of the fault situations, such as short-circuiting a single conductor to ground or earth (a one-phase short-circuit) or short-circuiting between two conductors (a two-phase short-circuit), the power system network becomes unsymmetrical. However, we shall start our study of transients in three-phase systems with a symmetrical three-phase fault, where all three conductors touch each other or fall to ground. Although this kind of fault occurs in only a very small percentage of cases, it is very severe for the system and its devices. The very extreme magnitudes of the fault currents in such faults give engineers the ratings of the circuit breakers and other equipment of the power network to be used.

In the case of a symmetrical three-phase fault in a symmetrical system, we can use a single-phase approach, which simplifies to a great degree the calculation of the short-circuit currents and performance of the transient analysis. By simplifying the system network, as was discussed in the previous sections, we may reduce it to the simplest circuit including a single source and a single impedance.

In the case of unsymmetrical faults, the most common method of analysis is to use symmetrical components (see further on), in which we attempt to find the symmetrical components of the voltages and the currents at the point of unbalance and connect the sequence networks, which are, in fact, symmetrical circuits. Hence, the following analysis can be made by again using a single-phase representation.

For a better understanding of the short-circuit phenomena in a three-phase system let us first consider the simple circuit, shown in Fig. 6.9, in which a

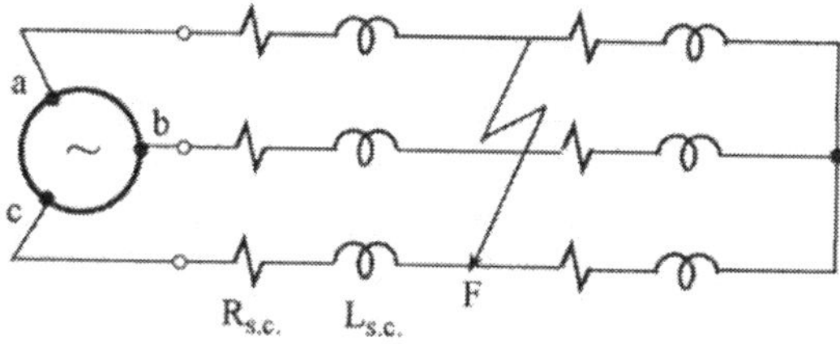

Figure 6.9 A simple three-phase circuit under a symmetrical three-phase fault.

symmetrical three-phase fault occurs. Following the classical approach in transient analysis (see Chaps. 1 and 2) we may represent the total fault current, say in phase "a", as the sum of a forced and a natural response

$$i_{sc} = i_f + i_n = I_{m,f} \sin(\omega t + \psi_v - \varphi_{sc}) + A e^{-(R_{sc}/L_{sc})t}, \tag{6.14}$$

where $I_{m,f} = V_m/Z_{sc}$ is an amplitude of the forced response, which is a steady-state short-circuit current,

$$Z_{sc} = \sqrt{R_{sc}^2 + (\omega L_{sc})^2}, \quad \varphi_{sc} = \tan^{-1}\frac{\omega L_{sc}}{R_{sc}}$$

are the magnitude and the angle of the total impedance up to the fault point F and ψ_v is an applied voltage phase angle at the moment of the short-circuiting.

Suppose that the current prior to short-circuiting was

$$i_{ld} = I_{m,ld} \sin(\omega t + \psi_v - \varphi_{ld}), \tag{6.15}$$

where $I_{m.ld} = V_m/Z_{ld}$ is the amplitude of the current under normal load conditions, just prior to short-circuiting,

$$Z_{ld} = \sqrt{R_{ld}^2 + (\omega L_{ld})^2} \quad \text{and} \quad \varphi_{ld} = \tan^{-1}\frac{\omega L_{ld}}{R_{ld}}$$

are the total impedance and the angle of a total impedance of the load and the system under normal operation. Then the integrating constant is

$$A = i_{n0} = i_{ld}(0) - i_f(0) = I_{m,ld} \sin(\psi_v - \varphi_{ld}) - I_{m,f} \sin(\psi_v - \varphi_{sc}), \tag{6.16}$$

and the time constant of the exponential term is

$$\tau = \frac{L_{sc}}{R_{sc}}. \tag{6.17}$$

Finally, we have the expressions of the natural and total responses:

$$i_n = A e^{-t/\tau} = [I_{m,ld} \sin(\psi_v - \varphi_{ld}) - I_{m,f} \sin(\psi_v - \varphi_{sc})] e^{-t/\tau}, \tag{6.18}$$

and

$$i_{sc} = I_{m,f} \sin(\omega t + \psi_v - \varphi_{sc}) - i_{n0} e^{-t/\tau}. \tag{6.19}$$

Since the current in phase "a" is determined, the rest of the currents in phases "b" and "c" may be found by replacing the current of phase "a" by $-120°$ for the current of phase "b" and by 120° for the current of phase "c". In Fig. 6.10 the three-phase phasor diagram of all three currents is given.

In accordance with the phasor concept, the phasors on the phasor diagram are vectors rotated in a counterclockwise direction at an angular velocity of ω, rad/s, and their projections on axis "t" (or on an axis of imaginary numbers) give the instantaneous values of the currents/voltages. Hence, the differences of two phasors $(I_{m,ld} - I_{m,f})$ in each of three phases (dashed phasors) represent the vectorized values of the integration constants, and their projection on axis t

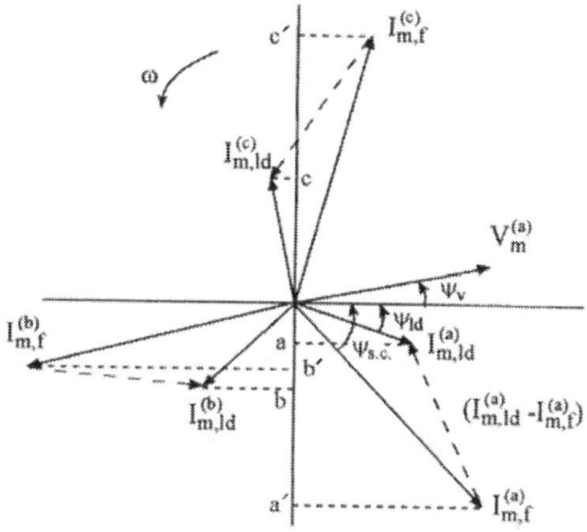

Figure 6.10 The phasor diagram of three-phase currents in a simple circuit at the three-phase short-circuit fault.

gives the initial values of the natural responses in the corresponding phase *a*, *b* and *c*. Such a representation clearly shows that the initial value of a natural response may vary from its maximal value, when the vector $(I_{m,ld} - I_{m,f})$, i.e., the dashed line, is parallel to axis *t*, to zero, when this vector is perpendicular to axis *t*. The position of this vector on the diagram is dependent on the applied voltage phase angle ψ_v at the moment of fault. In the latter case the exponential term is absent, which means that the forced current at the instant of switching is equal to the current prior to switching and no transient response takes place at all. It is obvious that such conditions may occur only in one of the phases. For the conditions of the phasor diagram, shown in Fig. 6.10, the short-circuit currents versus time in all three phases are shown in Fig. 6.11.

As can be seen from the current plots in Fig. 6.11, the transient currents in three phases, due to the aperiodic term, are different. Hence, we shall say that even the three-phase short circuit is not symmetrical. In one of the phases the instantaneous current might be much larger than in the others. However, after the aperiodical term decays, the short-circuit current becomes symmetrical.

The exponential term can be separated from the short-circuit current oscillogram, as shown in Fig. 6.11(c). As can be seen, the exponential term is a medium line in between two envelopes: an envelope of positive amplitudes and an envelope of negative amplitudes. We may also say that the exponential term represents the curve axis of a short-circuit current causing the current to be unsymmetrical.

The initial value of the exponential term also depends on the previous regime. It is easy to see that the largest value of the integration constant may be achieved, if in the previous regime the current was *leading* (Fig. 6.12(b)). Since the capacitance load in power systems is uncommon, the most severe case may occur if, prior to the fault, the network was under no load operation, Fig. 6.12(c).

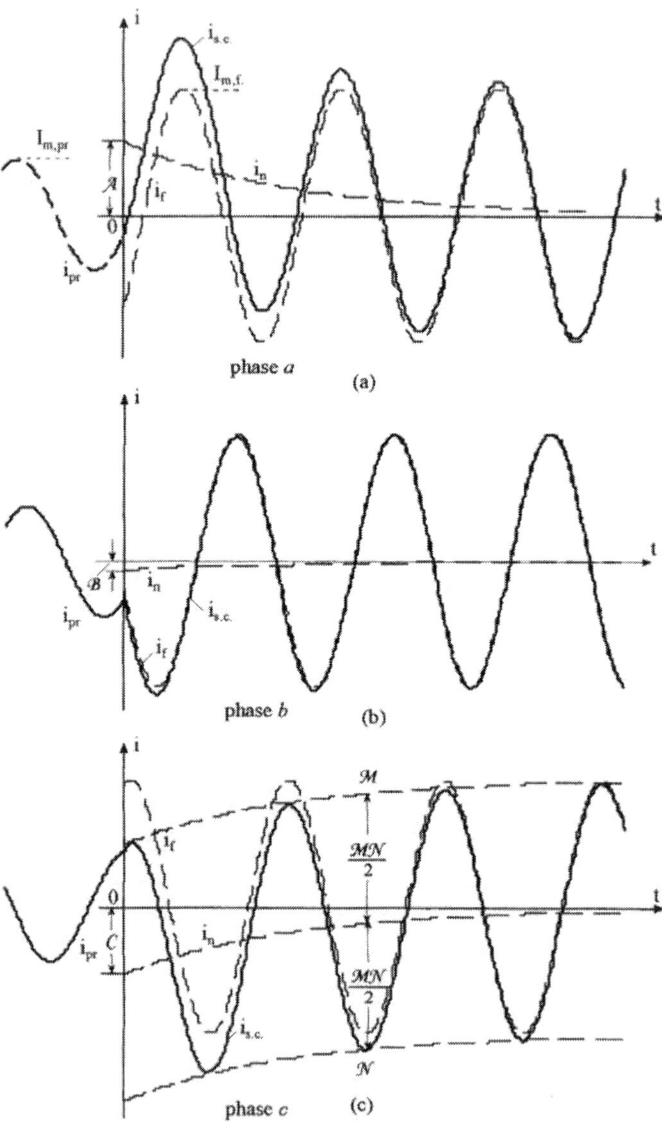

Figure 6.11 The short-circuit currents in a simple three-phase circuit.

The maximal value of the short-circuit current in the latter case will appear if the forced response current, at the instant of the fault, passes its maximum (positive or negative), so that $i_{n0} \cong I_{m,f}$ For the short-circuited network, which is primarily of inductive impedance, this takes place when the applied voltage

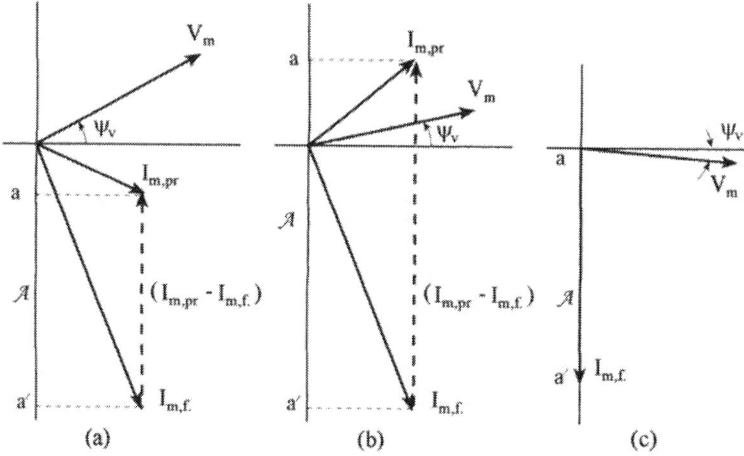

Figure 6.12 The most unfavorable conditions for the largest value of an aperiodic term to appear: 1) the lagging load (a), 2) the leading load (b) and 3) no-load operation (c).

passes its zero point. The plot of the short-circuit current under such conditions is shown in Fig. 6.13.

Note that the time constant T_a may be found experimentally from the short-circuit oscillogram, as shown in Fig. 6.13 (also refer to section 1.3.1). The time constant here is measured as an under-tangent, T_a, along axis t. To achieve good precision, using this method, point g must be taken at the beginning (the highest) part of the exponential curve.

We may estimate the highest peak (or just "peak") of a short-circuit current by using the "*peak-coefficient*". Since the highest peak is found to occur at about

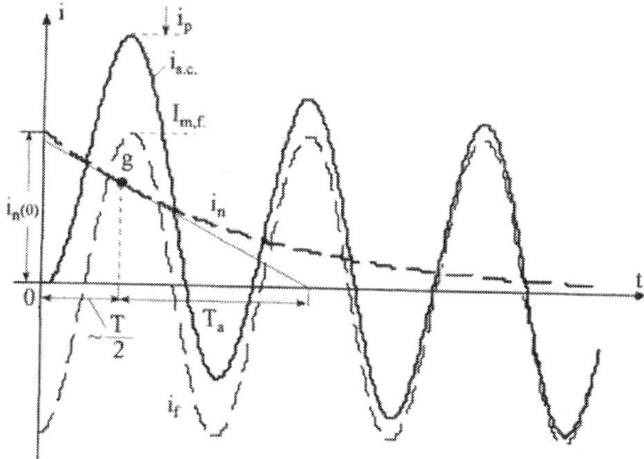

Figure 6.13 A plot of a short-circuit current having a maximal exponential component.

$t \cong T/2$ (i.e. at 50 Hz in 10 ms and at 60 Hz in 8 ms) after the incidence of the short-circuit, we have

$$i_{pk} = I_{m,f} + i_{n0} e^{-(T/2\tau)} = (1 + e^{-(T/2\tau)}) I_{m,f} = k_p I_{m,f}$$

where k_{pk} is the peak coefficient. Thus,

$$k_{pk} = 1 + e^{-(T/2\tau)}. \tag{6.20a}$$

The time constant, τ, changes between zero ($L = 0$) to infinity ($R = 0$), therefore the peak coefficient lies in the range

$$1 < k_p < 2 \tag{6.20b}$$

(except for the much less common case of the leading current, shown in Fig. 6.12(b)).

Due to the resistivity of the short-circuit network, the exponential term finally vanishes. Usually the time constant of power system networks is relatively large ($\tau = 0.01$–0.2 s), so that it takes a few periods for the exponential term to decay.

To check the *thermal stability* of electrical equipment under short-circuit fault conditions, the r.m.s value of the short-circuit current in its initial stage has to be estimated. Since this current is unsymmetrical, i.e., consisting of two components: sinusoidal, or a.c., and exponential, or d.c., we may calculate its r.m.s. value as

$$I_{sc} = \sqrt{I_f^2 + I_{\exp}^2} \tag{6.21}$$

where $I_f = I_{m,f}/\sqrt{2}$ is an r.m.s. value of a.c. and $I_{\exp}$ is an r.m.s. value of the exponential term. The *r.m.s. value of the exponential term* may be estimated as its average value in the interval of one period T or approximately, as the value in the middle point of the period, as shown in Fig. 6.14.

The highest *r.m.s. value of a short-circuit current* will appear at the first period

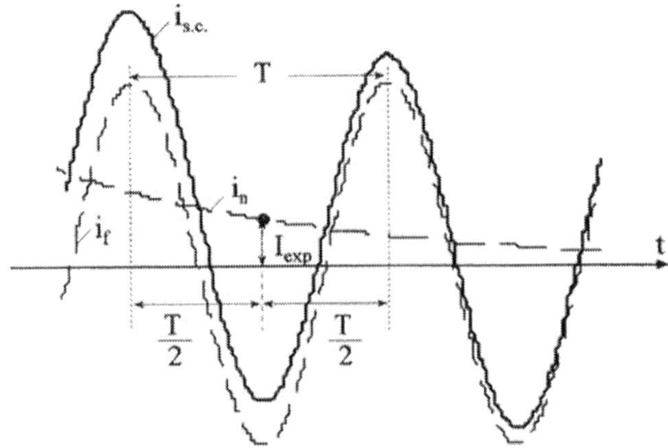

Figure 6.14 Calculation of an average value of the exponential term.

after the instant of fault. Thus, with equation 6.20a and in accordance with equation 6.21 we obtain

$$I_{sc,pk} = \sqrt{I_f^2 + [(k_{pk} - 1)\sqrt{2} I_f]^2} = I_f \sqrt{1 + 2(k_{pk} - 1)^2} \tag{6.22a}$$

and with equation 6.20b, the range limits of $I_{sc,pk}$ are

$$1 < \frac{I_{sc,pk}}{I_f} < \sqrt{3}. \tag{6.22b}$$

The value of i_{pk} is used by project engineers for checking the electrodynamic stability of electrical equipment under short-circuit fault conditions.

6.4 SWITCHING TRANSFORMERS

6.4.1 Short-circuiting of power transformers

The short-circuit phenomenon in any transformer must be analyzed as a transient response in mutual (magnetically interlinked) elements. Considering a three-phase transformer as a symmetrical element (which is an approximation of a three-phase core type transformer) we may reduce it to a single-phase circuit, as shown in Fig. 6.15. In this equivalent circuit a transformer is represented as two identical circuits. The resistance and inductance of the secondary winding are referred to the primary winding. Note that, as previously shown, p.u. impedances, resistances and inductances of a transformer are the same regardless of which winding they are referred to. This means that both the

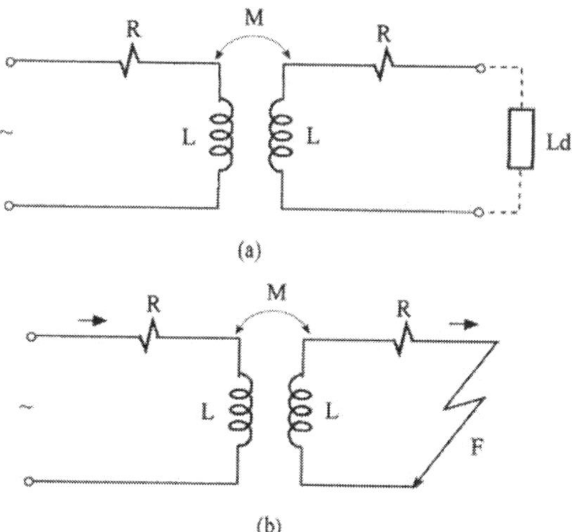

Figure 6.15 An equivalent single-phase transformer under the load (a) and under the short-circuit fault (b).

primary and secondary circuits have identical p.u. parameters. It should also be noted that the ratio of the winding inductance to its resistance is equal in both windings as is usual in power transformers (since the amounts of copper in the primary and secondary windings are nearly equal). Hence we may assume that the parameters of the secondary winding, which are referred to the primary, are about of the same values as the primary, and represent the transformer by two similar circuits, Fig. 6.15. Also note that this equivalent transformer has a unit turns ratio. We may also say that when the analysis is done in p.u. quantities, the actual values of the primary and secondary circuits may be obtained by simply multiplying the p.u. value of each current by an appropriate rated value.

By using the superposition properties discussed previously, we may separate the previous, i.e., the prior to short-circuiting, operation of the transformer and its transient behavior having zero initial conditions. To find its natural response we shall solve two homogeneous equations

$$\begin{aligned} L\frac{di_{1,n}}{dt} + Ri_{1,n} + M\frac{di_{2,n}}{dt} &= 0 \\ L\frac{di_{2,n}}{dt} + Ri_{2,n} + M\frac{di_{1,n}}{dt} &= 0. \end{aligned} \tag{6.23}$$

The characteristic equation has been developed in Example 1.2 (Chapter 1) and it roots are given by equation 1.34, which under the given conditions (that $L_1 = L_2 = L$ and $R_1 = R_2 = R$) yields

$$s_{1,2} = \frac{1}{L^2 - M^2}\left[RL \pm \sqrt{(RL)^2 - R^2(L^2 - M^2)}\right] = \frac{R(L \mp M)}{L^2 - M^2}, \tag{6.24a}$$

or

$$s_1 = -\frac{R}{L+M} = -\frac{1}{\tau_m}, \quad s_2 = -\frac{R}{L-M} = -\frac{1}{\tau_\ell}, \tag{6.24b}$$

and the time constants are

$$\tau_m = \frac{L+M}{R}, \quad \tau_\ell = \frac{L-M}{R}. \tag{6.24c}$$

Hence, the natural currents are

$$i_{1,n} = A_1 e^{-t/\tau_m} + A_2 e^{-t/\tau_\ell}, \quad i_{2,n} = B_1 e^{-t/\tau_m} + B_2 e^{-t/\tau_\ell}. \tag{6.25}$$

The transformer's equivalent circuit (Fig. 6.15) is of the second order and therefore both currents consist of two exponential terms, having two different time constants. The larger one τ_m is determined by the sum of the winding inductance L and the mutual inductance M and is related to the main magnetic flux linked to both windings. The smaller one τ_ℓ is determined by the difference between the inductances L and M and is related to the leakage flux. As is known from the power transformer theory, the difference between L and M represents the leakage inductance of the transformer windings and usually has a relatively small value. Thus,

$$L_\ell = L - M.$$

In the next step (step 2 of the classical approach) we shall find the forced response, i.e. the steady-state short-circuit current of a transformer. By neglecting the resistances and using the phasor approach: $i = I e^{j\omega t}$ and $v = Ve^{j\omega t}$, for the transformer in Fig. 6.15(b) we may write

$$\begin{aligned} j\omega L I_1 + j\omega M I_2 &= V_s \\ j\omega M I_1 + j\omega L I_2 &= 0. \end{aligned} \tag{6.26}$$

From the second equation we have

$$I_2 = -\frac{M}{L} I_1. \tag{6.27a}$$

Substituting this in the first equation (equation 6.26) yields (for the magnitudes)

$$I_1 = \frac{V_s L}{\omega(L^2 - M^2)} = \frac{V_s L}{\omega(L+M)(L-M)}. \tag{6.27b}$$

Because of the small leakage we may neglect in the sum $(L+M)$ the difference between inductance L and mutual inductance M $(L \cong M)$. Then the above expression simplifies to

$$I_{1,f} = I_{sc} = \frac{V_s}{\omega 2(L-M)} = \frac{V_s}{2\omega L_\ell} = \frac{V_s}{X_\ell}, \tag{6.27c}$$

where L_ℓ is the leakage inductance of one winding and X_ℓ is the leakage reactance of a transformer. The p.u. value of the steady-state short-circuit current, therefore, is

$$\frac{I_{sc}}{I_r} = \frac{V_s}{X_\ell I_r} = \frac{V_r}{V_{sc}},$$

i.e., as a ratio of the system voltage, which is usually the same as a rated voltage, and the voltage drop of the transformer caused by the short-circuit current (the voltage at the short-circuit test). Thus if, for instance, a relatively low power distribution transformer has a 4% short-circuit voltage, it will develop a steady-state short-circuit current

$$\frac{I_{sc}}{I_r} = \frac{100}{4} = 25,$$

i.e., 25 times the normal current in either of the transformer windings.

The next step is finding the independent initial conditions, i.e. the value of both currents at the instant of switching. For this reason we have to take into consideration that prior to switching the transformer carried a *magnetizing, or exciting, current* (the current at no-load), which is obtained from the first equation (equation 6.26). At zero secondary, it yields

$$I_M = \frac{V_s}{\omega L}. \tag{6.28}$$

The p.u. value of the magnetizing current for power transformers lies in the 0.5–3% range; the first number is appropriate for very large transmission transformers (200–300 MVA) and the second one is appropriate for relatively small distribution transformers. The magnetizing current, which is an open-circuit current, relates to the short-circuit current, with equation 6.27c and equation 6.28, as

$$\frac{I_M}{I_{sc}} = \frac{V_s}{\omega l} \Big/ \frac{V_s}{2\omega L_\ell} = \frac{2L_\ell}{L} \cong \frac{2L_\ell}{M}. \tag{6.29}$$

It is worthwhile to mention that the same results can be obtained by inspection of the equivalent circuit of a transformer with a cancelled mutual inductance, Fig. 6.16(a), and its common approximation with the magnetized branch moved to the transformer input, Fig. 6.16(b). As can be seen from Fig. 6.16(b), after neglecting the resistances and assuming $L \cong M$, the magnetized current I_M and short-circuit current I_{sc} become the expressions as in equations 6.27c and 6.28.

Let us consider the most unfavorable instant of the short-circuiting, when the steady-state primary current $i_{1,f}$ passes through its maximum $I_{1,f}$ (equation 6.27). Since both currents, the magnetizing and the short-circuit current, are

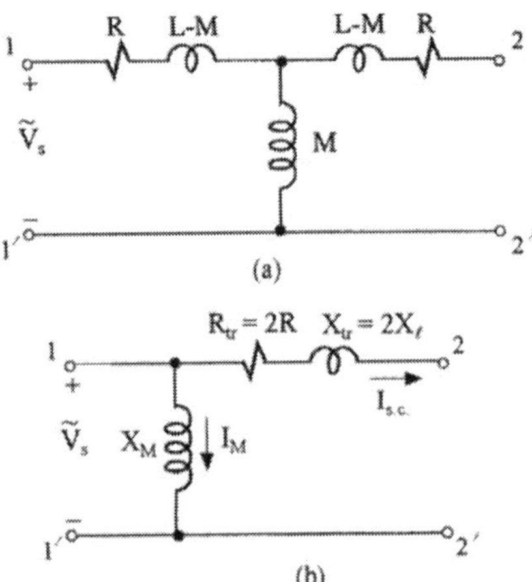

Figure 6.16 An equivalent circuit of a transformer with a cancelled mutual inductance (a) and an approximate circuit with the magnetized branch moved to the input (b).

almost purely inductive, and thus having nearly the same phase angle, we may write

$$i_{1,n}(0) = i_1(0) - i_{1,f}(0) = I_M - I_{1,f}, \tag{6.29a}$$

and

$$i_{2,n}(0) = 0 - i_{2,f}(0) = \frac{M}{L} i_{1,f}(0) = \frac{M}{L} I_{1,f}. \tag{6.30}$$

Next we shall find the dependent initial conditions, i.e., the derivatives of both natural currents at $t = 0$. The equations (equation 6.23) may be rewritten as

$$L \frac{di_{1,n}}{dt}\bigg|_{t=0} + M \frac{di_{2,n}}{dt}\bigg|_{t=0} = Ri_{1,n}(0) = I_m - I_{1,f}$$

$$M \frac{di_{1,n}}{dt}\bigg|_{t=0} + L \frac{di_{2,n}}{dt}\bigg|_{t=0} = Ri_{2,n}(0) = \frac{M}{N} I_{1,f}.$$

Solving these two equations for each of the derivatives yields

$$\begin{aligned} \frac{di_{1,n}}{dt}\bigg|_{t=0} &= -R\left[\frac{L}{L^2 - M^2} I_m - \frac{L^2 + M^2}{L(L^2 - M^2)} I_{1,f}\right] \\ \frac{di_{2,n}}{dt}\bigg|_{t=0} &= -R\left[\frac{-M}{L^2 - M^2} I_m + \frac{2M}{L^2 - M^2} I_{1,f}\right]. \end{aligned} \tag{6.31}$$

We can now obtain the integration constant by solving two simultaneous equations (see equation 1.55 and Example 2.3).

For the primary current $i_{1,n}$:

$$\begin{cases} A_1 + A_2 = I_m - I_{1,f} \\ \dfrac{-R}{L+M} A_1 + \dfrac{-R}{L-M} A_2 = -R\left[\dfrac{L}{L^2 - M^2} I_m - \dfrac{L^2 + M^2}{L(L^2 - M^2)} I_{1,f}\right], \end{cases}$$

which yields

$$\begin{aligned} A_1 &= \frac{1}{2} I_m - \frac{L - M}{2L} I_{1,f} = \frac{1}{2} I_m - \frac{1}{2}\frac{L_\ell}{L} I_{1,f}, \\ A_2 &= \frac{1}{2} I_m - \frac{L + M}{2L} I_{1,f} \cong \frac{1}{2} I_m - I_{1,f}. \end{aligned} \tag{6.32}$$

For the secondary current, $i_{2,n}$:

$$\begin{cases} B_1 + B_2 = \dfrac{M}{L} I_{1,f} \\ \dfrac{-R}{L+M} B_1 + \dfrac{-R}{L-M} B_2 = -R\left[\dfrac{M}{L^2 - M^2} I_m + \dfrac{2M}{L^2 - M^2} I_{1,f}\right], \end{cases}$$

which yields

$$B_1 = \frac{1}{2}I_m - \frac{L-M}{2L}I_{1,f} = \frac{1}{2}I_m - \frac{1}{2}\frac{L_\ell}{L}I_{1,f},$$
$$A_2 = -\frac{1}{2}I_m + \frac{L+M}{2L}I_{1,f} \cong -\frac{1}{2}I_m + I_{1,f}. \tag{6.33}$$

These results actually show that $B_1 = A_1$ and $B_2 = -A_2$. The expressions for $A_1(B_1)$ and $A_2(B_2)$ may be simplified: with equation 6.29 we have

$$A_1 = B_1 = \frac{1}{2}I_m - \frac{1}{2}\frac{L_\ell}{L}\frac{L}{2L_\ell}I_m = \frac{1}{4}I_m.$$

And, since I_m is negligibly small relative to I_{sc},

$$A_2 = -B_2 \cong -I_{1,f} = -I_{sc}.$$

Finally,

$$i_{1,n} = \frac{1}{4}I_m e^{-t/\tau_m} - I_{sc}e^{-t/\tau_\ell}, \quad i_{2,n} = \frac{1}{4}I_m e^{-t/\tau_m} + I_{sc}e^{-t/\tau_\ell}. \tag{6.34}$$

These expressions show that the short-circuiting of the transformer results in the appearance in both windings of two exponential (aperiodic) currents, which superimpose with the steady-state short-circuit currents. The first one decays relatively slowly with the large time constant τ_m, however it is insignificantly small and can be neglected. The second one decays much faster with the smaller time constant τ_ℓ, but its initial value is as large as the amplitude of the steady-state short-circuit current. Half a cycle after short-circuiting, the exponential term is added to the steady-state short-circuit current, which results in an almost double amplitude value. This means that a transformer having a leakage inductance in the order of 4–10% will develop a maximal short-circuit current of 50–20 times the rated value. A typical curve of such a short-circuit current versus time is shown in Fig. 6.17,

It should be noted that by neglecting the very small effect of the transient

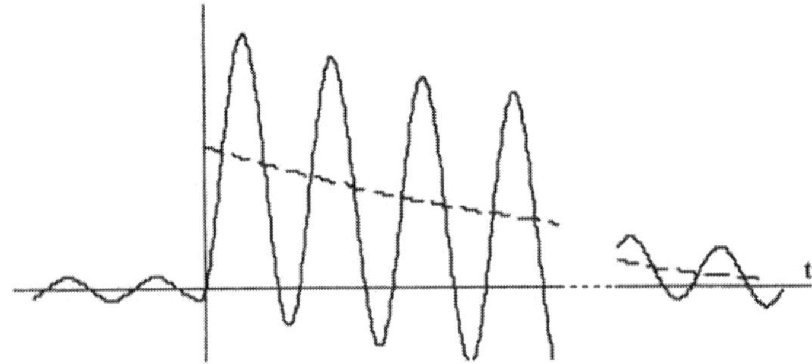

Figure 6.17 A typical waveform of a transformer's short-circuit current.

magnetizing currents in equation 6.34, the transient response to short-circuiting a transformer is similar that in the simple *RL* circuit having an inductance as a total leakage inductance of the transformer and the total resistance of both its windings. Hence, for the analysis of the short-circuit phenomena in any power network, we may replace every transformer by a single inductance in series with a resistance, both referred either to the high- or low-voltage side.

6.4.2 Current inrush by switching on transformers

Upon switching on a power transformer, an inrush of a magnetizing (exciting) current may initially reach a very high level of eight times the rated current, even under no-load conditions. From our previous study, we know that in *linear RL* circuits, even under the most unfavorable conditions, the transient current may not exceed the double value of its forced response. However, the magnetizing circuit of the transformer is non-linear due to its iron core. Hence, to analyze the transient phenomenon in the transformer we have to take into consideration the saturation of its *magnetizing characteristic*, i.e. $B = f(H)$.

The inrush is most severe when the transformer is switched on at the instant the voltage goes through zero with such polarity that the flux increases in the direction of the residual flux. For these conditions, we may write

$$v_s = \sqrt{2} V_s \sin \omega t = \frac{d\lambda}{dt} = N \frac{d\phi}{dt}.$$

The value of the flux is then found by integration:

$$\phi = \frac{\sqrt{2} V_s}{N} \int_0^t \sin \omega t \, dt + \phi(0), \qquad (6.35a)$$

where $\phi_{(0)} = \Phi_0$ is the residual flux. Thus

$$\phi = \frac{\sqrt{2} V_s}{\omega N} (1 - \cos \omega t) + \Phi_0 = -\Phi_m \cos \omega t + \Phi_m + \Phi_0. \qquad (6.35b)$$

Since we neglected all the resistances (representing the winding and core losses), the aperiodic (d.c.) component, $\Phi_m + \Phi_0$, is obtained as a constant quantity. However, due to these losses, the aperiodic term decays very slowly according to the large time constant of the magnetizing circuit. Then, at $\omega t = \pi$ (half a period after switching) the instantaneous flux will be

$$\phi_{\max} = 2\Phi_m + \Phi_0.$$

The magnetic flux density under steady-state conditions is $B_m \cong 1.3$ T. If Φ_0 is assumed to equal $0.6\Phi_m$, then the maximal flux density, which in a transformer is directly proportional to the flux, will be

$$B_{\max} = (2 + 0.6) \cdot 1.3 \cong 3.4 \, \text{T}.$$

This value is far beyond the rated range and according to the magnetizing

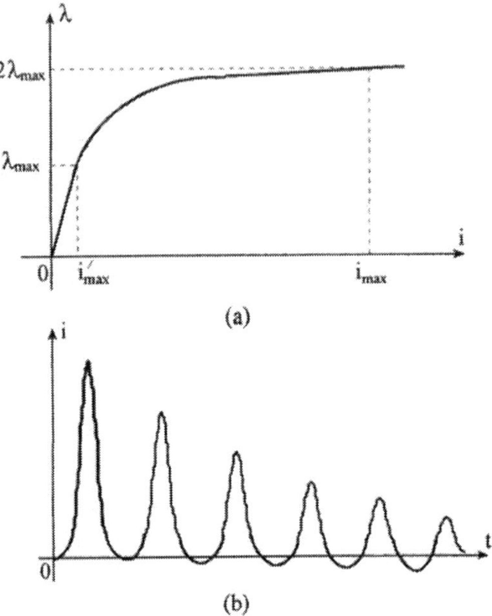

Figure 6.18 A magnetizing curve (a) and an inrush current of transformer (b).

curve, shown in Fig. 6.18(a), the magnetizing force, H (which is directly proportional to the magnetizing current) may reach as large a value as 8–10 times its rated value. The typical curve of an inrush current for a transformer switched at zero instantaneous voltage is shown in Fig. 6.18(b). Note that the waveform of this current is not sinusoidal due to the presence of high harmonics (as a result of the non-linearity of a transformer magnetize characteristic).

6.5 SHORT-CIRCUITING OF SYNCHRONOUS MACHINES

High-magnitude transient currents, or short-circuit currents, in the stator windings of a synchronous generator occur, particularly if the voltage at its terminals is suddenly changed by a considerable amount. This may happen as a result of a faulty switching operation, or by any other fault, which brings about short-circuiting, such as a result of bad synchronizing in the faulty position of the poles, by energizing a rotating machine by sudden connection to full voltage, etc. In such cases the transient currents may be much greater than the normal operating currents of the machine. Depending on the design of the machine and the process of switching excess currents, up to ten times the normal current may develop in the windings. In view of the large size of most modern generators, this would release an enormous amount of energy in the network, which might be dangerous for the normal operation of the network equipment.

The stator and rotor windings of a synchronous machine are mutually coupled, but in distinction to a transformer, due to the rotation of the rotor, they continuously change their relative position in the space. As a result of that, their mutual inductances are not constant, but vary in time. This leads to differential equations with variable coefficients, whose analysis and solution are very cumbersome.(*)

To simplify the practical approach to the calculation of short-circuit currents we shall make a few common assumptions. It should be noted that by any sudden change of the operation conditions of a synchronous machine, its revolution is disturbed and its angular velocity changes, which gives rise to mechanical oscillations. Obviously, the detailed analysis of the transient behavior of the synchronous generator becomes even more complicated. Thus, the first assumption is that the revolution of the generator does not change and remains constant during the transients.

As previously, we shall neglect the resistance of the generator windings and the short-circuit impedances of the generator will be considered approximately as an inductive reactance. The resistances will then be taken into consideration by determining the damping coefficients of decaying the transient currents.

To simplify the entire calculation of transients in a three-phase system and reduce it to a one-phase presentation, the generalized phasor of three-phase system currents will be introduced as well as the two-phase model of the synchronous machine.

6.5.1 Two-axis representation of a synchronous generator

Three-phase synchronous generators fall into two general classifications: 1) *cylindrical* (or *round*) *rotor* (high-speed turbogenerators) or 2) *salient-pole rotor* (low-speed hydrogenerators). While the air gap in the cylindrical rotor construction is practically of uniform length that of the salient-pole rotor is much longer in between the poles, Fig. 6.19.

We shall review here the two-axis representation of synchronous generators using the salient-rotor generator as an example rather than the cylindrical one, since the latter constitutes a particular case of the former. In Fig. 6.20 the schematic cross-section of a salient-rotor generator is given. Here the rotor has two axes: the *direct axis d*, which is in the direction of the magnetizing, or field flux, Φ_d and the *quadrature axis q*, which is perpendicular to axis d midway between the poles. Accordingly, the generator is represented by two reactances X_d and X_q and two EMF's, E_d and E_q in the direct axis and the quadrature axis respectively. The above two components of the EMF can always be combined in **one phasor** of a total generated EMF (or terminated voltage), E_{af}.

The stator three-phase winding carries three currents, which are displaced by 120° relative to each other. Following the idea of a one-phase representation of

(*)For a detailed discussion of this problem see, for example, C. Concordia, *Synchronous Machines*, John Wiley & Sons, New York, 1957.

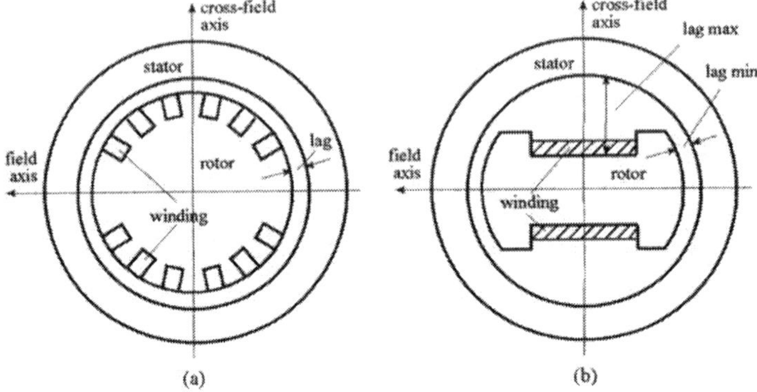

Figure 6.19 Two kinds of rotors: cylindrical rotor (a) and salient-pole rotor (b).

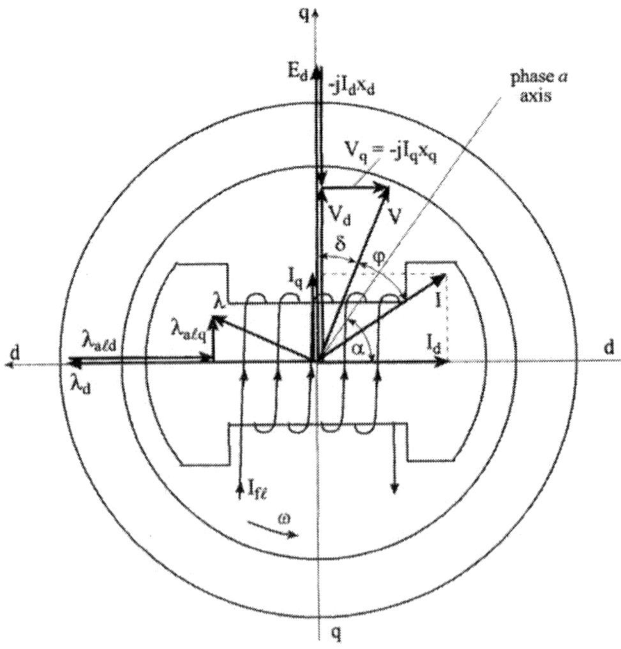

Figure 6.20 Salient rotor and phasor diagram in a two-axis representation.

a synchronous machine we shall transform the stator three-phase current system into one generalized current phasor I.

Consider the usual representation of a three-phase current by three phasors, as shown in Fig. 6.21a. The three instantaneous currents i_a, i_b and i_c can then be obtained as the projections of the three phasors on the time axis t, while the star of phasors is rotating with an angular velocity ω.

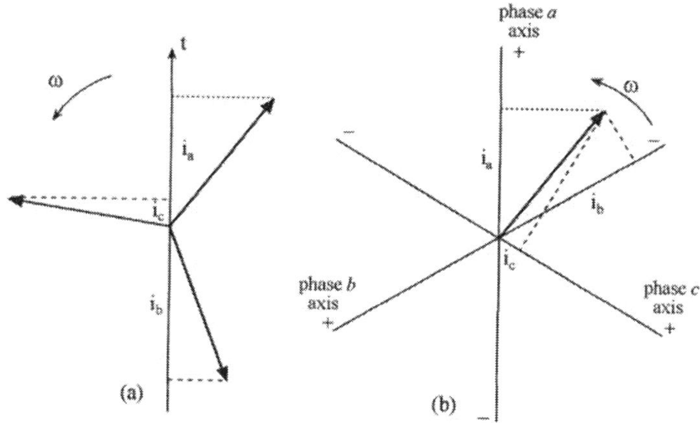

Figure 6.21 Determining instantaneous currents in a three-phase system: with three-phase phasors (a) and with one generalized phasor (b).

The same results may be derived using only one **rotating phasor**, or a so called **generalized phasor**, but with its projection on three time axes, which coincide with the axes of a three-phase stator winding, as shown in Fig. 6.21(b). If the generalized current phasor is rotated in the same direction as the phase-phasors, the sequence of the time axes should be taken as opposite to those of the phase-phasors, i.e., $a \to c \to b$.

Now, this single phasor can be expanded into two quadrature components, according to two rotor axes, I_d and I_q, as shown in Fig. 6.20. Here, λ_d is the flux linkage, produced by the field current, λ_{ald} and λ_{alq} are the armature reaction and stator winding leakage fluxes, produced by the currents I_d and I_q respectively and λ is the resultant flux linkage, which induces the terminal voltage V. In accordance with the phasor diagram for EMF's we may write

$$\tilde{V}_d = E_d - jX_d I_d, \quad \tilde{V}_q = -jX_q I_q, \tag{6.36a}$$

and

$$\tilde{V} = V_d + V_q \quad \text{or} \quad |V| = \sqrt{V_d^2 + V_q^2}, \tag{6.36b}$$

where X_d and X_q are the generator direct-axis and quadrature-axis reactances.

Finally, if the phasors I_d and I_q are known, and taking into consideration that $\tilde{I}_a + \tilde{I}_b + \tilde{I}_c = 0$, the phase-phasors can be expressed as

$$\begin{aligned} I_a &= I_d \cos\alpha + I_q \sin\alpha \\ I_b &= I_d \cos(\alpha + 2\pi/3) + I_q \sin(\alpha + 2\pi/3) \\ I_c &= I_d \cos(\alpha - 2\pi/3) + I_q \sin(\alpha - 2\pi/3), \end{aligned} \tag{6.38a}^{(*)}$$

(*)If the sum of the phase current phasors is not equal to zero, then each phase current consists of a zero sequence term.

where α is the angle between the rotor direct axis and the axis of the phase a winding.

In turn, the two components of a generalized current can be expressed by the phase currents:

$$I_d = \frac{2}{3}\left[I_a \cos\alpha + I_b \cos(\alpha + 2\pi/3) + I_c \cos(\alpha - 2\pi/3)\right]$$

$$I_d = \frac{2}{3}\left[I_a \sin\alpha + I_b \sin(\alpha + 2\pi/3) + I_c \sin(\alpha - 2\pi/3)\right]. \tag{6.38b}$$

Thus, the generalized current completely represents the three-phase stator currents and allows for the reduction of a three-phase generator to a one-phase machine, having constant mutual inductances between the stator and rotor, which is

$$M_{eq} = \frac{3}{2} M,$$

where M is the mutual inductance between the phase winding of the stator and rotor winding, when the axis of the stator winding coincides with the direct axis of the rotor.

As was previously mentioned, the cylindrical-rotor generator is a particular case of a salient-pole rotor. Thus, since the air gap lengths of both the d and q axes of the cylindrical rotor are the same, we have $X_d \cong X_q$ and all the expressions obtained for a salient-pole generator are valid for a cylindrical rotor generator.

6.5.2 Steady-state short-circuit of synchronous machines

As we know the steady-state regime, or the forced response, takes place after the natural responses decay, i.e., a few seconds after the moment of short-circuiting. However, for the sake of protecting all kinds of electrical equipment and providing the dynamic stability of synchronous generators operating in parallel, the short-circuit fault in present-day power systems is disconnected very fast (by means of modern relay protection and switch gears), Therefore, steady-state short-circuit conditions are very uncommon. We shall, however, start our analysis of the synchronous generators' behavior under short-circuit conditions with the steady-state short-circuit. In order to get the total response and estimate the maximal magnitudes of short-circuit currents in the first moments of the fault, we must know the forced responses, i.e., the steady-state short-circuit currents. In addition, the study of steady-state short-circuit behavior of a synchronous generator contributes largely to a better understanding of the whole process.

The steady-state short-circuit behavior of a synchronous generator depends to a greater degree on *the automatic voltage regulator* (AVR). Excitation of a synchronous generator is derived from a d.c. supply with a variable voltage.

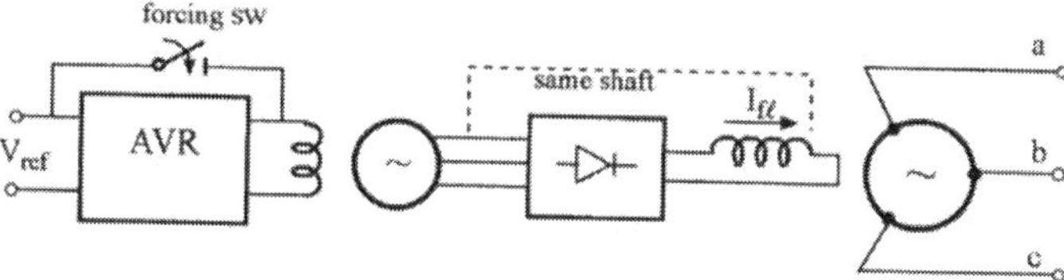

Figure 6.22 An excitation arrangement for a synchronous generator with AVR.

Originally, the main exciter consisted of an a.c. exciter with an integral diode or thyristor rectifiers rotating on the rotor (main) shaft, thus avoiding any brush gear, Fig. 6.22. In general, the AVR's are set out to control the output voltage of the synchronous generator, by controlling the exciter. The other important function of such regulators is to force the field current usually up to its maximal value at the event of a short-circuit fault, which requires a very fast-acting regulator. As a result of the AVR action, the steady-state short-circuit current might be larger than during the transients and even at the first moment of switching.

(a) *Short-circuit ratio (SCR) of a synchronous generator*

When short-circuiting occurs across the terminals of the generators or nearby, the magnetic saturation of their characteristics must be taken into consideration since the values of the voltages and of the inductances substantially depend on the magnetic saturation. The *open-circuit* (no-load) *characteristic* (OCC), or the magnetic curve, is the graph of the generated voltage against the *field current*, I_{fl}, of the machine on open circuit and running at synchronous speed. The typical OCC of turbo- and hydro-generators in p.u. are shown in Fig. 6.23. The air-gap line represents the linear part of the open-circuit characteristic and ignores saturation (Fig. 6.24).

For our further consideration, we will also need the *short-circuit characteristic* (SCC), which is the graph of a stator current against a field current with the terminals short-circuited. Both OCC and SCC are shown in Fig. 6.24.

With these two characteristics we may calculate, first of all, both the unsaturated and saturated (its approximate value) synchronous reactances of the generator. The p.u. **unsaturated reactance** is obtained with the air-gap (unsaturated) line as the ratio of the open-circuit voltage (length $\overline{ac}$) and the short-circuit current (length $\overline{ad}$), both produced by the same field current ($\overline{0a}$). Thus,

$$X_{du}=\frac{\overline{ac}}{\overline{ad}}\ \text{pu.} \tag{6.39a}$$

With the saturated air-gap line (of), also called the modified air gap line, and by the same procedure we may obtain the **saturated reactance**:

$$X_d=\frac{\overline{bc'}}{\overline{be}}\ \text{pu.} \tag{6.39b}$$

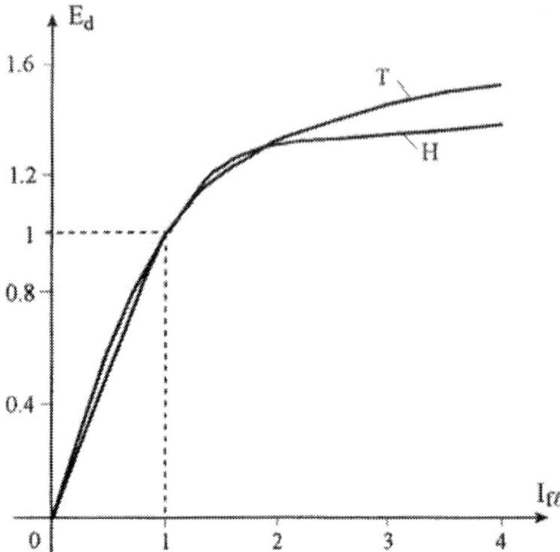

Figure 6.23 Typical open-circuit characteristics of turbogenerator (T) and hydrogenerator (H).

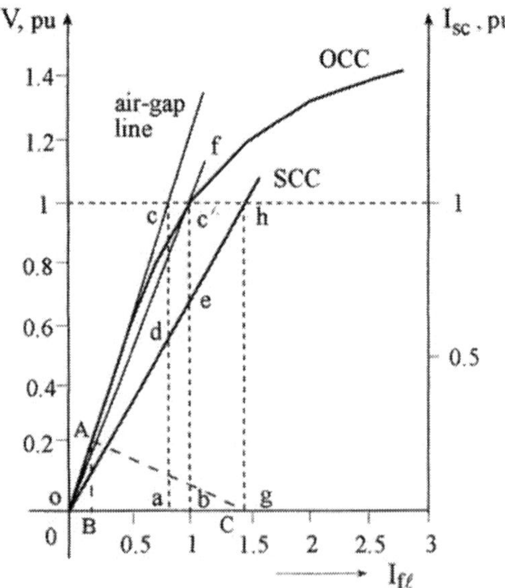

Figure 6.24 Open- and short-circuit characteristics of a synchronous generator.

As can be seen, the saturated reactance is less than the unsaturated reactance, and is usually taken as the active value of the synchronous reactance of a generator.(*) It is important to understand this feature of all magnetic circuits. The reactance is reciprocally proportional to the reluctance (or magnetic conductivity), which is a function of the permeability (μ) of the magnetic material. Since the permeability by saturation is getting larger, the reluctance subsequently decreases, which results in a lower reactance.

The second important parameter of a synchronous generator, which is obtained by the above two characteristics is the **short-circuit ratio** (SCR). It is defined as the ratio between the field current required for nominal open-circuit voltage and that required to circulate the full-load current in the armature winding when short-circuited. Thus with Fig. 6.24

$$SCR = \frac{\overline{ob}}{\overline{og}}. \tag{6.40}$$

With SCR the p.u. steady-state short-circuit current at the generator terminals will be

$$I_{sc,\infty} = SCR\ I_{fl}, \tag{6.41a}$$

where I_{fl} is a known *magnetizing, or field, current* in p.u. The steady-state short-circuit current in natural units (i.e., in amperes) will be

$$I_{sc,\infty} = SCR\ I_{fl} I_{\mathrm{r}}, \mathrm{A}$$

where I_{r} is a generator rated (nominal) current. The value of SCR in accordance with the OCC, and SCC in Fig. 6.24 is 0.67 (this value is typical for turbogenerators; for hydrogenerators it can be taken as 1.1).

Comparing triangles Δ ohg and Δ oeb and noting that $gh = bc'$ we have

$$\frac{gh}{be} = \frac{og}{ob} \quad \text{or} \quad \frac{bc'}{be} = \frac{1}{ob/og},$$

i.e.,

$$X_{d,pu} = \frac{1}{SCR}. \tag{6.42}$$

The *direct-axis synchronous reactance* X_d of a synchronous generator (it is often replaced by the so-called *synchronous reactance* X_s) includes the combined effect of the leakage reactance X_l and the *reactance* X_{ad} of the *armature reaction*. The value of the leakage reactances is usually in the range of 0.1–0.15 (for turbogenerators), 0.15–0.25 (for hydrogenerators).

As a reminder of the basic conditions at the terminal short-circuit (s.c.), the phasor diagram and the Potier triangle are shown in Fig. 6.25. The rotor current

(*)For more about the effect of saturation and calculation of the saturation value of X_{d}, see, for example, in McPerson, G. and Laramore, R. D. (1990) *Electrical Machines and Transformers*, Wiley & Sons.

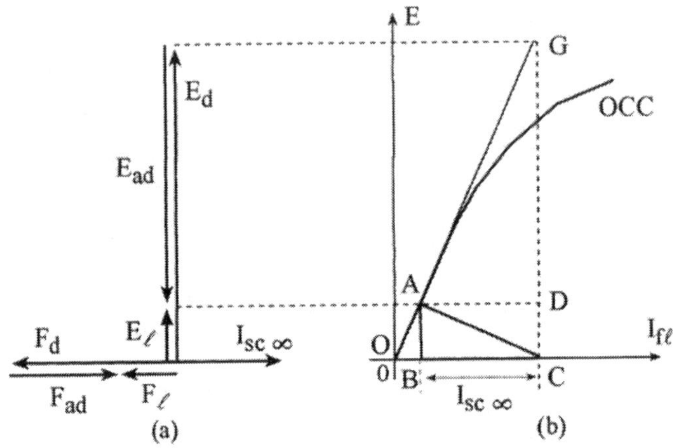

Figure 6.25 The phasor diagram (a) and OCC with a Potier triangle (b).

I_{fl} produces a *magnetomotive force* (MMF) F_d, which induces in the stator winding an *electromotive force* (EMF) E_d. Since the terminal voltage at short-circuiting is zero, this voltage is required to overcome the armature reaction $E_{ad} = jX_{ad}I_{sc}$ and the leakage reactance voltage drop $E_l = jX_lI_{sc}$, Fig. 6.25a. The corresponding fractions of F_d are also shown on the phasor diagram as leading the appropriate EMF by 90°. Note that the armature reaction F_{ad}, is in phase with F_d, since the short-circuit current is actually a zero-power-factor, or pure reactive, current.

With the known E_l point A, which is the upper vertex of the **Potier triangle** on the OCC, is determined. Point C, which is determined by the field current, required to produce a rated short-circuit current, gives the second vertex of the triangle. Point B, which is determined by the perpendicular drawn from vertex A to the abscissa, gives the third vertex. Length $\overline{BC}$ is the component of the field current required to overcome the MMF of the armature reaction and, therefore, is proportional to the stator current. The other component $\overline{OB}$ produces F_l, required for inducing E_l to overcome the leakage reactance voltage drop. Note that since point A is located on the linear part of OCC, the quantities E_d, E_{ad} and X_d are appropriate for an unsaturated generator.

Example 6.3

Use the open-circuit and short-circuit characteristics, shown in Fig. 6.26, for a 133.5 MVA three-phase 13.8 kV 60 Hz generator, to: a) find the unsaturated and saturated synchronous reactances in ohms and in p.u.; b) determine SCR; and c) draw the Potier triangle, if the leakage reactance is 0.145, and determine the scale of the stator current on the axis of the field current (abscissa).

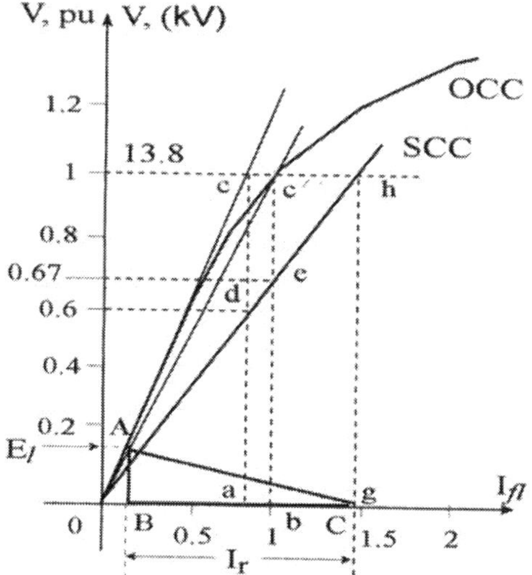

Figure 6.26 The OCC, SCC and Potier triangle for Example 6.3.

Solution

a) The unsaturated synchronous reactance (equation 6.39a) is (see Fig. 6.26)

$$X_{du} = \frac{\overline{ac}}{\overline{ad}} = \frac{1}{0.6} = 1.67 \text{ pu}$$

The rated impedance of the generator (equation 6.4) is

$$Z_r = \frac{13.8^2 \cdot 10^6}{133.5 \cdot 10^6} = 1.43\,\Omega.$$

Thus, the unsaturated reactance in ohms is

$$X_{du} = Z_r X_{du} = 1.43 \cdot 1.67 \cong 2.93\,\Omega.$$

The saturated reactance (equation 6.39b) is (see Fig. 6.26)

$$X_d = \frac{\overline{bc'}}{\overline{be}} = \frac{1}{0.67} = 1.49 \text{ pu}$$

or in ohms

$$X_d = Z_r X_d = 1.43 \cdot 1.49 \cong 2.13\,\Omega.$$

b) The short-circuit ratio (equation 6.40) with Fig. 6.21 is

$$SCR = \frac{\overline{ob}}{\overline{og}} = \frac{1}{1.49} \cong 0.67,$$

which is the reciprocal of X_d.

c) The vertex A is determined by the ordinate $E_{1,pu} = 0.145$ and the vertex C by the abscissa $\overline{OC} = 1.45$, which is the p.u. field current required to produce a rated short-circuit current (see Fig. 6.26). The rated current of the generator is

$$I_r = \frac{S_r}{\sqrt{3}\cdot 13.8\cdot 10^3} = 5580\ \text{A}.$$

Since the length $\overline{BC}$ determines the portion of the field current, which produces F_{ad} to overcome the armature current reaction ($F_{ad} = X_{ad} I_{sc}$), and therefore it is proportional to this current, we may determine the scale of the stator current on the abscissa as

$$m_I = \frac{I_r}{\overline{BC}} = \frac{5580}{1.35} \cong 4130\ \text{A/cm}.$$

The Potier triangle is shown in Fig. 6.26.

With the known p.u. field (magnetizing) current, I_{fl}, the steady-state short-circuit (s.c.) current can easily be found as

$$I_{sc,\infty} = SCR\ I_{fl} I_r.$$

However, this value of an s.c. current is valid only for an unsaturated generator, or for a linear OCC, i.e., in accordance with an air-gap line, which is, of course, only a rough approximation. For a more precise calculation of an s.c. current the graphical solution shall be introduced.

(b) *Graphical solution*

We shall start the graphical solution representation with a simple case of a short-circuit fault occurring on the main line fed by a single generator, as shown in Fig. 6.27. The generator is represented by the OCC and the leakage reactance X_l; the terminal voltage is V and X_F is the reactance of the external network (the resistances, as usual, are neglected).

The EMF of the generator required to overcome the leakage voltage and the terminal voltage is

$$E_g = E_l + V = (X_l + X_F) I_{sc} = X_{eq} I_{sc}. \tag{6.43}$$

This expression (since X_l and X_F are constants) can be represented graphically

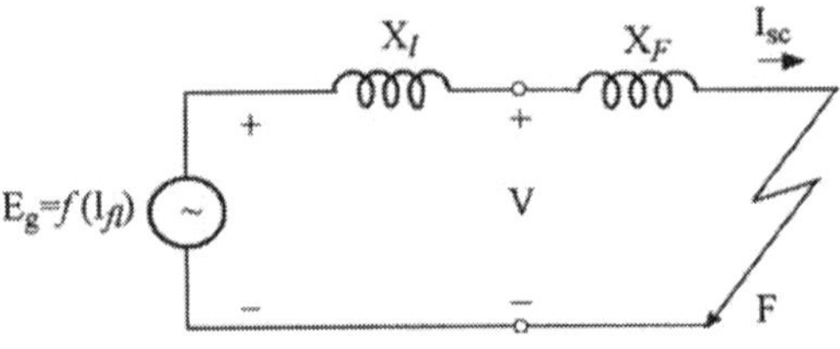

Figure 6.27 An equivalent circuit of a short-circuited synchronous generator through an external reactance.

as a straight line, as shown in Fig. 6.28. As has already been shown, the abscissa of the OCC can also be used as an axis of an armature current with an origin in point C and its positive direction opposes the positive direction of the axis of the field current. Hence, the straight line of $E_g = f(I_{sc})$ should be plotted from this point C with the slope angle

$$\alpha = \tan^{-1}(X_{l,pu} + X_{F,pu}),$$

i.e., line CM in Fig. 6.28.

The EMF of the generator is dependent on two quantities: 1) a magnetizing current I_{fl} (in accordance with its OCC) and 2) an s.c. current I_{sc} (in accordance with equation 6.43. Hence, the actual EMF will be given by the intersection, point M, of the two characteristics: the OCC and the straight line CM, as shown in Fig. 6.28.

The actual s.c. current will be determined by point N and can be expressed by the length $\overline{ON'}$ according to the scale of the axis I_{sc} (see Example 6.2). Note that this method of determining I_{sc} is actually a graphical solution of two equations (one of them is the OCC given as a curve and the second one is a straight line given by expression 6.43) on two unknowns: E_g and I_{sc}, i.e.,

$$E = f(I_{fl}), \quad E = X_{eq} I_{sc}. \tag{6.44}$$

(Also note that there is a relationship between I_{fl} and I_{sc}, e.g., given by equation 6.41a when the fault occurs at the generator terminals.)

Next we may separate the total EMF, induced by the stator winding, into two parts (in accordance with the circuit in Fig. 6.27): the leakage voltage drop E_l and the terminal voltage V. For this purpose we shall draw the Potier triangle

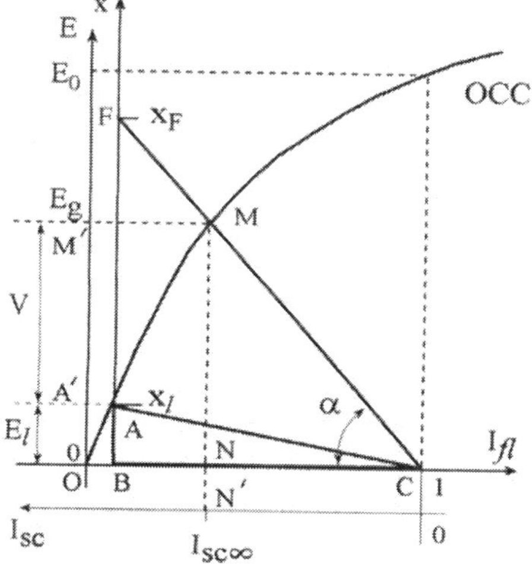

Figure 6.28 The graphical representation of two functions: 1) $E_g = f(I_{fl})$, which is the OCC and 2) $E_g = f(I_{sc})$ in accordance with equation 6.43).

ABC: vertex C is determined by the field current required for a rated s.c. current, i.e., $\overline{OC} = 1/SCR$ and vertex A (the triangle altitude) is determined by E_l.

We then plot the reactance axis as a continuation of the triangle leg AB (Fig. 6.28). The length $\overline{AB}$, which represents the leakage voltage E_l, is proportional to X_l ($E_l = X_l I_{sc}$) and therefore it determines the scale of reactance:

$$m_x = \frac{X_l}{\overline{AB}}, \text{ pu/cm.}$$

Then, length $\overline{AF}$ on the reactance axis will give X_F in the same scale, while the lengths $\overline{OA'}$ and $\overline{A'M'}$ on the voltage axis will give the leakage voltage E_l and the terminal voltage V respectively.

So far, in the above solution, the generator, previously to short-circuiting, was running under no-load conditions. Usually, short-circuits do not occur under no-load, but under the full operation of the power plant. Thus, the generators will carry a considerable current prior to the occurrence of a short-circuit, and in order to compensate for the armature reaction of the load current, the generator should be excited by a substantially higher field current than by the no-load field current in Fig. 6.28, which is often by a multiple of this value. If the field current under full load is known, we start the solution by indicating point C_1 according to the value of this current. Then we move the Potier triangle with the reactance axis, toward point C_1 so that its vertex C coincides with point C_1, as shown in Fig. 6.29.

Now, as in the previous case, we shall determine point F_1, in accordance with the value of X_F and plot the line C_1M_1 through F_1. The projection of M_1 on the I_{sc} axis, point N_1, gives the value of the steady-state short-circuit current, $I_{sc,\infty}$. To find the remaining terminal voltage V we must extend the hypotenuse

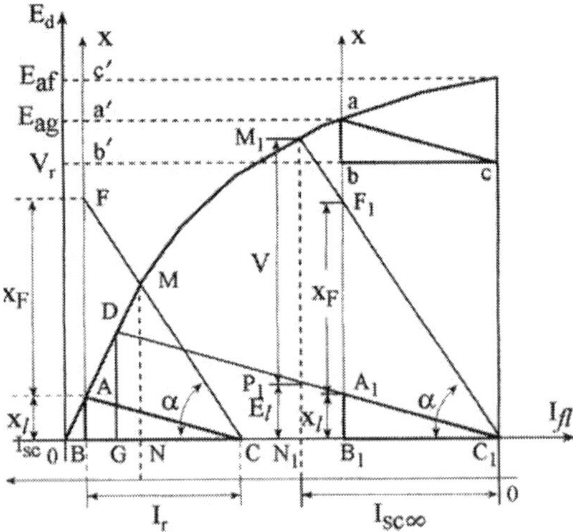

Figure 6.29 The graphical solution for finding the steady-state value of a short-circuit current.

A_1C_1 of the Potier triangle up to the OCC, point D. Then length $\overline{M_1P_1}$ will give the p.u. value of the generator terminal voltage. It is obvious that the above procedure can be performed for any given field (magnetizing) current and for any external reactance. However, if the field current is not known, we may determine it as follows. The leakage voltage E_l as length $\overline{ab} = \overline{AB}$ should be added to the rated voltage V_r level (dashed line $b'b$) so that point A reaches the OCC, as shown in Fig. 6.29. Then the length $\overline{bc} = 1/SCR$ should be plotted on the same line $b'b$ and the obtained triangle $\Delta\, abc$ is the Potier triangle. Point c, projected on the abscissa, as point C_1, will determine the required field current. On the OCC in Fig. 6.29: E_{ag} is the air-gap EMF and E_{af} is the total EMF generated by the field current I_{fl}. Let us now introduce the graphical solution in the following example.

Example 6.4

The synchronous generator, prior to short-circuiting, is operated under full load. Use the parameters and OCC of Example 6.3 to find the s.c. current and the terminal voltage of the generator if the fault is placed at the external reactances 1) 0.3 and 2) 0.9.

Solution

In Fig. 6.30 the OCC and the Potier triangle ABC from Example 6.3 are given. First we move the Potier triangle into the position of $\Delta\, abc$, so that $ab = AB$.

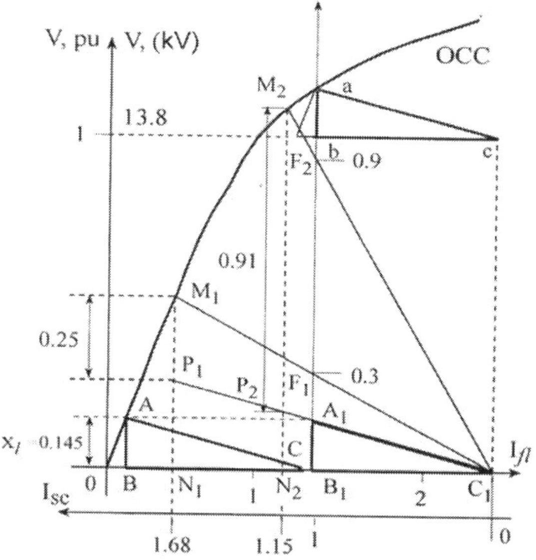

Figure 6.30 The graphical solution of Example 6.4.

Then, the field current of the generator under full load is given by point C_1, which is the projection of c on the abscissa, and the new position of the Potier triangle is $\Delta A_1B_1C_1$.

1) On the X-axis, which is an extension of A_1B_1, we determine point F_1 in accordance with the value of the external reactance 0.3. The intersection of the straight line, drawn from C_1 through F_1 to OCC, gives point M_1, which is the graphical solution of our problem. The length $\overline{C_1N_1}$, measured as 1.68, is the p.u. value of the s.c. current. Thus,

$$I_{sc,\infty} = 1.68I_r = 1.65 \cdot 5580 \cong 9.4 \text{ kA}.$$

The length $\overline{M_1P_1}$, measured as 0.25, is the p.u. value of the terminal voltage:

$$V = 0.25 \cdot 13.8 = 3.45 \text{ kV}.$$

2) *On the* X-axis we determine point F_2 in accordance with the second value 0.9. Then the intersection point M_2 gives the solution of the s.c. current (length $\overline{CN_2}$):

$$I_{sc,\infty} = 1.15I_r = 1.15 \cdot 5580 \cong 6.4 \text{ kA},$$

and of the terminal voltage (length $\overline{M_2P_2}$):

$$V = 0.91V_r = 0.91 \cdot 13.8 \cong 12.6 \text{ kV}.$$

Note that, in the first case, the intersection point M_1 lies on the straight part of the OCC. Therefore, the s.c. current can be found with the unsaturated reactance. Indeed, the total reactance up to the fault is

$$X_{tot} = (X_{du} + X_F)X_r = (1.67 + 0.3) \cdot 1.43 = 2.82\ \Omega,$$

and

$$I_{sc,\infty} = 13.8/2.82 = 4.89 \text{ kA}.$$

Since the generator prior to fault was under full load operation, its field current was about twice as large as under no-load (see the diagram in Fig. 6.30), the actual s.c. should be

$$I_{sc,\infty} = 4.89 \cdot 2.00 \cong 9.8 \text{ kA},$$

which is pretty close to the s.c estimated graphically.

It should also be noted that the actual PF of the generator load prior to the short-circuiting has not been taken into consideration, i.e. the armature reaction is considered as a pure reactive. However this approximation does not significantly change the final results.

In the above solution the field current has been kept constant, regardless of the distance to the fault (i.e. the value of the external reactance) and the level of the terminal voltage. However, as has already been mentioned, nowadays synchronous generators are equipped with an automatic voltage regulation system (AVR), which endeavors to hold the terminal voltage constant by changing the field current. Thus, if the short-circuit occurs far away from the power station, i.e. the external reactance is large enough so that the decrease in the terminal voltage will be unsubstantial, then the response of the AVR in increasing the field current will be low. On the other hand, if the short-circuit fault occurs close to the generator terminals, the drop in its voltage will be significant and the AVR response in increasing the field current will be very strong. It is also possible that the field current will reach its maximal value, but despite that the terminal voltage will remain lower than its normal level. Hence, we shall distinguish between two possible regimes:

a) the *maximal field current regime*, in which $I_{fl} = I_{fl,\max}$ and $V_\infty \leq V_r$; and

b) the *rated (nominal) voltage regime* $V_\infty = V_r$ and $I_{fl} \leq I_{fl,\max}$.

In order to determine in which of the two regimes the generator is operating, and to perform the graphical solution in these cases, let us consider the diagram shown in Fig. 6.31.

In this diagram $\overline{OC_m}$ represents the maximal field current $I_{fl,\max}$ and as previously $B_m a$ is a reactance axis. We shall now find the maximal value of X_F, in which the voltage drop, in the case of the maximal field current, will be equal to the rated voltage, $V_r = 1$. For this purpose we must plot a line from point R, which is positioned on the voltage axis at $V_r = 1$, parallel to the hypotenuse of the Potier triangle up to the intersection point K on the OCC. By connecting K with the origin C_m we obtain point k, on the x-axis, so that the length $\overline{A_m k}$ gives the desired reactance $X_{F,cr}$, which is called the *critical reactance*. Indeed, from the plotted diagram, it can be seen that the air-gap EMF, E_{ag}, at any

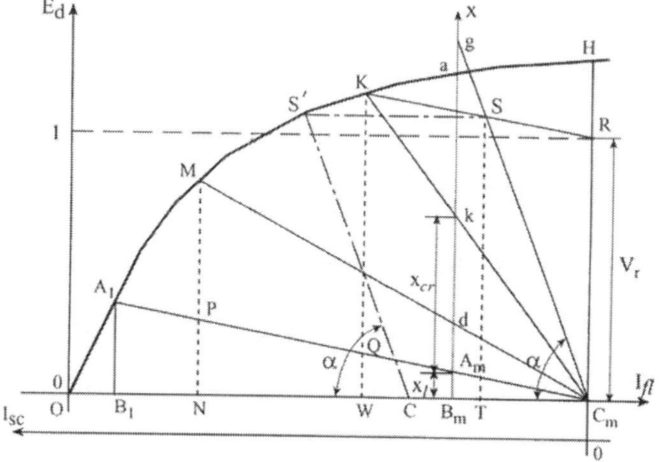

Figure 6.31 The graphical solution when the field current is not constant.

point on the OCC, which is lower than point K, less the leakage voltage E_l (at this point length $\overline{QW}$), will be smaller than unity (lower than at point R). At point K we have

$$E_{ag,k} - E_{l,Q} = \overline{QK} = 1.$$

For typical synchronous generators the critical fault reactance can be approximated as $X_{F,cr} \cong 0.5$ and therefore, $I_{\infty,cr} \cong 2$. As soon as $X_{F,cr}$ is found we may conclude that:

a) if $X_F \leq X_{F,cr}$ the regime of the maximal field current takes place,

b) if $X_F \geq X_{F,cr}$ the regime of the rated voltage takes place.

It is obvious that if $X_F = X_{F,cr}$, both regimes take place at the same time.

In the first case the graphical solution is held in the same way, which has been previously explained for the constant field current taking the maximal field current as a constant. The second case requires some additional discussion. Since the terminal voltage in this regime is the straight line KR, which represents this voltage as a function of the field current, it can be treated as an extension of the OCC (instead of the curve KH). Then, the s.c. current will be determined by point S on the intersection of $C_m g$ (g is given by X_F) and KR. The projection of RS on the abscissa, i.e., I_{sc}-axis, length $\overline{C_m T}$, gives the p.u. value of the steady-state s.c. current. The field current in this regime is smaller than $I_{fl,\max}$. To find its value we have to project point S on the OCC as point S^1 and to plot line CS^1 in parallel to $C_m S$. Then length $\overline{OC}$ will determine the actual field current.

Example 6.5

The generator of the previous example is equipped with the AVR, which ensures increasing the field current under the fault conditions up to $I_{fi,\max} = 4$ pu. Find $I_{sc,\infty}$ and the generator terminal voltage V_g, if the short-circuit fault occurred at 1) $X_F = 0.3$ and 2) $X_F = 0.9$. Determine the kind of regime: $I_{fl,\max}$ or $V_{g,r}$, for both cases.

Solution

In Fig. 6.32 the OCC of the generator and the Potier triangle are redrawn. Since the maximal field current at the full operation of AVR is $I_{fl,\max} = 4$ pu, the Potier triangle is moved to position $A_m B_m C_m$. Next we plot lines KR (point R is at the rated voltage V_r) and $C_m K$. The intersection of line $C_m K$ with the x-axis at point k gives the critical reactance X_{cr}, which is 0.68 pu. Thus the fault critical reactance is

$$X_{F,cr} = X_{cr} - X_l = 0.78 - 0.145 = 0.64.$$

1) Hence, at the fault of $X_F = 0.3$, which is less than critical, the generator operates under the regime of the first kind, i.e., the maximal field current. To find the s.c. in this case we determine the total reactance $X_{tot} = 0.145 + 0.3 \cong 0.45$

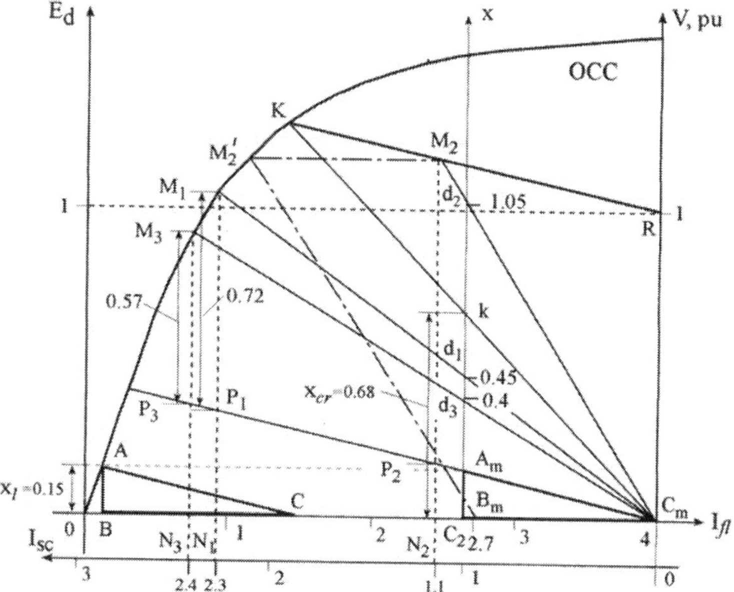

Figure 6.32 The graphical solution of Example 6.5.

on the x-axis at the point d_1 and plot line $C_m M_1$ through this point. The projection of point M_1 on the abscissa, i.e., point N_1, indicates the s.c. of the generator

$$I_{sc,\infty} = \overline{C_m N_1} \cdot I_r = 2.3 \cdot 5.58 = 12.8\ \text{kA},$$

and length $\overline{M_1 P_1}$ gives the terminal voltage

$$V_\infty = \overline{M_1 P_1} \cdot V_r = 0.72 \cdot 13.8 = 9.9\ \text{kV}.$$

2) The total fault reactance in this case is $X_{tot} = 0.145 + 0.9 = 1.05$, i.e., larger than the critical reactance and therefore the generator operates under the second kind of regime, in which the terminal voltage is of the rated value. The solution will be given by line $C_m M_2$ plotted through point d_2 on the x-axis at the value of 1.05. The s.c. current is determined by N_2, which is the projection of M_2. Thus,

$$I_{sc,\infty} = \overline{C_m N_2} \cdot I_r = 1.1 \cdot 5.58 = 6.2\ \text{kA},$$

or, since the terminal voltage is unity,

$$I_{sc,\infty} = \frac{1}{0.9} 5.58 = 6.2\ \text{kA},$$

and the terminal voltage

$$V_\infty = V_r = 13.8\ \text{kV}.$$

To find the field current at this regime we have to plot line M_2M_2' in parallel to the abscissa and line $M_2'C_2$ in parallel to M_2C_m. Point C_2 will indicate the field current, which, therefore, is $I_{fl}=2.7$ pu and less than the maximal.

(c) *Influence of the load*

The load of power systems, especially induction motors that compose 50–70% of the entire load, largely influence the transient behavior of the synchronous generators under short-circuit faults. Generally speaking, any load connected to the same node as the short-circuit line, Fig. 6.33, changes the current values and their flow in the affected network. Thus, by simplifying the network to get an equivalent circuit, we simply connect the load branch in parallel to the short-circuited branch, as shown in Fig. 6.33(b). This results in lowering the total fault reactance and consequently in decreasing the generators' voltages, which in turn results in decreasing the s.c. currents and changes their distribution in the whole network. Hence, the load connections must be taken into consideration by the short-circuit fault analysis. On the other hand the exact consideration of the load presents a lot of difficulties. The most typical kinds of loads: lightning, heating and mechanical operating (primarily induction motors), are not constant, but vary as a function of the voltage power ($V^{1.6}$ in the case of a lightning load and V^2 in the case of heating and induction motors). Furthermore, induction motors stop operating (their rotor speed reduces to zero), when the voltage is decreased 70%; and the motor turns into a short-circuited branch (this situation is very dangerous for induction motors and they would be disconnected by means of the protection relays).

Generally speaking, all kinds of loads also depend on frequency. However, information regarding the characteristics of composite loads with frequency is scarce. With the small frequency changes during most of the short-circuit faults, this effect is neglected in calculations.

A detailed analysis of the different ways of load considerations (which is not given here as it is beyond the scope of this book) shows that for the purpose

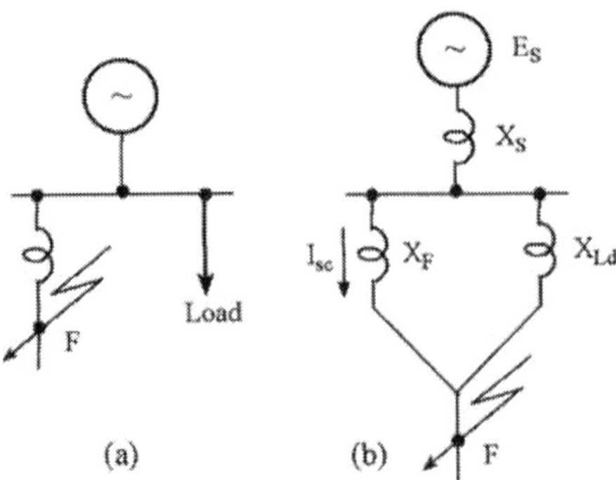

Figure 6.33 A simple network for illustrating the influence of the load on the short-circuit analysis.

of s.c. current calculations a good approximation can be achieved by considering the composite loads as a constant reactance of the 1.2 pu value.

Example 6.6

Suppose that the load of the generator of Example 6.5 is connected to its terminals as shown in Fig. 6.33. Find the generator's s.c. current if the value of the load is 85% of the generator rated power and the fault occurs at the p.u. reactance of 0.3 (the generator is equipped with AVR).

Solution

In accordance with the above recommendation, we shall represent the load as a 1.2 reactance. Hence, the reactance referred to the generator power is

$$X_{ld} = 1.2\frac{1}{0.85} = 1.41.$$

The equivalent fault reactance in this case, Fig. 6.33(b), will be

$$X_{eq} = 0.3//1.41 = 0.25,$$

and

$$X_{tot} = 0.25 + 0.145 \cong 0.4.$$

Since this reactance is less than critical, the regime of the generator is of maximal field current. Determining the above value on the x-axis, in Fig. 6.32, point d_3, and plotting line $C_m M_3$ through this point, we obtain the solution at point N_3. Thus

$$I_{sc,\infty} = \overline{C_m N_3} \cdot I_r = 2.4 \cdot 5.58 = 13.4 \text{ kA},$$

and

$$V_\infty = \overline{M_3 P_3} \cdot V_r = 0.57 \cdot 13.8 \cong 7.87 \text{ kV}.$$

As expected, consideration of the load results in increasing the s.c. current of the generator and in decreasing its terminal voltage (compare with the results of Example 6.5 for $X_F = 0.3$). Note that decreasing the terminal voltage results in decreasing the short-circuit current in the fault branch.

(d) *Approximate solution by linearization of the OCC*

A disadvantage of the graphical method is that its accuracy depends on the scale of the draft and experience of the performer of the graphical calculations. From this standpoint analytical methods are always preferable. However, to perform an analytical approximation of the short-circuit fault of a synchronous generator taking into consideration the saturation of its magnetic circuit, we need to know the analytical approximation of its OCC. The simplest one is a linearization of a given curve with a single straight line. It is obvious, however, that replacing the whole curve of the OCC with one straight line will give a very bad approximation. So, usually, only a specific part of the curve, which is considered as a working part, is replaced by a straight line. For generators having AVR (nowadays most synchronous generators are equipped with a voltage regulation system) the working part of the OCC, in accordance with Fig. 6.34, is A_mK' (note that the continuation of the generator characteristic in this case is also a straight line KR). This part of the OCC may be approximated by the straight line A_mN, which for a typical OCC is expressed as

$$E_g = 0.20 + 0.8I_{fl}. \tag{6.45}$$

(For different OCCs the numerical parameters in this expression may be different.)

The synchronous reactance of the generator, which is represented by a linear OCC, can also be estimated as a constant quantity. We shall obtain this value by considering a short-circuit fault at generator terminals. We then have

$$X_s = \frac{E_g}{I_{sc,\infty}} = \frac{E_g}{SCR\, I_{fl}}, \tag{6.46}$$

where E_g is in accordance with equation 6.45.

Since the position of point K' is different from those of point K on an actual OCC, we shall find a new value of the critical reactance $X'_{F,cr}$ for the linear characteristic. At point K' the terminal voltage is unity, the field current is still

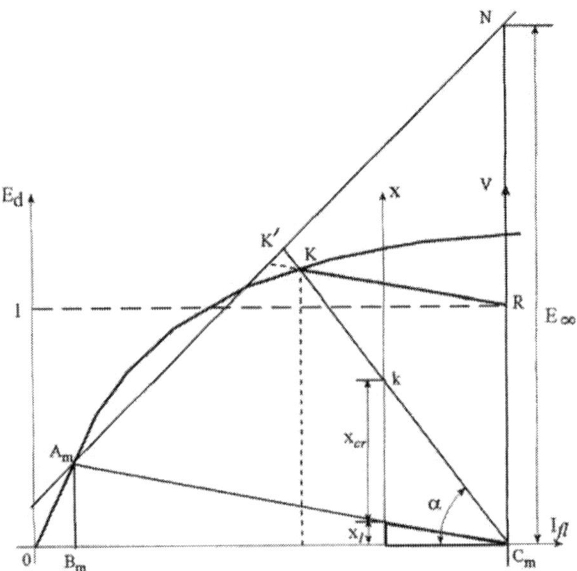

Figure 6.34 The linear approximation of a typical OCC.

maximal and therefore the induced EMF is also maximal. Hence

$$V_\infty = E_{g,\max} - X_s I_{\infty,cr} = 1,$$

which gives the critical short-circuit current

$$I_{\infty,cr} = \frac{E_{g,\max} - 1}{X_s}, \tag{6.47}$$

where

$$E_{g,\max} = 0.20 + 0.8 I_{fl,\max} \quad \text{and} \quad X_s = \frac{E_{g,\max}}{SCR\, I_{fl,\max}}.$$

With equation 6.44 the *critical reactance* is

$$X_{F,cr} = \frac{1}{I_{\infty,cr}}. \tag{6.48}$$

If the generator is operating under maximal field current, i.e., $X_F \leq X_{F,cr}$, then

$$I_\infty = \frac{E_{g,\max}}{X_s + X_F} \geq I_{\infty,cr}, \tag{6.49a}$$

and

$$V = X_F I_\infty \leq 1. \tag{6.49b}$$

If $X_F \geq X_{F,cr}$, which means that the generator operates under rated voltage, then

$$I_\infty = \frac{1}{X_F} \leq I_{\infty,cr} \quad \text{and} \quad V_\infty = 1. \tag{6.50}$$

Example 6.7

For the generator of Example 6.5 find the s.c. current using the linearization method.

Solution

First we shall estimate the critical reactance. With equation 6.45 through equation 6.48 we have (in p.u.)

$$E_{g,\max} = 0.2 + 0.8 \cdot 4 = 3.4, \quad X_s = \frac{3.4}{0.67 \cdot 4} = 1.27,$$

and

$$I_{\infty,cr} = \frac{3.4 - 1}{1.27} = 1.89, \quad X_{F,cr} = \frac{1}{1.89} = 0.53.$$

1) Since the fault reactance in the first case, $X_{F1} = 0.3$ pu, is less than the critical

reactance, by using equation 6.49, we have

$$I_\infty = \frac{3.4}{1.27+0.3} = 2.16 \quad \text{and} \quad V_\infty = 0.3 \cdot 2.16 = 0.648$$

or in natural units

$$I_\infty = 2.16 \cdot 5.58 = 12.1\,\text{kA} \quad \text{and} \quad V_\infty = 0.648 \cdot 13.8 = 8.94\,\text{kV}.$$

The generator in this case is operated under maximal field current.

2) Since the fault reactance in the second case $X_{F2} = 0.9\,\text{pu}$, which is greater than critical, we use equation 6.50. Thus,

$$I_\infty = \frac{1}{0.9} = 1.11 \quad \text{and} \quad V_\infty = 1,$$

or in natural units

$$I_\infty = 1.11 \cdot 5.58 = 6.2\,\text{kA} \quad \text{and} \quad V_\infty = 1 \cdot 13.8 = 13.8\,\text{kV}.$$

The generator in this case is operated under the nominal terminal voltage with less than maximal field current. The latter one may be estimated as

$$I_{fl} = \frac{I_\infty}{SCR} = \frac{1.11}{0.67} = 1.66.$$

Comparing the obtained results, for both cases of operation, with those of Example 6.5, we may conclude that the difference between them is less than 10% (note that the accuracy of all the engineering calculations is between 5–10%).

(e) *Calculation of steady-state short-circuit currents in complicated power networks*

As has been previously mentioned, most of the synchronous generators in a modern power system are equipped with AVR and, therefore, may operate under a short-circuit fault in one of two regimes: 1) maximal field current or 2) rated, i.e. normal terminal voltage. Depending on the kind of regime, each of the generators has to be represented by a different equivalent circuit: 1) in the first regime – with the OCC and X_l (using the graphical method) or with $E_{g,\max}$ and $X_{s,\max}$ (using the linearization method) and 2) in the second regime – as an ideal voltage source, i.e. with $E_g = 1$ and $X_s = 0$ (in both the graphical and linearization methods).

The determination of the kind of regime is made by comparing the actual short-circuit current of each of the generators with its critical value, $I_{\infty,cr}$, or the external reactance up to the fault with its critical value, $X_{F,cr}$. However, the s.c. currents of each of the generators are the goal of our solution and are not known at the first stage of the analysis, i.e. determining the equivalent circuit. To overcome this difficulty the iteration method, or method of successive approximations, may be applied. In accordance with this method, in the first calculation, as a starting point, the generators are represented by one of the two regimes, i.e. just by inspection of their location relative to the fault point. Those generators which are relatively "close" to the fault (by means of the estimated value of the reactance from the generators up to the fault) should be represented by an equivalent circuit as they operate under the regime of maximal field current, and those which are relatively "far" from the fault – as they operate under the regime of a normal voltage. Then, the results of this calculation, i.e., the first iterate, shall be compared with the critical ones, and the generator representation, which has been incorrectly chosen, should be changed. The calculation will be repeated and the results, i.e. the second iterate, shall be checked again and so on. In the final iterate all the generators will be represented in accordance with their actual behavior.

A straightforward method of s.c. fault analysis can also be applied to a complicated network, by means of a computer-aided calculation. With the superposition principle we may represent the s.c. current as a sum of the partial (or individual) currents caused by each generator acting alone:

$$I_{sc,\infty} = \sum_i I_{sc,i} = \sum_i B_{F,i} E_i, \tag{6.51}$$

where $B_{F,i}$ are the transfer susceptances (reciprocal of reactances) between a fault branch and each of the generator branches. These susceptances can be found by means of matrix analysis:

$$B_{F,i} = \frac{\Delta_{F,i}}{\Delta},$$

where Δ is the determinant of the network reactances' matrix, written in accordance with mesh analysis, and $\Delta_{F,i}$ is its appropriate cofactor. With these results an *equivalent circuit*, in which every generator is individually connected to the fault point, as shown in Fig. 6.35, may be obtained. Here, each generator is connected to the fault with the reactance $X_{F,i} = 1/B_{F,i}$. Now each of the reactances and/or currents (equation 6.51) can be compared with the critical reactance and/or critical currents and the correct representation of each generator

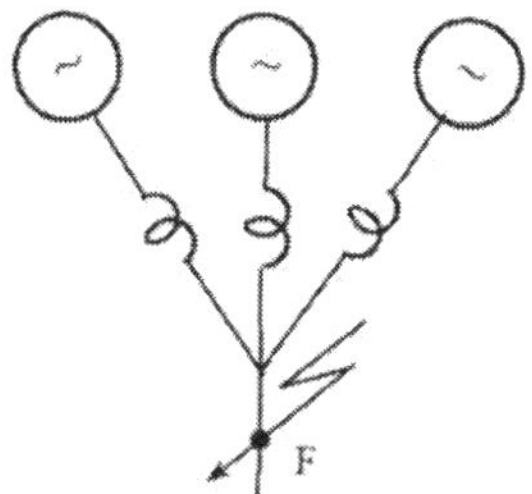

Figure 6.35 An equivalent circuit of a complicated network obtained by using the superposition principle.

will be chosen. An example of a short-circuit fault calculation in a complicated power network is given in Appendix III.

6.5.3 Transient performance of a synchronous generator

Mutually coupled stator and rotor windings of electrical machines, in distinction to the transformers, are in motion with respect to each other. The d.c. winding of the rotor of the synchronous generator moves with respect to the a.c. three-phase stator winding so that the mutual inductances between these two windings, and even between different phases of the stator winding, change with time. This leads to differential equations with variable coefficients, which results in a very cumbersome analysis and difficult understanding of the whole transient process. However, we may describe and analyze the transient behavior of the synchronous generator using an artifice. We shall use the generalized current phasor for the stator windings and the two-axis representation of a synchronous machine (see section 6.5.1), which reduce the three-phase system to a single one. Furthermore, we will make a couple of common assumptions, which allow us to not only simplify the analysis, but also to obtain results that are still close to the actual ones.

Firstly, we assume that the rotor angular speed ω stays constant during the whole transient process. For the machine with a damper winding in the rotor poles, we assume that the influence of the damper currents on the transient process can be obtained by superimposing the appropriate calculations on the transient results obtained first for the generator without damper windings. In the first stages of the transients we shall also neglect the winding resistances, being very small compared to their reactances (less than 10% for the stator winding, even when including the external network, and about 1% for the rotor winding). The influence of the resistances on the entire process will be taken into consideration as the cause of decaying all the natural responses. Transient analysis of synchronous generators will be given for a salient pole generator, as a more general case. For the round rotor generator the results may then be obtained by equaling the reactances on both axes.

(a) *Transient EMF, transient reactance and time constant*

It should be noted that the transient equivalent circuit of a generator differs from those representing it in the steady-state regime. This is shown in Fig. 6.36(a) and its simplification in Fig. 6.36(b). Here E_d is the EMF, induced by the magnetizing or field current, and

$$X_d = X_{ad} + X_l \tag{6.52}$$

is the *synchronous reactance* of a generator (sometimes designated as X_s). The value of E_d is obtained by OCC in accordance with I_{fl} (or I_μ).

However, since by the sudden interruption of a synchronous machine, the total magnetic flux has to be considered and kept constant, the rotor leakage

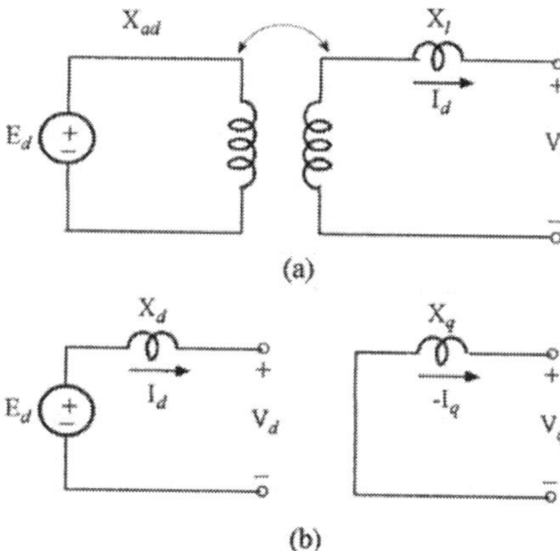

Figure 6.36 An equivalent circuit of a synchronous generator for steady-state operation (a) and its simplification in two axes (b).

reactance must be taken into account and added to the equivalent circuit (just like the leakage reactance of a transformer primary).

The equivalent circuit of the synchronous generator for the transient behavior is shown in Fig. 6.37(a). In this circuit the rotor and stator windings on the direct axes including the rotor winding leakage reactance are shown. Since the rotor and the stator magnetic fields are rotated with the same speed (as we have previously assumed), i.e., they remain fixed with respect to one another; we may treat this circuit as a transformer. Here, as in the previous circuit, X_{ad} is an *armature reaction reactance* (sometimes it is called the magnetizing reactance), and X_{rl} and X_{sl} are the *leakage reactances* of the rotor and the stator respectively. Note that in p.u. notation the magnetizing reactance expresses the relation between the rotor and stator currents:

$$I_d = \frac{I_{fl}}{X_{ad}}. \tag{6.53}$$

This circuit can be transformed into those shown in Fig. 6.37(b), in which the mutual inductance is illuminated. (Note that the p.u. values of the inductances and their corresponding reactances are equal and, therefore, the reactances can be used in place of inductances and vice versa.) We may now apply the Thévenin theorem to get the circuit in (c) and finally a very simple circuit including a voltage source and a single reactance, as shown in (d):

$$E_d' = E_{Th} = E_{fl}\frac{X_{ad}}{X_{rl} + X_{ad}}, \tag{6.54a}$$

and

$$X_d' = X_{Th} + X_{sl} = \frac{X_{rl}X_{ad}}{X_{rl} + X_{ad}} + X_l, \tag{6.54b}$$

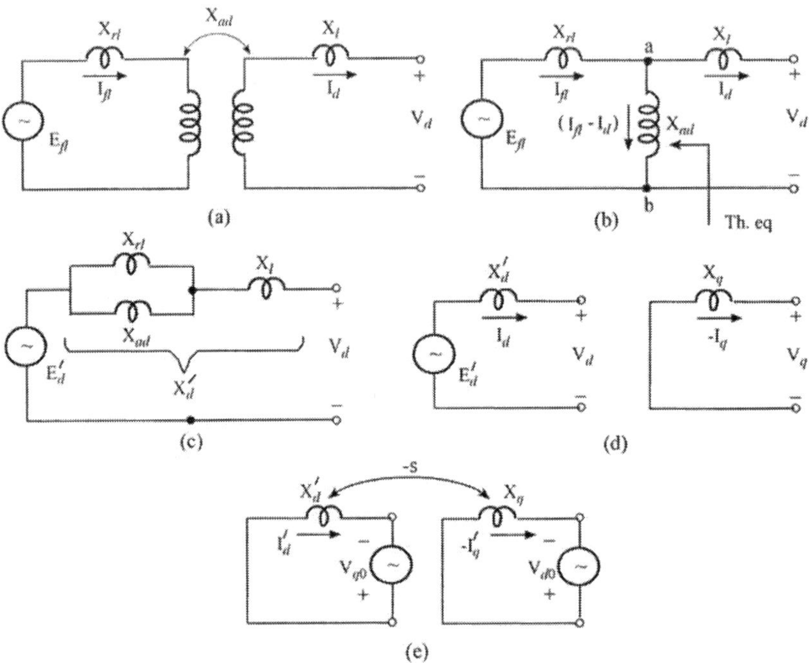

Figure 6.37 The equivalent circuit of the synchronous generator for transient behavior: in the *d*-axis as for a two-winding transformer (a), its simplification (b), after applying the Thévenin theorem (c), its Laplace transform equivalent (d) and the interconnection between the equivalent circuits in the d- and q-axes (e).

where E'_d and X'_d are the *transient EMF* (generated voltage) and the *transient reactance*.

In some technical books the transient reactance X'_d is given in the form

$$X'_d = X_d - \frac{X_{ad}^2}{X_{rl} + X_{ad}} = X_d - (1 - \sigma_{fd})X_{ad},$$

where $\sigma_{fd} = X_{rl}/(X_{rl} + X_{ad})$ is the leakage coefficient of the rotor winding. Then

$$X'_d = X_d - X_{ad} + \sigma_{fd}X_{ad} = X_l + \frac{X_{rl}X_{ad}}{X_{rl} + X_{ad}},$$

which is as was previously obtained.

Recall that similar results were obtained for power transformers (see section 6.4.1), i.e. the entire magnetic circuit of a transformer can be represented by only a single reactance, which incorporates all the magnetic fluxes of both windings in a total flux. In accordance with the principle of a constant flux linkage, this total flux must be kept constant at the instant of switching. Hence, the equivalent circuit in Fig. 6.37(d) represents the synchronous generator at the moment of short-circuiting. The equivalent circuit in the q-axis remains the same, since no additional winding in this axis is present (in distinction to the rotor equipped with the damper winding – see further on). Electromagnetic force E'_d, being induced by a total flux linkage, also stays constant at the moment of fault, which allows us to calculate the first moment short-circuit current of the generator.

Knowing the above two transient parameters E'_d and X'_d of a synchronous generator we may readily obtain the *first moment a.c.* (*periodic*) and *d.c.* (*aperiodic*) *components* of the *short-circuit current*. Thus, after short-circuiting in Fig. 6.37(d), we have

$$I'_{d0} = \frac{E'_{d0}}{X'_d}, \tag{6.55a}$$

or as an instantaneous value:

$$i_d = \frac{E'_{d0}}{X'_d} \cos \omega t, \tag{6.55b}$$

where the initial, or switching, angle ψ_i is taken as zero.

The d-axis component of the terminal voltage prior to switching will then be

$$V_{d0} = E'_{d0} - X'_d I'_{d0}. \tag{6.56}$$

The steady-state s.c. current can be found from the circuit for the steady-state analysis, Fig. 6.36. After short-circuiting $V_d = 0$ we have

$$I_{d,\infty} = \frac{E_d}{X_d}. \tag{6.57}$$

Next we shall find the quadrature-axis component of the voltage. Consider the *phasor diagram* shown in Fig. 6.38, which is drawn for a round-pole generator with $X_d = X_q$. From this diagram we obtain

$$\tilde{E}_d = \tilde{V}_d + jX_d \tilde{I}_d = \tilde{V} - \tilde{V}_q + jX_d \tilde{I}_d. \tag{6.58a}$$

Since $\tilde{V}_q = -jX_q \tilde{I}_q = -jX_d \tilde{I}_q$ we have

$$\tilde{E}_d = \tilde{V} + jX_d \tilde{I}. \tag{6.58b}$$

(Here and further on the tilde sign ~ stands for phasor quantities.)

Considering triangle ABC, we may determine angle δ, between E_d and V, which is known as the load angle:

$$\delta = \tan^{-1} \frac{X_d I \cos \varphi}{V + X_d I \sin \varphi}. \tag{6.59}$$

With this angle the V_{q0} component is

$$V_{q0} = V_0 \cos \delta_0. \tag{6.60a}$$

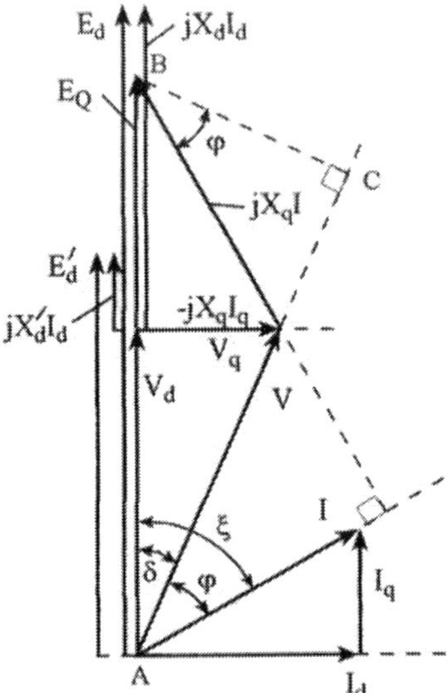

Figure 6.38 Phasor diagram for a round-pole generator.

and the q-axis component of the current is

$$I_{q0} = \frac{V_{q0}}{X_q}. \tag{6.60b}$$

The two components of the s.c. current (equations 6.55a and 6.60b) are actually the initial values of the two components of the s.c. current. To find the transient s.c. current we must solve the differential equations for each of the two components. To simplify the solution we will use the superposition principle and consider each of the currents as a sum of the current prior to switching, and a transient current due to switching. This current is found as a result of applying, to the passive circuit, the voltages equal and opposed to those which existed at the fault point prior to switching. In doing so we must remember that the two equivalent circuits in the d- and q-axes in transient behavior are not unconnected any more. Thus, any change of current in one will induce voltage in the second just like in transformer windings, Fig. 6.37(e). With the Laplace transform technique we may write

$$I_d' X_d' - sI_q' X_q = V_{q0},$$
$$-I_q' X_q - sI_d' X_d' = V_{d0}.$$

(Note that here, as in some technical books, the voltage induced by current I_q

is also assigned as V_q, although it is directed on the d-axis, and subsequently the voltage induced by current I_d is assigned as V_d.) Solving the above equations yields

$$I'_d = \frac{1}{X'_d}\left(-V_{d0}\frac{s}{s^2+1} - V_{q0}\frac{1}{s^2+1}\right)$$

$$I'_q = \frac{1}{X_q}\left(V_{q0}\frac{s}{s^2+1} - V_{d0}\frac{1}{s^2+1}\right).$$

Taking the inverse transform (using the table of the Laplace transform pairs) we may obtain

$$\begin{aligned} i'_d &= -\frac{V_{d0}}{X'_d}\cos\omega t - \frac{V_{q0}}{X'_d}\sin\omega t \\ i'_q &= \frac{V_{q0}}{X'_q}\cos\omega t - \frac{V_{d0}}{X_q}\sin\omega t. \end{aligned} \tag{6.61}$$

Next we can find the s.c. current in each of the three phases of a stator winding. Thus, for instance, for phase a, by substituting equation 6.61 in the first equation of 6.38a (note that in order to obtain the *instantaneous value of the short-circuit currents*, the argument ωt must be added to angle α under the cos and sin functions), and after algebraic simplifications we may obtain

$$\begin{aligned} i_a = {} & \frac{E'_{dm,0}}{X'_d}\cos(\omega t+\alpha) + I_{d,2\omega}\cos(2\omega t+\alpha) + I_{da}\cos\alpha \\ & + I_{q,2\omega}\sin(2\omega t+\alpha) + I_{qa}\sin\alpha, \end{aligned} \tag{6.62}$$

where

$$I_{d,2\omega} = \frac{V_{dm,0}(X'_d - X_q)}{2X'_dX_q} \quad \text{and} \quad I_{q,2\omega} = \frac{V_{qm,0}(X'_d - X_q)}{2X'_dX_q} \tag{6.63}$$

are the *double frequency component* amplitudes, and

$$I_{da} = -\frac{V_{dm,0}(X'_d + X_q)}{2X'_dX_q} \quad \text{and} \quad I_{qa} = -\frac{V_{qm,0}(X'_d - X_q)}{2X'_dX_q} \tag{6.64}$$

are the d.c., or *aperiodic components*, and α is the angle between the a-phase axis and d-axis. (Here the subscript "m" in EMF and voltages stands for the amplitude values.)

The appearance of the double frequency term in the stator transient current may be explained by considering the transient current in the rotor field winding. The sudden change of the stator currents in turn results in the appearance of a transient current in the rotor winding to keep the total magnetic flux constant. This current is of two components. As the three-phase stator current of basic frequency abruptly increases, the armature reaction on the d-axis also increases respectively. The transient current will appear in the field winding to compensate for the rise of this reaction. Since it is proportional to the change of the stator current, we may write its p.u. expression as

$$I_{fl,n} = \frac{X_{ad}}{X_{ad} + X_{rl}}(I'_d - I_{d,0}) = \frac{X_d - X'_d}{X_{ad}} \frac{V_{d,0}}{X'_d}. \tag{6.65}$$

The magnetic field produced by the d.c. component of the stator currents remains fixed in the air gap space and therefore is rotated with respect to the rotor with synchronous speed. This results in inducing an a.c. component in the field transient current. Since the field current, at the first moment, does not change, the initial value of an a.c. component is oppositely equal to the d.c. component. Subsequently, the resulting *current in the field winding* is

$$i_{fl} = I_{fl,f} + i_{fl,n} = I_{fl,0} - \frac{X_d - X'_d}{X_{ad}} \frac{V_{dm,0}}{X'_d} \cos \omega t, \tag{6.66}$$

where $I_{fl,0} = E_{d0}/X_{ad}$.

The a.c. term in the rotor produces a pulsating magnetic flux, which can be resolved into two components, having equal (half of the original) amplitudes, and revolving with synchronous speed in opposite directions. The component, which revolves in the direction opposite to that of the rotor, is actually fixed in the space and interferes, therefore, with the magnetic field of the d.c. component of the stator current, decreasing it slightly.

The component, which revolves in the same direction as the rotor, produces a magnetic field rotating in space with double the speed of the rotor and inducing in the stator windings the *double frequency component*. This component, however, is relatively small and when using engineering calculations is usually neglected.

The resistances of the generator windings, which have been neglected when determining the magnitudes of the transient currents, are responsible for decaying all these currents so that only the steady-state a.c. term in the stator and the d.c. term in the rotor remain invariable. The decay process is of an exponential form and the damping factors are mainly determined by the ratio of the resistance to the leakage inductance of the circuits. Since large synchronous machines have very small resistances compared to their considerable leakage reactances, their transient currents decrease very slowly and may predominantly determine the transients in a few seconds.

We may recognize two kinds of currents and the fields related to them, one of which adheres to the stator windings and the other to the rotor field winding. Each of them has a different damping factor or time constant, which primarily depends on the value of their resistances, related to the reactances. Thus, the stator resistances (including the external network) may be roughly estimated to be 10% of its leakage reactance, and the rotor circuit resistance may be about 1% of its leakage reactance.

The time constant of the rotor circuit is usually known and is given as a generator catalogue parameter. This time constant is related to the mutual flux linkage between the rotor and stator windings and is determined by an open-circuit test. This time constant determines the rate of increasing the field current (and therefore the generator open-circuit terminal voltage), when the constant voltage V_{fl} is suddenly applied to the field winding. Since the field winding is a simple *LR* circuit, the differential equation for the rotor circuit may be written as

$$\frac{d\lambda_{fl}}{dt} + I_{fl}R_{fl} = V_{fl},$$

or

$$\frac{1}{R_{fl}}\frac{d\lambda_{fl}}{dt} + I_{fl} = I_{fl,\infty}, \tag{6.67}$$

where $I_{fl,\infty} = V_{fl}/R_{fl}$ is a long-term (steady-state) field current. Since the open-circuit terminal voltage is approximately proportional to the field current: $E_d = X_{ad}I_{fl}$, where X_{ad} is the mutual reactance/inductance, we have

$$E_{d,\infty} = E_d + \frac{X_{ad}}{R_{fl}}\frac{d\lambda_{fl}}{dt}. \tag{6.68}$$

Using the relation between the transient EMF E'_d and λ_{fl}: $E'_d = (X_{ad}/[X_{fl} + X_{ad})]\lambda_{fl}$, equation 6.68 becomes

$$E_{d,\infty} = E_d + T_{d0}\frac{dE'_d}{dt}, \tag{6.69}$$

where

$$T_{d0} = \frac{X_{fl} + X_{ad}}{R_{fl}} \tag{6.70}$$

is the *open-circuit time constant* in p.u. (as the basic time is $t_b = 1/\omega$ s), or in seconds

$$T_{d0} = \frac{X_{fl} + X_{ad}}{\omega R_{fl}}. \tag{6.71}$$

Since $E_d = X_d I_d$ and $E'_d = X'_d I_d$, then $E'_d = (X'_d/X_d)E_d$, and substituting E'_d in equation 6.69 with $(X'_d/X_d)E_d$, we have

$$T'_d\frac{dE_d}{dt} + E_d = E_{d,\infty}, \tag{6.72}$$

where the *transient time constant* is

$$T'_d = \frac{X'_d}{X_d}T_{d0}. \tag{6.73}$$

For the short-circuit fault, remote from the generator, the transient time constant is given by the equation

$$T'_d = \frac{X'_d + X_F}{X_d + X_F} T_{d0}, \tag{6.74}$$

where X_F is the external reactance. If the system (external) impedance contains a relatively high resistance R_{ex}, the transient time constant is given by the extended relationship

$$T'_d = \frac{R_F^2 + (X'_d + X_F)(X_d + X_F)}{R_F^2 + (X_d + X_F)^2} T_{d0}. \tag{6.75}$$

The transient time constant T'_d lies in the range of 0.4 to 2 s for high power, high voltage turbogenerators and 0.7 to 2.55 s for salient-pole hydrogenerators. In low power generators T'_d may be less than 0.2 s.

The d.c. or aperiodical terms of both the stator and rotor windings are actually exponential functions and each of them decays with an appropriate time constant, which is determined by the parameters of that winding to which they are linked. Thus, the d.c. term of the rotor current and, adherent to it, the a.c. transient term of the stator current, decay at the rate of the above transient time constant T'_d, which is determined primarily by the time constant of the rotor winding, T_{d0}.

The d.c. (aperiodic) and double-frequency components in the stator currents die out with the *armature time constant* T_a. Although the initial value of the d.c. components in different phases is determined by the switching moment, or by the initial phase angle of the prior to switching phase currents, the total MMF, produced by these currents, is stationary in space and of a magnitude which is independent of the initial phase angle. This stationary MMF reacts with the rotating rotor alternately on the d- and q-axes. Therefore, the inductance associated with the d.c. component may be regarded as a sort of average of X'_d and X_q, More precisely, by observing equation 6.64 we may conclude that the reactance, which determines the d.c. (aperiodic) term, is $X_2 = X'_d X_q/(X'_d + X_q)$ (also known as the negative-sequence of a synchronous machine – see further on). Thus, by using this reactance, the stator winding transient time constant may be determined as:

in p.u.
$$T_a = \frac{2X'_d X_q}{R_a(X'_d + X_q)} = \frac{X_2}{R_a} \text{ pu} \tag{6.76}$$

or in seconds
$$T_a = \frac{2X'_d X_q}{R_a \omega (X'_d + X_q)} = \frac{X_2}{R_a \omega} \text{ s} \tag{6.77}$$

For high-voltage generators, T_a is in the range of 0.07 to 0.5 s and for low-voltage generators its value lies in the range of 0.01 to 0.1 s. If the short-circuit occurs at a distance from the generator, then the time constant is given as

$$T_a = \frac{X_2 + X_{ex}}{R_a}. \tag{6.78}$$

The rotor rotation relative to the fixed MMF of the d.c. component in the stator causes the a.c. component of the fundamental frequency in the field current to appear. This a.c. component in the field current, as has been previously mentioned, is responsible for the double-frequency component of the armature short-circuit current. The time constant T_a therefore also applies to the a.c. component in the rotor current and to the double-frequency component of the stator current.

The a.c. component of s.c. current, which appears at the first moment of switching (equation 6.62) differs from the steady-state s.c. current, which can be approximately determined by the saturated reactance X_d, or by using the linearization method (see section 6.5.2), as E_d/X_d (equation 6.57). This difference may then be expressed as

$$\Delta I'_d = \frac{E'_{d0}}{X'_d} - \frac{E_d}{X_d} = I'_{d0} - I_{d,\infty}. \tag{6.79}$$

The first component is a *transient current*, which decays at the rate of the transient time constant; therefore, the difference, $\Delta I'_d$, also decays at the same rate, so that the short-circuit a.c. current at fundamental frequency falls off from its initial value I'_{d0} to its final value of the steady-state short-circuit current with the time constant T'_d.

In conclusion, the *total transient response current* of a synchronous generator to the short-circuit fault at its terminal (phase *a*) and *total rotor field current*

$$\begin{aligned} i_a = {} & \left[\left(\frac{E'_{d0m}}{X'_d} - \frac{E_{dm}}{X_d}\right) e^{-t/T'_d} + \frac{E_{dm}}{X_d}\right] \cos(\omega t + \alpha) \\ & - \frac{V_{d0m}(X'_d + X_q)}{2X'_d X_q} e^{-t/T_a} \cos\alpha + \frac{V_{d0m}(X'_d - X_q)}{2X'_d X_q} e^{-t/T_a} \cos(2\omega t + \alpha) \\ & + \frac{V_{q0m}(X'_d + X_q)}{2X'_d X_q} e^{-t/T_a} \sin\alpha + \frac{V_{q0m}(X'_d - X_q)}{2X'_d X_q} e^{-t/T_a} \sin(2\omega t + \alpha), \end{aligned} \tag{6.80}$$

$$i_{fl} = I_{fl0} + \frac{V_{dm,0}}{X'_d} \frac{X_d - X'_d}{X_{ad}} e^{-t/T'_d} - \frac{V_{dm,0}}{X'_d} \frac{X_d - X'_d}{X_{ad}} e^{-t/T_a} \cos\omega t. \tag{6.81}$$

Both currents and their components are plotted in Fig. 6.39.

The initial value of the aperiodic term may be obtained by combining two components: of axes *d* and *q*

$$A = \frac{V_{d0m}\cos\alpha + V_{q0m}\sin\alpha}{2X'_d X_q/(X'_d + X_q)}. \tag{6.82}$$

(b) *Transient effects of the damper windings: subtransient EMF, subtransient reactance and time constant*

Nowadays synchronous machines are usually equipped with **damper windings**, which consist of short-circuited turns, or bars of copper strip set in poles. The

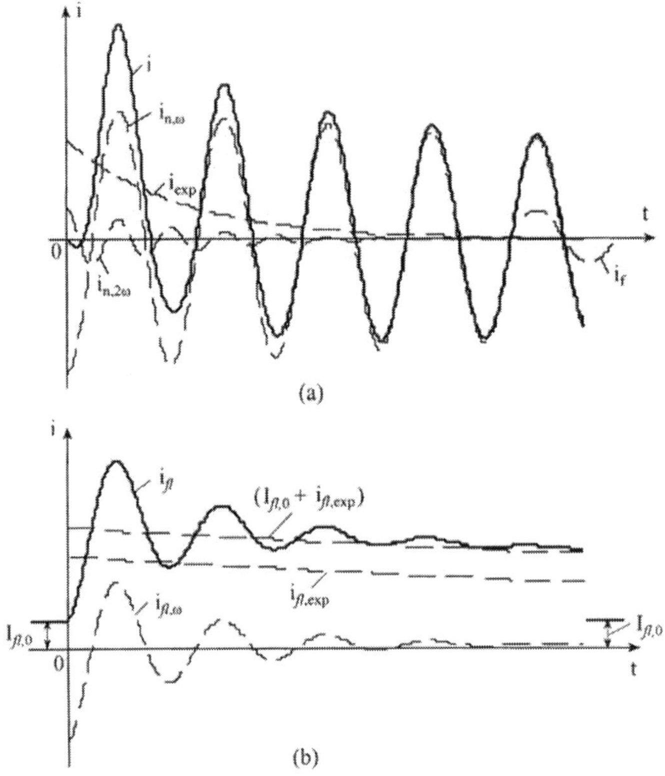

Figure 6.39 The short-circuit currents of a synchronous generator at sudden fault at its terminals: stator current (a) and rotor current (b).

reason for using the damper windings is to aid in starting and to reduce (to damp) mechanical oscillatory tendencies, which may arise under different faults, and thereby to increase the dynamic stability of the generator. The damper windings are placed in both axes, *d* and *q*, as can be seen from Fig. 6.40.

The damper winding does not change in principle the nature of the transients. Its influence results in increasing the short-circuit current magnitudes and in the appearance of an additional component on the *q*-axis, which is a subtransient EMF E''_q. Presenting an additional winding on a rotor makes the straightforward analysis of the generator transients even more complicated. However, analyzing the generator equivalent circuit in both axes, shown in Figure 6.41, will allow us to get the final results in a much easier way.

As can be seen, the equivalent circuit in the *d*-axis differs from those in Fig. 6.37 by an additional mutual winding (damper winding). By illuminating the mutual inductance (in a similar way as for a three-winding power transformer) we may obtain the circuit shown in (b), and by applying the Thévenin

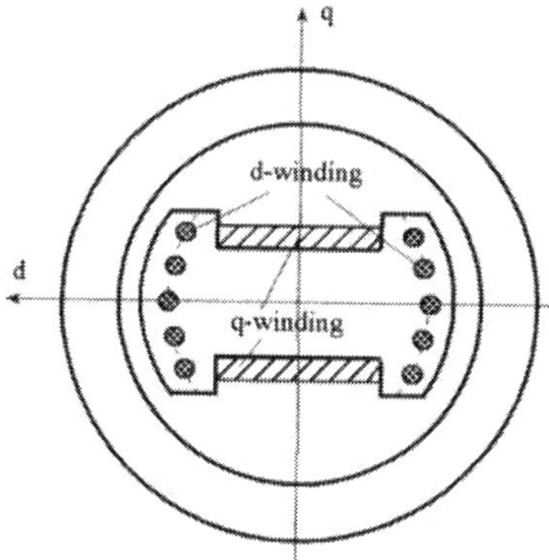

Figure 6.40 The damper windings in axes d and q.

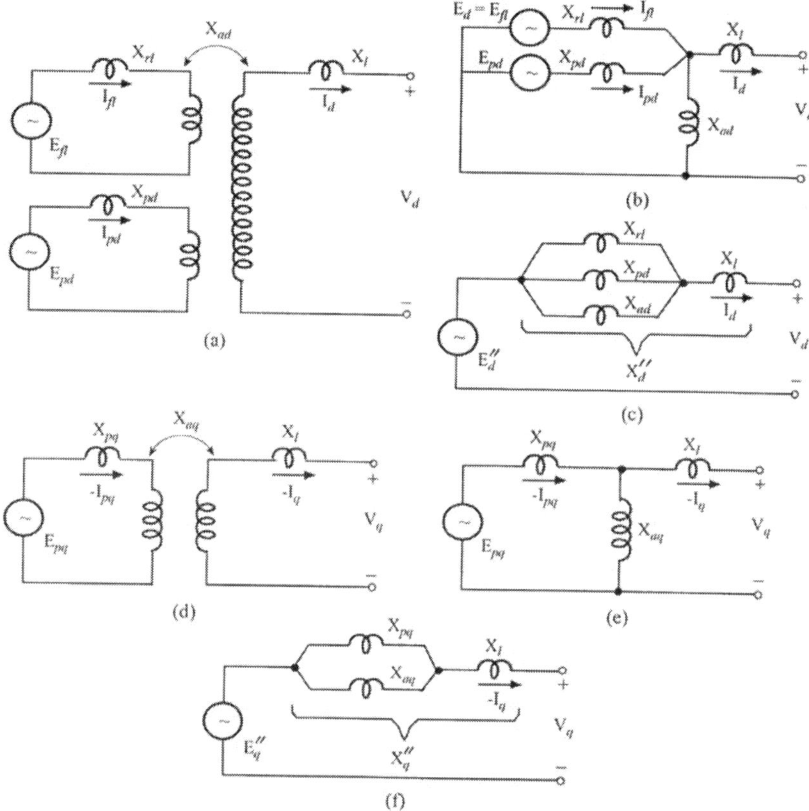

Figure 6.41. An equivalent circuit of a synchronous machine having damper windings and its simplification: in the d-axis (a, b and c) and in the q-axis (d, e and f).

theorem the one shown in (c). In the resulting circuit E''_d is called a *subtransient EMF* and X''_d is a *subtransient reactance*:

$$E''_d = \frac{E_d/X_{fl} + E_{pd}/X_{pd}}{1/X_{fl} + 1/X_{pd} + 1/X_{ad}} \tag{6.83}$$

and

$$X''_d = \frac{1}{1/X_{rl} + 1/X_{pd} + 1/X_{ad}} + X_l. \tag{6.84}$$

The subtransient EMF E''_d may also be determined by using the known terminal voltage and load current prior to short-circuiting:

$$E''_{d0} = V_{d0} + X''_d I_{d0}. \tag{6.85a}$$

The phasor diagram for a synchronous generator having damper windings is shown in Fig. 6.42.

For the generators having $X''_d = X''_q$, the initial subtransient EMF can be easily found from the simplified *phasor diagram of a synchronous generator with a damper winding*, shown in Fig. 6.42(b). Thus,

$$E''_0 = \sqrt{(V_0 \cos\varphi_0)^2 + (V_0 \sin\varphi_0 + X'' I_0)^2}, \tag{6.85b}$$

or approximately as a projection on V_0,

$$E''_0 \cong V_0 + X'' I_0 \sin\varphi_0. \tag{6.85c}$$

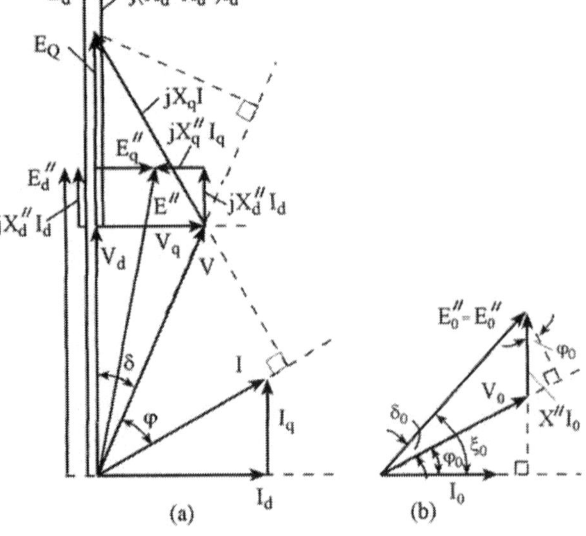

Figure 6.42 Phasor diagram for a synchronous generator with a damper winding (a) and a simplified phasor diagram for the generator having $X''_d = X''_q$ (b).

The subtransient parameters on the q-axis may be obtained by using the equivalent circuit in Fig. 6.41 (d, e and f). After simplification we have

$$E''_q = \frac{E_{pq}/X_{pq}}{1/X_{pq} + 1/X_{aq}} \tag{6.86}$$

and

$$X''_q = \frac{1}{1/X_{pq} + 1/X_{aq}} + X_l. \tag{6.87}$$

Similar to equation 6.85a we have

$$E''_{q0} = V_{q0} - X''_q I_{q0}. \tag{6.88}$$

The *subtransient time constants* are found as

$$T''_d = \frac{X''_d}{X'_d} T''_{d0} \tag{6.89}$$

and usually

$$T''_q \cong T''_d, \tag{6.90}$$

where T''_{d0} is a subtransient open-circuit d-axis time constant of a generator having a damper winding. The subtransient time constant is relatively small, $T''_d < T'_d$, and is in the range of 20 to 50 ms.

For the generator with damper windings the magnitudes of the fundamental frequency subtransient currents at the first moment of the fault are given by expressions

$$I''_{d0} = \frac{E''_{d0}}{X''_d}, \quad I''_{q0} = \frac{E''_{q0}}{X''_q} \tag{6.91}$$

and

$$I''_0 = \sqrt{I''^2_{d0} + I''^2_{q0}}. \tag{6.92}$$

Similar to equation 6.80, the *total short-circuit current versus time for a generator with damper windings is*

$$\begin{aligned} i_a = {} & \left[\left(\frac{E''_{d0m}}{X''_d} - \frac{E'_{d0m}}{X'_d}\right) e^{-t/T''_d} + \left(\frac{E'_{d0m}}{X'_d} - \frac{E_{dm}}{X_d}\right) e^{-t/T'_d} + \frac{E_{dm}}{X_d}\right] \cos(\omega t + \alpha) \\ & - \frac{V_{d0m}(X''_d + X''_q)}{2X''_d X''_q} e^{-t/T_a} \cos\alpha + \frac{V_{d0m}(X''_d - X''_q)}{2X''_d X''_q} e^{-t/T_a} \cos(2\omega t + \alpha) \\ & - \frac{E''_{q0m}}{X''_q} e^{-t/T''_q} \sin(\omega t + \alpha) + \frac{V_{q0m}(X''_d + X''_q)}{2X''_d X''_q} e^{-t/T_a} \sin\alpha \\ & + \frac{V_{q0m}(X''_d - X''_q)}{2X'_d X_q} e^{-t/T_a} \sin(2\omega t + \alpha). \end{aligned} \tag{6.93}$$

When the s.c. fault occurs after some external reactance X_F in all the previous expressions, this reactance must be added to the generator reactances on both axes.

As previously mentioned, this short-circuit current differs from that of a generator without damper windings (equation 6.80) by the presence of the subtransient term E''_{d0}/X''_d and the term on the q-axis E''_{q0}/X''_q. Both of these terms, however, decay very fast, at the rate of the time constants T''_d and T''_q.

After these two components die out, the instantaneous s.c. current is practically similar to those of a generator without damper windings. (Precisely speaking the damper winding also influences the transient process after decaying the subtransient currents: as an additional short-circuited winding on the d-axis it results in an increase in the aperiodic component in the field current to a slightly higher value than in the first moment of the fault. However, in practice this phenomenon is usually neglected. It should also be noted that for turbogenerators having $X''_d \cong X''_q$, the double frequency component in the s.c. currents is practically absent.)

In conclusion, the change of the r.m.s. or amplitude values, the *envelope curve* of an a.c. short-circuit (fundamental frequency) versus time is plotted in Fig. 6.43. As can be seen, this curve consists of three stages of the transient process: subtransient, transient and steady-state. The *subtransient stage* is given by the difference

$$\Delta I''_{d0} = I''_{d0} - I'_{d0},$$

and the *transient stage* is given by the difference

$$\Delta I'_{d0} = I'_{d0} - I'_{d\infty}.$$

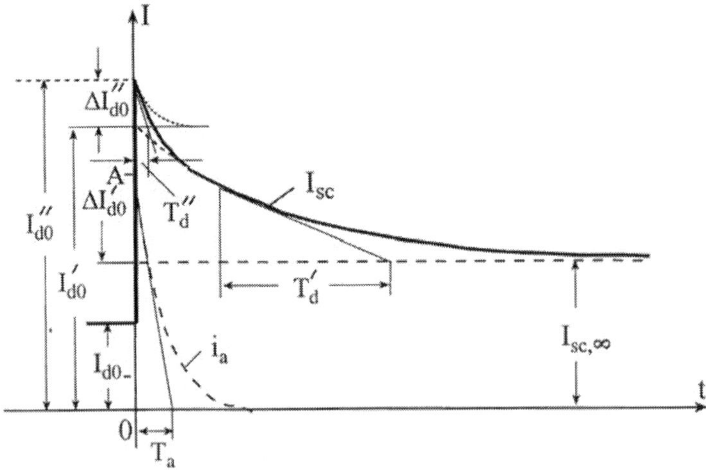

Figure 6.43 The r.m.s. (envelope) curve of the periodic term and exponential term of a short-circuit current for a generator having a damper winding.

The aperiodic (or exponential) component, which is decaying from its initial value A with the time constant T_a, is also plotted in Fig. 6.43. The curves in Fig. 6.43 can be used for experimentally determining the generator time constants, as shown in the figure.

The decaying process of the a.c. term of the short-circuit current can also be explained by increasing the generator reactances gradually from X''_d to X'_d and to X_d (remember that $X''_d < X'_d < X_d$) during the transient process. This phenomenon is opposed to the one in the first moment of short-circuiting: the armature reaction flux Φ_{ad} being suddenly increased, is opposed to the damper windings flux and is forced out of the poles to some extent, thereby increasing the reluctance and yielding a reduced synchronous reactance X''_d. As the time of the transients progresses, Φ_{ad} moves back on through the pole, which yields a relatively low reluctance path, and therefore the reactance will increase.

(c) *Transient behavior of a synchronous generator with AVR*

If the generator is equipped with AVR (automatic voltage regulator), the voltage supplied to the field winding does not keep constant, but is increased at the moment of short-circuiting. It can be approximately assumed that this rise of the supplied voltage is exponential. Hence, equation 6.72 may now be written in the form

$$T'_d \frac{dE_d}{dt} + E_d = E_{d,\max} - (E_{d,\max} - E_{d0\_})e^{-t/T_{ff}}. \tag{6.94}$$

Here on the right side of the differential equation is an exponential function having its prior to switching value $E_{d0\_}$ and the final, steady-state value $E_{d,\max}$; T_{ff} is the time constant of the supplied voltage circuit (exciter and/or power supply circuit). The range of this constant is 0.4 to 1 s.

The natural solution of this equation as we already know is a simple exponent

$$E_{d,n} = Ae^{-t/T'_d}. \tag{6.95}$$

The forced solution should be of the same form as the forced function

$$E_{d,f} = B + Ce^{-t/T_{ff}}. \tag{6.96}$$

The integration constants A, B and C might be found by applying the known quantities: prior to switching ($t = 0_-$) value $E_{d0\_}$, the initial ($t = 0_+$) value E_{d0_+} $[E_{d0_+} = E'_{d0}(X_d/X'_d)]$ and the final or steady-state value $E_{d,\max}$ (in accordance with the known $I_{ff,\max}$). Omitting all the algebraic calculations we may obtain the integration constants as

$$B = E_{d,\max}, \quad C = \Delta E_d \frac{T_{ff}}{T'_d - T_{ff}}, \tag{6.97a}$$

where $\Delta E_d = E_{d,\max} - E_{d0\_}$ and

$$A = -(E_{d,\max} - E_{d0_+}) - \Delta E_d \frac{T_{ff}}{T'_d - T_{ff}}. \tag{6.97b}$$

Thus, we finally have

$$E_d(t) = E_{d,f} + E_{d,n} = E_{d,\max} + \Delta E_d \frac{T_{ff} e^{-t/T_{ff}}}{T'_d - T_{ff}}$$

$$- \left[E_{d,\max} - E_{d0_+} + \Delta E_d \frac{T_{ff}}{T'_d - T_{ff}} \right] e^{-t/T'_d}, \tag{6.98}$$

which results in E_{d0_+} at $t = 0$ and in $E_{d,\max}$ at $t \to \infty$, i.e. in accordance with the given initial and steady-state conditions.

For the generator without AVR the time constant T_{ff} should be infinite, i.e., $T_{ff} = \infty$, and the differential equation 6.94 turns into

$$T'_d \frac{dE_d}{dt} + E_d = E_{d0_-}.$$

The solution of this equation is

$$E_d(t) = (E_{d0_+} - E_{d0_-}) e^{-t/T'_d} + E_{d0_-}. \tag{6.99}$$

By rearranging the terms in equation 6.98 and after performing the appropriate algebraic calculations, we may obtain

$$E_d(t) = [(E_{d0_+} - E_{d0_-}) e^{-t/T'_d} + E_{d0_-}] + (E_{d,\max} - E_{d0_-}) F(t)$$

$$= E_{d(\text{without AVR})} + \Delta E_{d(\text{with AVR})}, \tag{6.100a}$$

where

$$F(t) = 1 - \frac{T'_d e^{-t/T'_d} - T_{ff} e^{-t/T_{ff}}}{T'_d - T_{ff}}. \tag{6.100b}$$

The above expression clearly shows that due to AVR the EMF of the generator increases gradually during the transients by $\Delta E_{d(\text{with AVR})}$ relatively to the EMF of the generator without AVR. This in turn results in increasing the transient a.c. term of the short-circuit current. Dividing equation 6.100a by X_d and taking into consideration that $E_d = (X_d / X'_d) E'_d$ and $E_{d0_-} / X_d = I_\infty$, we have

$$I'(t) = [(I'_d - I_\infty) e^{-t/T'_d} + I_\infty] + \Delta I_\infty F(t), \tag{6.101}$$

where $\Delta I_\infty = I_{\infty,\max} - I_\infty$ and $I_{\infty,\max}$ is the steady-state short-circuit current due to the maximal increased by AVR field current. The term in the brackets on the right hand side in equation 6.101 is the transient a.c. component of the short-circuit current without AVR, and ΔI_∞ is its increase due to AVR. The short-circuit current, given by equation 6.101, which is actually the r.m.s. envelope curve of an a.c. component, is plotted in Fig. 6.44 for two cases: with and without AVR. Note that, in the very beginning of the transients, the two curves are practically the same, which means that the subtransient current is not influenced by AVR. This current decays before the AVR has had time to affect the generator EMF.

As was earlier shown, operating regime of the AVR depends on the fault

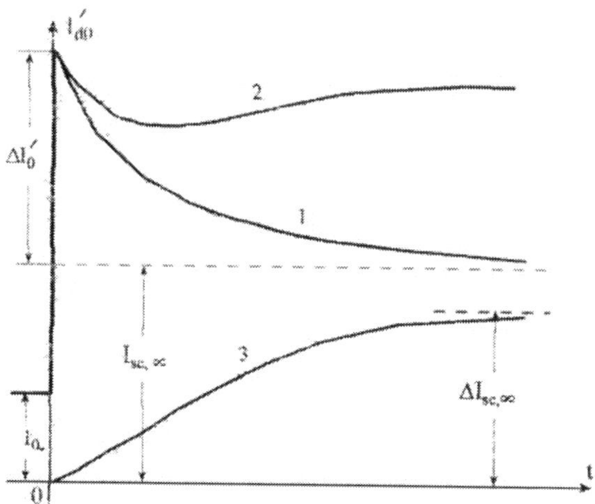

Figure 6.44 An envelope curve of the transient current for two cases: 1, without AVR; 2, with AVR and 3, the increase on the current due to AVR.

point location. Namely, if the external reactance is lower than the critical one. i.e., $X_F < X_{cr}$, then the AVR operates under the condition of maximal field current. If, however, $X_F > X_{cr}$, then the AVR operates under the condition of nominal/rated terminal voltage, $V = V_{nom}$. Hence, in the first case the short-circuit current varies in accordance with equation 6.101, but in the second case the current is limited by the value of $V_{nom}/(X'_d + X_F)$.

(d) *Peak values of a short-circuit current*

For some conditions (such as to check the dynamic stability of electric equipment under short-circuit conditions or for the proper design of relay protection), it is necessary to know the maximum instantaneous or *peak value* of the *short-circuit current*. In highly inductive circuits the peak current appears nearly half a period after the occurrence of the short-circuit.

Neglecting the double-frequency term, we shall take into consideration the subtransient a.c. term I''_d and the exponential term A (equation 6.82), so that the peak value will be

$$i_{pk} = \sqrt{2}I''_d + Ae^{-0.01/T_a}, \tag{6.102}$$

or by approximating $A \cong \sqrt{2}I''_d$, and with $T_a = 0.05$ (or $X/R = 1.5$) we have

$$i_{pk} = \sqrt{2}I''_d(1 + e^{-0.01/T_a}) = \sqrt{2}k_{pk}I''_d \cong \sqrt{2}\cdot 1.8 I''_d, \tag{6.103}$$

where, as previously (equation 6.20a),

$$k_{pk} = 1 + e^{-0.01/T_a}$$

is the *peak coefficient*.

Furthermore, because of very small damping, the exponential term approaches unity and the maximal peak is simply

$$i_{pk} = 2\sqrt{2}I_d'' \tag{6.104}$$

The *r.m.s. value* of the *peak current* can be calculated in accordance with equation 6.22a

$$I_{pk} = I_d''\sqrt{1 + 2(k_{pk} - 1)^2},$$

and for $T_a = 0.05$ ($k_{pk} = 1.8$) we have

$$I_{pk} = 1.52I_d''.$$

As can be seen, the peak coefficient depends on the value of the time constant. Thus, for $T_a = 0.008$ ($X/R = 2.5$) the peak coefficient decreases to 1.3 and

$$i_{pk} = \sqrt{2}\cdot 1.3I_d'' \quad \text{and} \quad I_{pk} = 1.1I_d''.$$

However, the aperiodic component should be taken into consideration for the periods of less than $t = 0.15$ s after the short-circuit fault occurs.

Example 6.8

For a synchronous generator having the following p.u. parameters: $X_\ell = 0.1$, $X_d = 1.2$, $X_d' = 0.25$, $X_q = 0.6$, $R = 0.005$ and $T_{do} = 8.5$ s: a) find all the components of the transient short-circuit current at $t = 0$, b) write the expression of the s.c. current in phase a and in the field (rotor) winding versus time, c) write the expression of the s.c. current envelope and plot the phase a current and the envelope curve and d) calculate the peak value of the s.c. current. Prior to short-circuiting the generator has been operated at the rated voltage $V = 1$ and 0.8 of the rated current with PF = 0.85 ($\varphi = 31.8°$) and $f = 50$ Hz (the AVR is absent).

Solution

a) To find E_d and E_d' we must calculate the power angle δ (see the phasor diagram in Fig. 6.38). For this purpose we first calculate the angle ξ:

$$\xi = \tan^{-1}\frac{V\sin\varphi + X_q I}{V\cos\varphi} = \tan^{-1}\frac{1\cdot\sin 31.8° + 0.6\cdot 0.8}{1\cdot 0.85} = 49.8°$$

and $\delta = \xi - \varphi = 49.8° - 31.8° = 18.0°$.

Now we may find

$$E_d = V\cos\delta + X_d I\sin\xi = 1\cdot\cos 18.0° + 1.2\cdot 0.8\sin 49.8° = 1.70$$

$$E_d' = V\cos\delta + X_d' I\sin\xi = 1\cdot\cos 18.0° + 0.25\cdot 0.8\sin 49.8° = 1.1.$$

Then the s.c. current components at $t = 0$ are:

a.c. (or periodic)

$$I'_{d0} = \frac{E'_{d0}}{X'_d} = \frac{1.1}{0.25} = 4.4, \quad I_\infty = \frac{E_{d0}}{X_d} = \frac{1.70}{1.2} = 1.42,$$

and

$$\Delta I_d = I'_{d0} - I_\infty = 4.4 - 1.42 = 2.98;$$

d.c. (or aperiodic)

$$I_{da} = \frac{V_{d0}(X_q + X'_d)}{2X_q X'_d} = \frac{0.951\cdot(0.6+0.25)}{2\cdot 0.6\cdot 0.25} = 2.69$$

$$I_{qa} = \frac{V_{q0}(X_q + X'_d)}{2X_q X'_d} = \frac{0.310\cdot(0.6+0.25)}{2\cdot 0.6\cdot 0.25} = 0.878,$$

where

$$V_{d0} = V\cos\delta = 1\cdot\cos 18.0^\circ = 0.951$$

$$V_{q0} = X_q I_q = X_q I\cos\xi = 0.6\cdot 0.8\cos 49.8^\circ = 0.310;$$

2ω (or double frequency)

$$I_{d,2\omega} = \frac{V_{d0}(X_q - X'_d)}{2X_q X'_d} = \frac{0.951\cdot(0.6-0.25)}{2\cdot 0.6\cdot 0.25} = 1.11$$

$$I_{q,2\omega} = \frac{V_{q0}(X_q - X'_d)}{2X_q X'_d} = \frac{0.310\cdot(0.6-0.25)}{2\cdot 0.6\cdot 0.25} = 0.362.$$

b) The time constants are:

$$T_a = \frac{X_2}{\omega R} = \frac{0.353}{314\cdot 0.005} = 0.225,$$

where

$$X_2 = \frac{2X'_d X_q}{X'_d + X_q} = \frac{2\cdot 0.25\cdot 0.6}{0.25+0.6} = 0.353,$$

and

$$T'_d = T_{d0}\frac{X'_d}{X_d} = 8.5\cdot\frac{0.25}{1.2} = 1.77.$$

The s.c. current of phase a versus time for the initial angle $\alpha = 30^\circ$ will be

$$\begin{aligned} i_a(t) = \sqrt{2}[&(1.42 + 2.98e^{-t/1.77})\cos(\omega t + 30^\circ) - 2.69e^{-t/0.225}\cos 30^\circ \\ &- 1.11e^{-t/0.225}\cos(2\omega t + 30^\circ) + 0.878e^{-t/0.225}\sin 30^\circ \\ &- 0.362e^{-t/0.225}\sin(2\omega t + 30^\circ)], \end{aligned}$$

or, after simplification,

$$i_a(t) = (2.01 + 4.21e^{-t/1.77})\cos(\omega t + 30^\circ) - 2.67e^{-t/0.225} - 1.65e^{-t/0.225}\cos(2\omega t + 11.9^\circ).$$

Note that at $t = 0$, the d- and q-components are

$$I_{d0} = 4.4 - 2.69 - 1.11 = 0.60, \quad I_{q0} = 0.878 - 0.362 = 0.516,$$

and $I_0 = \sqrt{0.60^2 + 0.516^2} = 0.79 \cong 0.8$, as it is given.

The rotor winding current, i.e., the field current, is calculated with equation 6.81:

$$i_{fl}(t) = I_{fl0} + i_{fl,a0} + i_{fl,\omega 0} = 1.55 + 4.65e^{-t/1.77} - 4.65e^{-t/0.225}\cos\omega t,$$

where

$$I_{fl0} = \frac{E_{d0}}{x_{ad}} = \frac{1.70}{1.1} = 1.55, \quad X_{ad} = X_d - X_\ell = 1.2 - 0.1 = 1.1$$

and

$$I_{fl,a0} = -I_{fl,\omega 0} = \frac{\sqrt{2}V_{d0}}{X'_d}\frac{X_d - X'_d}{X_{ad}} = \frac{\sqrt{2}\cdot 0.951}{0.25}\cdot\frac{1.2 - 0.25}{1.1} = 4.65.$$

c) The envelope curve may be obtained (by neglecting the 2ω-component) as

$$I_d(t) = \sqrt{2}(1.42 + 2.98e^{-t/1.77} - 2.69e^{-t/0.225}), \quad I_q(t) = \sqrt{2}\cdot 0.878e^{-t/0.225}.$$

The phase a s.c. current versus time as well as the envelope curve are given in Fig. 6.45.

d) The peak value of the s.c. current (which arises after about $t = 0.01$ s) is

$$I_{pq} \cong 2.01 + 4.21^{-0.01/1.77} + 2.67e^{-0.01/0.225} = 8.8,$$

or by using the approximate formula 6.103

$$i_{pk} = \sqrt{2}\cdot 1.8\cdot 4.4 = 11.2.$$

The difference in the above results is because the approximate formula is given for a no-loaded generator and therefore the initial value of an aperiodic component is as high as the subtransient s.c. current. However, in our example the generator prior to switching was operated under load and the aperiodic component is much lower.

Example 6.9

A turbogenerator is connected to the system through a power transformer. The parameters of these two apparatuses are: 1) turbo-generator – 125 MVA, 15.8 kV, PF = 0.8, and $X_\ell = 0.1$, $X_d = X_q = 1.35$, $X'_d = 0.2$, $X''_d = 0.13$, $X''_q = 0.15$, $T_a =$ 0.1 s, $T'_{d0} = 11.45$ s, $T''_{d0} = 0.25$ s, $T''_{q0} = 0.55$ s; 2) transformer – 120 MVA, 242/15.8 kV, $v_{sc} = 11.5\%$. For the three-pole short-circuit fault at the secondary of the transformer, find the subtransient s.c. current as its r.m.s. value versus

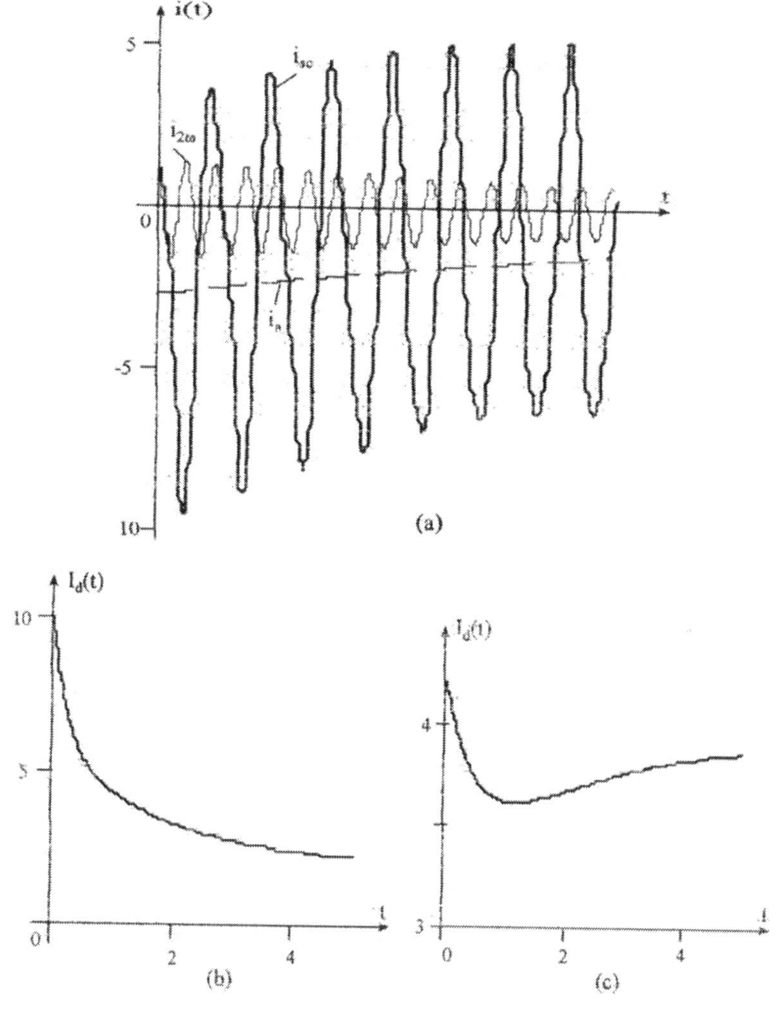

Figure 6.45 Short-circuit current of phase *a* versus time (a) and envelope curve of the short-circuit current without AVR (b) and with AVR (see Example 6.10) (c).

time (an envelope curve). The prior to switching operating conditions of the generator were as follows: $P = 100$ MVA, $\mathrm{PF} = 0.8$, $V_t = 16.5$ kV (without AVR).

Solution

The rated and load currents of the generator are

$$I_r = \frac{125}{\sqrt{3}\cdot 15.8} = 4.6 \text{ kA}, \quad I_{ld} = \frac{100}{\sqrt{3}\cdot 15.8} = 3.5 \text{ kA}.$$

Thus, the generator current, prior to switching, in p.u. is

$$I_0 = \frac{3.5}{4.6} = 0.76,$$

and the operating voltage in p.u. is

$$V_0 = \frac{16.6}{15.8} = 1.05.$$

In accordance with the phasor diagram (see Fig. 6.42)

$$E_{d0} = \sqrt{(1.05 \cdot 0.8)^2 + (1.05 \cdot 0.6 + 1.35 \cdot 0.76)^2} = 1.85,$$

where $\cos \varphi_0 = 0.8$ and $\sin \varphi_0 = 0.6$.

The angle between the current I_0 and the EMF E_{d0} is

$$\xi = \tan^{-1} = \frac{1.05 \cdot 0.6 + 1.35 \cdot 0.76}{1.05 \cdot 0.8} = \tan^{-1} 1.97 = 63.1^\circ.$$

Thus,

$$\cos \xi = 0.452 \quad \text{and} \quad \sin \xi = 0.883.$$

The power angle is

$$\delta_0 = 63.1^\circ - \cos^{-1} 0.8 = 26.2^\circ,$$

and $\cos \delta_0 = 0.897$, $\sin \delta_0 = 0.442$.

The d- and q-components of the initial current and voltage can now be calculated as

$$I_{d0} = 0.76 \cdot 0.883 = 0.671 \qquad I_{q0} = 0.76 \cdot 0.452 = 0.344$$

$$V_{d0} = 1.05 \cdot 0.897 = 0.942 \qquad V_{q0} = 1.05 \cdot 0.442 = 0.464.$$

The d- and q-components of the subtransient EMF are

$$E''_{d0} = V_{d0} + X''_d I_{d0} = 0.942 + 0.13 \cdot 0.671 = 1.03$$

$$E''_{q0} = V_{q0} - X''_q I_{q0} = 0.464 - 0.15 \cdot 0.344 = 0.412.$$

For further calculation we need to know the p.u. reactance of the transformer referred to the generator power $X_T = 0.115(125/120) = 0.12$. The first moment subtransient current components, therefore, will be

$$I''_{d0} = \frac{E''_{d0}}{X''_d + X_T} = \frac{1.03}{0.13 + 0.12} = 4.12$$

$$I''_{q0} = \frac{E''_{q0}}{X''_q + X_T} = \frac{0.412}{0.15 + 0.12} = 1.53,$$

and

$$I''_0 = \sqrt{I''^2_{d0} + I''^2_{q0}} = \sqrt{4.12^2 + 1.53^2} = 4.40.$$

The steady-state s.c. current will be

$$I_\infty = \frac{E_{d0}}{X_d + X_T} = \frac{1.85}{1.35 + 0.12} = 1.26.$$

To find the transient current we must first determine the transient EMF

$$E'_{d0} = V_{d0} + X'_d I_{d0} = 0.942 + 0.2 \cdot 0.671 = 1.08.$$

Hence, the transient current is

$$I'_{d0} = \frac{E'_{d0}}{X'_d + X_T} = \frac{1.08}{0.2 + 0.12} = 3.38,$$

and the subtransient and transient stages are given by

$$\Delta I''_d = 4.12 - 3.38 = 0.74, \quad \Delta I' = 3.38 - 1.26 = 2.12.$$

The aperiodic components (at $t = 0$) are

$$I_{da0} = \frac{V_{dF0}(X''_d + X''_q + 2X_T)}{2(X''_d + X_T)(X''_q + X_T)} = \frac{0.861 \cdot (0.12 + 0.15 + 0.24)}{2 \cdot 0.25 \cdot 0.27} = 3.32$$

$$I_{qa0} = \frac{V_{qF0}(X''_d + X''_q + 2X_T)}{2(X''_d + X_T)(X''_q + X_T)} = \frac{0.50 \cdot (0.12 + 0.15 + 0.24)}{2 \cdot 0.25 \cdot 0.27} = 1.92,$$

where the voltages at the fault point are

$$V_{dF0} = V_{d0} - X_T I_{d0} = 0.942 - 0.12 \cdot 0.671 = 0.861$$

$$V_{qF0} = V_{q0} + X_T I_{q0} = 0.464 + 0.12 \cdot 0.344 = 0.50.$$

Note that since the subtransient reactances in the d- and q-axes are almost equal, the double-frequency terms are neglected.

The subtransient and transient time constants are

$$T''_d = \frac{X''_d + X_T}{X'_d + X_T} T''_{d0} = \frac{0.13 + 0.12}{0.2 + 0.12} 0.25 = 0.195 \text{ s}$$

$$T''_q = \frac{X''_q + X_T}{X_q + X_T} T''_{q0} = \frac{0.15 + 0.12}{1.35 + 0.12} 0.55 = 0.101 \text{ s},$$

and

$$T'_d = \frac{X'_d + X_T}{X_d + X_T} T'_{d0} = \frac{0.2 + 0.12}{1.35 + 0.12} 11.45 = 2.49 \text{ s}.$$

Thus, the r.m.s. value of the a.c. component versus time (the envelope curve) is

$$I_d = 0.74 e^{-t/0.195} + 2.12 e^{-t/2.49} + 1.26, \quad I_q = -1.53 e^{-t/0.101}.$$

The aperiodic terms in both axes are

$$I_{da} = -3.32 e^{-t/0.1}, \quad I_{qa} = 1.92 e^{-t/0.1}.$$

The initial value of the subtransient current is

$$I''_0 = \sqrt{4.12^2 + 1.53^2} = 4.4.$$

The initial value of the entire current is

$$I_0 = \sqrt{(I''_{d0} - I_{da0})^2 + (-I_{q0} + I_{qa0})^2} = \sqrt{(4.12 - 3.32)^2 + (-1.53 + 1.92)^2} \cong 0.88.$$

(Note that this value varies slightly from the actual initial current because we neglected the double-frequency terms.)

Example 6.10

For the generator of Example 6.8 find the a.c. component of the short-circuit current (an envelope curve) if the generator is equipped with an AVR, having $I_{fl,\max} = 4.3$ and $T_{ff} = 0.55$ s.

Solution

Since the subtransient current decays very fast, practically before the AVR substantially affects the field current increasing, we shall take into consideration only the transient and steady-state currents. The steady-state s.c. current under the maximal field current will be

$$I_{\infty,\max} = \frac{I_{fl,\max}}{X_{ad}} = \frac{4.3}{1.1} = 3.91,$$

where $X_{ad} = X_d - X_\ell = 1.2 - 0.1 = 1.1$.
Thus,

$$\Delta I'_0 = I'_0 - I_\infty = 4.4 - 1.42 = 2.98, \quad \Delta I_\infty = I_{\infty,\max} - I_\infty = 3.91 - 1.42 = 2.49.$$

The increasing function will be

$$F(t) = 1 - \frac{T'_d e^{-t/T'_d} - T_{ff} e^{-t/T_{ff}}}{T'_d - T_{ff}} = 1 - \frac{1.77 e^{-t/1.77} - 0.55 e^{-t/0.55}}{1.77 - 0.55}$$

$$= 1 + 0.451 e^{-t/0.55} - 1.45 e^{-t/1.77}.$$

From Example 6.8 we have

$$I_{d(\text{without AVR})} = 2.98 e^{-t/1.77} + 1.42.$$

Therefore, we may now obtain

$$I_d(t) = I_{d(\text{without AVR})} + \Delta I_\infty F(t) = 3.91 + 1.12 e^{-t/0.55} - 0.63 e^{-t/1.77},$$

which at $t = 0$ again gives 4.4. This curve is also shown in Fig. 6.45.

6.6 SHORT-CIRCUIT ANALYSIS IN INTERCONNECTED (LARGE) NETWORKS

In general an electric system is supplied by a number of generators of different designs and different ratings, which are interconnected in complicated networks.

In practice the operation of a single synchronous generator in an isolated system, as has been discussed so far, is limited. The short-circuit analysis in interconnected systems is very complicated. The short-circuit currents of each of the generators are dependent on each other. The operation conditions of the AVR of each of the generators will depend on the distance to the location point of the fault. The mechanical oscillations of some of the generators will almost always follow the short-circuit faults. All this makes the precise calculation extremely complicated, if not impossible.

Therefore, in practical calculations it is common to make a few additional simplifications:

1) Each of the generators has a round (cylindrical) rotor, which allows us to neglect the double-frequency component and operate with only one current, alleviating the need of dividing the current and voltage onto two axes;

2) The periodic a.c. and aperiodic exponential terms are of the same form and obey the same law of behavior, as for a single generator;

3) All the generators operate under a constant speed of rotation.

These assumptions allow us to obtain the final results of a short-circuit fault in an interconnected system relatively easily and with accuracy, satisfying the practical needs. We shall now illustrate our study with the following examples.

Example 6.11

Find the subtransient s.c. current and its peak value in the network, shown in Fig. 6.46(a), if the three-pole fault occurs at the secondary (2) terminals of transformer T_1, point F. The two transformers T_1 and T_2 are identical and the circuit breaker Br is closed.

Solution

Let the basic power be 600 MVA. Then the p.u. reactances (see Fig. 6.46(b)) are calculated as

$$X_1 = 0.20\frac{600}{380} = 0.32, \quad X_2 = 0.093\frac{600}{160} = 0.35,$$

$$X_3 = 0.38 \cdot 200\frac{600}{230^2} = 0.86,$$

$$X_4 = X_5 = 0.5(0.181 + 0.123 - 0.058)\frac{600}{60} = 1.23, \quad X_6 = X_7 \cong 0,$$

$$X_8 = X_9 = 0.5(0.181 - 0.123 + 0.058)\frac{600}{60} = 0.58.$$

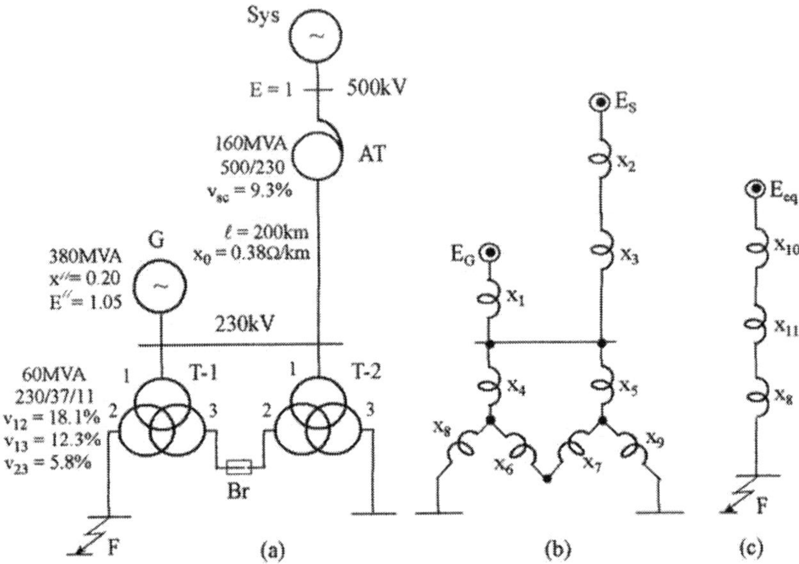

Figure 6.46 A network diagram for Example 6.11 (a) and its simplification (b) and (c).

To simplify the network, we find

$$X_{10} = 0.32//(0.35 + 0.86) = 0.25, \quad X_{11} = X_4/2 = 0.62,$$

$$X_{eq} = 0.25 + 0.62 + 0.58 = 1.45,$$

and

$$E''_{eq} = \frac{1.05 \cdot 1.21 + 1 \cdot 0.32}{1.21 + 0.32} = 1.04.$$

Therefore, the subtransient current will be

$$I'' = \frac{E''_{eq}}{X_{eq}} = \frac{1.04}{1.45} = 0.72, \quad \text{or in amperes} \quad I_{sc} = 0.72\frac{600}{\sqrt{3} \cdot 37} = 6.74\ \text{kA}.$$

The peak current (equation 6.103) will be

$$i_{pk} = \sqrt{2} \cdot 1.8 I_{sc} = 2.6 \cdot 6.74 = 17.2\ \text{kA}.$$

Example 6.12

The power network, shown in Fig. 6.47(a), consists of two identical generators G_1 and G_2, two identical transformers T_1 and T_2 and a power station G_3, which are connected by a 161 kV transmission line. All the circuit parameters are given in Fig. 6.47(a). If a three-pole fault occurs at point F, find the s.c. current at the moments of 0.1 s and 1 s. All the generators are equipped with AVR and the circuit breaker Br is opened.

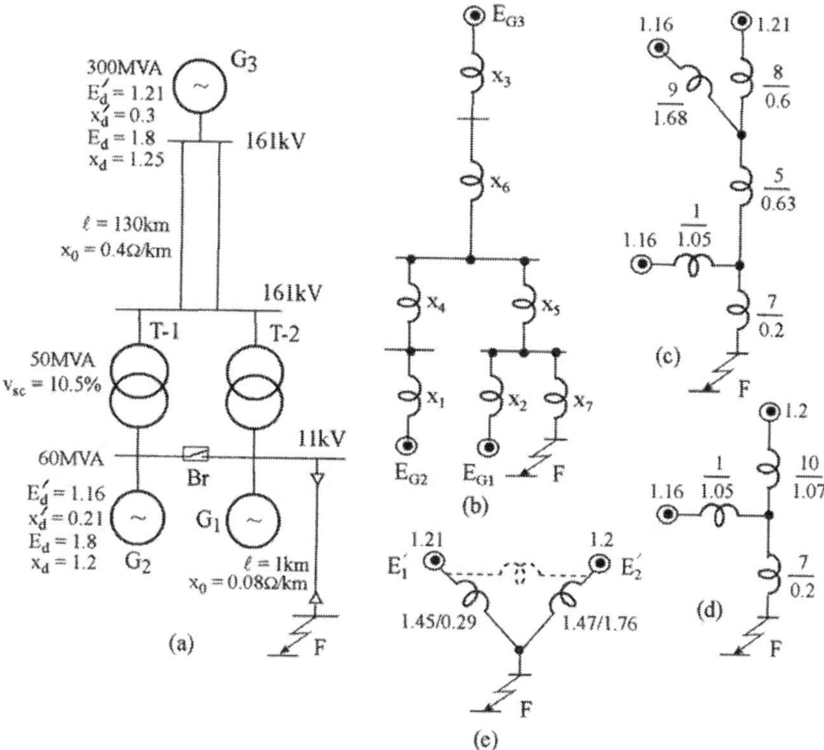

Figure 6. 47 A network for Example 6.12 (a) and the stages of its simplification (b)–(e).

Solution

Since at the time of 0.1 s the subtransient currents are already decayed, we shall represent the generators by their transient parameters. By choosing the basic power of 300 MVA, the circuit reactances are calculated as follows (see Fig. 6.47(b)):

$$X_1 = X_2 = 0.21\,\frac{300}{60} = 1.05, \quad X_3 = 0.3, \quad X_4 = X_5 = 0.105\,\frac{300}{50} = 0.63,$$

$$X_6 = \frac{1}{2}\,0.4 \cdot 130\,\frac{300}{161^2} = 0.3, \quad X_7 = 0.08 \cdot 1 \cdot \frac{300}{11^2} = 0.2.$$

The obvious simplification of the circuit is then performed in three stages, as shown in Fig. 6.47(c), (d) and (e). Generator G_1 is relatively "close" to the fault point; therefore it is treated separately, as shown in Fig. 6.47(e). First we find the s.c. current flowing from this generator. The transfer reactance between the generator and the fault point is found by transforming the Y-configuration in

Fig. 6.47(d) to the Δ-configuration in Fig. 6.47(e):

$$X'_{trs1} = 1.05 + 0.2 + \frac{1.05 \cdot 0.2}{1.07} = 1.45,$$

and referred to the rated power of the generator $X^{(n)}_{trs1} = 1.45(60/300) = 0.29$. Hence, the transient current is

$$I'_{01} = \frac{E'_1}{X_{trs1}} = \frac{1.16}{0.29} = 4.0.$$

Next we shall calculate the steady-state s.c. current. To do this, the generators must be represented by their synchronous reactances, X_d, and steady-state EMF, E_d. Performing the same steps of circuit simplification we will obtain

$$X^{(n)}_{trs1} = 1.37 \quad \text{and} \quad I_{\infty,1} = \frac{E_{d1}}{X^{(n)}_{trs1}} = \frac{1.8}{1.37} = 1.31.$$

Suppose that the maximal field current and time constant of the generators are: $I_{fl,\max} = 4.7$, $T_{ff} = 0.55$ s and $T'_{d0} = 8.5$ s. Then we may find the maximal steady-state s.c. current (equation 6.53)

$$I_{\infty,1\max} = \frac{I_{fl,\max}}{X_{ad}} = \frac{4.7}{1.37 - 0.1} = 3.7,$$

where $X_{ad} = X_d - X_\ell$. Then

$$\Delta I'_{01} = I'_{01} - I_{\infty,1} = 4 - 1.31 = 2.69$$

$$\Delta I'_{\infty,1} = I'_{\infty,1\max} - I_{\infty,1} = 3.62 - 1.31 = 2.31.$$

The transient time constant is

$$T'_{d1} = T'_{d0}\frac{X'_{trs1}}{X_{trs1}} = 8.5\frac{0.29}{1.37} = 1.8.$$

Now we may find the increasing function (equation 6.100b)

$$F(t) = 1 - \frac{1.8e^{-t/1.8} - 0.55e^{-t/0.55}}{1.8 - 0.55} = 1 + 0.44e^{-t/0.55} - 1.44e^{-t/1.8}.$$

The r.m.s. short-circuit current versus time without AVR is

$$I_{sc1(\text{without AVR})} = 2.69e^{-t/1.8} + 1.31,$$

and the total current due to the AVR action will be

$$\begin{aligned} I_{sc1}(t) &= I_{sc1(\text{without AVR})} + \Delta I_{\infty,1}F(t) \\ &= 2.69e^{-t/1.8} + 1.3 + 2.31(1 + 0.44e^{-t/0.55} - 1.44e^{-t/1.8}) \\ &= 3.7 + 1.05e^{-t/0.55} - 0.75e^{-t/1.8}. \end{aligned}$$

Thus, the s.c. current of generator G_1 is:

at $t=0.1$ $\quad I_{sc1}(0.1)=3.9 \quad \text{or} \quad 3.9\frac{60}{\sqrt{3}\cdot 11}=12.3\ \text{kA},$

at $t=1.0$ $\quad I_{sc1}(1.0)=3.4 \quad \text{or} \quad 3.4\frac{60}{\sqrt{3}\cdot 11}=10.7\ \text{kA}.$

(Note that without AVR the s.c. current at 1 s would be 2.8 or 8.8 kA, i.e., less than with the AVR.) Next we find the s.c. current flowing from the power station and generator 2. The equivalent EMF in Fig. 6.47(d) is

$$E'_2=\frac{1.16\cdot 0.6+1.21\cdot 1.68}{1.68+0.6}=1.2.$$

The transfer reactance in Fig. 6.47(e) is

$$X^{(n)}_{trs2}=1.07+0.2+\frac{1.07\cdot 0.2}{1.05}=1.47,$$

which as referred to the rated power will be

$$X^{(n)}_{trs2}=1.47\frac{360}{300}=1.76.$$

Hence, the transient current flowing from the rest of the network will be

$$I'_{02}=\frac{1.2}{1.76}=0.674.$$

To determine in which regime the AVR is operated, we shall calculate the terminal voltage of the equivalent generator in Fig. 4.67(d):

$$V_{ter2}=X_{F2}I'_{02}=1.55\cdot 0.674=1.05>1,$$

where $X_{F2}=X_{trs2}-X'_d\cong 1.76-0.21=1.55.$

Since $V_{ter2}>1$, the generators of the power station and G_2 operate under constant voltage and, therefore, the s.c. current flowing from the rest of the network is almost constant. Its ampere value is

$$I_{sc2}=0.674\frac{360}{\sqrt{3}\cdot 11}=12.7\ \text{kA}.$$

Thus, the total s.c. currents are

$$I_{sc2}(0.1)=12.3+12.7=25.0\,\text{kA},\quad I_{sc2}(1.0)=10.7+12.7=23.4\,\text{kA}.$$

6.6.1 Simple computation of short-circuit currents

The simplest calculation of short-circuit transients is based on the assumption that the fault circuit is connected to a system of infinite power. In this case the inner impedance of such a system is taken as zero and its voltage is unity. The change of the short-circuit current in this case is only due to the aperiodic component and can be approximated by using the peak coefficient. The periodic component of the short-circuit current, therefore, may be found with just the total reactance between the fault point and the system on the equivalent circuit

$$I_{sc} = \frac{1}{X_{tot}}. \tag{6.105a}$$

The elements of the equivalent circuit are usually transformers, cables and/or transmission lines. The short-circuit currents in such a calculation become a little bit larger than in reality. However, because of its simplicity, this way of calculating is widely used for a quick estimation of the s.c. currents and the results might be appropriate for solving some of the practical problems. This method is also used when the system configuration and its parameters are unknown.

Up to this point in our transient analysis, power circuits, which have been under consideration, consisted primarily of pure reactances, i.e., their very small resistances have been neglected. It can be shown that if $R \leq (1/3)X$, then neglecting such resistances results in increasing the periodic component of the s.c. current only at a rate of less than 5%, which anyway is within the accuracy of engineering calculations.

However, in the distribution networks the value of the resistances might be much higher. In such cases the resistances should be taken into consideration:

1) by the correction of the time constant of the aperiodic component:

$$T_a = \frac{X}{\omega R},$$

and respectively of the peak coefficient

$$k_{pk} = \sqrt{2}(1 - e^{-0.01/T_a}),$$

2) if the ratio of $R/X \geq 1/3$, by the replacement of X_{tot} with $Z_{tot} = \sqrt{R_{tot}^2 + X_{tot}^2}$ in the formula

$$I_{sc} = \frac{1}{Z_{tot}}. \tag{6.105b}$$

Finally, the influence of the load, such as big motors and high power composed loads, can be considered by equivalent parameters

$$X''_{Ld} = 0.35, \quad E''_{Ld} = 0.8. \tag{6.106}$$

A very rough approximation of the initial value of a subtransient s.c. can be made by

$$I''_{sc} = \frac{V_{F0}}{X''_{tot}}, \tag{6.107}$$

where V_{F0} is the voltage prior to switching at the fault point F, and (if the generator and/or system are represented by their subtransient reactances) X''_{tot} is the total subtransient reactance of the circuit up to the fault point.

6.6.2 Short-circuit power

The product of the initial subtransient s.c. current I''_{sc} and the rated voltage with the factor $\sqrt{3}$ gives the short-circuit power:

$$S''_{sc} = \sqrt{3}V_r I''_{sc}. \tag{6.108}$$

This power is used for characterizing the rate of the fault disturbance, which includes both the s.c. current and the voltage at the fault point. The s.c. power is primarily used for determining the breaking capacity, which is given in MVA, and is included in the information which manufacturers of circuit breakers are required to provide.

Sometimes the short-circuit power is given for the s.c. current at the switching instant, i.e., at the moment that the circuit breaker opens its contacts, rather than at $t = 0$, and which is called the breaking current.

Example 6.13

In the network shown in Fig. 6.48(a), find the peak and r.m.s value of the s.c. current when the three-pole fault occurs at points F_1 and F_2.

Solution

Assuming $S_B = 100$ MVA, the p.u. reactances will be as shown in Fig. 6.48(b):

$$X_1 = 0.4\cdot 140\frac{100}{161^2} = 0.22, \quad X_2 = 0.105\frac{100}{50} = 0.21, \quad X_3 = 0.04\frac{5.24}{0.5} = 0.42,$$

where

$$I_B = \frac{100}{\sqrt{3}\cdot 11} = 5.24\ \text{kA} \quad \text{and} \quad X_4 = 0.08\cdot 2\frac{100}{11^2} = 1.45.$$

The s.c current at point F_1 will be

$$I_{sc1} = \frac{1}{(0.22 + 0.21)} = 2.32, \quad \text{or} \quad I_{sc1} = 2.32\cdot 5.24 = 12.2\ \text{kA}.$$

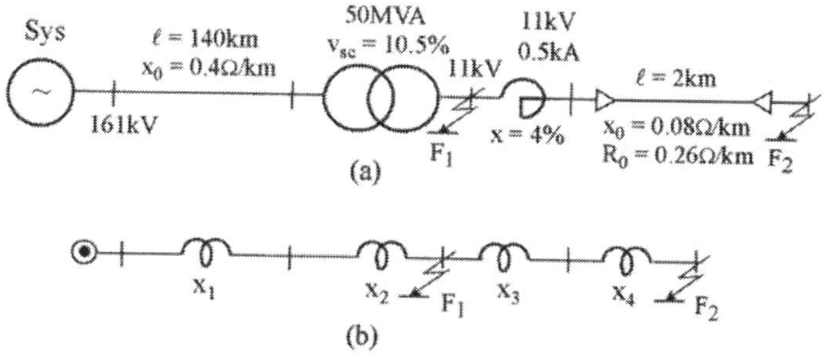

Figure 6.48 A circuit diagram for Example 6.13 (a) and its equivalent circuit (b).

Assuming $T_a = 0.05$ s ($k_{pk} = 1.8$), we have (see equation 6.103)

$$i_{pk} = 1.8 \cdot \sqrt{2} \cdot 12.2 = 31.1 \text{ kA},$$

and r.m.s. value is

$$I_{pk} = \sqrt{1 + 2(1.8 - 1)^2} I_{sc} = 1.52 \cdot 12.2 = 18.5 \text{ kA}.$$

For the short-circuiting at point F_2 we have

$$X_{tot} = 0.43 + 0.42 + 1.45 = 2.23,$$

and

$$I_{sc2} = \frac{1}{2.33} = 0.429, \quad \text{or} \quad I_{sc2} = 0.429 \cdot 5.24 = 2.45 \text{ kA}.$$

The peak values with a 1.8 peak coefficient will be

$$i_{pk} = 1.8 \cdot \sqrt{2} \cdot 2.45 = 6.25 \text{ kA} \quad \text{and} \quad I_{pk} = 1.52 \cdot 2.45 = 3.72 \text{ kA}.$$

However, the resistance of the cable is relatively high:

$$R_{tot} = 0.260 \cdot 2 \frac{100}{11^2} = 0.43,$$

and by taking it into consideration we can calculate the s.c. current more precisely. Thus, the time constant of the aperiodic component will be

$$T_a = \frac{X_{tot}}{\omega R_{tot}} = \frac{2.33}{314 \cdot 0.43} = 0.02,$$

and

$$k_{pk} = 1 + e^{-0.01/0.02} = 1.6.$$

Hence,

$$i_{pk} = 1.6 \cdot \sqrt{2} \cdot 2.45 = 5.54 \text{ kA},$$

and

$$I_{pk} = \sqrt{1 + 2(1.6 - 1)^2} \cdot 2.45 = 3.21 \text{ kA}.$$

If we now consider that the transformer is connected straight to the system (by neglecting the transmission line), the s.c. current at point F_1 will increase by 35%, but at point F_2 only by 7%.

Example 6.14

The power network shown in Fig. 6.49a includes a generator, synchronous condenser (SC) and three compound loads. Taking into consideration SC and all the loads, (a) find the first moment s.c. current and its peak value and (b) calculate the short-circuit power.

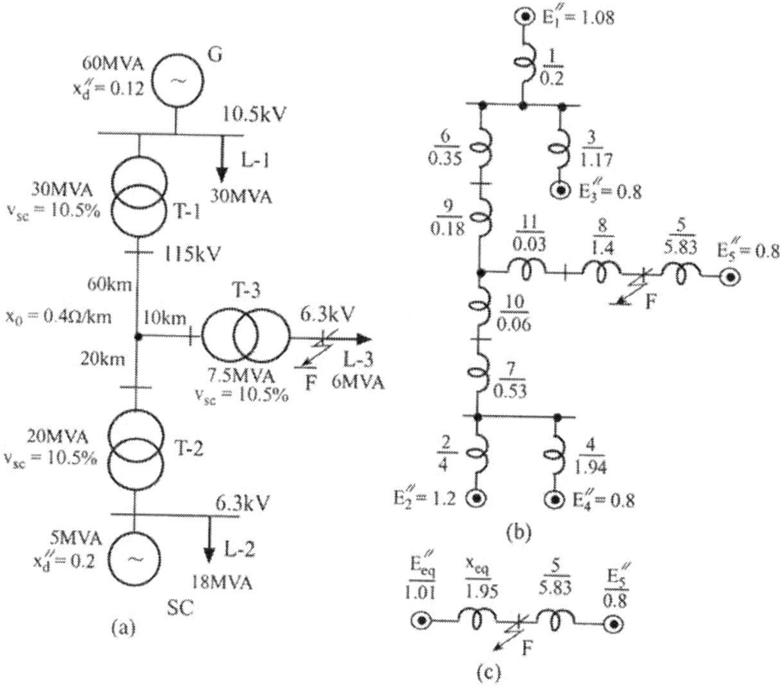

Figure 6.49 A network diagram for Example 6.14 (a), its equivalent circuit (b) and simplified circuit (c).

Solution

Assuming $S_B = 100$ MVA, the p.u. reactances of all the circuit elements are calculated and shown in Fig. 6.49(b). The loads are represented by $X''_{ld} = 0.35$ and $E''_{ld} = 0.8$. The given circuit is then simplified in a few obvious steps:

$$X_{12} = 0.2//1.17 = 0.18, \quad E_6 = \frac{1.08 \cdot 1.17 + 0.8 \cdot 0.2}{1.17 + 0.2} = 1.04,$$

$$X_{13} = 0.18 + 0.35 + 0.18 = 0.71, \quad X_{14} = 1.94//4 = 1.31,$$

$$E_7 = \frac{1.2 \cdot 1.94 + 0.8 \cdot 4}{1.94 + 4} = 0.93, \quad X_{15} = 1.31 + 0.53 + 0.06 = 1.9,$$

$$X_{16} = 1.9//0.71 = 0.52,$$

$$X_{eq} = 0.52 + 0.03 + 1.4 = 1.95, \quad E_{eq} = \frac{1.04 \cdot 1.9 + 0.93 \cdot 0.71}{1.9 + 0.71} = 1.01.$$

The s.c. current flowing from the system through transformer T_3 is

$$I''_S = \frac{1.01}{1.95} = 0.52 \quad \text{or in amperes} \quad I''_S = 0.52 \cdot 9.2 = 4.8 \text{ kA},$$

where $I_B = (100/\sqrt{3}\cdot 6.3) = 9.2$ kA is the basic current.

The s.c. current flowing from the load is

$$I''_{Ld} = \frac{0.8}{5.83} 9.2 = 1.26 \text{ kA}.$$

Thus, the total short-circuit current is

$$I''_{sc} = I''_S + I''_{Ld} = 4.8 + 1.26 = 6.06 \text{ kA}.$$

The total peak current will be

$$i_{pk} = 1.8\sqrt{2}\cdot 4.8 + \sqrt{2}\cdot 1.26 = 14.0 \text{ kA}.$$

Note that the s.c. current from the load can be calculated straightforwardly, without referring its parameters to the basic quantities. Indeed,

$$I''_{Ld} = \frac{0.8}{0.35} I_r = 2.29 \frac{6}{\sqrt{3}\cdot 6.3} = 1.26 \text{ kA},$$

which is the same as it was calculated previously.

(b) With the first moment s.c. current the short-circuit power at the load *L*-3 bus is

$$S''_{sc} = \sqrt{3} V_{L3} I''_{sc} = \sqrt{3}\cdot 6.3\cdot 6.06 = 66.1 \text{ MVA}.$$

6.7 METHOD OF SYMMETRICAL COMPONENTS FOR UNBALANCED FAULT ANALYSIS

Earlier we mentioned that truly balanced three-phase systems exist only in theory. Actually many real systems are very nearly balanced and for practical purposes can be analyzed as balanced systems, i.e., on per-phase basis. However, sometimes the degree of unbalance cannot be neglected. Such cases may occur during emergency conditions like unsymmetrical faults (one- or two-phase short-circuiting), unbalanced loads, open conductors, unsymmetrical operation of rotating machines, etc. Of course straightforward methods for the application of Kirchhoff's laws might be used for such three-phase circuit analysis. However, such cases may be calculated without difficulty by an indirect method in which the unbalanced or unsymmetrical system is replaced by equivalent component systems, each of which is symmetrical and balanced. The calculation of the currents and voltages in these symmetrical systems is a simple process (since it can be provided on a per-phase basis), and the superposition or vector addition of these currents and voltages is then easily carried out to obtain the actual results for the original unbalanced system.

This method, called the **method of symmetrical components**, was proposed by Charles L. Fortescue in 1913[(*)] and was developed by C.F. Wagner and R.D.

[(*)]This method was published by Fortescue, C.L. (1918), "Method of symmetrical co-ordinates applied to the solution of polyphase networks", *AIEE Transactions*, 37.

Evans later to apply it to the analysis of unsymmetrical faults in three-phase systems. Today, the symmetrical component method is widely used in studying unbalanced systems. Many electrical devices have been developed to operate on the basis of the concept of symmetrical components. In this section we shall briefly introduce this method followed by a few examples of its application.

6.7.1 Principle of symmetrical components

(a) *Positive-, negative- and zero-sequence systems*

Any unbalanced (unsymmetrical) three-phase system of phasors can be resolved into three balanced systems of phasors: (1) positive-sequence system, (2) negative-sequence system, and (3) zero-sequence system, as shown in Fig. 6.50, as an example of a set of three unbalanced voltages.

The **positive-sequence system** is represented by a balanced system of phasors

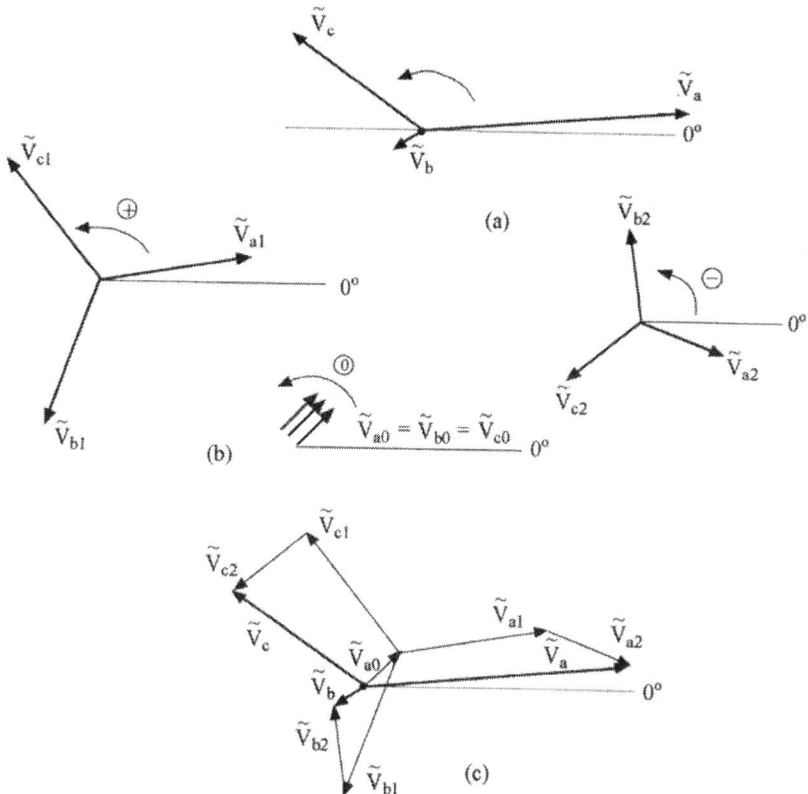

Figure 6.50 The symmetrical components of three unbalanced voltages: given system of unbalanced phasors (a); positive (+), negative (−) and zero (0) sequence components (b) and the graphical addition of the symmetrical components to obtain the given set of unbalanced phasors (c).

having the same phase sequence as the original unbalanced system. This set consists of three-phase currents and three-phase line-to-neutral voltages supplied by the power system generator and therefore of **positive** or counterclockwise **phase rotation**. Thus, the phasors of the positive-sequence system are equal in magnitude and displaced from each other by 120°, as shown by set "+" in Fig. 6.50(b).

The **negative-sequence system** is represented by a balanced system of phasors having the opposite phase sequence from the original system and, therefore, a **negative phase rotation**. The phasors of the negative-sequence system are also equal in magnitude and displaced from each other by 120°, set "−" in Fig. 6.50(b). Thus, if a positive sequence is *abc*, a negative sequence will be *acb*.

The **zero-sequence system** is represented by three single phasors that are equal in magnitude and are in phase, as shown by set "0" in Fig. 6.50(b). Note that the zero-sequence system is also a set of rotating phasors.

Using subscripts 0, 1 and 2 to denote the zero, positive and negative sequences we may write

$$\begin{aligned}\tilde{V}_a &= \tilde{V}_{a1} + \tilde{V}_{a2} + \tilde{V}_{a0}\\ \tilde{V}_b &= \tilde{V}_{b1} + \tilde{V}_{b2} + \tilde{V}_{b0}\\ \tilde{V}_c &= \tilde{V}_{c1} + \tilde{V}_{c2} + \tilde{V}_{c0},\end{aligned} \tag{6.109}$$

i.e., three voltage phasors $\tilde{V}_a$, $\tilde{V}_b$, $\tilde{V}_c$ of an unbalanced set can be expressed in terms of their symmetrical components as shown in Fig. 6.50(c).

With a **unit phasor operator** a ($a = \angle 120°$; $a^2 = \angle 240°$; $a^3 = 1$; $a^{-1} = \angle -120°$, etc.) the positive-sequence set can be designated

$$\tilde{V}_{a1} = V_1 \angle \psi_{a1}, \quad \tilde{V}_{b1} = a^2\tilde{V}_{a1}, \quad \tilde{V}_{c1} = a\tilde{V}_{a1} \tag{6.109a}$$

$$\tilde{V}_{a2} = V_2 \angle \psi_{a2}, \quad \tilde{V}_{b2} = a\tilde{V}_{a2}, \quad \tilde{V}_{c2} = a^2\tilde{V}_{a2} \tag{6.109b}$$

$$\tilde{V}_{a0} = V_0 \angle \psi_{a0}, \quad \tilde{V}_{b0} = \tilde{V}_{a0}, \quad \tilde{V}_{c0} = \tilde{V}_{a0}. \tag{6.109c}$$

Substituting the above equations into equation 6.109, the phase voltages can be expressed in terms of the *sequence voltages* as

$$\begin{aligned}\tilde{V}_a &= \tilde{V}_{a0} + \tilde{V}_{a1} + \tilde{V}_{a2}\\ \tilde{V}_b &= \tilde{V}_{a0} + a^2\tilde{V}_{a1} + a\tilde{V}_{a2}\\ \tilde{V}_c &= \tilde{V}_{a0} + a\tilde{V}_{a1} + a^2\tilde{V}_{a2},\end{aligned} \tag{6.110a}$$

and in matrix form as

$$\begin{bmatrix}\tilde{V}_a\\ \tilde{V}_b\\ \tilde{V}_c\end{bmatrix} = \begin{bmatrix}1 & 1 & 1\\ 1 & a^2 & a\\ 1 & a & a^2\end{bmatrix}\begin{bmatrix}\tilde{V}_{a0}\\ \tilde{V}_{a1}\\ \tilde{V}_{a2}\end{bmatrix} \tag{6.110b}$$

or

$$[V_{abc}] = [a][\tilde{V}_{012}], \tag{6.110c}$$

where

$$[\tilde{V}_{abc}] = \begin{bmatrix} \tilde{V}_a \\ \tilde{V}_b \\ \tilde{V}_c \end{bmatrix}, \quad [\tilde{V}_{012}] = \begin{bmatrix} \tilde{V}_{a0} \\ \tilde{V}_{a1} \\ \tilde{V}_{a2} \end{bmatrix}$$

and the *operator matrix* **a** is

$$\mathbf{a} = [a] = \begin{bmatrix} 1 & 1 & 1 \\ 1 & a^2 & a \\ 1 & a & a^2 \end{bmatrix}.$$

To shorten the writing of the symmetrical components, later on we will ignore the subscript "a" for phase a, which means that the symmetrical components $\tilde{V}_0$, $\tilde{V}_1$ and $\tilde{V}_2$ and $\tilde{I}_0$, $\tilde{I}_1$ and $\tilde{I}_2$ belong to the phase a voltages and currents.

Equations 6.110 are also known as the **synthesis equations** since they synthesize the set of unbalanced phasors from three sets of symmetrical components. These equations may be solved to find the symmetrical components of a known three-phase system of unbalanced voltages or currents:

$$[\tilde{V}_{012}] = [a]^{-1}[\tilde{V}_{abc}] \tag{6.111}$$

or

$$\begin{bmatrix} \tilde{V}_{a0} \\ \tilde{V}_{a1} \\ \tilde{V}_{a2} \end{bmatrix} = \frac{1}{\det[a]} \begin{bmatrix} a^4 - a^2 & a - a^2 & a - a^2 \\ a - a^2 & a^2 - 1 & 1 - a \\ a - a^2 & 1 - a & a^2 - 1 \end{bmatrix} \begin{bmatrix} \tilde{V}_a \\ \tilde{V}_b \\ \tilde{V}_c \end{bmatrix}. \tag{6.112}$$

By performing the appropriate computations with the phasor operators in equation 6.112, as with complex numbers, we may simplify the inverse of matrix $[a]$ as follows

$$[a]^{-1} = \frac{1}{3\sqrt{3}j} \begin{bmatrix} \sqrt{3}j & \sqrt{3}j & \sqrt{3}j \\ \sqrt{3}j & \sqrt{3}\angle -150^\circ & \sqrt{3}\angle -30^\circ \\ \sqrt{3}j & \sqrt{3}\angle -30^\circ & \sqrt{3}\angle -150^\circ \end{bmatrix} = \frac{1}{3} \begin{bmatrix} 1 & 1 & 1 \\ 1 & a & a^2 \\ 1 & a^2 & a \end{bmatrix}$$

where $\det[a] = 3(a - a^2) = 3\sqrt{3}j$. Therefore,

$$\begin{bmatrix} \tilde{V}_{a0} \\ \tilde{V}_{a1} \\ \tilde{V}_{a2} \end{bmatrix} = \frac{1}{3} \begin{bmatrix} 1 & 1 & 1 \\ 1 & a & a^2 \\ 1 & a^2 & a \end{bmatrix} \begin{bmatrix} \tilde{V}_a \\ \tilde{V}_b \\ \tilde{V}_c \end{bmatrix}, \tag{6.113}$$

i.e. the sequence voltages can be expressed in terms of phase voltages as

$$\tilde{V}_{a0} = \frac{1}{3}(\tilde{V}_a + \tilde{V}_b + \tilde{V}_c) \tag{6.114a}$$

$$\tilde{V}_{a1} = \frac{1}{3}(\tilde{V}_a + a\tilde{V}_b + a^2\tilde{V}_c) \tag{6.114b}$$

$$\tilde{V}_{a2} = \frac{1}{3}(\tilde{V}_a + a^2\tilde{V}_b + a\tilde{V}_c). \tag{6.114c}$$

These equations are also known as the **analysis equations**.

Of course, the synthesis and analysis equations can also be used for current phasors

$$[\tilde{I}_{abc}] = [a][\tilde{I}_{012}] \tag{6.115a}$$

$$[\tilde{I}_{012}] = [a]^{-1}[\tilde{I}_{abc}]. \tag{6.115b}$$

Example 6.15

Determine the symmetrical components for the line voltages $\tilde{V}_a = 220\angle 0°$ V, $\tilde{V}_b = 200\angle -150°$ V, and $\tilde{V}_c = 180\angle 120°$ V, Fig. 6.51(a), and construct their phasor diagram.

Solution

In accordance with equation 6.114a we have

$$\tilde{V}_{a0} = \tilde{V}_{b0} = \tilde{V}_{c0} = \frac{1}{3}(\tilde{V}_a + \tilde{V}_b + \tilde{V}_c)$$

$$= \frac{1}{3}(220\angle 0° + 200\angle -150° + 180\angle 120°)$$

$$= -14.3 + j18.6 = 23.5\angle 127.7°\ \text{V}.$$

Applying equation 6.114b, the positive components are

$$\tilde{V}_{a1} = \frac{1}{3}(\tilde{V}_a + a\tilde{V}_b + a^2\tilde{V}_c)$$

$$= \frac{1}{3}[220\angle 0° + (1\angle 120°)(200\angle -150°) + (1\angle 240°)(180\angle 120°)]$$

$$= 191.1 - j33.3 = 194.0\angle -9.9°\ \text{V},$$

and

$$\tilde{V}_{b1} = a^2\tilde{V}_{a1} = (1\angle 240°)(194.0\angle -9.9°) = 194.0\angle -129.9°\ \text{V}$$

$$\tilde{V}_{c1} = a\tilde{V}_{a1} = (1\angle 120°)(194.0\angle -9.9°) = 194.0\angle 110.1°\ \text{V}.$$

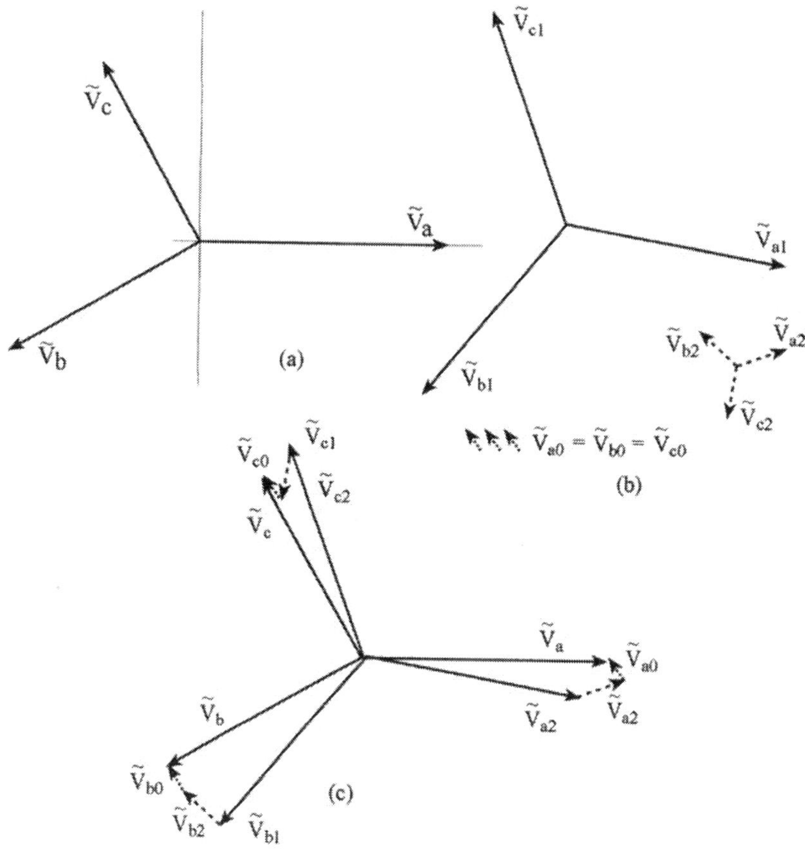

Figure 6.51 Three unbalanced voltages (a), their symmetrical components (b) and the original phasor system as composed of the symmetrical components (c).

Applying equation 6.114c, the negative components are

$$\tilde{V}_{a2} = \frac{1}{3}(\tilde{V}_a + a^2\tilde{V}_b + a\tilde{V}_c)$$

$$= \frac{1}{3}[220\angle 0^\circ + (1\angle 240^\circ)(200\angle -150^\circ) + (1\angle 120^\circ)(180\angle 120^\circ)]$$

$$= 43.3 + j14.8 = 45.8\angle 18.9^\circ \text{ V},$$

and

$$\tilde{V}_{b2} = a\tilde{V}_{a2} = (1\angle 120^\circ)(45.8\angle 18.9^\circ) = 45.8\angle 138.9^\circ \text{ V},$$

$$\tilde{V}_{c2} = a^2\tilde{V}_{a2} = (1\angle 240^\circ)(45.8\angle 18.9^\circ) = 45.8\angle -101.1^\circ \text{ V}.$$

Using the above results the phasor diagrams for positive and negative symmetrical components are constructed in Fig. 6.51(b). The diagram in Fig. 6.51(c) shows that the original phasor system is obtained when the symmetrical components are compounded either numerically or also graphically.

In the general case of an unsymmetrical three-phase, three-wire system, i.e. when the neutral line is absent, the vector sum of three line currents is also always (like sum of line voltages) zero. Therefore, the zero-sequence components for these unbalanced currents as well as for line voltages are zero. Furthermore, we may conclude that in a four-wire system, since the neutral-wire current in every case is the sum of line currents, the zero-sequence components, equation 6.114a, are equal to one-third of this current.

In the next example we shall show the resolving of an unbalanced set of phase voltages into symmetrical components.

Example 6.16

A synchronous generator, which is connected to an infinite busbar system Fig. 6.52(a), is subjected to a short-circuit line-to-line fault at its terminals. The generator's p.u. short-circuit currents are found (see further on) to be $\tilde{I}_a = -2.1$, $\tilde{I}_b = 3.37\angle -71.8°$ and $\tilde{I}_c = 3.37\angle 71.8°$. Find the symmetrical components of these currents and construct their phasor diagram.

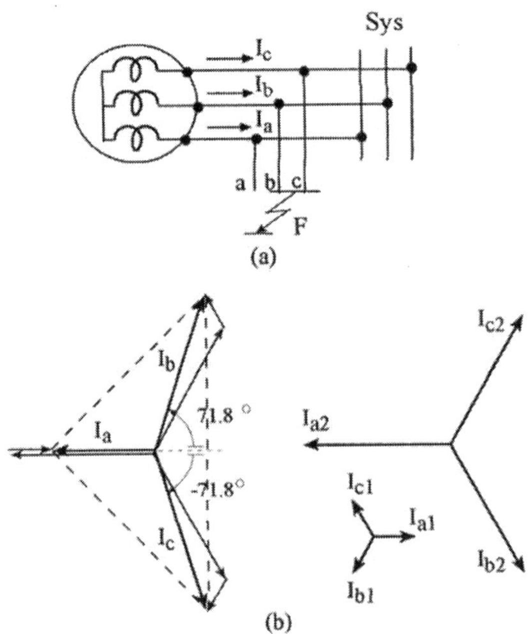

Figure 6.52 A circuit diagram for Example 6.15 (a) and the phasor diagram of the current sequences (b).

Solution

In accordance with equation 4.114a we obtain

$$\tilde{I}_{a0} = \tilde{I}_{b0} = \tilde{I}_{c0} = \frac{1}{3}(-2.1 + 3.37\angle -71.8^\circ + 3.37\angle 71.8^\circ) = 0.$$

This result should be expected since the sum of three-phase system currents (without a neutral line) is always zero (as the sum of three vectors, which form a triangle, see Fig. 6.52(b)).

In accordance with equation 4.114b we obtain the positive components:

$$\tilde{I}_{a1} = \frac{1}{3}[-2.1 + (1\angle 120^\circ)(3.37\angle -71.8^\circ) + (1\angle -120^\circ)(3.37\angle 71.8^\circ)] = 0.8$$

$$\tilde{I}_{b1} = a^2\tilde{I}_{a1} = 0.8\angle -120^\circ, \quad \tilde{I}_{c1} = a\tilde{I}_{a1} = 0.8\angle 120^\circ,$$

and in accordance with equation 4.114c, the negative components are

$$\tilde{I}_{a2} = \frac{1}{3}[-2.1 + (1\angle -120^\circ)(3.37\angle -71.8^\circ)$$

$$+ (1\angle 120^\circ)(3.37\angle 71.8^\circ)] = -2.9$$

$$\tilde{I}_{b2} = a\tilde{I}_{a2} = -2.9\angle 120^\circ, \quad \tilde{I}_{c2} = a^2\tilde{I}_{a2} = -2.9\angle -120^\circ.$$

In checking the results, we have $\tilde{I}_a = \tilde{I}_{a1} + \tilde{I}_{a2} = 0.8 - 2.9 = -2.1$. The phasor diagram of the currents is shown in Fig. 6.52(b).

If the set of line voltages is balanced, it is obvious that the negative-sequence for these voltages is zero; hence, the negative-sequence for the phase voltages will also be zero. That is, the set of unbalanced phase voltages, forming a balanced set of line voltages, resolves into positive- and zero-components, as shown, for example, in Fig. 6.53. As can be seen the negative-sequence voltage, $\tilde{V}_2$, is absent and each of the phase voltages is equal to the sum of the positive- and zero-sequences. Their values can then be easily found:

$$V_{a,ph} = V_1 + V_0$$

$$V_{b,ph} = V_{c,ph} = \sqrt{(V/2)^2 - [V/(2\cdot\sqrt{3}) - V_0]^2},$$

where V is the given line voltage.

(b) *Sequence impedances*

Consider first the circuit of Fig. 6.54(a), which represents a three-phase, three-wire, Y-connected, generally unbalanced system, i.e., $Z_a \neq Z_b \neq Z_c$. The matrix equation for phase voltages, across these three impedances, will be

$$\begin{bmatrix} \tilde{V}_a \\ \tilde{V}_b \\ \tilde{V}_c \end{bmatrix} = \begin{bmatrix} Z_a & 0 & 0 \\ 0 & Z_b & 0 \\ 0 & 0 & Z_c \end{bmatrix} \begin{bmatrix} \tilde{I}_a \\ \tilde{I}_b \\ \tilde{I}_c \end{bmatrix} \tag{6.116a}$$

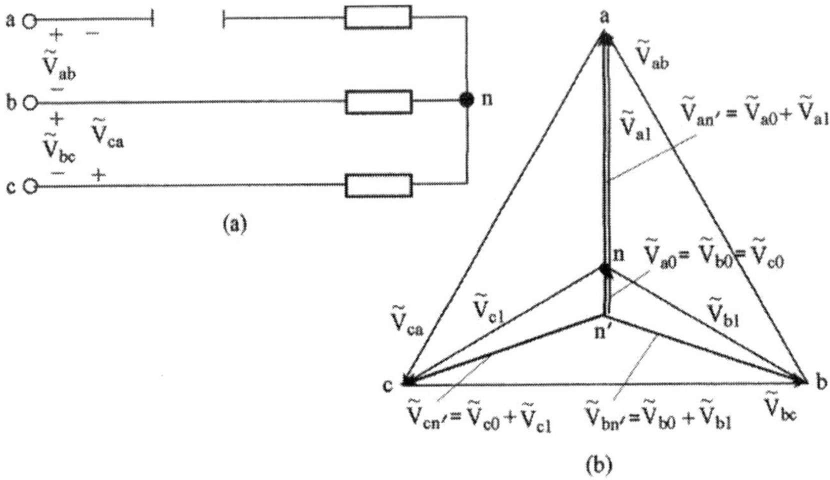

Figure 6.53 A faulted network with balanced line voltages but unbalanced phase voltages (a) and the phasor diagram of the symmetrical components (b).

or

$$[\tilde{V}_{abc}] = [Z_{abc}][\tilde{I}_{abc}]. \tag{6.116b}$$

Here both the voltages and currents are unsymmetrical. Multiplying both sides of equation 6.116b by $[a]^{-1}$ and also substituting equation 6.115a, we obtain

$$[a]^{-1}[\tilde{V}_{abc}] = [a]^{-1}[Z_{abc}][a][\tilde{I}_0 \tilde{I}_1 \tilde{I}_2],$$

or, with equation 6.111,

$$[\tilde{V}_{012}] = [Z_{012}][\tilde{I}_{012}], \tag{6.117}$$

where the *matrix transformation* is defined as

$$[Z_{012}] = [a]^{-1}[Z_{abc}][a]. \tag{6.118a}$$

Performing the matrix multiplication and upon simplification this transformation results in a *sequence impedance matrix* of an unbalanced load

$$[Z_{012}] = \begin{bmatrix} Z_{00} & Z_{01} & Z_{02} \\ Z_{10} & Z_{11} & Z_{12} \\ Z_{20} & Z_{21} & Z_{22} \end{bmatrix} = \begin{bmatrix} Z_0 & Z_2 & Z_1 \\ Z_1 & Z_0 & Z_2 \\ Z_2 & Z_1 & Z_0 \end{bmatrix} \tag{6.118b}$$

where by definition the **zero-sequence impedance** is

$$Z_0 = \frac{1}{3}(Z_a + Z_b + Z_c) \quad (Z_{00} = Z_{11} = Z_{22} = Z_0), \tag{6.119a}$$

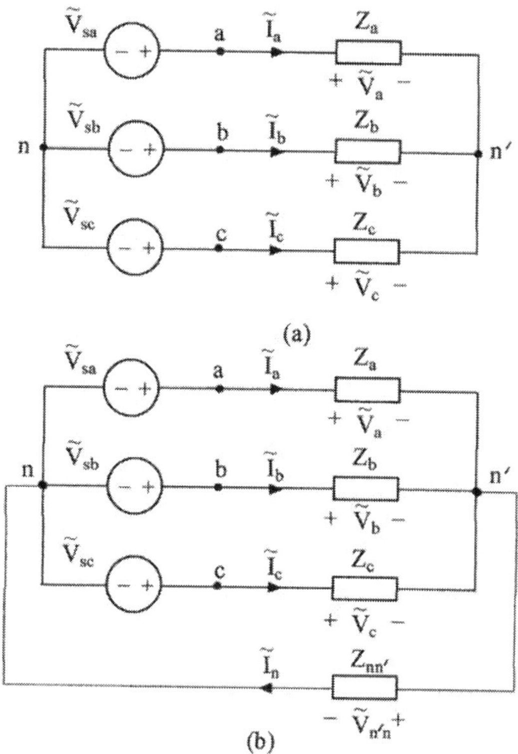

Figure 6.54 Three-phase, three-wire unbalanced system ($\tilde{I}_n = 0$) (a), three-phase, four-wire unbalanced system ($\tilde{I}_n \neq 0$) (b).

the **positive-sequence impedance** is

$$Z_1 = \frac{1}{3}(Z_a + aZ_b + a^2 Z_c) \quad (Z_{02} = Z_{10} = Z_{21} = Z_1) \tag{6.119b}$$

and the **negative-sequence impedance** is

$$Z_2 = \frac{1}{3}(Z_a + a^2 Z_b + aZ_c) \quad (Z_{01} = Z_{12} = Z_{20} = Z_2). \tag{6.119c}$$

These component impedances have little physical significance but they are useful in a general mathematical formulation of symmetrical-component theory. It should be noted in this respect that the real parts of the component impedances may possess negative signs even though the real parts of Z_a, Z_b and Z_c are all positive.

Providing the matrix multiplication in equation 6.117 yields

$$\begin{aligned}\tilde{V}_0 &= Z_0\tilde{I}_0 + Z_2\tilde{I}_1 + Z_1\tilde{I}_2\\ \tilde{V}_1 &= Z_1\tilde{I}_0 + Z_0\tilde{I}_1 + Z_2\tilde{I}_2\\ \tilde{V}_2 &= Z_2\tilde{I}_0 + Z_1\tilde{I}_1 + Z_0\tilde{I}_2.\end{aligned} \tag{6.120}$$

Note that the sum of the sequence indexes (0, 1, or 2) of Z and $\tilde{I}$ in the voltage drops $Z_i\tilde{I}_j$ in each of these equations gives the index of the voltage-sequence to which these voltage drops belong. Therewith, $2+1=3$ is considered as 0 $(3-3=0)$ and $2+2=4$ is considered as 1 $(4-3=1)$, since there are only three sequences (0, 1 and 2). This simple rule is known as the **sequence rule**.

Recall that the above symmetrical components (equation 6.120) are of the a-phase voltage, i.e.,

$$\tilde{V}_a = \tilde{V}_0 + \tilde{V}_1 + \tilde{V}_2.$$

Applying, for example, the second equation of equation 6.120 to phase b and making appropriate substitutions we may write

$$\begin{aligned}\tilde{V}_b &= Z_{b1}\tilde{I}_{b0} + Z_{b0}\tilde{I}_{b1} + Z_{b2}\tilde{I}_{b2}\\ &= (Z_{a1}\angle -120^\circ)\tilde{I}_{a0} + Z_{a0}(I_{a1}\angle -120^\circ) + (Z_{a2}\angle 120^\circ)(I_{a2}\angle 120^\circ)\\ &= (Z_1\tilde{I}_0 + Z_0\tilde{I}_1 + Z_2\tilde{I}_2)\angle -120^\circ = \tilde{V}_{a1}\angle -120^\circ.\end{aligned}$$

This result shows that $\tilde{V}_{b1}$ is equal in magnitude to $\tilde{V}_{a1}$ and 120° displaced behind $\tilde{V}_{a1}$, as, of course, it should be for a positive-sequence system. An opportunity is given to the reader to check in the above manner that

$$\tilde{V}_{c1} = \tilde{V}_{a1}\angle 120^\circ \quad \text{and} \quad \tilde{V}_{c0} = \tilde{V}_{b0} = \tilde{V}_{a0}.$$

The sequence currents can be found by solving equation 6.117, i.e.,

$$[\tilde{I}_{012}] = [Y_{012}][\tilde{V}_{012}], \tag{6.121}$$

where $[Y_{012}]$ is the associated *sequence admittance matrix*

$$[Y_{012}] = [Z_{012}]^{-1}.$$

This sequence admittance matrix may be found in the same manner as the impedance sequence matrix (equation 6.118a), i.e.,

$$[Y_{012}] = [a^{-1}][Y_{abc}][a]. \tag{6.122}$$

Indeed, applying the reversal rule to find the inverse of the product of the matrixes, we obtain

$$[Z_{012}]^{-1} = ([a]^{-1}[Z_{abc}][a])^{-1} = [a]^{-1}[Z_{abc}]^{-1}[a],$$

where

$$[Z_{abc}]^{-1} = [Y_{abc}] = \begin{bmatrix} Y_a & 0 & 0\\ 0 & Y_b & 0\\ 0 & 0 & Y_c\end{bmatrix}$$

and $Y_a = 1/Z_a$, $Y_b = 1/Z_b$, $Y_c = 1/Z_c$.

Therefore, similar to equation 6.119 we observe that:

the **zero-sequence admittance** is

$$Y_0 = \frac{1}{3}(Y_a + Y_b + Y_c), \tag{6.123a}$$

the **positive-sequence admittance** is

$$Y_1 = \frac{1}{3}(Y_a + aY_b + a^2 Y_c), \tag{6.123b}$$

and the **negative-sequence admittance** is

$$Y_2 = \frac{1}{3}(Y_a + a^2 Y_b + aY_c). \tag{6.123c}$$

When the applied voltage sequence-components are known, the sequence-components of the a-phase current may be readily found according to equation 6.121. Thus,

$$\begin{aligned} \tilde{I}_0 &= Y_0 \tilde{V}_0 + Y_2 \tilde{V}_1 + Y_1 \tilde{V}_2 \\ \tilde{I}_1 &= Y_1 \tilde{V}_0 + Y_0 \tilde{V}_1 + Y_2 \tilde{V}_2 \\ \tilde{I}_2 &= Y_2 \tilde{V}_0 + Y_1 \tilde{V}_1 + Y_0 \tilde{V}_2. \end{aligned} \tag{6.124}$$

Example 6.17

Let the line-to-line voltages and the phase impedances of the Y-connected, three-wire, load, as shown in Fig. 6.54(a), be as follows:

$$\tilde{V}_{ab} = 200\angle 0^\circ \,\text{V}, \quad \tilde{V}_{bc} = 141.4\angle -135^\circ \,\text{V}, \quad \tilde{V}_{ca} = 141.4\angle 135^\circ \,\text{V}$$

$$Z_a = 6\,\Omega, \quad Z_b = 6\angle -30^\circ \,\Omega, \quad Z_c = j12\,\Omega.$$

Find the a-phase current symmetrical components.

Solution

The phase admittances are

$$Y_a = 0.1667\,\text{S}, \quad Y_b = 0.1667\angle 30^\circ \,\text{S}, \quad Y_c = -j0.08333\,\text{S}.$$

Then, employing equation 6.123, the sequence-component admittances are

$$Y_0 = \frac{1}{3}(0.1667 + 0.1443 + j0.08333 - j0.08333) = 0.1037 \text{ S}$$

$$Y_1 = \frac{1}{3}[0.1667 + (1\angle 120^\circ)(0.1667\angle 30^\circ) + (1\angle 240^\circ)(0.08333\angle -90^\circ)]$$

$$= 0.04485\angle 111.71^\circ \text{ S}$$

$$Y_2 = \frac{1}{3}[0.1667 + (1\angle 240^\circ)(0.1667\angle 30^\circ) + (1\angle 120^\circ)(0.08333\angle -90^\circ)]$$

$$= 0.08987\angle -27.63 \text{ S.}$$

Resolving the above line-to-line voltages into symmetrical components yields

$$\tilde{V}_{ab0} = \frac{1}{3}[200\angle 0^\circ + 141.4\angle -135^\circ + 141.4\angle 135^\circ] = 0$$

$$\tilde{V}_{ab1} = \frac{1}{3}[200\angle 0^\circ + 141.4(\angle -135^\circ + \angle 120^\circ) + 141.4(\angle 135^\circ + \angle 240^\circ)]$$

$$= 157.7 \text{ V}$$

$$\tilde{V}_{ab2} = \frac{1}{3}[200\angle 0^\circ + 141.4(\angle -135^\circ + \angle 240^\circ) + 141.4(\angle 135^\circ + \angle 120^\circ)]$$

$$= 42.3 \text{ V.}$$

The positive- and negative-components of the phase voltages are

$$\tilde{V}_{a1} = \tilde{V}_1 = \frac{157.7}{\sqrt{3}}\angle -30^\circ = 91.1\angle -30^\circ \text{ V}$$

$$\tilde{V}_{a2} = \tilde{V}_2 = \frac{42.3}{\sqrt{3}}\angle 30^\circ = 24.4\angle 30^\circ \text{ V.}$$

Note that even if $\tilde{I}_0 = 0$ (since the neutral wire is absent) $\tilde{V}_0$ will possess a finite value, which may be calculated in accordance with the first equation of 6.124:

$$\tilde{V}_0 = \frac{1}{Y_0}(-Y_2\tilde{V}_1 - Y_1\tilde{V}_2)$$

$$= \frac{1}{0.1037}[-(0.08987\angle -27.63^\circ)(91.1\angle -30^\circ)$$

$$-(0.04485\angle 111.71^\circ)(24.4\angle 30^\circ)] = 69.08\angle 119.47^\circ \text{ V.}$$

Now, the positive- and negative-sequence currents may be calculated in accordance with equation 6.124

$$\tilde{I}_1 = (0.4485\angle 111.71^\circ)(69.08\angle 119.47^\circ) + 0.1037(91.1\angle -30^\circ)$$

$$+(0.08987\angle -27.63^\circ)(24.4\angle 30^\circ) = 8.42 - j7.04 = 10.98\angle -39.9^\circ \text{ A}$$

$$\tilde{I}_2 = (0.08987\angle -37.63^\circ)(69.08\angle 119.47^\circ) + (0.04485\angle 111.71^\circ)(91.1\angle -30^\circ)$$
$$+0.1037(24.4\angle 30^\circ) = 2.58 - j11.51 = 11.80\angle 77.36^\circ \text{ A}.$$

Note that for a balanced load, i.e., $Z_a = Z_b = Z_c = Z_L$, the positive- and negative-sequence impedances are zero and $Z_0 = Z_L$. Thus,

$$\begin{bmatrix} \tilde{V}_0 \\ \tilde{V}_1 \\ \tilde{V}_2 \end{bmatrix} = \begin{bmatrix} Z_L & 0 & 0 \\ 0 & Z_L & 0 \\ 0 & 0 & Z_L \end{bmatrix} \begin{bmatrix} \tilde{I}_0 \\ \tilde{I}_1 \\ \tilde{I}_2 \end{bmatrix}. \tag{6.125a}$$

This matrix equation indicates that there is no mutual coupling among the three sequences and it can be separated into three independent equations

$$\tilde{V}_0 = Z_L \tilde{I}_0, \quad \tilde{V}_1 = Z_L \tilde{I}_1, \quad \tilde{V}_2 = Z_L \tilde{I}_2. \tag{6.125b}$$

Consider next the circuit of Fig. 6.54(b), in which, for more generality, a neutral wire is represented by the impedance $Z_{nn'}$. In this case the matrix equation for phase voltages (equation 6.116a) shall be written as

$$\begin{bmatrix} \tilde{V}_a \\ \tilde{V}_b \\ \tilde{V}_c \end{bmatrix} = \begin{bmatrix} Z_a & 0 & 0 \\ 0 & Z_b & 0 \\ 0 & 0 & Z_c \end{bmatrix} \begin{bmatrix} \tilde{I}_a \\ \tilde{I}_b \\ \tilde{I}_c \end{bmatrix} + \begin{bmatrix} Z_{nn'} \\ Z_{nn'} \\ Z_{nn'} \end{bmatrix} [\tilde{I}_n].$$

Substituting equations 6.110c and 6.115a into this equation, and since $\tilde{I}_n = 3\tilde{I}_0$, we obtain

$$[a][\tilde{V}_{012}] = [Z_{abc}][a][\tilde{I}_{012}] + [Z_{nn'}][3\tilde{I}_0].$$

Performing the matrix multiplication and upon simplification this equation becomes

$$[a]\begin{bmatrix} \tilde{V}_0 \\ \tilde{V}_1 \\ \tilde{V}_2 \end{bmatrix} = \begin{bmatrix} Z_a + 3Z_{nn'} & Z_a & Z_a \\ Z_b + 3Z_{nn'} & a^2 Z_b & aZ_b \\ Z_c + 3Z_{nn'} & aZ_c & a^2 Z_c \end{bmatrix} \begin{bmatrix} \tilde{I}_0 \\ \tilde{I}_1 \\ \tilde{I}_2 \end{bmatrix}.$$

Multiplying both sides of this equation by $[a]^{-1}$ yields

$$[\tilde{V}_{012}] = [Z'_{012}][\tilde{I}_0 \tilde{I}_1 \tilde{I}_2],$$

where the sequence-impedance matrix can be expressed as

$$[Z'_{012}] = \begin{bmatrix} Z_0 + 3Z_{nn'} & Z_2 & Z_1 \\ Z_1 & Z_0 & Z_2 \\ Z_2 & Z_1 & Z_0 \end{bmatrix} \tag{6.126a}$$

and

$$Z_{00} = Z_0 + 3Z_{nn'}. \tag{6.126b}$$

Note again that for a balanced load, i.e., $Z_a = Z_b = Z_c = Z_L$, the positive- and negative-sequence impedances are zero, $Z_1 = Z_2 = 0$, and $Z_0 = Z_L$. Thus,

$$\begin{bmatrix} \tilde{V}_0 \\ \tilde{V}_1 \\ \tilde{V}_2 \end{bmatrix} = \begin{bmatrix} Z_L = 3Z_{nn'} & 0 & 0 \\ 0 & Z_L & 0 \\ 0 & 0 & Z_L \end{bmatrix} \begin{bmatrix} \tilde{I}_0 \\ \tilde{I}_1 \\ \tilde{I}_2 \end{bmatrix}. \tag{6.127a}$$

This matrix equation may also be separated into three independent equations

$$\tilde{V}_0 = (Z_L + 3Z_{nn'})\tilde{I}_0, \quad \tilde{V}_1 = Z_L\tilde{I}_1, \quad \tilde{V}_2 = Z_L\tilde{I}_2. \tag{6.127b}$$

In accordance with these equations three sequence networks of the balanced load may be drawn, as shown in Fig 6.55. Therefore, the positive- and negative-network impedances for a balanced load are equal to each other and simply equal the load phase impedance, but the zero-sequence network includes, in addition, the triplicate neutral line impedance. It is important to mention that with $Z_{nn'} = \infty$, i.e., for a three-wire system, the zero-sequence current $\tilde{I}_0$ is zero ($\tilde{I}_0 = 0$). It follows from the first equation in 6.127b.

It is worthwhile to note that the **impedances of sequence networks**, Fig 6.55, are not the same as the **sequence impedances** (equation 6.119) in equation 6.117. To make this sentence clearer we shall consider the sequence networks' impedances as the ratio of the voltage and the current of the same sequence, which

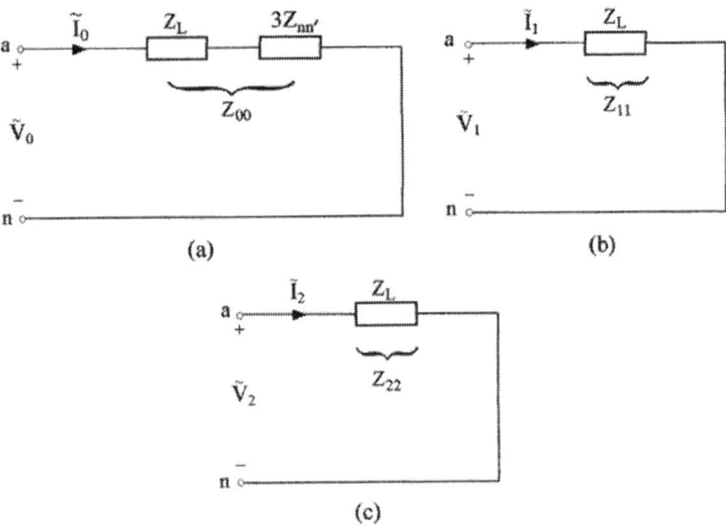

Figure 6.55 *Sequence networks*: zero-sequence network (a), positive sequence network (b) and negative-sequence network (c).

also are called the **impedances to positive-, negative-**, and **zero-sequence currents**. Thus, if a balanced system of positive-sequence voltages is applied to a balanced three-phase network, then the currents are also balanced and of positive-sequence. So, the ratio of the positive-sequence phase voltages and the appropriate positive-sequence phase currents gives the *positive-sequence network impedance*

$$Z_{11}=\frac{\tilde{V}_{a1}}{\tilde{I}_{a1}}=\frac{\tilde{V}_{b1}}{\tilde{I}_{b1}}=\frac{\tilde{V}_{c1}}{\tilde{I}_{c1}}. \tag{6.128a}$$

In a similar manner we shall define the *negative-* and *zero-sequence network impedances*

$$Z_{22}=\frac{\tilde{V}_{a2}}{\tilde{I}_{a2}}=\frac{\tilde{V}_{b2}}{\tilde{I}_{b2}}=\frac{\tilde{V}_{c2}}{\tilde{I}_{c2}}, \tag{6.128b}$$

and, since $\tilde{V}_{a0}=\tilde{V}_{b0}=\tilde{V}_{c0}=\tilde{V}_0$ and $\tilde{I}_{a0}=\tilde{I}_{b0}=\tilde{I}_{c0}=\tilde{I}_0$.

$$Z_{00}=\frac{\tilde{V}_0}{\tilde{I}_0}. \tag{6.128c}$$

In other words, positive-sequence currents flowing in a balanced network produce **only** positive-sequence voltage drops, negative-sequence currents will produce **only** negative-sequence voltage drops and zero-sequence currents will produce only zero-sequence voltage drops, as follows from equations 6.125a and 6.127a.

Therefore, as has already been mentioned, for a balanced load the three-sequence networks, Fig 6.55, can be separated and treated independently. It is important to recall at this point that the zero-sequence system is not a three-phase system but a single-phase system, i.e. the zero-sequence currents and voltages are equal in magnitude and are in phase at any point in all the phases of the system. Thus, the zero-sequence currents can only exist in a circuit if there is a complete path for their flow.

Figure 6.56 shows zero-sequence networks for Y- and Δ-connected three-phase loads. As can be seen from Fig 6.56(a), in a Y-connected load with an open neutral wire, there is no return path to zero-sequence currents, hence the zero-sequence impedance is infinite (in a zero-sequence network drawing this infinite impedance is indicated by an open circuit). In the circuit of Fig 6.56(b) the fourth wire, connecting the neutrals, provides a return path for the zero-sequence currents, so that their sum, $3\tilde{I}_0$, flows through this wire. If the neutral wire impedance is $Z_{nn'}$, the zero-sequence voltage drop of $3Z_{nn'}\tilde{I}_0$ will be produced, across this impedance, by a triple zero-sequence current $3\tilde{I}_0$. For this reason an impedance of $3Z_{nn'}$ should be inserted in the zero-sequence network, as shown in Fig 6.56(b). This result is in full agreement with those achieved previously (equation 6.126b) by the mathematical treatment. Note that in the particular case of $Z_{nn'}=0$ (solidly grounded neutrals) no potential difference exists between neutral points n-n', so they should be short-circuited. A

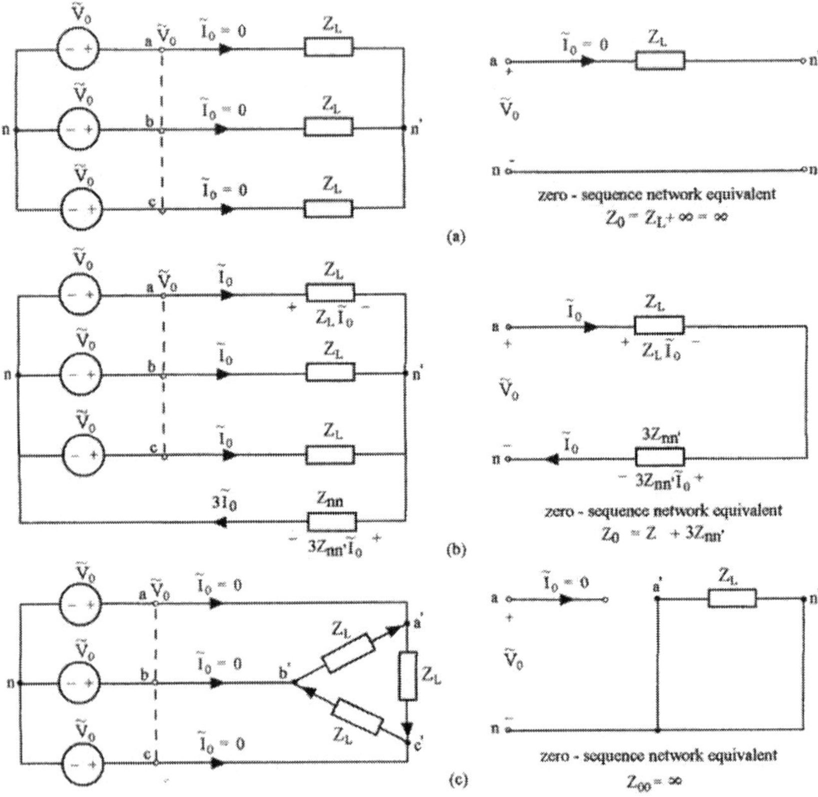

Figure 6.56 Zero-sequence network equivalent for a Y-connected three-wire load (a), Y-connected four-wire load (b) and Δ-connected load (c).

Δ-connected load, as shown in Fig 6.56(c), provides no path for zero-sequence currents flowing in the line wires. Therefore, the zero-sequence impedance, as seen from the source terminals, is infinite (open circuit). However it is possible to have zero-sequence currents circulating within the delta circuit, if zero-sequence voltages are applied independently, or by induction, in the delta circuit.

Consider, for example, the network shown in Fig. 6.57(a), in which the single-pole short-circuit to earth occurs on the transmission line between transformers T_1 and T_2. The arrows in each of the generator and transformer windings show the circulation paths of the zero-sequence currents. In accordance with these possible zero-sequence currents flowing, the zero-sequence equivalent circuit is formed and shown in Fig. 6.57(b).

Consider next, a more general case representing a circuit with unequal mutual impedances (e.g. transmission lines, transformers and tree-core cables). We also assume that there is mutual coupling between the phase branches and the neutral

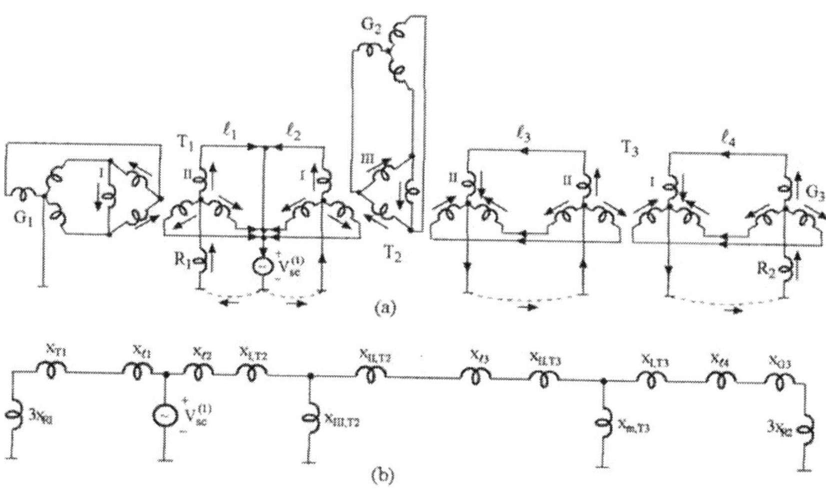

Figure 6.57 A network with a single-pole short-circuit to earth (a) and its zero-sequence equivalent (b).

line (e.g., as in transmission lines with overhead ground wire), as shown in Fig 6.58.

The KVL equation for phase a may be written as

$$\tilde{V}_a = Z_{aa}\tilde{I}_a + Z_{ab}\tilde{I}_b + Z_{ac}\tilde{I}_c - Z_{an}\tilde{I}_n + \tilde{V}_{nn'},$$

where $\tilde{V}_{nn'} = -Z_{na}\tilde{I}_a - Z_{nb}\tilde{I}_b - Z_{nc}\tilde{I}_c + Z_{nn'}\tilde{I}_n$ and, as shown earlier, $\tilde{I}_n = 3\tilde{I}_0$.

For three phases, in matrix form, these equations can be expressed as

$$\begin{bmatrix} \tilde{V}_a \\ \tilde{V}_b \\ \tilde{V}_c \end{bmatrix} = \begin{bmatrix} Z_{aa} & Z_{ab} & Z_{ac} \\ Z_{ba} & Z_{bb} & Z_{bc} \\ Z_{ca} & Z_{cb} & Z_{cc} \end{bmatrix} \begin{bmatrix} \tilde{I}_a \\ \tilde{I}_b \\ \tilde{I}_c \end{bmatrix} - \begin{bmatrix} Z_{na} & Z_{nb} & Z_{nc} \\ Z_{na} & Z_{nb} & Z_{nc} \\ Z_{na} & Z_{nb} & Z_{nc} \end{bmatrix} \begin{bmatrix} \tilde{I}_a \\ \tilde{I}_b \\ \tilde{I}_c \end{bmatrix}$$

$$+ \begin{bmatrix} 3(Z_{nn'} - Z_{an}) & 0 & 0 \\ 3(Z_{nn'} - Z_{bn}) & 0 & 0 \\ 3(Z_{nn'} - Z_{cn}) & 0 & 0 \end{bmatrix} \begin{bmatrix} \tilde{I}_0 \\ \tilde{I}_1 \\ \tilde{I}_2 \end{bmatrix} \qquad (129a)$$

where Z_{aa}, Z_{bb}, Z_{cc} and $Z_{nn'}$ are the phase and neutral self-impedances, and $Z_{ab} = Z_{ba}$, $Z_{bc} = Z_{cb}$, $Z_{ca} = Z_{ac}$ and $Z_{an} = Z_{na}$, $Z_{bn} = Z_{nb}$, $Z_{cn} = Z_{nc}$ are the *phase-phase* and *phase-neutral mutual impedances*. In reduced notation equation 6.129a may be written as

$$[\tilde{V}_{abc}] = [Z_{abc}][\tilde{I}_{abc}] - [Z_{nabc}][\tilde{I}_{abc}] + 3[Z_{nn'}][\tilde{I}_{012}]. \qquad (6.129b)$$

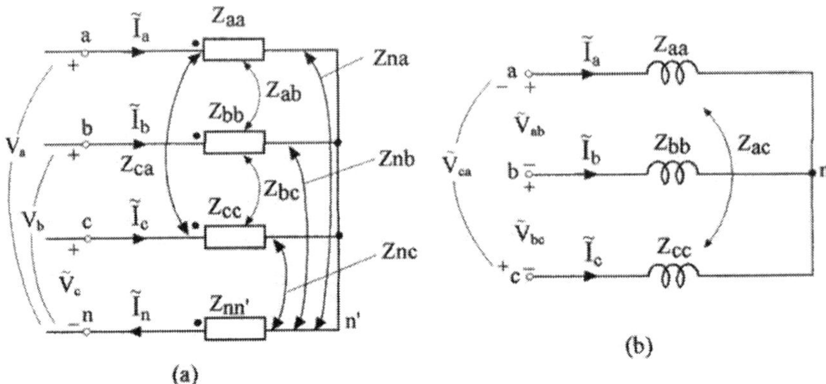

Figure 6.58 Mutually coupled three-phase network (a) and network for Example 6.18 (b).

Substituting equations 6.110c and 6.115a into this equation we obtain

$$[a][\tilde{V}_{012}] = [Z_{abc}][a][\tilde{I}_{012}] - [Z_{nabc}[a][\tilde{I}_{012}] + 3[Z_{nn'}][\tilde{I}_{012}],$$

or

$$[a][\tilde{V}_{012}] = (\{[Z_{abc}] - [Z_{nabc}]\}[a] + 3[Z_{nn'}])[\tilde{I}_{012}] = [Z_{nabc}]_{eq}[\tilde{I}_{012}]. \tag{6.130}$$

Solving this equation for $[\tilde{V}_{012}]$ yields

$$[\tilde{V}_{012}] = [a]^{-1}(\{[Z_{abc}] - [Z_{nabc}]\}[a] + 3[Z_{nn'}])[\tilde{I}_{012}], \tag{6.131a}$$

or

$$[\tilde{V}_{012}] = [Z_{012}^{(M)}][\tilde{I}_{012}]. \tag{6.131b}$$

Here $[Z_{012}^{(M)}]$ is the *sequence impedance matrix* of a three-phase load with *mutual coupling*. Performing all the matrix operations on the right side of equation 6.131a and after simplification the sequence impedance matrix can be expressed as

$$[Z_{012}^{(M)}]$$

$$= \begin{bmatrix} (Z_0 + 2Z_{M0} + 3Z_{nn'} - 6Z_{n0}) & (Z_2 - Z_{M2} - 3Z_{n2}) & (Z_1 - Z_{M1} - 3Z_{n1}) \\ (Z_1 - Z_{M1} - 6Z_{n1}) & (Z_0 - Z_{M0}) & (Z_2 + 2Z_{M2}) \\ (Z_2 - Z_{M2} - 6Z_{n2}) & (Z_1 + 2Z_{M1}) & (Z_0 - Z_{M0}) \end{bmatrix} \tag{6.132}$$

where, in addition to equation 6.119, we defined *sequence mutual impedances*:

$$Z_{M0} = \frac{1}{3}(Z_{bc} + Z_{ca} + Z_{ab}) \tag{6.133a}$$

as a zero-sequence mutual phase impedance,

$$Z_{M1} = \frac{1}{3}(Z_{bc} + aZ_{ca} + a^2 Z_{ab}) \tag{6.133b}$$

as a positive-sequence mutual phase impedance,

$$Z_{M2} = \frac{1}{3}(Z_{bc} + a^2 Z_{ca} + aZ_{ab}) \tag{6.133c}$$

as a negative-sequence mutual phase impedance,

$$Z_{n0} = \frac{1}{3}(Z_{nc} + Z_{nb} + Z_{nc}) \tag{6.133d}$$

as a zero-sequence mutual neutral impedance,

$$Z_{n1} = \frac{1}{3}(Z_{na} + aZ_{nb} + a^2 Z_{nc}) \tag{6.133e}$$

as a positive-sequence mutual neutral impedance,

$$Z_{n2} = \frac{1}{3}(Z_{na} + a^2 Z_{nb} + aZ_{nc}) \tag{6.133f}$$

as a negative-sequence mutual neutral impedance, and $Z_{nn'}$ as the impedance of a neutral line.

If neither self- nor mutual-impedances are equal, the application of equation 6.131b will show that there is a mutual coupling among three sequences and, therefore, the sequence networks cannot be separated. The sequence currents can be found by solving equation 6.131b:

$$[\tilde{I}_{012}] = [Y_{012}^{(M)}][\tilde{V}_{012}], \tag{6.134}$$

where $[Y_{012}^{(M)}]$ is the associated *sequence admittance matrix*

$$[Y_{012}^{(M)}] = [Z_{012}^{(M)}]^{-1} = \begin{bmatrix} Y_{00} & Y_{01} & Y_{02} \\ Y_{10} & Y_{11} & Y_{12} \\ Y_{20} & Y_{21} & Y_{22} \end{bmatrix}. \tag{6.135}$$

If a balanced voltage is applied to an unbalanced load (as frequently happens), then the symmetrical component voltage matrix $[\tilde{V}_{012}]$ reduces to only a positive-sequence component $\tilde{V}_1$. Indeed, if the applied voltage is balanced, then

$$\tilde{V}_b = a^2 \tilde{V}_a, \quad \tilde{V}_c = a\tilde{V}_a.$$

Substituting this in equation 6.113 yields

$$\begin{bmatrix} \tilde{V}_0 \\ \tilde{V}_1 \\ \tilde{V}_2 \end{bmatrix} = \frac{1}{3} \begin{bmatrix} 1 & 1 & 1 \\ 1 & a & a^2 \\ 1 & a^2 & a \end{bmatrix} \begin{bmatrix} \tilde{V}_a \\ a^2 \overline{V}_a \\ a\tilde{V}_a \end{bmatrix} = \begin{bmatrix} 0 \\ \tilde{V}_a \\ 0 \end{bmatrix}.$$

In other words, the three-phase balanced system consists only of the positive-sequence components, $\tilde{V}_1 = \tilde{V}_a$. However, if the system of line-to-line voltages is balanced, then in general zero-sequence voltages may also be present. Thus, the current sequences can be expressed as

$$\begin{bmatrix} I_{a0} \\ I_{a1} \\ I_{a2} \end{bmatrix} = \begin{bmatrix} Y_{00} & Y_{01} & Y_{02} \\ Y_{10} & Y_{11} & Y_{12} \\ Y_{20} & Y_{21} & Y_{22} \end{bmatrix} \begin{bmatrix} \tilde{V}_0 \\ \tilde{V}_1 \\ 0 \end{bmatrix}. \tag{6.136}$$

However, if only a few mutual inductances are present, the solution may be simplified, as can be seen from the following example.

Example 6.18

Consider a particular case of a mutually coupled three-phase network as shown in Fig. 6.58(b). Let the self-impedances be $Z_{aa} = j1\,\Omega$, $Z_{bb} = 2\,\Omega$, $Z_{cc} = j3\,\Omega$ and only the mutual-impedances be $Z_{ac} = Z_{ca} = -j0.5\,\Omega$. Find the current $\tilde{I}_a$, if the system of line-to-line voltages is balanced and given as $\tilde{V}_{ab} = 100\angle 0°$, $\tilde{V}_{bc} = 100\angle -120°$, $\tilde{V}_{ca} = 100\angle 120°$ V.

Solution

The impedance matrix is

$$[Z_{abc}] = \begin{bmatrix} j1 & 0 & -j0.5 \\ 0 & 2 & 0 \\ -j0.5 & 0 & j3 \end{bmatrix}$$

and we first calculate $[Y_{abc}]$:

$$[Y_{abc}] = [Z_{abc}]^{-1} = \frac{1}{-5.5} \begin{bmatrix} j6 & 0 & j1 \\ 0 & -2.75 & 0 \\ j1 & 0 & j2 \end{bmatrix}.$$

Then, as in equation 4.132 and taking into consideration that $Y_{n1} = Y_{n2} = Y_{n0} = Y_{nn'} = 0$, we may obtain the sequence admittance matrix as

$$[Y_{012}] = \begin{bmatrix} Y_0 + 2Y_{M0} & Y_2 - Y_{M2} & Y_1 - Y_{M1} \\ Y_1 - Y_{M1} & Y_0 - Y_{M0} & Y_2 + 2Y_{M2} \\ Y_2 - Y_{M2} & Y_1 + 2Y_{M1} & Y_0 - Y_{M0} \end{bmatrix} = \begin{bmatrix} Y_{00} & Y_{01} & Y_{02} \\ Y_{10} & Y_{11} & Y_{12} \\ Y_{20} & Y_{21} & Y_{22} \end{bmatrix},$$

where

$$Y_0 = \frac{1}{3(-5.5)}(j6 - 2.75 + j2) = 0.1667 - j0.4849 = 0.5128\angle -71.03^\circ \text{ S}$$

$$Y_1 = \frac{1}{3(-5.5)}(j6 - 2.75\angle 120^\circ + j2\angle 240)$$

$$= -0.1883 - j0.1587 = -0.2463\angle 40.12^\circ \text{ S}$$

$$Y_3 = \frac{1}{3(-5.5)}(j6 - 2.75\angle 240^\circ + j2\angle 120^\circ)$$

$$= 0.0216 - j0.4474 = 0.4479\angle 87.23^\circ \text{ S}$$

and as in equations 4.133

$$Y_{MO} = \frac{1}{3(-5.5)}(0 + j1 + 0) = -j0.06061 \text{ S}$$

$$Y_{M1} = \frac{1}{3(-5.5)}(0 + j1\angle 120^\circ + 0) = 0.06061\angle 30^\circ \text{ S}$$

$$Y_{M2} = \frac{1}{3(-5.5)}(0 + j1\angle 240^\circ + 0) = 0.06061\angle 150^\circ \text{ S.}$$

Now the elements of matrix $[Y_{012}]$ are

$$Y_{00} = Y_0 + 2Y_{MO} = 0.6286\angle -74.62^\circ \text{ S}$$

$$Y_{11} = Y_{22} = Y_0 - Y_{MO} = 0.4559\angle -68.55^\circ \text{ S}$$

$$Y_{10} = Y_{02} = Y_1 - Y_{M1} = -0.3061\angle 38.13^\circ \text{ S}$$

$$Y_{01} = Y_{20} = Y_1 - Y_{M2} = 0.4834\angle -81.18^\circ \text{ S}$$

$$Y_{21} = Y_1 + 2Y_{M1} = -0.1287\angle 49.66^\circ \text{ S.}$$

Next we determine the sequence-voltages:

$$\tilde{V}_1 = \frac{100}{\sqrt{3}}\angle -30^\circ = 57.4\angle -30^\circ \text{ V}, \quad \tilde{V}_2 = 0.$$

Since $\tilde{I}_0 = 0$, in accordance with the first equation in 4.136 we may calculate $\tilde{V}_0$

$$\tilde{V}_0 = -\frac{Y_{01}}{Y_{00}}\tilde{V}_1 = -\frac{0.4834\angle -81.18^\circ}{0.6286\angle -74.62^\circ}57.74\angle -30^\circ = -44.4\angle -36.56^\circ \text{ V.}$$

With the two other equations in 6.136 we have

$$\tilde{I}_1 = Y_{10}\tilde{V}_0 + Y_{11}\tilde{V}_1 = 9.67 - j25.66 \text{ A}$$

$$\tilde{I}_2 = Y_{20}\tilde{V}_0 + Y_{21}\tilde{V}_1 = 2.99 + j16.50 \text{ A}$$

and

$$\tilde{I}_a = \tilde{I}_1 + \tilde{I}_2 = 12.66 - j9.16 = 15.63\angle -35.89^\circ \text{ A}.$$

When the load is balanced, i.e., the mutual impedances are equal to each other: $Z_{bc} = Z_{ca} = Z_{ab} = Z_M$ and $Z_{na} = Z_{nb} = Z_{nc} = Z_{np}$ and so the self-impedances: $Z_{aa} = Z_{bb} = Z_{cc} = Z_L$, the sequence impedance matrix (equation 6.132) simplifies to

$$[Z_{012}^{(M)}] = \begin{bmatrix} (Z_L + 2Z_M + 3Z_{nn'} - 6Z_{n0}) & 0 & 0 \\ 0 & (Z_L - Z_M) & 0 \\ 0 & 0 & (Z_L - Z_M) \end{bmatrix}. \tag{6.137}$$

As can be seen, there is no mutual coupling among the three sequences in this case either, and the sequence circuit impedances are

$$Z_{00} = Z_L + 2Z_M + 3Z_{nn'} - 6Z_{np}, \quad Z_{11} = Z_{22} = Z_L - Z_M. \tag{6.138}$$

Thus,

$$\tilde{V}_0 = Z_{00}\tilde{I}_0, \quad \tilde{V}_1 = Z_{11}\tilde{I}_1, \quad \tilde{V}_2 = Z_{22}\tilde{I}_2, \tag{6.139}$$

The *degree of current or voltage unbalances* is usually estimated as:

for the **zero sequence**

$$m_{0i} = \frac{I_{a0}}{I_{a1}}, \quad m_{0v} = \frac{V_{a0}}{V_{a1}}; \tag{6.140a}$$

for the **negative sequence**

$$m_{2i} = \frac{I_{a2}}{I_{a1}}, \quad m_{2v} = \frac{V_{a2}}{V_{a1}}. \tag{6.140b}$$

Example 6.19

The Y-connected load, having self- and mutual-impedances $Z_L = 1 + j22\,\Omega$ and $Z_M = j6\,\Omega$; and the self- and mutual-impedance of a neutral line $Z_{nn'} = 2 + j18\,\Omega$ and $Z_{np} = j2\,\Omega$ is supplied by an unbalanced three-phase system with the phase voltages being $\tilde{V}_a = 100\angle -30^\circ$, $\tilde{V}_b = 150\angle 180^\circ$ and $\tilde{V}_c = 75\angle 60^\circ$ V. Calculate the current in each branch of the load.

Solution

The first step is to calculate the sequence impedance matrix (in accordance with equation 6.137):

$$\begin{aligned} Z_{01} &= Z_L + 2Z_M + 3Z_{nn'} - 6Z_{np} \\ &= 1 + j22 + 2\cdot j6 + 3(2 + j8) - 6\cdot j2 = 7 + j46 = 46.5\angle 81.3^\circ \text{ V} \end{aligned}$$

and

$$Z_{11} = Z_{22} = Z_L - Z_M = 1 + j22 - j6 = 1 + j16 = 16.03\angle 86.4^\circ \text{ V}.$$

Next we shall calculate the phase-sequence components of the unbalanced voltages

$$\tilde{V}_0 = \frac{1}{3}(100\angle -30^\circ + 150\angle 180^\circ + 75\angle 60^\circ) = 10.0\angle 150^\circ \text{ V}$$

$$\tilde{V}_1 = \frac{1}{3}(100\angle -30^\circ + (1\angle 120^\circ)(150\angle 180^\circ) + (1\angle 240^\circ)(75\angle 60^\circ)]$$

$$= 105\angle -51^\circ \text{ V}$$

$$\tilde{V}_2 = \frac{1}{3}(100\angle -30^\circ + (1\angle 240^\circ)(150\angle 180^\circ) + (1\angle 120^\circ)(75\angle 60^\circ)]$$

$$= 39\angle 43^\circ \text{ V}.$$

Since there is no mutual coupling along the three sequences, the second step is to calculate the phase-sequence components of the current $\tilde{I}_a$. Thus,

$$\tilde{I}_{a0} = \tilde{I}_0 = \frac{\tilde{V}_0}{Z_{00}} = \frac{10.0\angle 150^\circ}{46.5\angle 81.3^\circ} = 0.215\angle 68.7^\circ \text{ A}$$

$$\tilde{I}_{a1} = \tilde{I}_1 = \frac{\tilde{V}_1}{Z_{11}} = \frac{105\angle -51^\circ}{16.03\angle 86.4^\circ} = 6.55\angle -137.4^\circ \text{ A}$$

$$\tilde{I}_{a2} = \tilde{I}_2 = \frac{\tilde{V}_2}{Z_{22}} = \frac{39.0\angle 43^\circ}{16.03\angle 86.4^\circ} = 2.43\angle -43.4^\circ \text{ A}.$$

Therefore,

$$\tilde{I}_a = \tilde{I}_{a0} + \tilde{I}_{a1} + \tilde{I}_{a2} = 0.215\angle 68.7^\circ + 6.55\angle -137.4^\circ + 2.43\angle -43.4^\circ$$

$$= 6.60\angle -116.7 \text{ A}$$

$$\tilde{I}_b = \tilde{I}_{a0} + a^2\tilde{I}_{a1} + a\tilde{I}_{a2}$$

$$= 0.215\angle 68.7^\circ + (1\angle 240^\circ)(6.55\angle -137.4^\circ) + (1\angle 120^\circ)(2.43\angle -43.4^\circ)$$

$$= 8.98\angle 85^\circ \text{ A}$$

$$\tilde{I}_c = \tilde{I}_{a0} + a\tilde{I}_{a1} + a^2\tilde{I}_{a2}$$

$$= 0.215\angle 68.7^\circ + (1\angle 120^\circ)(6.55\angle -137.4^\circ) + (1\angle 240^\circ)(2.43\angle -43.4^\circ)$$

$$= 4.69\angle -31.5^\circ \text{ A}.$$

In the above treatment of three-phase loads, and the development of the phase-sequence networks' equivalent, it was derived that the values of these network impedances are the same for currents of positive-, and negative-sequences. In practice, such a result is quite in order in the case of "static" circuits, such as transformers, transmission lines and the like, in which the mutual inductances between the circuits of different phases are bilateral. The phase sequence, positive or negative, of the currents flowing in static circuits does not change the impedances, so the same values of impedances in both the positive-, and negative-sequence networks are used.

With rotating machinery, e.g. alternators, induction motors, synchronous motors, etc., the impedance will have different values for currents of positive and negative phase-sequences. Indeed, the negative-network impedance, Z_{22}, can be determined by applying the negative-sequence voltages and measuring the negative-sequence currents, when the machine is run at specified speed and direction. Since the negative phase-sequence field (also called the *backward field*) rotates in the direction opposite to the positive phase-sequence field (also called the *forward field*), it will also rotate opposite to the rotor. Thus, for instance, for asynchronous machines the difference in speed between the backward field and rotor is $n_s + n$, where n_s and n are rotating speeds of the field and rotor respectively. This results in a slip for the backward field $s_2 = 2 - s$. Since the regular slip s (i.e. slip for the forward field) is very small ($s = 0.02$–0.05), the slip s_2 equals approximately 2, so that it is much larger than s. As a result the negative-sequence currents will be larger than the positive-sequence currents and, therefore, the impedance to currents of negative phase-sequence, Z_{22}, will be lower than that to currents of positive phase-sequence, Z_{11}. To develop the mathematical representation of rotating machine symmetrical-component impedances we shall assume that the mutual inductances between the phases of these machines are not bilateral, as shown in Fig 6.59. Thus, two different values Z_p and Z_q are the **mutual impedances of rotating machines** (clockwise and counterclockwise respectively) between the phases. The impedance matrix in

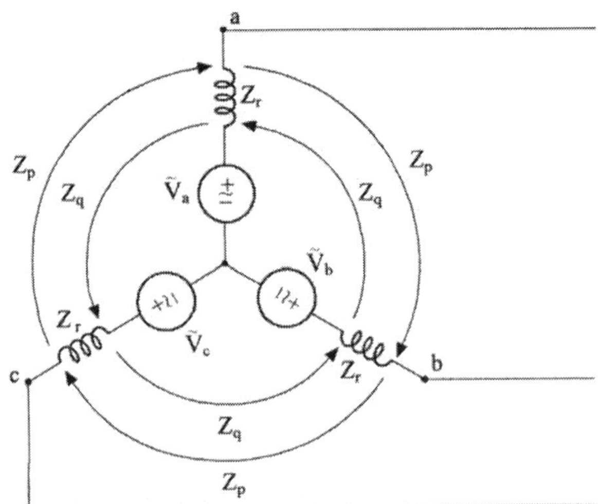

Figure 6.59 Equivalent circuit of a rotating machine.

this case is of a circular form:

$$[Z_{rpq}] = \begin{bmatrix} Z_r & Z_p & Z_q \\ Z_q & Z_r & Z_p \\ Z_p & Z_q & Z_r \end{bmatrix} \tag{6.141}$$

where Z_r is the self-impedance of each phase and $Z_p \neq Z_q$. Applying the matrix transformation of the form of equation 6.118a yields

$$[Z_{012}^{(r)}] = \begin{bmatrix} Z_{00} & 0 & 0 \\ 0 & Z_{11} & 0 \\ 0 & 0 & Z_{22} \end{bmatrix} \tag{6.142}$$

where

$$\begin{aligned} Z_{00} &= Z_r + Z_p + Z_q \\ Z_{11} &= Z_r + a^2 Z_p + a Z_q \\ Z_{22} &= Z_r + a Z_p + a^2 Z_q \end{aligned} \tag{6.143}$$

are the *zero-, positive- and negative-sequence impedances of the machine*. Thus, the sequence matrix equation for a rotating machine will be

$$[\tilde{V}_{012}] = [Z_{012}^{(r)}][\tilde{I}_{012}]. \tag{6.144}$$

Since the matrix in equation 6.142 is diagonal, also in this case, this matrix equation may be separated into three independent equations, each for each sequence:

$$\tilde{V}_0 = Z_0 \tilde{I}_0, \quad \tilde{V}_1 = Z_1 \tilde{I}_1, \quad \tilde{V}_2 = Z_2 \tilde{I}_2. \tag{6.145}$$

(For the sake of simplicity here and further on single subscripts, 0, 1 and 2, are used to indicate sequence-network impedances.) However, in distinction to the "static" load (see equations 6.125, 6.127 and 6.139) here the positive- and negative-network impedances are unequal, with the negative-network impedance lower than the positive-network impedance, $|Z_2| < |Z_1|$(*).

Example 6.20

A three-phase, Y-connected, induction motor, having the positive- and negative-sequence network impedances: $Z_1 = 3.6 + j3.6\ \Omega$ and $Z_2 = 0.15 + j0.5\ \Omega$, is supplied from an unsymmetrical three-wire system. The line voltages being $V_{ab} = V_{ca} = 365\,\mathrm{V}$ and $V_{bc} = 312\,\mathrm{V}$, calculate the current in each phase of the motor.

Solution

The first step is to calculate the phase voltages. Drawing the triangle of line

(*)There is more about symmetrical components in Gönen, T. (1988) *Electric Power Transmission System Engineering*. Wiley & Sons, New York, Chichester, Brisbane, Toronto, Singapore.

voltages as shown in Fig 6.60(a), we assumed that the neutral point n is located at the midpoint of the line voltage V_{bc} (it was already shown that the location of a neutral point does not influence the positive- and negative-sequence voltages, but only the zero-sequence; however the zero-sequence currents anyway are zero, since $Z_{nn'} = \infty$). Choosing $\tilde{V}_a$ as a reference phasor, we have

$$\tilde{V}_b = -j156 \text{ V}, \quad \tilde{V}_c = j156 \text{ V}$$

and

$$\tilde{V}_a = \sqrt{365^2 - 156^2} = 330 \text{ V}.$$

The next step is to calculate the positive- and negative-sequence components of the phase voltages

$$\tilde{V}_1 = \frac{1}{3}(\tilde{V}_a + a\tilde{V}_b + a^2\tilde{V}_c) = \frac{1}{3}[330 + j156(-a + a^2)] = 200 \text{ V}$$

and

$$\tilde{V}_2 = \frac{1}{3}(\tilde{V}_a + a^2\tilde{V}_b + a\tilde{V}_c) = \frac{1}{3}[330 + j156(a - a^2)] = 20 \text{ V}$$

Now, from positive- and negative-sequence circuits, Fig 6.60(b), we obtain

$$\tilde{I}_1 = \frac{\tilde{V}_1}{Z_1} = \frac{200}{3.6\sqrt{2}\angle 45^\circ} = 39.3\angle -45^\circ \text{ A}$$

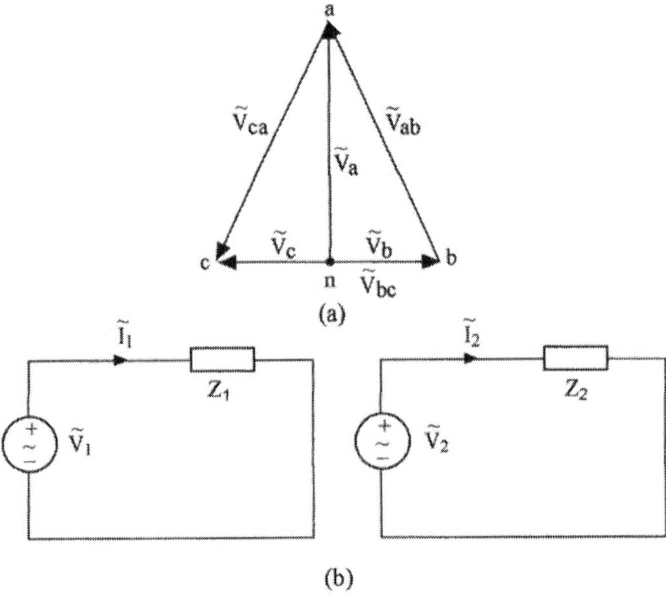

Figure 6.60 Phasor diagram (a) and sequence network for Example 6.20 (b).

$$\tilde{I}_2 = \frac{\tilde{V}_2}{Z_2} = \frac{20}{0.522\angle 73.3^\circ} = 38.3\angle -73.3^\circ \text{ A}.$$

Finally,

$$\tilde{I}_a = \tilde{I}_1 + \tilde{I}_2 = 38.8 - j64.5 = 75.3\angle -60^\circ \text{ A}$$

$$\tilde{I}_b = a^2\tilde{I}_1 + a\tilde{I}_2 = -11.7 - j17.7 = 21.2\angle 123.5^\circ \text{ A}$$

$$\tilde{I}_c = a\tilde{I}_1 + a^2\tilde{I}_2 = -27.1 - j46.8 = 54.1\angle 120^\circ \text{ A}.$$

6.7.2 Using symmetrical components for unbalanced three-phase system analysis

As we have already mentioned, the symmetrical components method is very useful for analyzing and solving the unbalanced faults of power systems. To illustrate this let us consider the most frequently occurring **single line-to-ground fault**, which occurs when one conductor contacts the ground or the neutral wire. Fig. 6.61 shows the general representation of a single line-to-ground fault at a fault point F with fault impedance Z_F. Usually, the fault impedance Z_F is ignored in fault studies. In general the voltage-current sequences' relationship for an unbalanced system is given by the matrix equation 6.117:

$$\tilde{\mathbf{V}}_{012} = \mathbf{Z}_{012}\tilde{\mathbf{I}}_{012}. \tag{6.146}$$

Here the elements of the sequence-impedance matrix usually are known. However, neither the voltage nor current symmetrical components are known. The remaining equations, called constraint equations, may be obtained using the relationship between the symmetrical components in accordance with a kind of unsymmetrical fault. Thus for the fault under consideration we have

$$\tilde{I}_b = \tilde{I}_c = 0.$$

Then, using equation 6.115b, we have

$$\tilde{I}_0 = \tilde{I}_1 = \tilde{I}_2 = \tfrac{1}{3}\tilde{I}_a. \tag{6.147}$$

Now, the current-sequence matrix can be written in terms, for instance, of $\tilde{I}_0$ as

$$\mathbf{I}_{012} = \begin{bmatrix} \tilde{I}_0 \\ \tilde{I}_1 \\ \tilde{I}_2 \end{bmatrix} = \begin{bmatrix} 1 \\ 1 \\ 1 \end{bmatrix} \tilde{\mathbf{I}}_0 = \mathbf{C}\tilde{\mathbf{I}}', \tag{6.148a}$$

where

$$\tilde{\mathbf{I}}' = [\tilde{I}_0] = \tilde{I}_0 \tag{6.148b}$$

(note that in this particular case matrix $\tilde{\mathbf{I}}'$ is reduced to just a scalar I_0), and

$$\mathbf{C} = \begin{bmatrix} 1 \\ 1 \\ 1 \end{bmatrix} \tag{6.148c}$$

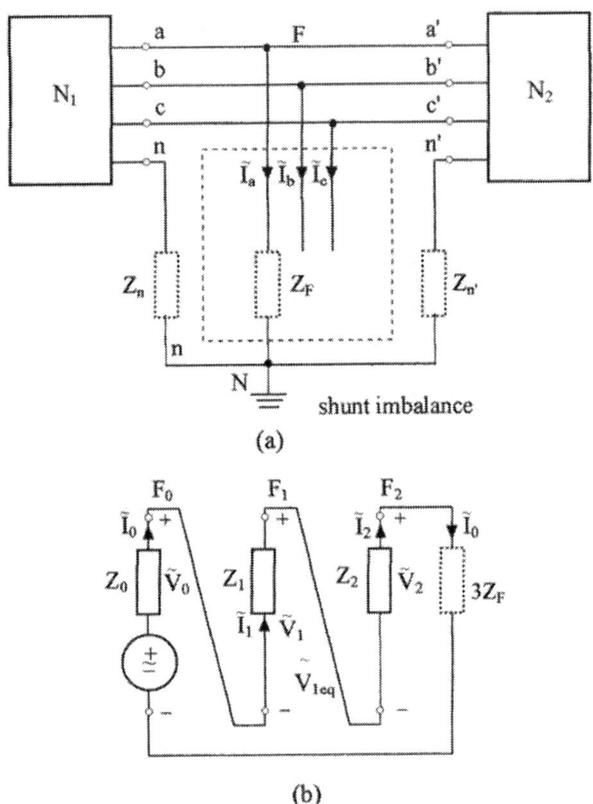

Figure 6.61 Single line-to-ground fault (a) and interconnection of sequence networks (b).

Then equation 6.148a may be written in the general form as

$$\tilde{\mathbf{I}} = \mathbf{C}\tilde{\mathbf{I}}', \tag{6.149}$$

where matrix $\mathbf{C}$ is called a **constraint matrix**. This *matrix transformation* is similar to those we used in mesh and nodal analysis, when a new set of variables, currents and voltages were chosen, for some reason, instead of a previous one. The new set of voltages' matrix is then given as

$$\tilde{\mathbf{V}}' = \mathbf{C}^{\mathrm{T}}\tilde{\mathbf{V}}. \tag{6.150}$$

(This also follows from the fact that the power volt-amperes of the network calculated in terms of the old voltage and current matrixes ($\mathbf{V}^{\mathrm{T}}\mathbf{I}^*$) must be the same as when calculated in terms of the new voltage and current matrixes ($\mathbf{V}'^{\mathrm{T}}\mathbf{I}'^*$). Then the corresponding impedance matrix is given by

$$\mathbf{Z}' = \mathbf{C}^{\mathrm{T}}\mathbf{Z}\mathbf{C}, \tag{6.151}$$

and the equation in terms of the new set of variables is denoted by

$$\mathbf{V}' = \mathbf{Z}\mathbf{I}' \tag{6.152}$$

Continuing with the above example, we apply equation 6.150 to yield

$$\mathbf{V}' = \tilde{\mathbf{V}}'_{012} = \mathbf{C}^{\mathrm{T}}\mathbf{V}_{012} = [1 \quad 1 \quad 1]\begin{bmatrix} 0 \\ \tilde{V}_1 \\ 0 \end{bmatrix} = \tilde{V}_1, \tag{6.153}$$

where the symmetrical components of the applied voltages consist only of a positive sequence since the three-phase sources of the network N_1 and/or network N_2 in Fig. 6.61(a), which actually represent the power system generators, are symmetrical. The *transformed impedance matrix* (equation 6.151) is

$$\tilde{\mathbf{Z}}'_{012} = [1 \quad 1 \quad 1][Z_{012}]\begin{bmatrix} 1 \\ 1 \\ 1 \end{bmatrix}. \tag{6.154}$$

The sequence impedances of matrix $\mathbf{Z}_{012}$ are viewed from the fault point F and, since the system generally is balanced and consists of the rotating loads as well as the static ones, this impedance matrix $[Z_{012}]$ is diagonal with unequal positive- and negative-sequence impedances

$$Z_{012} = \begin{bmatrix} Z_0 & 0 & 0 \\ 0 & Z_1 & 0 \\ 0 & 0 & Z_2 \end{bmatrix}. \tag{6.155}$$

Substituting this matrix into equation 6.154 and performing the multiplication we easily obtain

$$\mathbf{Z}' = [Z_0 + Z_1 + Z_2]. \tag{6.156}$$

Substituting equations 6.153, 6.156 and 6.148b into equation 6.152 yields

$$\tilde{V}_1 = [Z_0 + Z_1 + Z_2]\tilde{I}_0. \tag{6.157}$$

Thus the matrix equation 6.152 in this case becomes the single scalar equation. Then

$$\tilde{I}_0 = \frac{\tilde{V}_1}{Z_0 + Z_1 + Z_2}. \tag{6.158}$$

These equations 6.157 and 6.158 are appropriate for Fig. 6.61(b), where, to meet this relationship, the symmetrical component networks have to be connected in series. The fault current of phase a, therefore, is (equation 6.147)

$$\tilde{I}^{(1)}_{sc} = 3\tilde{I}_0. \tag{6.159}$$

Recall that the superscript indicates the following kind of faults: (1) a single-pole ground fault, (2) two-pole fault and (1,1) two-pole-ground fault. The numerical examples follow.

Example 6.21

The faulted network and all the parameters are given in Fig. 6.62(a). Form the sequence networks and calculate the steady-state single-pole-to-ground short-circuit current.

Solution

The p.u. reactances referred to $S_B = 120$ MVA and to the average basic voltages are shown in Fig. 6.62(b), where the three sequence networks are also given.

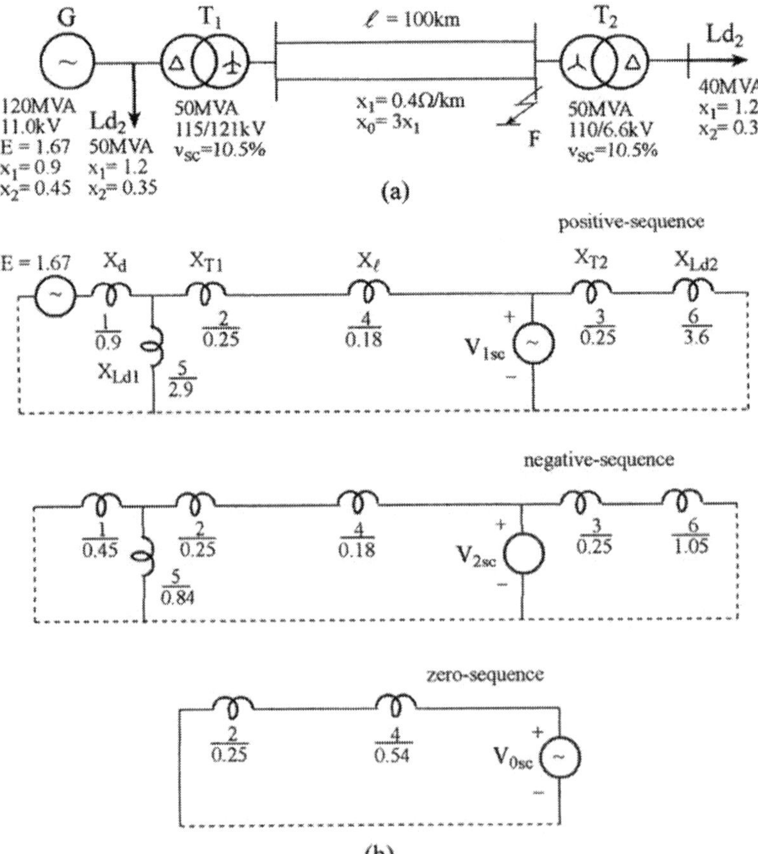

Figure 6.62 A given network for Example 6.21 (a) and the sequence networks (b).

By simplifying the positive-sequence network we have

$$X_{1,7} = 0.9//2.9 = 0.69, \quad X_{1,8} = 0.25 + 3.6 = 3.85,$$

$$X_{1,9} = 0.69 + 0.25 + 0.18 = 1.12,$$

and the equivalent reactance is

$$X_{1,eq} = 1.12//3.85 = 0.87.$$

The equivalent EMF is found as follows

$$E_2 = \frac{1.67 \cdot 2.9}{0.9 + 2.9} = 1.27 \quad \text{and} \quad E_{1eq} = \frac{1.27 \cdot 3.85}{1.12 + 3.85} = 0.98.$$

The simplification of the negative-sequence network gives

$$X_{2,7} = 0.45//0.84 = 0.29, \quad X_{2,8} = 0.25 + 1.05 = 1.3,$$

$$X_{2,9} = 0.29 + 0.25 + 0.18 = 0.72,$$

and

$$X_{2,eq} = 0.72//1.3 = 0.46.$$

Finally, the simplification of the zero-sequence network results in

$$X_{0,eq} = 0.25 + 0.54 = 0.79.$$

The zero sequence current will be (equation 6.158)

$$I_0^{(1)} = \frac{0.98}{0.87 + 0.46 + 0.79} = 0.45.$$

And the short-circuit current in a single-pole-ground fault, therefore, is (equation 6.159)

$$I_{sc}^{(1)} = 3I_0^{(1)} = 3 \cdot 0.45 = 1.35,$$

or

$$I_{sc}^{(1)} = 1.35 \frac{120}{\sqrt{3} \cdot 115} = 0.81.$$

Example 6.22

Consider the low power system shown in Fig. 6.61(a) and assume that there is a single-line-to-ground solid (i.e., $Z_F = 0$) fault involving phase a at the end of a transmission line, i.e. at point a'. Let the network N_1 represent a generator having phase voltages 240 V and the sequence impedances $Z_{s1} = j4\,\Omega$, $Z_{s2} = j2\,\Omega$ and $Z_{s0} = j1\,\Omega$; and the network N_2 represents a load with $Z_{L1} = 22.22\angle 25.84^\circ = 20 + j9.69\,\Omega$ (which corresponds to 0.9 PF), $Z_{L2} = (8 + j5)\Omega$ and $Z_{L0} = (2.5 + j1)\Omega$. The transmission line sequence impedances are $Z_{l1} = Z_{l2} = j1\ \Omega$ and $Z_{l0} = j1.5\ \Omega$. Also assume that the neutral wire/ground impedances are $Z_n = Z_{n'} = 0.5\ \Omega$, and the fault impedance Z_F is zero. At the fault point F, determine: (a) the sequence and phase currents and (b) the sequence and phase voltages.

Solution

(a) Figure 6.63(a) shows the corresponding positive-, negative- and zero-sequence networks which are interconnected in series. To reduce them we first find the equivalent impedances:

$$Z_{1eq} = \frac{(Z_{s1} + Z_{l1})Z_{L1}}{Z_{s1} + Z_{l1} + Z_{L1}} = \frac{(j4 + j1)(20 + j9.69)}{20 + j14.69} = 4.47\angle 79.54^\circ\ \Omega$$

$$Z_{2eq} = \frac{(Z_{s2} + Z_{l2})Z_{L2}}{Z_{s2} + Z_{l2} + Z_{L2}} = \frac{(j2 + j1)(8 + j5)}{8 + j8} = 2.50\angle 77.01^\circ\ \Omega$$

$$Z_{0eq} = \frac{(Z_{s0} + Z_{l0} + 3Z_n)(Z_{L0} + 3Z_{n'})}{Z_{s0} + Z_{l0} + Z_{L0} + 3(Z_n - Z_{n'})} = \frac{(j1 + j1.5 + 1.5)(2.5 + j1 + 1.5)}{5.5 + j3.5}$$

$$= 1.84\angle 40.61^\circ\ \Omega,$$

and the equivalent voltage source seen at the fault point (which is the Thévenin

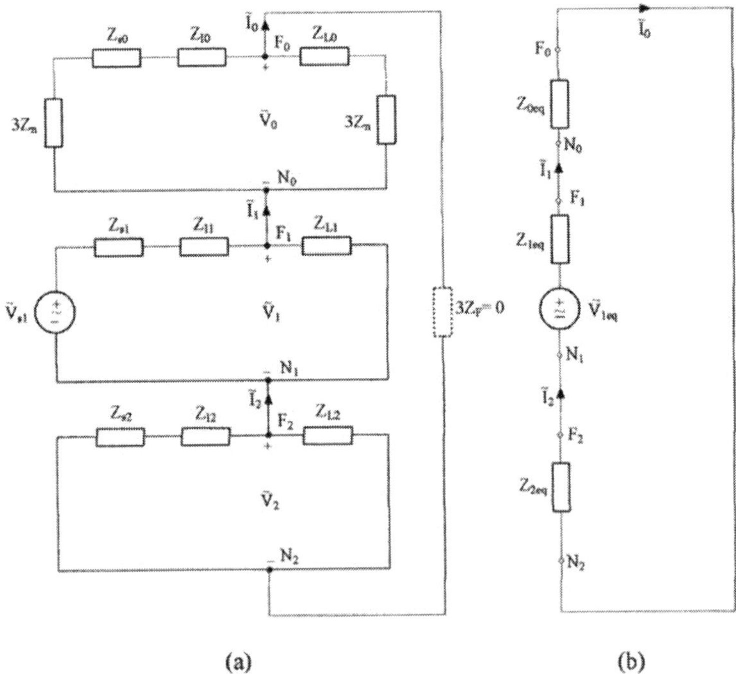

Figure 6.63 Interconnection of zero-, positive- and negative-sequence networks (a) and an equivalent network for Example 6.22 (b).

voltage)

$$\tilde{V}_{1eq} = V_{s1}\frac{Z_{L1}}{Z_{s1}+Z_{l1}+Z_{L1}} = 240\frac{20+j9.69}{20+j14.69} = 215.0\angle -10.46^\circ \text{ V.}$$

The resulting equivalent sequence network interconnection is shown in Fig. 6.63(b). Thus, the sequence currents of phase a are (equation 6.158)

$$\tilde{I}_0 = \tilde{I}_1 = \tilde{I}_2 = \frac{V_{1eq}}{Z_{0eq}+Z_{1eq}+Z_{2eq}} = 25.3\angle -81.42^\circ \text{ A,}$$

and the phase currents (Fig. 6.61a)

$$\begin{bmatrix} \tilde{I}_{a'F} \\ \tilde{I}_{b'F} \\ \tilde{I}_{c'F} \end{bmatrix} = \begin{bmatrix} 1 & 1 & 1 \\ 1 & a^2 & a \\ 1 & a & a^2 \end{bmatrix}\begin{bmatrix} 25.3\angle -81.4^\circ \\ 25.3\angle -81.4^\circ \\ 25.3\angle -81.4^\circ \end{bmatrix} = \begin{bmatrix} 75.8\angle -81.4^\circ \\ 0 \\ 0 \end{bmatrix} \text{ A.}$$

(b) The sequence voltages are (in matrix representation)

$$\tilde{\mathbf{V}}_{F,012} = \tilde{\mathbf{V}}_{s,012} - \mathbf{Z}_{012}\tilde{\mathbf{I}}_{012},$$

or

$$\begin{bmatrix} \tilde{V}_0 \\ \tilde{V}_1 \\ \tilde{V}_2 \end{bmatrix} = \begin{bmatrix} 0 \\ 215.0\angle -10.46^\circ \\ 0 \end{bmatrix} - \begin{bmatrix} 1.84\angle 40.61^\circ & 0 & 0 \\ 0 & 4.48\angle 79.54^\circ & 0 \\ 0 & 0 & 2.50\angle 77.01^\circ \end{bmatrix}$$

$$\times \begin{bmatrix} 25.3\angle -81.42^\circ \\ 25.3\angle -81.42^\circ \\ 25.3\angle -81.42^\circ \end{bmatrix} = \begin{bmatrix} -35.3+j30.44 \\ 98.3-j35.30 \\ -63.0+j4.86 \end{bmatrix} = \begin{bmatrix} 46.6\angle 139.19^\circ \\ 104.4\angle -19.76^\circ \\ 63.2\angle 175.59^\circ \end{bmatrix} \text{ V,}$$

and the phase voltages are

$$\begin{bmatrix} \tilde{V}_{a'F} \\ \tilde{V}_{b'F} \\ \tilde{V}_{c'F} \end{bmatrix} = \begin{bmatrix} 1 & 1 & 1 \\ 1 & a^2 & a \\ 1 & a & a^2 \end{bmatrix}\begin{bmatrix} 46.6\angle 139.19^\circ \\ 104.4\angle -19.76^\circ \\ 63.2\angle 175.59^\circ \end{bmatrix}$$

$$= \begin{bmatrix} 0.02+j0.00 \\ -87.7-j94.0 \\ -18.1+j185.4 \end{bmatrix} = \begin{bmatrix} \approx 0 \\ 129\angle -133.0^\circ \\ 186\angle 95.6^\circ \end{bmatrix} \text{ V.}$$

For the **line-to line fault**, shown in Fig. 6.64(a), if, for example, the fault occurs on phases b and c, the constraint equations are

$$\tilde{I}_a = 0 \quad \text{and} \quad \tilde{I}_b = -\tilde{I}_c. \tag{6.160}$$

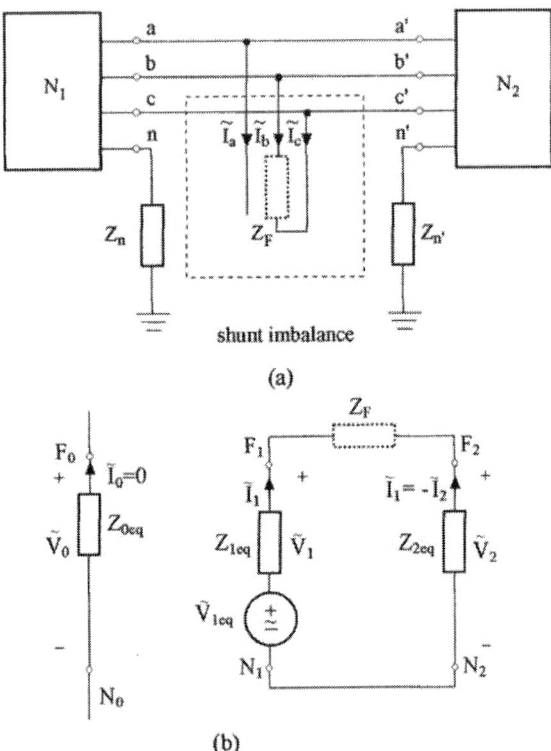

Figure 6.64 Line-to-line fault (a) and the interconnection of the sequence networks (b).

Applying now equation 6.115b we have

$$\begin{bmatrix} \tilde{I}_0 \\ \tilde{I}_1 \\ \tilde{I}_2 \end{bmatrix} = \frac{1}{3} \begin{bmatrix} 1 & 1 & 1 \\ 1 & a & a^2 \\ 1 & a^2 & a \end{bmatrix} \begin{bmatrix} 0 \\ \tilde{I}_b \\ -\tilde{I}_b \end{bmatrix} = \frac{j}{\sqrt{3}} \begin{bmatrix} 0 \\ \tilde{I}_b \\ -\tilde{I}_b \end{bmatrix}$$

or

$$\tilde{I}_0 = 0 \quad \text{and} \quad \tilde{I}'_1 = -\tilde{I}'_2 = \frac{j}{\sqrt{3}} I_b. \tag{6.161}$$

Note that the absence of I_0 can also be recognized from the fact that the zero-sequence current fault path in the circuit of Fig. 6.64(a) is open, which is indicated in Fig. 6.64(b) by ignoring the zero-sequence network.

Hence the constraint matrix in terms of $\tilde{I}_1$ would be written as

$$\mathbf{C} = \begin{bmatrix} 0 \\ 1 \\ -1 \end{bmatrix}. \tag{6.162}$$

Substituting this constraint equation into equations 6.150 and 6.151 and remembering that only the positive-sequence source voltage is non-zero, we easily obtain

$$[\tilde{V}_{012}] = \tilde{V}_1 \quad \text{and} \quad [Z_{012}] = [Z_1 + Z_2].$$

Thus, with equation 6.152 we have

$$\tilde{V}_1 = [Z_1 + Z_2]\tilde{I}_1 \tag{6.163a}$$

and

$$\tilde{I}'_1 = \frac{\tilde{V}'_1}{Z_1 + Z_2}. \tag{6.163b}$$

These equations are appropriate for Fig. 6.64(b), where only two symmetrical component networks, positive and negative, are connected in series.

In accordance with equation 6.161 the short-circuit current in phase b is

$$\tilde{I}_{sc,b} = -j\sqrt{3}I'_1. \tag{6.163c}$$

Since at the fault point the voltage is zero we have

$$\tilde{V}_{F,b} = \tilde{V}_{F,c},$$

which gives

$$\tilde{V}_F = \tilde{V}_{F,b} - \tilde{V}_{F,c} = 0.$$

Then

$$\tilde{V}_{F1} = \frac{1}{3}(\tilde{V}_{F,a} + a\tilde{V}_{F,b} + a^2\tilde{V}_{F,c}) = \frac{1}{3}[\tilde{V}_{F,a} + (a + a^2)\tilde{V}_{F,b}]$$

$$\tilde{V}_{F2} = \frac{1}{3}(\tilde{V}_{F,a} + a^2\tilde{V}_{F,b} + a\tilde{V}_{F,c}) = \frac{1}{3}[\tilde{V}_{F,a} + (a + a^2)\tilde{V}_{F,b}],$$

and

$$\tilde{V}_{F1} = \tilde{V}_{F2}. \tag{6.163d}$$

Example 6.23

Repeat Example 6.22 assuming that there is a line-to-line fault, involving phases b and c at the end of the transmission line, i.e. at points b' and c'.

Solution

(a) In accordance with Fig. 6.64(b), where $\tilde{V}_{s1} \equiv \tilde{V}_{1eq}$, $Z_1 \equiv Z_{1eq}$ and $Z_2 \equiv Z_{2eq}$ and with the other data of the previous example, the sequence currents are

$$\tilde{I}_0 = 0$$

$$\tilde{I}_1 = -\tilde{I}_2 = \frac{\tilde{V}_{1eq}}{Z_{1eq} + Z_{2eq}} = \frac{215\angle -10.46^\circ}{4.48\angle 79.54^\circ + 2.50\angle 77.01^\circ}$$

$$= 30.8\angle -89.1^\circ \text{ A},$$

and the phase currents are

$$\begin{bmatrix} \tilde{I}_{a'F} \\ \tilde{I}_{b'F} \\ \tilde{I}_{c'F} \end{bmatrix} = \begin{bmatrix} 1 & 1 & 1 \\ 1 & a^2 & a \\ 1 & a & a^2 \end{bmatrix} \begin{bmatrix} 0 \\ 30.8\angle -89.1^\circ \\ 30.8\angle 90.9^\circ \end{bmatrix} = \begin{bmatrix} 0 \\ 53.3\angle -179.1^\circ \\ 53.3\angle 0.9^\circ \end{bmatrix} \text{ A}.$$

(b) The sequence and phase voltages are

$$\begin{bmatrix} \tilde{V}_0 \\ \tilde{V}_1 \\ \tilde{V}_2 \end{bmatrix} = \begin{bmatrix} 0 \\ 215\angle -10.46^\circ \\ 0 \end{bmatrix} - \begin{bmatrix} 1.84\angle 40.61^\circ & 0 & 0 \\ 0 & 4.48\angle 79.54^\circ & 0 \\ 0 & 0 & 2.50\angle 77.01^\circ \end{bmatrix}$$

$$\times \begin{bmatrix} 0 \\ 30.8\angle -89.1^\circ \\ 30.8\angle 90.9^\circ \end{bmatrix} = \begin{bmatrix} 0 \\ 77.1\angle -12.07^\circ \\ 77.0\angle -12.09^\circ \end{bmatrix} \text{ V},$$

i.e., $\tilde{V}_1 \approx \tilde{V}_2$, as can also be seen from Fig. 6.64(b), since $Z_F = 0$, and

$$\begin{bmatrix} \tilde{V}_{a'F} \\ \tilde{V}_{b'F} \\ \tilde{V}_{c'F} \end{bmatrix} = \begin{bmatrix} 1 & 1 & 1 \\ 1 & a^2 & a \\ 1 & a & a^2 \end{bmatrix} \begin{bmatrix} 0 \\ 77.06\angle -12.1^\circ \\ 77.06\angle -12.1^\circ \end{bmatrix} = \begin{bmatrix} 154.1\angle -12.1^\circ \\ 77.06\angle 107.9^\circ \\ 77.06\angle 107.9^\circ \end{bmatrix} \text{ V}.$$

Example 6.24

Repeat Example 6.21 and find the line-to-line short-circuit current.

Solution

The resulting sequence-network in this kind of fault is formed by a series connection of positive- and negative-sequences. Therefore, the positive-sequence current is

$$I_1 = \frac{0.98}{0.87 + 0.46} = 0.71,$$

and the short-circuit current (equation 6.163c) is

$$I_{sc}^{(2)} = \sqrt{3}{\cdot}0.71 = 1.24,$$

or (in amperes)

$$I_{sc}^{(2)} = 1.24\frac{120}{\sqrt{3}{\cdot}115} = 0.74\ \text{kA}.$$

Example 6.25

In the power network shown in Fig. 6.65(a) a power station is connected through an equivalent reactance to an infinite busbar. Assume that there is a line-to-line fault involving phases b and c and find 1) the steady-state value of a short-circuit current and 2) the currents flowing from the generator.

Solution

In accordance with the sequence networks shown in Fig. 6.65(b) we find that

$$X_{1eq} = 1.23//0.25 = 0.21 \quad \text{and} \quad E_{1eq} = \frac{1.7{\cdot}0.25 + 1{\cdot}1.23}{1.23 + 0.25} = 1.12,$$

$$X_{2eq} = 0.17//0.25 = 0.1.$$

Thus, the positive- and negative-sequence of the fault current are

$$\tilde{I}_1 = -\tilde{I}_2 = \frac{1.12}{j(0.21 + 0.1)} = -j3.61.$$

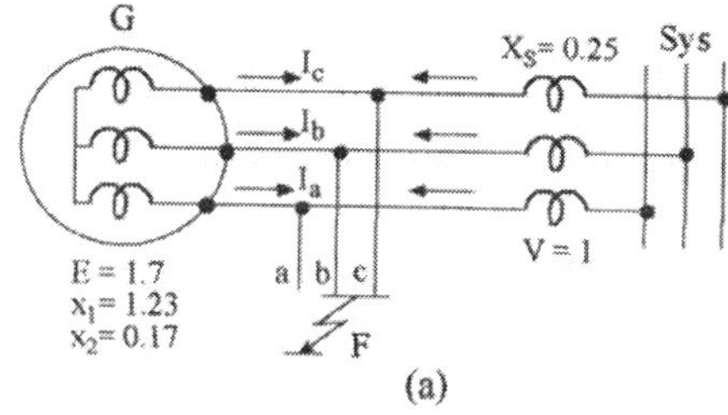

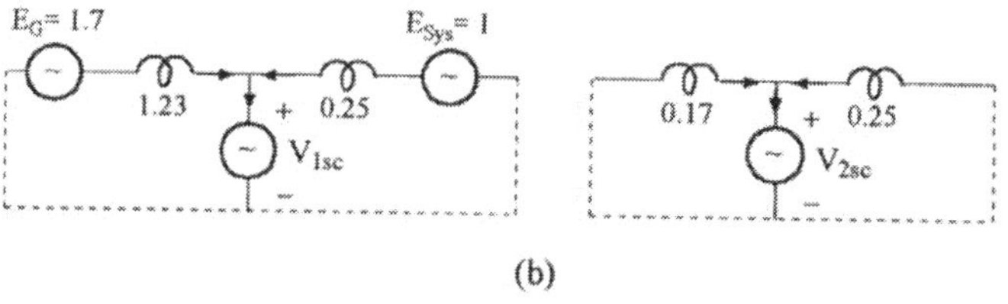

Figure 6.65 A network diagram for Example 6.25 (a) and the sequence networks (b).

The phase currents of the fault point will be

$$\tilde{I}_{sc,b} = -j\sqrt{3}(-j3.61) = -6.25 \quad \text{and} \quad I_{sc,c} = 6.25.$$

2) The sequence voltages of the fault point will then be (equation 6.163d)

$$\tilde{V}_{1F} = \tilde{V}_{2F} = -\tilde{I}_2(jX_{2eq}) = -j3.61(j0.1) = 0.361.$$

We can now find the generator current sequence

$$\tilde{I}_{1G} = \frac{1.7 - 0.361}{j1.23} = -j1.09, \quad \tilde{I}_{2G} = -\frac{0.361}{j0.17} = j2.12,$$

and the phase currents are

$$\tilde{I}_{aG} = -j1.09 + j2.12 = j1.03$$

$$\tilde{I}_{bG} = a^2(-j1.09) + a(j2.12) = -2.78 - j0.51 \quad \text{or} \quad I_{bG} = 2.83$$

$$\tilde{I}_{cG} = a(-j1.09) + a^2(j2.12) = -2.78 - j0.515 \quad \text{or} \quad I_{cG} = 2.83.$$

Note that the current in the non-faulted phase is about 40% of the current in the faulted phases. This means that the short-circuit current flows not only through the faulted phases, but also through a non-faulted phase.

Finally, consider the **double line-to-ground fault** on a transmission system, as shown in Fig. 6.66(a). This fault occurs when two conductors are connected through ground, Z_G, or directly, to the neutral of a three-phase grounded, or four-wire, system. If the fault is between phases b and c then

$$\tilde{I}_a = \tilde{I}_0 + \tilde{I}_1 + \tilde{I}_2 = 0, \tag{6.164}$$

and the current-sequence matrix could be written in terms, for instance, of $\tilde{I}_0$ and $\tilde{I}_1$

$$\tilde{I}_{012} = \begin{bmatrix} 1 & 0 \\ 0 & 1 \\ -1 & -1 \end{bmatrix} \begin{bmatrix} \tilde{I}_0 \\ \tilde{I}_1 \end{bmatrix}, \quad \text{i.e.,} \ \mathbf{C} = \begin{bmatrix} 1 & 0 \\ 0 & 1 \\ -1 & -1 \end{bmatrix}.$$

By determining $\mathbf{V}'$ (equation 6.150) and $\mathbf{Z}'$ (equation 6.151) and substituting them into equation 6.152 we obtain

$$\begin{bmatrix} 0 \\ \tilde{V}_1 \end{bmatrix} = \begin{bmatrix} Z_0 + Z_2 & Z_2 \\ Z_2 & Z_1 + Z_2 \end{bmatrix} \begin{bmatrix} \tilde{I}_0 \\ \tilde{I}_1 \end{bmatrix}$$

and

$$\tilde{I}_1 = \frac{\begin{vmatrix} Z_0 Z_2 & 0 \\ Z_2 & \tilde{V} \end{vmatrix}}{(Z_0 + Z_2)^2 - Z_2^2} = \frac{\tilde{V}_1}{Z_1 + Z_2 Z_0/(Z_2 + Z_0)} \tag{6.165}$$

which is interpreted as for the equivalent circuit shown in Fig. 6.66(b). Note

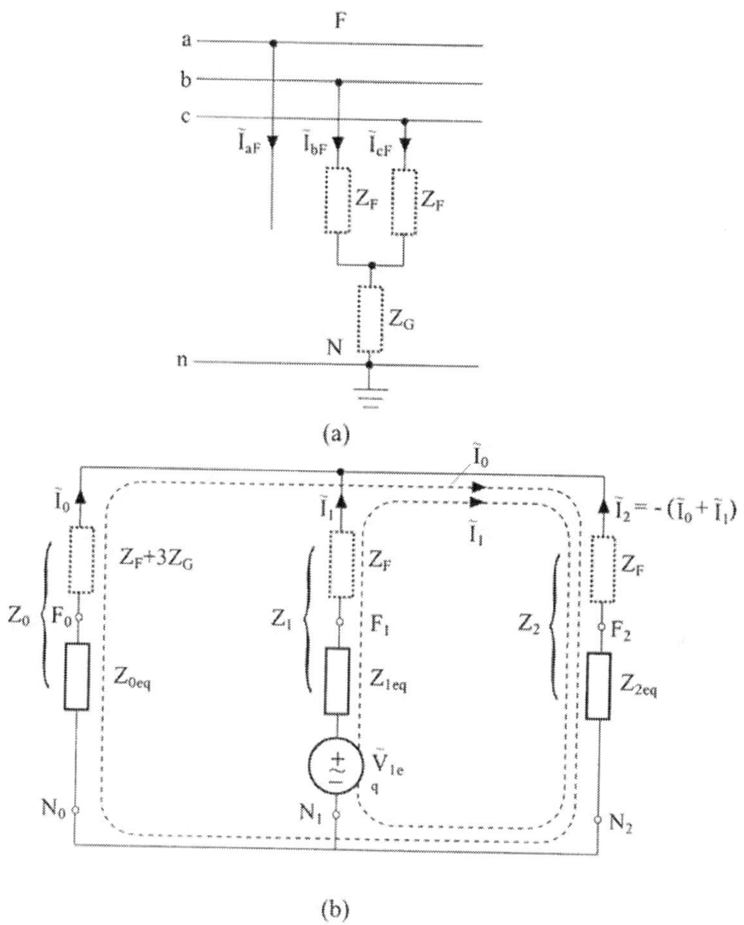

Figure 6.66 Double line-to-line ground fault: general representation (a) and interconnection of sequence networks (b).

that if $Z_G = \infty$, i.e., there is a line-to-line fault only, this circuit will reduce to the circuit in Fig. 6.64(b).

The faults, considered above, are commonly called **shunt faults**. A variety of series imbalances that occur in a power system are called **series faults**. A common one is a **broken or open conductor fault**, as shown in Fig. 6.67(a). The constraint equation for this fault is

$$\tilde{I}_a = \tilde{I}_0 + \tilde{I}_1 + \tilde{I}_2 = 0, \tag{6.166}$$

or

$$\tilde{I}_2 = -\tilde{I}_0 - \tilde{I}_1.$$

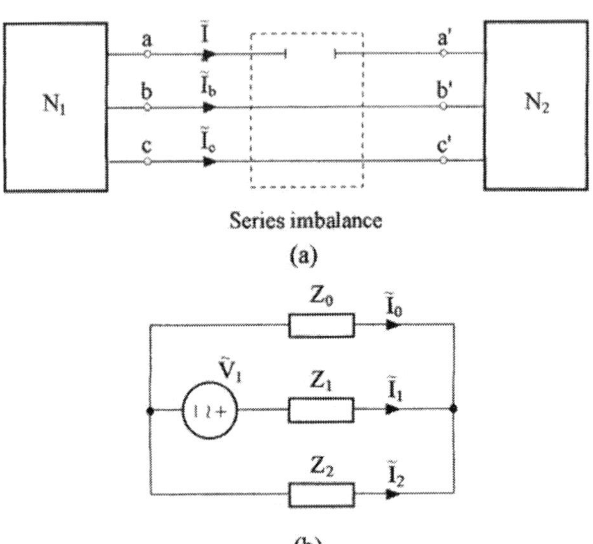

Figure 6.67 Single-phase open-fault (a) and interconnection of sequence networks (b).

Hence the constraint matrix in terms of $\tilde{I}_0$ and $\tilde{I}_1$ may be determined from the equation

$$\begin{bmatrix} \tilde{I}_0 \\ \tilde{I}_1 \\ \tilde{I}_2 \end{bmatrix} = \begin{bmatrix} 1 & 0 \\ 0 & 1 \\ -1 & -1 \end{bmatrix} \begin{bmatrix} \tilde{I}_0 \\ \tilde{I}_1 \end{bmatrix}$$

i.e.,

$$\mathbf{C} = \begin{bmatrix} 1 & 0 \\ 0 & 1 \\ -1 & -1 \end{bmatrix}. \tag{6.167}$$

With this constraint matrix the transformed voltage-sequence (equation 6.150) and impedance-sequence (equation 6.151) matrixes can be expressed as

$$\mathbf{V}' = \begin{bmatrix} 0 \\ \tilde{V}_1 \end{bmatrix}$$

and

$$\mathbf{Z}' = \begin{bmatrix} Z_0 + Z_2 & Z_2 \\ Z_2 & Z_1 + Z_2 \end{bmatrix}.$$

Substituting these expressions into equation 6.152 yields

$$\begin{bmatrix} Z_0+Z_2 & Z_2 \\ Z_2 & Z_1+Z_2 \end{bmatrix}\begin{bmatrix} \tilde{I}_0 \\ \tilde{I}_1 \end{bmatrix}=\begin{bmatrix} 0 \\ \tilde{V}_1 \end{bmatrix} \tag{6.168}$$

and solving it for $\tilde{I}_1$ (using, for instance, Kramer's rule) we obtain

$$\tilde{I}_1=\frac{\tilde{V}_1(Z_0+Z_2)}{Z_1Z_0+Z_1Z_2+Z_0Z_2}=\frac{\tilde{V}_1}{Z_1+Z_0Z_2/(Z_0+Z_2)}. \tag{6.169}$$

This result could be obtained straightforwardly from the parallel interconnection of the three sequence-networks as shown in Fig. 6.67(b). Note that this kind of sequence-network interconnection is actually the same as for a double line-to-ground fault. The difference, however, is that here the interconnection circuit refers to the line currents, whereas in the double line-to-ground case it refers to the fault currents.

Example 6.27

An induction motor is supplied by a three-phase three-wire balanced system. With the fault being the phase a conductor open, Fig. 6.68(a), find the line currents of the remaining phases $\tilde{I}_b$ and $\tilde{I}_c$ and phase voltages across the load $\tilde{V}_{a'n'}$, $\tilde{V}_{b'n'}$ and $\tilde{V}_{c'n'}$. Also find the voltages $\tilde{V}_{aa'}$ and $\tilde{V}_{nn'}$. The supplied line voltage is 400 V and the motor impedances are: positive-sequence $Z_{M1}=3.6+j3.6\,\Omega$ and negative-sequence $Z_{M2}=0.15+j0.5\,\Omega$. The line sequence impedances are $Z_{l1}=Z_{l2}=0.1+j0.1\,\Omega$. (The system impedances might be neglected, being relatively very small.)

Solution

Since the neutral wire is absent, $Z_{nn'}=\infty$, only two sequence networks (positive- and negative-sequence networks) are connected in parallel, as shown in Fig. 6.68(b). Thus, the positive-sequence current is found as

$$\tilde{I}_1=\frac{\tilde{V}_{s1}}{Z_1+Z_2}=\frac{231}{3.7+j3.7+0.25+j0.6}=\frac{231}{5.84\angle 47.4^\circ}=39.6\angle-47.4^\circ\text{ A},$$

where $\tilde{V}_{s1}=400/\sqrt{3}=231$ V.

Therefore,

$$\tilde{I}_b=a^2\tilde{I}_1+a\tilde{I}_2=(a^2-a)\tilde{I}_1=-j\sqrt{3}\tilde{I}_1=68.5\angle-137.4^\circ\text{ A}$$

$$\tilde{I}_c=-\tilde{I}_b=68.5\angle 42.6^\circ\text{ A}.$$

The phase voltages are found as

$$\tilde{V}_{a'n'}=Z_{M1}\tilde{I}_1+Z_{M2}\tilde{I}_2=(Z_{1M}-Z_{2M})\tilde{I}_1$$

$$=(3.45+j3.10)(39.6\angle-47.4^\circ)=183.7\angle-5.4^\circ\text{ V}$$

$$\tilde{V}_{b'n'}=a^2Z_{M1}\tilde{I}_1+aZ_{M2}\tilde{I}_2=(a^2Z_{M1}-aZ_{M2})\tilde{I}_1$$

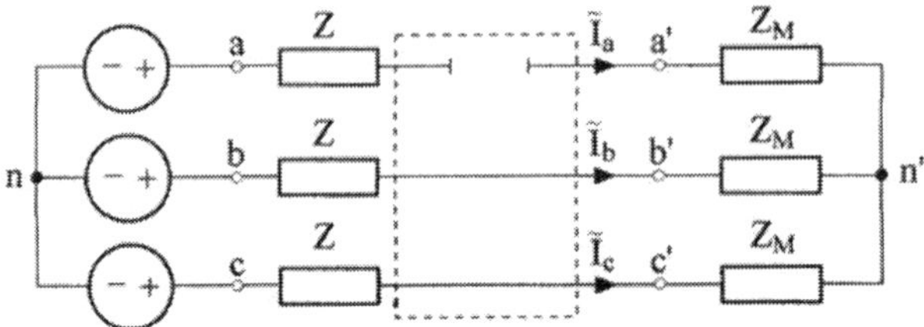

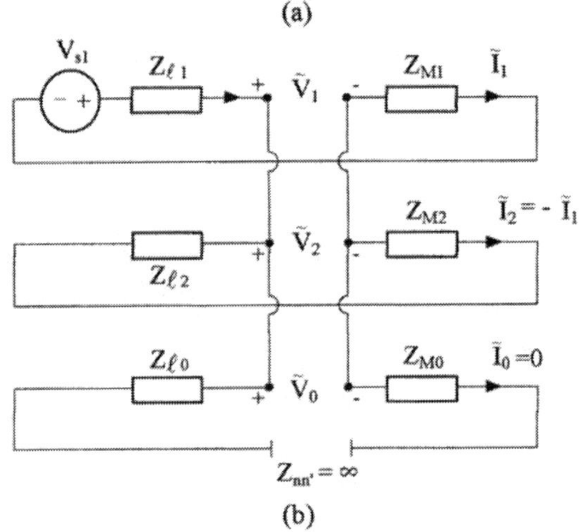

Figure 6.68 An open-fault circuit (a) and the interconnection of sequence networks (b) for Example 6.27.

$$= (5.09\angle -75^\circ - 0.522\angle 193.3^\circ)(39.6\angle -47.4^\circ) = 203\angle -116.6^\circ \text{ V}$$

$$\tilde{V}_{c'n'} = (aZ_{M1} - a^2 Z_{M2})\tilde{I}_1 = 220\angle 114.8^\circ \text{ V}.$$

To find the fault voltage $\tilde{V}_{aa'}$ we shall first analyze the series unbalanced voltages:

$$\begin{aligned} \tilde{V}_{aa'} &= \tilde{V}_0 + \tilde{V}_1 + \tilde{V}_2 \\ \tilde{V}_{bb'} &= \tilde{V}_0 + a^2\tilde{V}_1 + a\tilde{V}_2 \\ \tilde{V}_{cc'} &= \tilde{V}_0 + a\tilde{V}_1 + a^2\tilde{V}_2. \end{aligned} \tag{6.170}$$

The constraint voltage equations are

$$\tilde{V}_{bb'} = 0 \quad \tilde{V}_{cc'} = 0. \tag{6.171}$$

Solving equation 6.170 with equation 6.171 yields

$$\tilde{V}_1 = \tilde{V}_2 = \tilde{V}_0. \tag{6.172}$$

The second step is to determine $\tilde{V}_2$ in accordance with the negative-sequence network, which is the part of the equivalent circuit shown in Fig. 6.68(b). Thus,

$$\tilde{V}_2 = -Z_2\tilde{I}_2 = -(0.25 + j0.6)(-39.6\angle -47.5^\circ)$$
$$= 0.65\angle 67.4^\circ \cdot 39.6\angle -47.5^\circ = 25.7\angle 19.9^\circ \text{ V}.$$

Therefore, with the first equation of (1.170) we have

$$\tilde{V}_{aa'} = 3\tilde{V}_2 = 77.1\angle 19.9^\circ \text{ V}.$$

Since the neutral line is open the potential difference between neutral points n and n' equals zero-sequence voltage. Thus,

$$\tilde{V}_{nn'} = \tilde{V}_0 = 25.7\angle 19.9^\circ \text{ V}.$$

In the final example of this section let us consider the influence of AVR on the unsymmetrical faults.

Example 6.28

A two-pole-ground-fault occurs in the network shown in Fig. 6.69. Find at $t =$ 0.5 s the short-circuit currents at the fault point F. The generators and transformers are identical and both generators are equipped with an AVR.

Solution

The network reactances, referred to the basic power $S_B = 100$ MVA, are shown in Fig. 6.69(b) and (c). Note that the reactances of the high voltage winding of the transformers are not taken into account due to the symmetrical properties of the network relative to the fault point position.

By simplification of the positive-sequence circuit, we have

$$X_{1eq} = X_{2eq} = \frac{0.24 + 0.06}{2} + 0.23 = 0.38,$$

and for the zero-sequence circuit

$$X_{11} = 0.06//(2\cdot 0.12 + 0.06) = 0.5, \quad X_{12} = 0.05 + 0.8 = 0.85$$

and

$$X_{0eq} = 0.85//0.5 = 0.31.$$

In accordance with equation 6.165 we may calculate the positive-sequence of the short-circuit current

$$\tilde{I}_1 = \frac{1.15}{j(0.38 + 0.38//0.31)} = -j2.1.$$

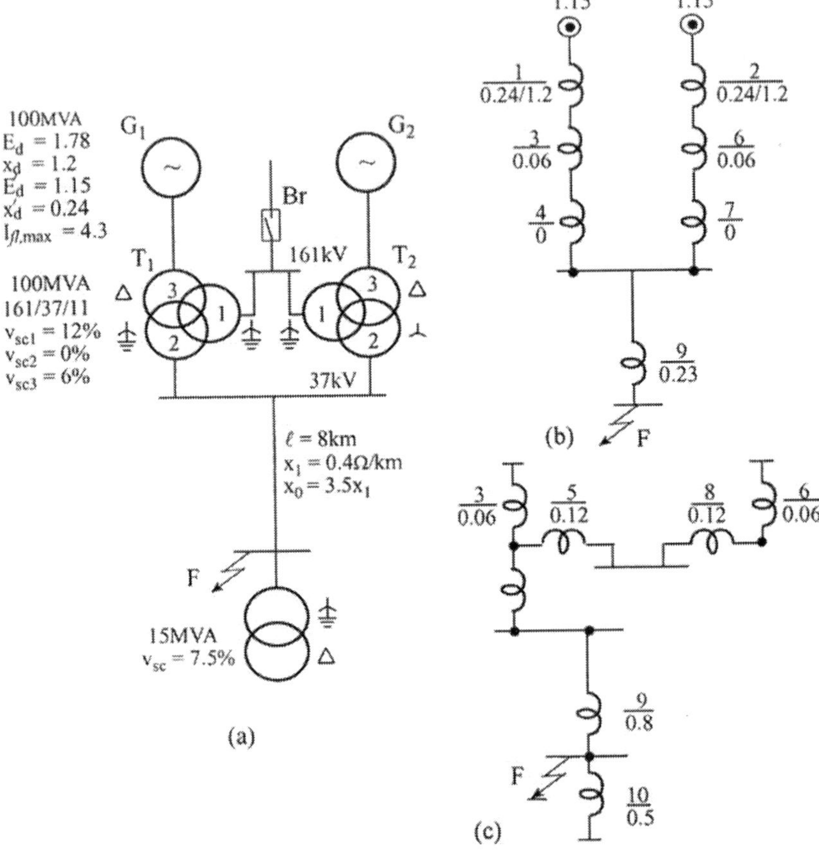

Figure 6.69 A network diagram for Example 6.28 (a), positive- and negative-sequence circuit (b) and zero-sequence circuit (c).

In accordance with Fig. 6.66(b) we have

$$\tilde{I}_2 = -\tilde{I}_1 \frac{X_{0eq}}{X_{2eq} + X_{0eq}} = j2.1 \frac{0.31}{0.38 + 0.31} = j0.94,$$

and

$$\tilde{I}_0 = -(\tilde{I}_1 + \tilde{I}_2) = -(-j2.1 + j0.94) = j1.16.$$

Thus, the first moment short-circuit current (phase b) is

$$\tilde{I}_{sc} = a^2\tilde{I}_1 + a\tilde{I}_2 + \tilde{I}_0 = (-0.5 - j0.866)(-j2.1) + (-0.5 + j0.866)(j0.94) + j1.16$$
$$= -2.63 + j1.74 \quad \text{or} \quad I_{sc} = 3.15.$$

Performing the same calculation for the steady-state s.c. yields

$$X_{1eq} = X_{2eq} = \frac{1.2 + 0.06}{2} + 0.23 = 0.86.$$

The zero-sequence resistances do not change, therefore

$$X_{0eq} = 0.31.$$

The positive-sequence of the steady-state s.c. current can now be calculated as

$$\tilde{I}_{1,\infty} = \frac{1.8}{j(0.86 + 0.86//0.31)} = -j1.65.$$

The negative- and zero-sequences of the s.c. current are

$$\tilde{I}_{2,\infty} = -\left(-j1.65\frac{0.31}{0.86 + 0.31}\right) = j0.44,$$

$$\tilde{I}_{0,\infty} = -(-j1.65 + j0.44) = j1.21.$$

The short-circuit current (phase b) is then found as

$$I_{\infty} = 2.57.$$

With the maximal field current, $I_{fl,\max} = 4.3$, we have

$$I_{\infty,\max} = \frac{4.3}{1.2 - 0.1} = 3.91,$$

and

$$\Delta I'_0 = 3.15 - 2.57 = 0.58, \quad \Delta I_{\infty} = 3.91 - 2.57 = 1.34.$$

Suppose that the transient time constants (see example 6.12) are $T'_d = 1.8$ s and $T_{ff} = 0.55$ s. Then the s.c. current at $t = 0.5$ s will be

$$I_{sc}^{(1,1)}(0.5) = 0.58e^{-0.5/1.8} + 2.57 + 0.08 \cdot 1.34 = 3.1,$$

where (see Example 6.12) $F(0.5) = 1 + 0.44e^{-0.5/0.55} - 1.44e^{-0.5/1.8} = 0.08$. As can be seen, the s.c. current, due to AVR action, has almost not changed.

6.7.3 Power in terms of symmetrical components

In general, the three-phase complex power of an unbalanced three-phase system can be expressed as the sum of three complex powers of each phase

$$\bar{S}_{3ph} = P_{3ph} + jQ_{3ph} = \bar{S}_a + \bar{S}_b + \bar{S}_c = \tilde{V}_a\tilde{I}_a^* + \tilde{V}_b\tilde{I}_b^* + \tilde{V}_c\tilde{I}_c^*. \tag{6.173}$$

The above in matrix notation will be

$$\bar{S}_{3ph} = [\tilde{V}_a \quad \tilde{V}_b \quad \tilde{V}_c]\begin{bmatrix}\tilde{I}_a\\ \tilde{I}_b\\ \tilde{I}_c\end{bmatrix}^* = \begin{bmatrix}\tilde{V}_a\\ \tilde{V}_b\\ \tilde{V}_c\end{bmatrix}^T\begin{bmatrix}\tilde{I}_a\\ \tilde{I}_b\\ \tilde{I}_c\end{bmatrix}^* \tag{6.174}$$

or

$$\bar{\mathbf{S}}_{3ph} = \tilde{\mathbf{V}}_{abc}^T \tilde{\mathbf{I}}_{abc}^*.$$

Using the matrix transformations

$$\tilde{\mathbf{V}}_{abc} = \mathbf{a}\tilde{\mathbf{V}}_{012}, \quad \tilde{\mathbf{I}}_{abc} = \mathbf{a}\tilde{\mathbf{I}}_{012}.$$

we may write

$$\tilde{\mathbf{V}}_{abc}^T = \tilde{\mathbf{V}}_{012}^T \mathbf{a}^T, \quad \tilde{\mathbf{I}}_{abc}^* = \mathbf{a}^* \tilde{\mathbf{I}}_{012}^*.$$

Substituting these equations into equation 6.174 we obtain

$$\bar{\mathbf{S}}_{3ph} = \tilde{\mathbf{V}}_{012}^T \mathbf{a}^T \mathbf{a}^* \tilde{\mathbf{I}}_{012}^*,$$

where

$$\mathbf{a}^T\mathbf{a}^* = \begin{bmatrix} 1 & 1 & 1 \\ 1 & a^2 & a \\ 1 & a & a^2 \end{bmatrix} \begin{bmatrix} 1 & 1 & 1 \\ 1 & a & a^2 \\ 1 & a^2 & a \end{bmatrix} = 3 \begin{bmatrix} 1 & 0 & 0 \\ 0 & 1 & 0 \\ 0 & 0 & 1 \end{bmatrix}.$$

Therefore

$$\bar{\mathbf{S}}_{3ph} = 3\tilde{\mathbf{V}}_{012}^T \tilde{\mathbf{I}}_{012}^* = 3[\tilde{V}_0 \quad \tilde{V}_1 \quad \tilde{V}_2] \begin{bmatrix} \tilde{I}_0 \\ \tilde{I}_1 \\ \tilde{I}_2 \end{bmatrix}^* \tag{6.175a}$$

or

$$\bar{S}_{3ph} = 3(\tilde{V}_0 \tilde{I}_0^* + \tilde{V}_1 \tilde{I}_1^* + \tilde{V}_2 \tilde{I}_2^*). \tag{6.175b}$$

This significant result means that there are no **cross terms** (e.g., $\tilde{V}_0 I_1^*$ or $\tilde{V}_1 I_2^*$) in the expression of a power (equation 6.175). In other words, there is no **coupling** of power among three sequences. It is also important to mention that the symmetrical components of three-phase voltages and currents belong to the same phase, i.e., in equation 6.175 all the sequence components are of phase a (the subscript of phase a here is just ignored).

Example 6.29

For the motor operated under unbalanced conditions of Example 6.27 determine the power delivered to the motor. Perform the calculations in two ways: (a) using the symmetrical components of currents and voltages; and (b) straightforwardly by calculating each phase power.

Solution

(a) First we shall calculate the sequence voltages across the motor. Since the zero-sequence current is zero, the zero-sequence voltage is also zero. The positive-sequence voltage may be calculated as

$$\tilde{V}_{1M} = Z_{1M}\tilde{I}_1 = (\sqrt{2}\cdot 3.6\angle 45^\circ)(39.6\angle -47.4^\circ)$$
$$= 202\angle -2.4^\circ \text{ V},$$

and similarly, the negative-sequence voltage is

$$\tilde{V}_{2M} = Z_{2M}\tilde{I}_2 = (0.15 + j0.5)(-39.6\angle -47.4^\circ)$$
$$= 20.7\angle -154.1^\circ \text{ V}.$$

Therefore, in accordance with equation 6.175b, we have

$$\bar{S}_M = 3[(202\angle -2.4^\circ)(39.6\angle 47.4^\circ) + (20.7\angle -154.1)(-39.6\angle 47.4^\circ)]$$
$$= (24.0\angle 45^\circ + 2.46\angle 73.3^\circ)\cdot 10^3 = 17.68 + j19.33 \text{ kW}.$$

(b) In accordance with equation 6.173 and substituting the results of the previous example, we have

$$\bar{S}_M = 0 + (203\angle -116.6^\circ)(68.5\angle 137.4^\circ) + (220\angle 114.8)(68.5\angle -42.6^\circ)$$
$$= (13.91\angle 20.8^\circ + 15.07\angle 72.2^\circ)\cdot 10^3 = 17.61 + j19.29 \text{ kW}.$$

Note that the minor differences (less than 0.5%) in results (a) and (b) are due to rounding off the calculated numbers.

6.8 TRANSIENT OVERVOLTAGES IN POWER SYSTEMS

Transients occurring in a power system, primarily as a result of switching and lightning strokes, cause overvoltages whose peak values can be much in excess of the normal operating voltage. The first kind of overvoltage, caused by switching, is considered an inner overvoltage, and the second kind, which is caused by lightning, is considered an outer overvoltage. Estimation and/or calculation of such overvoltages is of importance in the design of a power system, particularly in consideration of the insulation requirements and the protective equipment for the lines, transformers, generators etc.

Until recently outer overvoltages have been largely determined by the insulation requirements. However, with much higher operating voltages now in use of 500 kV and 750 kV, and the projected range likely to be 1000–1500 kV, the inner overvoltages due to switching have become the major consideration.

The outer overvoltages appear on an overhead conductor of transmission lines, caused by a lightning stroke, which can be as high as 200 kA (although an average value is in the order of 20 kA). When such a current stroke arrives on an overhead conductor, two equal current surges propagate in both directions away from the point of impact. The magnitude of each voltage surge is estimated therefore as $(1/2)Z_c i_{peak}$, where Z_c is the conductor surge impedance, usually of the value of 350 Ω to 400 Ω. Thus, the average voltage surges on a 400Ω transmission line will have a peak of $(400/2)\cdot 20\cdot 10^3 = 4000$ kV. A detailed analysis of the traveling current and voltage waves, and the different methods of calculating the overvoltage in transmission lines, is given in the next chapter. We will now continue with our consideration of inner overvoltages.

6.8.1 Switching surges

From our previous considerations (see section 2.7.4) we know that when a.c. circuits are to be interrupted, as in the case of a switching short-circuit in any line (Fig. 6.70(a)), the arc between the circuit-breaker contacts occurs and when breaking the arc the recovery voltage suddenly appears across the open gap.

In our previous analysis of this circuit, however, we had assumed an instantaneous switching, i.e. that the air gap resistance was increased from zero to infinity in zero time. Omitting the detailed analysis of the *phenomenon of the burning arc* caused by an interrupted a.c. current (which is beyond the scope of this book), we may perform our analysis under the assumption that the arc has a constant length and possesses a rectangular volt-ampere characteristic, shown in Fig. 6.71.

However, during the *quashing period of the arc*, the arc voltage does not

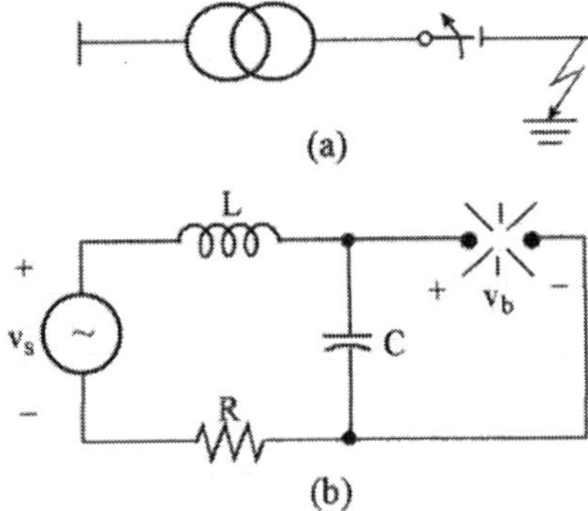

Figure 6.70 Switching of an s.c. fault: a network diagram (a) and an equivalent circuit (b).

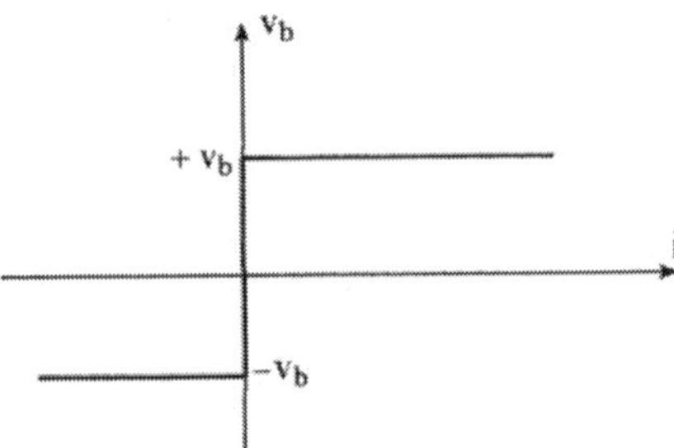

Figure 6.71 A rectangular characteristic of a burning a.c. arc.

remain constant, but gradually increases. After the interruption in such circuits, like Fig. 6.70, transient oscillations occur, which have been analyzed in sections 2.7.3 and 2.7.4. With equations 2.62, 2.63 and 2.95, 2.96 they are

$$i_n = I_n e^{-\alpha t} \sin(\omega_n t + \beta), \quad v_{C,n} \cong V_{C,n} e^{-\alpha t} \sin(\omega_n t + \beta - 90°), \qquad (6.176a)$$

where $\omega_n = 1/\sqrt{LC}$, $\alpha = R/2L$, $\tan\beta = (\omega/\omega_n)\tan\psi_i$,

$$V_{C,n} \cong V_S \sqrt{\left(\frac{\omega_n}{\omega}\right)^2 \sin^2\psi_i + \cos^2\psi_i}, \quad I_n = \sqrt{\frac{C}{L}}\, V_{C,n}. \qquad (6.176b)$$

Such oscillations will occur, after the circuit-breaker contacts start to move, at every zero passage of the current. At a few of the first passages, however, since the *restriking voltage* is higher than the electric strength of the arc, the arc will reignite, Fig. 6.72. As can be seen this happens four times at the reversal of the current during the gradual separation of the contacts. The last time, however, the transient voltage of the *restriking oscillation* does not succeed in igniting the arc again and the circuit is ultimately switched. The number of reignitions depends on the speed of the contact's separation and the electric strength of the arc, which in turn depends on the deionization process, i.e. on the diffusion and recombination of the ions and the temperature of the arc and the electrodes(*).

To analyze the overvoltages under the influence of a burning arc, we shall derive the differential equations for the circuit, shown in Fig. 6.73, in which for simplicity sake the relatively small resistances are neglected. Thus, with Kirchhoff's two laws we have

$$L\frac{di}{dt} + v_B = v_S, \quad v_B = v_C = \frac{1}{C}\int i_C, \quad i = i_C + i_B, \qquad (6.177)$$

where v_B is the voltage across and i_B is the current through the arc.

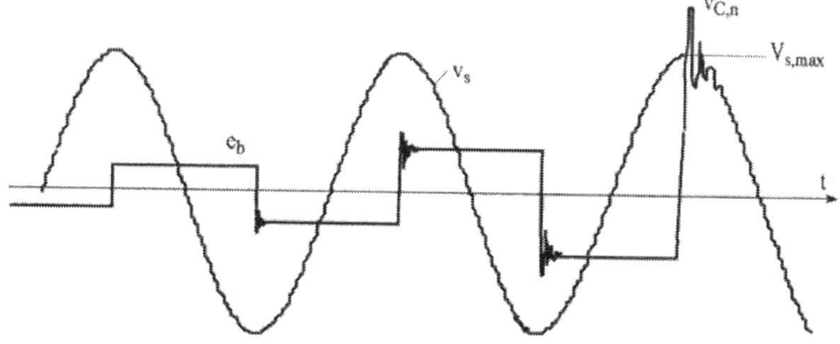

Figure 6.72 Transient oscillations during the contact separation.

(*)For a more detailed analysis of the restriking voltage after interruption see in R. Rudenberg (1969), *Transient Performance of Electric Power Systems*, MIT Press.

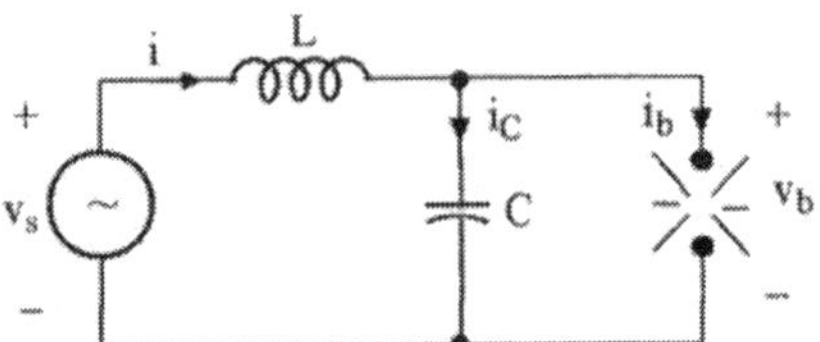

Figure 6.73 An equivalent circuit for analyzing the influence of a burning arc.

Or, substituting the third equation into the first one and expressing the capacitance current by the second yields

$$LC\frac{d^2 v_B}{dt^2} + v_B + L\frac{di_B}{dt} = v_s. \tag{6.178}$$

Since the arc has a resistive characteristic, the voltage and current will have the same form, and it is suitable to assume for the voltage (and current) the form of the exponent

$$v_B = v_{B,0}e^{(t-t_0)/T_B} = v_{B,0}e^{-t'/T_B}, \tag{6.179}$$

where t_0 is the time at which the quenching starts and T_B is the *quenching time constant*, i.e. of the deionization process. This time constant is different for different types of quenching agents: for air it is about 10^{-3} s, for gases, as in oil breakers, 10^{-4} s and for pure hydrogen 10^{-5} s.

The voltage change in accordance with equation 6.179 is shown in Fig. 6.74. Before the time t_0 the arc voltage within every half period will be nearly constant, as in Fig. 6.71, and after t_0 it will rise according to equation 6.179, which means that with arc quenching its electric strength will increase. As can be seen, the *quenching curves of the current* (starting at t_0) depend on the capacitance in parallel to the arc: from $t'=0$ when $C=\infty$, to $t'=\pi/\omega$ when $C=0$ (see further on).

Substituting equation 6.179 into equation 6.178 yields

$$L\frac{di_B}{dt} = v_s - \frac{1}{\omega_n^2 T_B^2}v_{B,0}e^{t'/T_B} - v_{B,0}e^{t'/T_B}, \tag{6.180}$$

and by straightforward integration we have

$$i_B = i_{B,0} - v_{B,0}\frac{T_B}{L}\left(1+\frac{1}{\omega_n^2 T_B^2}\right)(e^{t'/T_B}-1), \tag{6.181}$$

where $i_{B,0}$ is the current at t_0, Fig. 6.74. This current can be found by integrating only the first equation in 6.177 (the capacitance prior to interruption does not act), and the solution is

$$i_B = \frac{V_s}{\omega L}\sin(\omega t + \theta) + \frac{v_B}{\omega L}(\pi/2 - \omega t). \tag{6.182}$$

This current consists of two components: the first represents the steady-state

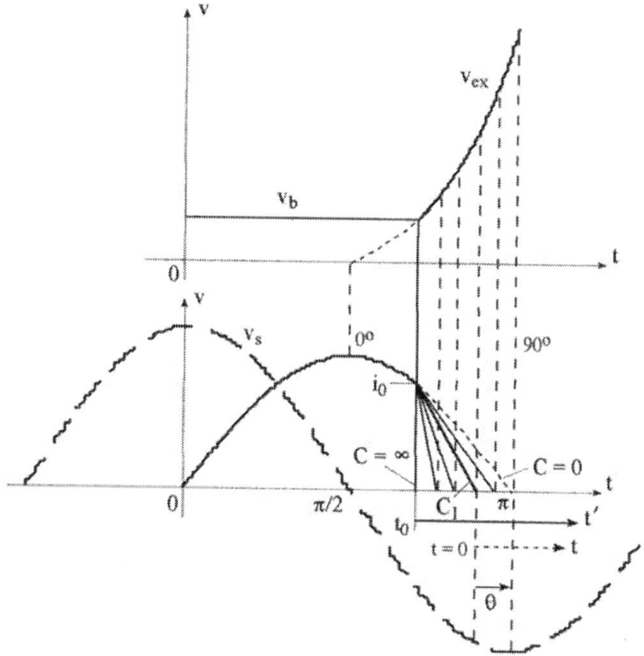

Figure 6.74 Exponentially increasing quenching voltage and current curve.

inductive current in the circuit with a closed circuit breaker and lags behind the applied voltage by 90°; the second component represents the linearly changing arc current. Note that the arc voltage, which opposes the current flow and thereby changes the phase of the current, brings it more into phase with the supply voltage, so that it will be less than 90°.

The displacement angle θ may be found, using the condition that in the quasi-steady-state regime the current must pass each half period through zero. Thus,

$$\sin\theta = \frac{\pi}{2}\frac{v_B}{V_s}. \tag{6.183a}$$

Knowing θ, the initial current i_0 can be found from its expression 6.182, or from the plot in Fig. 6.74, and may be approximated as

$$i_0 = I\sin\theta. \tag{6.183b}$$

In accordance with equation 6.181, under the effect of an exponentially increasing quenching voltage, the current will decrease exponentially passing through zero. This will happen when $i_B = 0$, then in accordance with equation 6.181 and since $i_{B,0} = i_0$ we have

$$e^{t'/T_B} = \frac{i_0 L}{v_{B,0} T_B(1 + 1/\omega_n^2 T_B^2)} + 1. \tag{6.184}$$

The amplitude of the current prior to interruption can be determined approximately as

$$I \cong \frac{V_s}{\omega L}, \tag{6.185a}$$

then

$$i_0 L = \frac{i_0 V_s}{I\omega}. \tag{6.185b}$$

Solving equation 6.184 for t' by using equation 6.185b, we have

$$t' = T_B \ln\left[1 + \frac{i_0}{I}\frac{V_s}{v_{B,0}}\frac{1}{\omega T_B(1 + 1/\omega_n^2 T_B^2)}\right]. \tag{6.186}$$

Due to the relatively small capacitance, the natural frequency is very high so that the quantity $1/\omega_n^2 T_B^2$ might be approximated as a unity. Then, with time constant $T_B \cong 10^{-4}$ and with the most common ratios $v_{B,0}/V_s \cong 1/20$ and $i_0/I \cong 1/5$, we have

$$t' = T_B \ln\left(1 + \frac{1}{5}\cdot 20\cdot \frac{1}{377\cdot 10^{-4}\cdot 2}\right) \cong 4T_B.$$

Thus, the approximate quenching time is about four times the quenching time constant.

The *extinction voltage* in p.u. at the moment that the current attains zero is determined by substituting condition 6.184 into equation 6.179, and using equation 6.185b

$$v_{B,pu}|_{i=0} = v_{ex,pu} = \frac{v_{B,0}}{V_s} + \frac{i_0/I}{\omega T_B(1 + 1/\omega_n^2 T_B^2)}. \tag{6.187a}$$

Now, equations 6.186 and 6.187a are giving better insight into the part of the capacitance, in parallel to the arc. Thus, without the capacitance, the natural frequency would be $\omega_n \to \infty$ and the quenching time will be maximal, while the p.u. extinction voltage will reach the value:

$$v_{ex,pu} = \frac{v_{B,0}}{V_s} + \frac{i_0/I}{\omega T_B}. \tag{6.187b}$$

With the previously used data, this voltage will be

$$v_{ex,pu} = 1/20 + \frac{1/5}{377\cdot 10^{-4}} \cong 5.4,$$

On the other hand, by using a very large capacitance the natural frequency approaches zero, so that the second term under the logarithm in equation 6.186 disappears and the quenching time therefore reduces to zero. The physical explanation of this result is that the current shifts instantaneously from the arc to the capacitance. The extinction p.u. voltage in this case with a moderate capacitance, giving a natural frequency of 10^3 Hz, will be much lower:

$$v_{ex,pu} = \frac{1}{20} + \frac{1/5}{377 \cdot 10^{-4}[1 + 1/(2\pi \cdot 10^3 \cdot 10^{-4})^2]} \cong 1.5.$$

Our next goal is to derive the actual values of the restriking voltage, in accordance with (6.176) and applying the switching laws. The charging current, prior to the passage of the arc current through zero, is

$$i_C = C\frac{d}{dt}(v_{B,0}e^{t'/T_B}) = \frac{C}{T_B}v_{B,0}e^{t'/T_B}, \tag{6.188}$$

which increases exponentially as does the voltage. At the end of the quenching period this current, by substituting equation 6.184, is

$$i_C(t') = \frac{C}{T_B}v_{B,0} + \frac{i_0}{1 + (\omega_n T_B)^2}. \tag{6.189}$$

If the extinction voltage, at the moment of passing the current zero, has risen to a value above the burning voltage of the arc, there is an expectation that with the reversal of the current the electric strength of the arc will withstand the appearing restriking voltage. For the sake of simplicity, we shall neglect the small burning voltage $v_{B,0}$ in equations 6.187 and 6.189, and also the very small current of fundamental frequency through the capacitance in the steady state. Note also that at the moment of the appearance of the restriking voltage the arc is extinct and $i_C = i_L$.

The initial conditions with equations 6.187a and 6.189, at $t = 0$, will now be

$$\begin{aligned} V_{C,n}\cos\beta' &= v_C(0) - v_{C,f}(0) = \frac{V_s(i_0/I)}{\omega T_B(1 + 1/\omega_n^2 T_B^2)} - V_s(-\cos(-\theta)) \\ I_n \sin\beta' &= i_C(0) - i_{C,f}(0) = \frac{i_0}{1 + \omega_n^2 T_B^2} - 0, \quad \text{where } \beta' = 90^\circ - \beta. \end{aligned} \tag{6.190}$$

(Note that at $t = 0$, the moment of passing the current zero, the forced capacitance voltage, i.e., the applied voltage v_s, has the initial angle $-\theta$, as shown in Fig. 6.74, for instance, for capacitance C_1.)

Dividing the first equation in 6.190 by the second one and noting that $I_n/V_{C,n} = \sqrt{C/L}$ (see equation 6.176b), we obtain for the initial phase angle β' of the restriking oscillations

$$\cot\beta' = \sqrt{\frac{C}{L}}\left[\frac{V_s(1 + \omega_n^2 T_B^2)}{\omega I T_B(1 + 1/\omega_n^2 T_B^2)} + \frac{V_s\cos\theta(1 + \omega_n^2 T_B^2)}{i_0}\right]. \tag{6.191}$$

Using equations 6.183b and 6.185a, and substituting the natural frequency from equation 6.176b, simplifies the above expression to

$$\cot\beta' = \omega_n T_B + (1 + \omega_n^2 T_B^2)\frac{\omega}{\omega_n}\cot\theta, \tag{6.192}$$

where θ is related to the end of the quenching period, Fig. 6.74.

Note that for instantaneous switching, i.e. with $T_B = 0$, this expression reduces to the previous one, as in equation 6.176. However, by comparing equation 6.192 with those in 6.176a one should take into consideration that here θ comes instead of ψ_i and the current is taken as a basic phasor, which means that β' must be replaced by $(90° - \beta)$.

In power networks the value of ω/ω_n is always about 10^{-2} or lower, which in turn results in values of $\omega_n T_B$ greater than unity, and equation 6.192 then simplifies to

$$\cot \beta' = \omega_n T_B(1 + \omega T_B) \cot \theta. \tag{6.193}$$

Furthermore, if the product $\omega_n T_B$ is substantially larger than unity, as with most of the circuit breakers in practice, $\cot^2 \beta'$ will also be large compared to unity, so thus

$$1/\sin \beta' = \sqrt{1 + \cot^2 \beta'} \cong \cot \beta',$$

and with equation 6.193 for the current in equation 6.190 we have

$$I_n = i_0 \frac{\cot \beta'}{1 + \omega_n^2 T_B^2} \cong i_0 \left(\frac{1}{\omega_n T_B} + \frac{\omega}{\omega_n} \cot \theta \right). \tag{6.194}$$

Now the amplitude of the transient voltage in equation 6.190 with equations 6.194, 6.183b and 6.185a becomes

$$V_{C,n} = \sqrt{\frac{L}{C}} I_n = \sqrt{\frac{L}{C}} \frac{V_s}{\omega L} \sin \theta \left(\frac{1}{\omega_n T_B} + \frac{\omega}{\omega_n} \cot \theta \right),$$

or, after simplification

$$V_{C,n} = V_s \left(\frac{1}{\omega T_B} \sin \theta + \cos \theta \right). \tag{6.195}$$

Comparing these results with (6.176b), given for instantaneous interruption, we see that in the first term the reciprocal value of the quenching time constant T_B has taken the place of the natural frequency ω_n. Checking these results numerically we may obtain for a medium network frequency of 10 kHz, with instantaneous interruption, that the restriking voltage would be (wherein the insignificant term with $\cos \theta$ is omitted):

$$V_{C,n} = \frac{\omega_n}{\omega} V_s \sin \theta = \frac{2\pi \cdot 10 \cdot 10^3}{377} V_s \sin \theta = 167 V_s \sin \theta.$$

By gradual interruption with the quenching time constant of 10^{-4} s, an amplitude develops of only

$$V_{C,n} = \frac{1}{\omega T_B} V_s \sin \theta = \frac{1}{377 \cdot 10^{-4}} \sin \theta = 26.5 V_s \sin \theta.$$

Considering now the premature extinction by 10°, the interruption angle will be $\theta = 10°$, as is often found with the interruption of s.c. currents, the p.u. restriking voltage amplitude will be

$$V_{C,n,pu} = \frac{V_{C,n}}{V_s} = 26.5 \sin 10° = 4.6.$$

The restriking voltage will then be reduced by the damping effect at the rate of the damping coefficient α (equation 6.176a). Note that such restriking oscillations as shown in Fig. 6.75 start from the last extinction voltage v_{ex}.

Once again recall that the physical reason for the much smaller restriking voltage amplitude and the more favorable initial phase angle is the fact that, by increasing the arc voltage, the current is shifted away from the arc to the shunt capacitance before the final interruption is established. A resistance connected across the contacts of the circuit breaker can significantly increase the damping effect. With such a resistance the oscillation may be critically damped when $R_{dam} \geq \frac{1}{2}\sqrt{L/C}$, so that the severity of the transient will be reduced.

6.8.2 Multiple oscillations

This kind of oscillation will occur if the circuit breaker is located not at the place of the short-circuit, but rather at some distance away, as shown in Fig. 6.76. This may represent a case in which the circuit breaker is located in between a generator-fed bus and a current-limiting reactor.

The voltages across the two circuit meshes before the interruption are

$$V_1 = \frac{L_1}{L_1 + L_2} V_s, \quad V_2 = \frac{L_2}{L_1 + L_2} V_s.$$

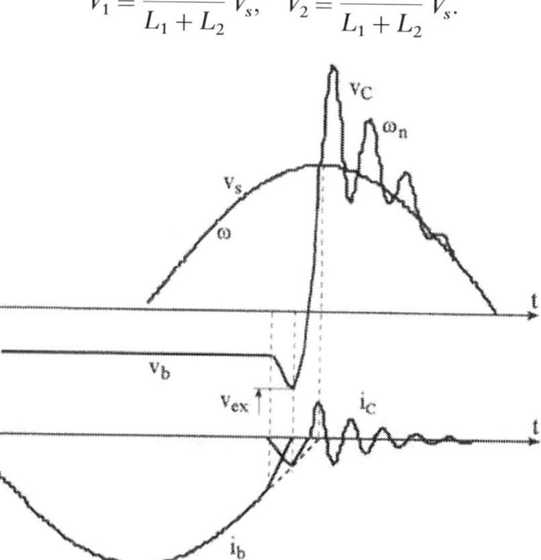

Figure 6.75 Restriking oscillations of the capacitance voltage and current.

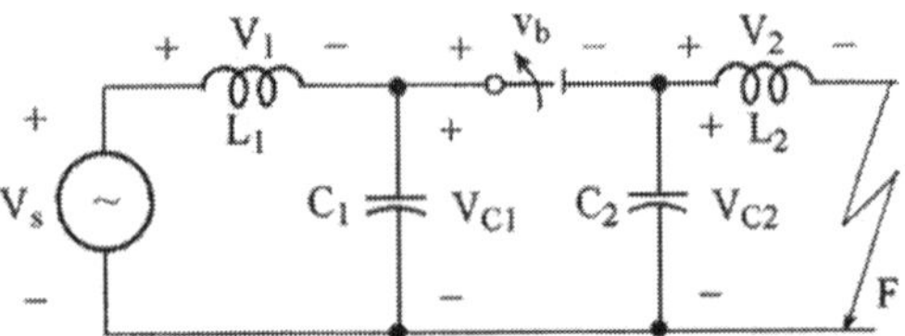

Figure 6.76 A case where a circuit breaker is connected between two meshes.

After the fault, from the instant at which the arc is finally extinct, the two circuits are separated and each oscillates at its own natural frequency.

$$\omega_{n1} = \frac{1}{\sqrt{L_1 C_1}} \quad \text{and} \quad \omega_{n2} = \frac{1}{\sqrt{L_2 C_2}}.$$

The voltage across the circuit breaker is then given by the difference between the two capacitive voltages

$$v_B = v_{C1} - v_{C2} = V_s \cos \omega t - V_1 e^{-t/T_1} \cos \omega_{n1} t - V_2 e^{-t/T_2} \cos \omega_{n2} t.$$

Figure 6.77 shows the above voltages after interruption at zero current. As can be seen from this figure, the restriking voltage across the circuit breaker has a more complicated form due to the summation of two oscillations at different frequencies.

Example 6.30

Determine the overvoltage surge set up on a 66 kV cable fed through a bulk-oil circuit breaker, when the breaker opens on a short-circuit fault. The network and breaker parameters are $R = 7.8\ \Omega$, $L = 6.5$ mH, $C = 0.16\ \mu$F and $T_B = 10^{-4}$ s. In order to increase the damping effect the shunt resistor R_{sh} is connected in parallel to the capacitance. What should its value be in order to damp the oscillations during 2 or 3 natural periods?

Solution

The natural frequency is

$$\omega_n = \frac{1}{\sqrt{LC}} = \frac{1}{\sqrt{6.5 \cdot 10^{-3} \cdot 0.16 \cdot 10^{-6}}} = 3.1 \cdot 10^4 \text{ rad/s}.$$

The $\omega_n T_B$ product is

$$v = \omega_n T_B = 3.1 \cdot 10^4 \cdot 10^{-4} = 3.1.$$

With the assumption of a premature extinction by 10° as is often found with the interruption of short-circuit currents, we will have $\theta = 10°$ and $i_0/I = \sin 10° = 0.173$. The quenching time, with equation 6.186 and with the previously

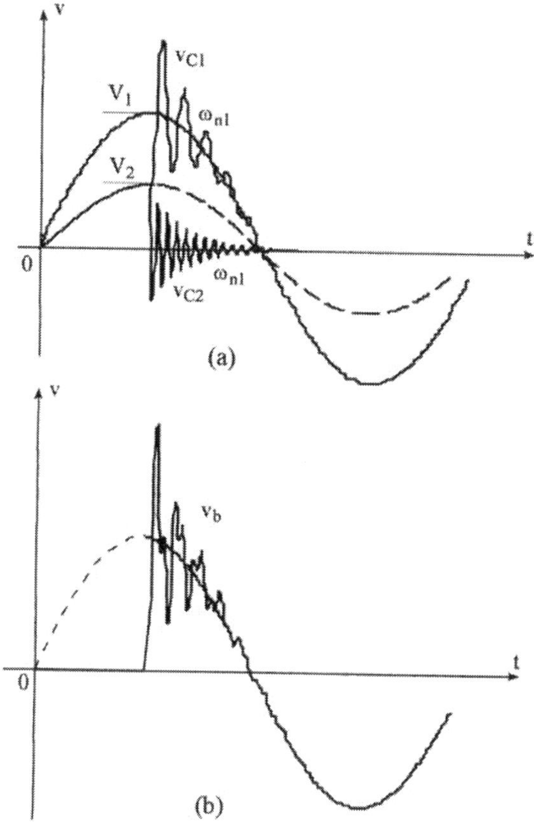

Figure 6.77 Development of the voltages across two capacitances (a) and the voltage across the circuit breaker (b).

assumed ratio $v_B/V_s = 1/120$, will be

$$t' = 10^{-4}\ln\left(1 + \frac{0.173\cdot 20}{377\cdot 10^{-4}(1 + 1/3.1^2)}\right) = 10^{-4}\ln 82.7 = 0.442 \text{ ms},$$

or $\omega t' = 377\cdot 0.442\cdot 10^{-3} = 0.167$ rad and the quenching angle will be $\theta = 0.167\cdot 57.3 = 9.6^\circ$ (as about what was assumed). Next, we determine the initial angle β' (equation 6.193):

$$\cot\beta' = v(1 + \omega T_B)\cot\theta = 0.31(1 + 377\cdot 10^{-4})\cot 10^\circ = 18.2,$$

and $\beta' = 3.15^\circ$. The p.u. extinction voltage (equation 6.187a) is

$$v_{ex,pu} = \frac{1}{20} + \frac{0.173}{377\cdot 10^{-4}(1 + 1/3.1^2)} = 4.1.$$

The amplitude of the transient voltage becomes (equation 6.195)

$$V_{C,n} = V_s\left(\frac{1}{377 \cdot 10^{-4}} \sin 10^\circ + \cos 10^\circ\right) \cong 5.6V_s,$$

and the amplitude of the capacitance current is (equation 6.194)

$$I_{C,n} = i_0 \frac{\cot \beta'}{1 + \omega_n^2 T_B^2} = i_0 \frac{18.2}{1 + 3.1^2} = 1.71 \cdot 0.173I = 0.3I.$$

Note that for the same circuit the amplitude of the voltage oscillation, in accordance with equation 6.176b, i.e., ignoring the arc and quenching time as happens with an instantaneous switching at the same premature angle of 10°, should be

$$V_{C,n} = V_s\sqrt{\left(\frac{\omega_n}{\omega}\right)^2 \sin^2 \psi_i + \cos^2 \psi_i}$$

$$= V_s\sqrt{\left(\frac{3.1 \cdot 10^4}{377}\right)^2 \sin^2 10^\circ + \cos^2 10^\circ} = 14.2V_s,$$

which is almost three times higher than in this example.

The desired damping may be derived by using a 2000 Ω resistor. Indeed, the reader may easily convince himself that in this case the damping coefficient is

$$\alpha = \frac{R}{2L} + \frac{1}{2R_{sh}C} = \frac{7.8}{2 \cdot 6.5 \cdot 10^{-3}} + \frac{1}{2 \cdot 2000 \cdot 0.16 \cdot 10^{-6}}$$

$$= 4.62 \cdot 10^3 \text{ 1/s}, \quad \text{or} \quad \tau \cong 0.2 \text{ ms},$$

which is about $2.3T_n$.

Finally, we have:

$$v_{C,n}(t) = 5.6V_s e^{-2.16 \cdot 10^3 t} \sin(3.1 \cdot 10^4 t - 3.15^\circ)$$

$$i_{C,n}(t) = 0.3Ie^{-9.6 \cdot 10^3 t} \cos(3.1 \cdot 10^4 t + 86.9^\circ).$$

Here the initial angle β' is negative as the interruption is assumed to have occurred when the current changes from negative to positive values, as shown in Fig. 6.78.

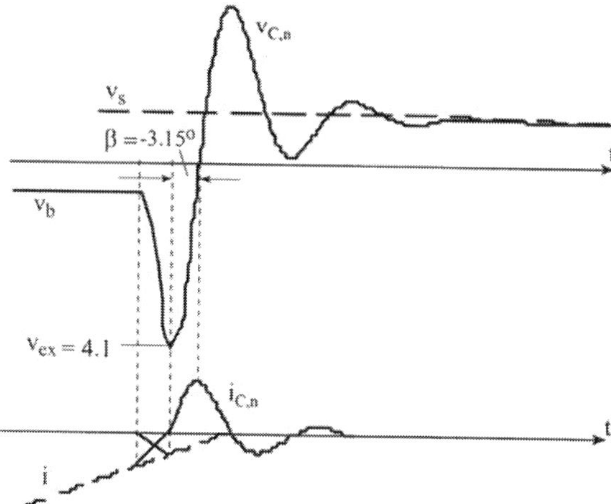

Figure 6.78 Restriking oscillation of the capacitance voltage and current.

Chapter #7

TRANSIENT BEHAVIOR OF TRANSMISSION LINES (TL)

7.1 INTRODUCTION

Transient phenomena in TL occur, like in networks with bulk parameters, when any change in their parameters, driving sources and/or configuration takes place. In general, the transients are caused by lightning, switching or faults in TL. Studies of transient disturbances on a transmission system have shown that changes are followed by traveling waves, which at first approximation can be treated as step front waves. For example, when the lightning's strike influences a line conductor, the induced voltage wave tends to divide into two halves, with the two halves *going in opposite directions.* When a voltage wave reaches a power transformer, for example, it causes a stress distribution, which is not uniform and may lead to the breakdown of the insulation system. Transient phenomena also occur in communication systems when signals of different forms are transmitted along the transmission line.

As the transmission line is a network with distributed parameters, its transient analysis, like the steady-state behavior, has to be based on partial differential equations.

7.2 THE DIFFERENTIAL EQUATIONS OF TL AND THEIR SOLUTION

Let R, G, L and C be the uniformly distributed parameters of the homogeneous line throughout its length (i.e. related to the unit of line length). Then we can represent the long line as a chain of an infinite number of incremental sections dx with the parameters: resistance Rdx, inductance Ldx, conductance Gdx and capacitance Cdx connected in series and parallel as shown in Fig. 7.1i. Let x be the distance from the sending-end to the considered section of the line; v and i be the voltage and the current at the beginning of section dx and $v + (\partial v/\partial x)dx$ and $i + (\partial i/\partial x)dx$ at the end of section dx.

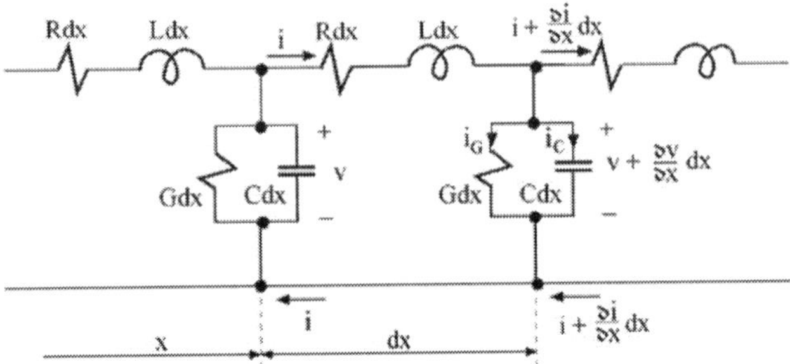

Figure 7.1i The incremental section of a transmission line.

Note that the voltage and current in a transmission line are functions of two variables x and t. We can now write two equations for this section by applying Kirchhoff's two laws:

$$v = \left(v + \frac{\partial v}{\partial x} dx\right) + Rdx\, i + Ldx \frac{\partial i}{\partial t}$$

$$i = \left(i + \frac{\partial i}{\partial x} dx\right) + Gdx\left(v + \frac{\partial v}{\partial x} dx\right) + Cdx \frac{\partial}{\partial t}\left(v + \frac{\partial v}{\partial x} dx\right).$$

Combining similar terms, dividing by dx and neglecting the quantities of second order infinitesimality, we obtain two *differential equations of partial derivatives*:

$$\begin{aligned} -\frac{\partial v}{\partial x} &= Ri + L\frac{\partial i}{\partial t}, \\ -\frac{\partial i}{\partial x} &= Gv + C\frac{\partial v}{\partial t}. \end{aligned} \tag{7.1}$$

Equations 7.1 are known in classical physics as the equations of telegraphy. They reduce to wave equations if R and G are set equal to zero. The solution of equation 7.1 with known initial and boundary (terminal) conditions allows for obtaining the line current and voltage in any point of the line as a function of time and distance from a terminal point.

The influence of resistance R and conductance G relative to L and C in transmission lines is negligible (especially for fast running processes like high frequency signals or transient phenomena). In addition, since the traveling time of waves is relatively small, the influence of losses is scarcely significant. So to simplify the analysis the line will be assumed to be loss-less. Therefore

$$\frac{\partial v}{\partial x} = -L\frac{\partial i}{\partial t} \tag{7.2a}$$

$$\frac{\partial i}{\partial x} = -C\frac{\partial v}{\partial t}. \tag{7.2b}$$

Note that the negative signs in equations 7.2 are due to the fact that both voltage v and current i decrease as x increases (the direction at which distance x advances along the line).

Taking the partial derivative of equation 7.2a with respect to x and the derivative of equation 7.2b with respect to t, we obtain

$$\frac{\partial^2 v}{\partial x^2} = -L\frac{\partial^2 i}{\partial x\,\partial t} \tag{7.3}$$

$$\frac{\partial^2 i}{\partial x\,\partial t} = -C\frac{\partial^2 v}{\partial t^2}. \tag{7.4}$$

Substituting equation 7.3 into equation 7.4, current i can be eliminated, so that

$$\frac{\partial^2 v}{\partial x^2} = LC\frac{\partial^2 v}{\partial t^2}. \tag{7.5}$$

Similarly, voltage v can be eliminated, so that

$$\frac{\partial^2 i}{\partial x^2} = LC\frac{\partial^2 i}{\partial t^2}. \tag{7.6}$$

Equations 7.5 and 7.6 are known as wave equations – they are identical for both v and i. When one of these functions is found, the other can be found by applying either equation 7.2a or 7.2b.

The solution of the wave equation can be determined intuitively. Paying attention to the fact that the second derivatives of the voltage v and current i functions, with respect to t and x, have to be directly proportional to each other, means that the solution can be any function as long as both independent variables t and x appear in the form:

$$w_{1,2} = x \pm vt. \tag{7.7}$$

Therefore, usually the solution of equation 7.5 will be

$$v(x,t) = v_1 + v_2 = f_1(x - vt) + f_2(x + vt), \tag{7.8}$$

which satisfies equation 7.5.

In order to ensure this and determine the meaning of v, let us substitute one of the functions (equation 7.8), for example f_1, in equation 7.5. Its first derivative with respect to x is:

$$\frac{\partial v_1}{\partial x} = \frac{\partial f_1}{\partial w_1}\frac{\partial w_1}{\partial x} = \frac{\partial f_1}{\partial w_1}, \tag{7.9a}$$

and the second derivative is

$$\frac{\partial^2 v_1}{\partial x^2} = \frac{\partial^2 f_1}{\partial w_1^2}. \tag{7.9b}$$

The first derivative of equation 7.8 with respect to t is:

$$\frac{\partial v_1}{\partial t} = \frac{\partial f_1}{\partial w_1}\frac{\partial w_1}{\partial t} = \frac{\partial f_1}{\partial w_1}(-v), \tag{7.10a}$$

and the second derivative is

$$\frac{\partial^2 v_1}{\partial t^2} = v^2 \frac{\partial^2 f_1}{\partial w_1^2}. \tag{7.10b}$$

Substituting equations 7.9b and 7.10b in equation 7.5 yields

$$\frac{\partial^2 f_1}{\partial w_1^2} = LC\, v^2 \frac{\partial^2 f_1}{\partial w_1^2}.$$

This equation becomes an equality, if $LC\, v^2 = 1$, or

$$v = \frac{1}{\sqrt{LC}} \text{ (m/s)}. \tag{7.11}$$

Hence, v having unit meters per second represents the velocity and, as will be shown in the following paragraph, it is the *velocity of the voltage and current wave propagation* along the line. Similarly, it can be shown that the second term (f_2) in equation 7.8 satisfies equation 7.5 with the same meaning of v.

Now the current function i may be found in accordance with equations 7.2a and 7.9a. Indeed, substituting first $\partial/\partial w_1$ (where f_1 is the first function of equation 7.8) into equation 7.2a for $\partial v/\partial x$ gives

$$\frac{\partial f_1}{\partial w_1} = -L\frac{\partial i_1}{\partial t},$$

and after integration, with respect to t

$$\int \frac{\partial f_1}{\partial w_1} dt = -L \int \frac{\partial i_1}{\partial t} dt$$

yields $[1/(-v)]f_1 = -Li_1$, since $\partial w_1/\partial t = -v = \text{const}$, or

$$i_1 = \frac{1}{vL} f_1(x - vt) = \frac{1}{Z_c} v_1, \tag{7.12}$$

where

$$Z_c = vL = \frac{L}{\sqrt{LC}} = \sqrt{\frac{L}{C}} \tag{7.13}$$

is the *characteristic impedance* of a loss-less transmission line.

Following the same steps, the second part of the current, i.e., i_2, may be obtained with only a difference in the sign. Indeed, after the integration of equation 7.13 for f_2 and i_2 we obtain $(1/v)f_2 = -Li_2$ or

$$i_2 = -\frac{1}{vL} f_2(x + vt) = -\frac{1}{Z_c} v_2. \tag{7.14}$$

Therefore, the entire current function is

$$i(x, t) = \frac{1}{Z_c} [f_1(x - vt) - f_2(x + vt)] = i_1 + i_2. \tag{7.15}$$

In conclusion, it must be mentioned that the actual shape of the voltage and current functions and their components f_1 and f_2 is defined by the initial and boundary (or terminal) conditions of a given problem, and also by the activating sources.

7.3 TRAVELING-WAVE PROPERTY IN A TRANSMISSION LINE

The behavior of the voltage and current functions of equations 7.8 and 7.15 can be understood by selecting some particular point on the wave (zero-crossing, maximum/minimum etc.) and checking (following) it for different instances of time. This result may be achieved by keeping the argument of v_1 (or i_1) constant, for example, for point A of $v_1 = 0$ in Fig. 7.1(a).

$$w_A = x - vt = \text{const.} \tag{7.16a}$$

This means that when t increases, x increases too, so $\Delta x = v\Delta t$ and this particular point A moves a distance of Δx, as shown in Fig. 7.1(a). Thus, the voltage function v_1, if plotted as a function of x for consecutive values of time as shown in Fig. 7.1(a) (bold line), appears to move in the positive ($+x$) direction (broken line). Hence, v_1 and i_1 are said to be the **forward-traveling waves** v_f and i_f (or **incident waves**).

Similarly, checking v_2 (or i_2) and keeping

$$w_B = x + vt = \text{const.} \tag{7.16b}$$

causes x to decrease as t increases, i.e., $\Delta x = -v\Delta t$, which means that a particular point (for example, point B) on the v_2 wave shown in Fig. 7.1(c) appears to move in the negative ($-x$) direction.

Hence, v_2 and i_2 are said to be the **backward-traveling waves** v_b and i_b (or **reflected waves**). In both cases, v represents the velocity of the voltage and current wave propagation, or simply the **velocity of propagation.**

In loss-less transmission lines the waves of voltage and current propagate without changing their shape. The measurement instrument, such as an oscilloscope, which is connected for example at the point x_1 of the line, will show the voltage wave as a function of time as shown in Fig. 7.1(b) (bold line). Note that the viewed curve is similar (although in a different scale) to the voltage distribution on the line, i.e., as a function of x. The oscilloscope connected at the next point x_2 will show the same curve (broken line) but with a time delay of $\Delta t =$

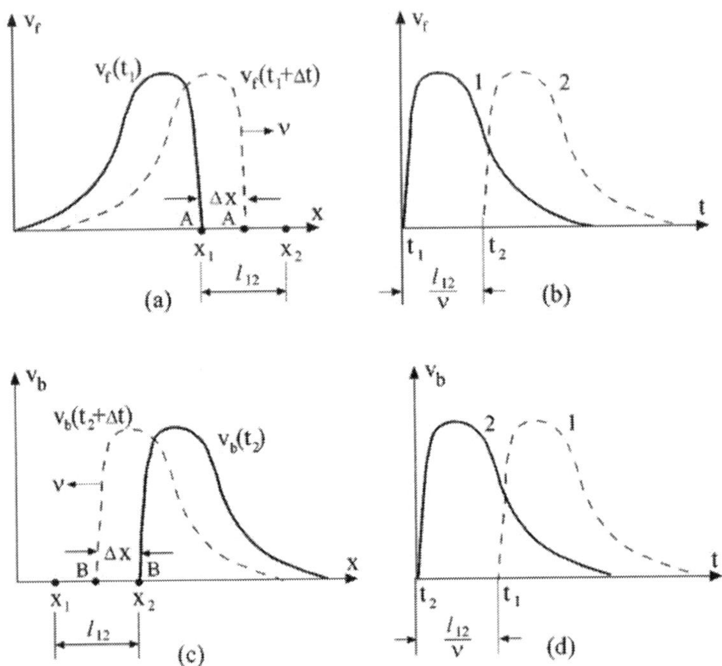

Figure 7.1 Traveling voltage wave as a function of distance x and as a function of time t: (a) and (b) forward-traveling wave, (c) and (d) backward-traveling wave.

$(x_2 - x_1)/v = \ell_{12}/v$, where ℓ_{12} is the distance between the points x_1 and x_2 (Fig. 7.1(b). The backward-traveling wave pattern for different values of time is shown in Fig. 7.1(d).

In conclusion, it must be mentioned that at any point on the line, including points of discontinuity (i.e., at the end of the line at the point of the connection of two different lines, etc.), the instantaneous voltage and current can be expressed as

$$v = v_f + v_b \tag{7.17a}$$

$$i = i_f + i_b, \tag{7.17b}$$

where the voltage and current traveling wave pair is connected by the characteristic impedance of the line Z_c:

$$i_f = \frac{v_f}{Z_c}, \quad i_b = -\frac{v_b}{Z_c}. \tag{7.18}$$

The negative sign in the relation between the voltage and current of the backward-traveling waves is important. It is not dependent on how either the coordinate system or the positive polarity/direction of voltage/current may be chosen. It is understandable in terms of the power shown that the power of a backward-traveling wave is always negative, which indicates a movement of energy in the negative direction of x, i.e. in the direction of travel of the $v_b(x, t)$ and $i_b(x, t)$ waves.

It can be shown that in the transient behavior of TL, like in the steady state regime, the power of a forward-traveling wave for example can be expressed in terms of energy content and wave propagation velocity

$$P_f = W_f v, \tag{7.19}$$

where $P_f = v_f i_f$ is the *power of the forward traveling wave.*

Indeed, the energies stored in electric (C) and magnetic (L) fields per unit length of TL are

$$W_e = \frac{1}{2} C v_f^2, \quad W_m = \frac{1}{2} L i_f^2. \tag{7.20}$$

Since the two components of energy storage are equal, the *total energy content stored per unit length* is

$$W_f = W_e + W_m = C v_f^2 = L i_f^2. \tag{7.21}$$

Therefore, the above power can be expressed as

$$P_f = v_f i_f = \frac{v_f^2}{Z_c} = \frac{C v_f^2}{\sqrt{LC}} = W_f v. \tag{7.22a}$$

Of course, the same result can be obtained for a backward-traveling wave

$$P_b = W_b v. \tag{7.22b}$$

Note that the *total transient power* is a sum of these two components, i.e., the forward- and backward-traveling waves:

$$P = vi = (v_f + v_b)(i_f - i_b) = v_f i_f + v_b i_b + (v_b i_f - v_f i_b) = P_f + P_b$$

since

$$(v_b i_f - v_f i_b) = \left(\frac{v_b v_f}{Z_c} - \frac{v_f v_b}{Z_c} \right) = 0.$$

Example 7.1

The surge voltage of 1000 kV caused by lightning propagates along the transmission line, having the distributed parameters $L = 1.34$ mH/km and $C = 8.6$ nF/km.

Determine: (a) The surge power in the line, (b) The surge current in the line.

Solution

(a) The total energy stored in an electromagnetic field per unit length of the line is

$$W = Cv^2 = 8.6 \cdot 10^{-9}(1000 \cdot 10^3)^2 = 8.6 \cdot 10^3 \text{ J/km}$$

The surge velocity is

$$v = \frac{1}{\sqrt{LC}} = \frac{1}{\sqrt{1.34 \cdot 10^{-3} \cdot 8.6 \cdot 10^{-9}}} = 295 \cdot 10^3 \text{ km/s}$$

Therefore, the surge power is

$$P = 8.6 \cdot 10^3 \cdot 295 \cdot 10^3 \cong 2500 \text{ MW}.$$

(b) The characteristic impedance of the line is

$$Z_c = \sqrt{\frac{L}{C}} = \sqrt{\frac{1.34 \cdot 10^{-3}}{8.6 \cdot 10^{-9}}} = 395\ \Omega.$$

Therefore the surge current is

$$i = \frac{v}{Z_c} = \frac{1000 \cdot 10^3}{395} = 2.53 \text{ kA}.$$

It should be noted that for the transient analysis of some problems, when the voltage and current change versus time is needed, it is more convenient to express the voltage and current-traveling waves in the form

$$v = v_f + v_b = \phi_1\left(t - \frac{x}{v}\right) + \phi_2\left(t + \frac{x}{v}\right) \tag{7.23a}$$

$$i = i_f - i_b = \frac{1}{Z_c}\left[\phi_1\left(t - \frac{x}{v}\right) - \phi_2\left(t - \frac{x}{v}\right)\right]. \tag{7.23b}$$

7.4 WAVE FORMATIONS IN TL AT THEIR CONNECTIONS

In practice, all kinds of transmission lines are necessarily terminated by sources or by loads. In addition, a lumped impedance or lumped admittance network may be inserted in tandem between sections of a line, or two (or more) different lines may be connected in a network junction. As has been already mentioned, to determine traveling-wave functions, the boundary conditions of line terminations must be taken into consideration. In other words, at any point of such non-uniformness or discontinuity, i.e., transition points, Ohm's and Kirchhoff's law equations must be obeyed in addition to traveling-wave equations.

Therefore, if the voltage and current at such a transition point are known, it can be written as

$$v_T = v_f + v_b \tag{7.24}$$

$$i_T = i_f + i_b. \tag{7.25}$$

Taking into consideration the relation between the voltage and current traveling waves (equation 7.18), according to equation 7.25 we obtain

$$Z_c i_T = v_f - v_b. \tag{7.26}$$

Adding equations 7.24 and 7.26 we have

$$v_T + Z_c i_T = 2v_f, \tag{7.27}$$

or the forward-traveling wave will be

$$v_f = \frac{1}{2}(v_T + Z_c i_T). \tag{7.28}$$

Similarly, by subtracting equation 7.26 from equation 7.24 we obtain

$$v_b = \frac{1}{2}(v_T - Z_c i_T), \tag{7.29}$$

also from equation 7.26.

$$v_b = v_f - Z_c i_T. \tag{7.30}$$

7.4.1 Connecting the TL to a d.c./a.c. voltage source

Consider the a.c. source, which is connected at $t = 0$ to a power transmission line. Such a source of industrial frequency of 50–60 Hz hardly changes during the time which is needed for a wave to propagate in hundreds of kilometres (note that the wave propagation time along a line of 1000 km length is about 30 μs as the period of the 50 Hz voltage source is $20 \cdot 10^3$ μs). Therefore, in the initial stage of wave formation, the a.c. source can be treated as a d.c. source.

Now consider the transmission line connecting at the time $t = 0$ to the d.c. source V_s having input impedance Z_s as shown in Fig. 7.2(a). As the line was not initially charged, no backward (reflection) waves exist at the first moment after the connection. Therefore equations 7.24 and 7.25 yield

$$v_T = v_f, \quad i_T = i_f. \tag{7.31}$$

Applying the boundary condition

$$v_T = V_s - Z_s i_T \tag{7.32}$$

and solving equations 7.28 and 7.32 with equation 7.31 yields

$$v_f = \frac{Z_c}{Z_s + Z_c} V_s = \rho_s V_s, \tag{7.33}$$

where

$$\rho_s = \frac{Z_c}{Z_s + Z_c} \tag{7.33a}$$

is the **source transmission coefficient**.

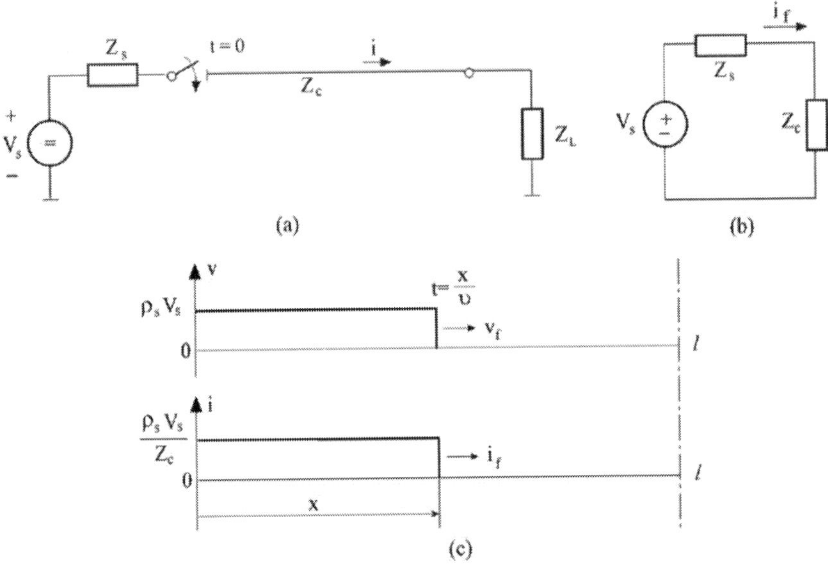

Figure 7.2 Waves traveling on a line by connecting to a source: circuit diagram (a), equivalent circuit (b), voltage and current distribution (c).

Note that the forward-traveling wave voltage (equation 7.33) might be determined in accordance with the equivalent circuit shown in Fig. 7.2(b) as the voltage across the characteristic impedance Z_c.

If the connecting source is ideal ($Z_s = 0$), the forward-traveling wave is simply $v_f = V_s$. Assuming at this point that the source input impedance is pure resistive, we can conclude that the voltage distribution along the line will be just the voltage at the sending-end of the line with the time delay $t = x/v$ as is shown in Fig. 7.2(c), i.e., a step function wave.

Disconnecting a transmission line from the source also causes step function waves to appear (Fig. 7.3). Assume that, at the disconnecting moment, the current in the line was I_s and the voltage was V_s. Since after the disconnection the current at the sending-end becomes zero $i_T = i_f + i_b + I_s = 0$ and noting that no reflection wave yet exists, we obtain

$$i_f = -I_s, \quad v_f = -Z_c I_s. \tag{7.34}$$

The voltage distribution will be a sum of the previous voltage V_s and the forward-traveling wave

$$v(x, t) = V_s - Z_c I_s \left(t - \frac{x}{v} \right), \tag{7.35}$$

and it will be positive if the current I_s is smaller than the charge current V_s/Z_c (i.e., $I_s < (V_s/Z_c)$) as shown in Fig. 7.3(b) or negative if vice versa.

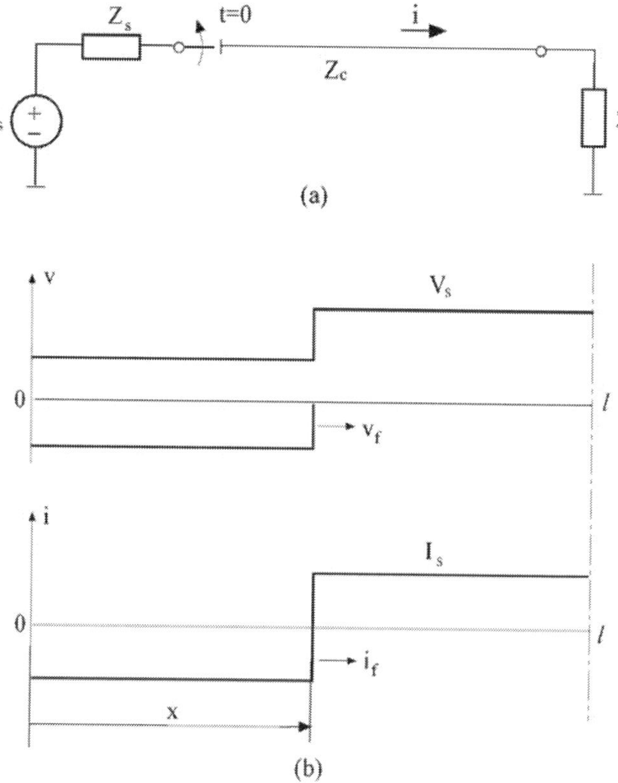

Figure 7.3 Waves traveling on line by disconnecting from a source: circuit diagram (a), voltage and current distribution (b).

7.4.2 Connecting the TL to load

Consider the transmission line shown in Fig. 7.4(a). After turning on the switch, the load impedance terminates the TL and the backward-traveling wave will appear. To determine it, Ohm's law must be obeyed for the receiving terminal of the line:

$$v_{\mathrm{T}} = Z_L i_{\mathrm{T}}. \tag{7.36}$$

Taking into consideration that the line was charged by voltage, say V_s, the equations 7.24 and 7.25 with equation 7.18 yield

$$v_{\mathrm{T}} = v_{\mathrm{b}} + V_s, \quad i_{\mathrm{T}} = i_{\mathrm{b}} = -\frac{v_{\mathrm{b}}}{Z_{\mathrm{c}}}. \tag{7.37}$$

Solving equations 7.36 and 7.37 gives

$$v_{\mathrm{b}} = -\frac{Z_{\mathrm{c}}}{Z_L + Z_{\mathrm{c}}} V_s, \quad i_{\mathrm{b}} = \frac{V_s}{Z_L + Z_{\mathrm{c}}}. \tag{7.38}$$

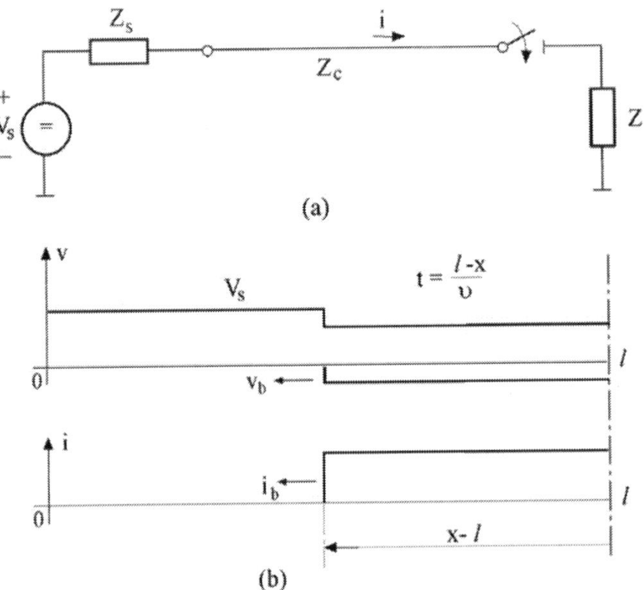

Figure 7.4 Traveling waves by load connection: circuit diagram (a): voltage and current distribution (b).

Assuming that load impedance Z_L is pure resistive ($Z_L = R_L$), the back-traveling wave will be a step-wave as shown in Fig. 7.4(b):

$$v_b(x, t) = \rho_\tau V_s\left(t + \frac{x - \ell}{v}\right), \tag{7.39}$$

where $\rho_\tau = -Z_c/(R_L + Z_c)$ is the **load transmission coefficient**.

The step-wave also appears at the time when the resistive load is disconnecting just as in disconnecting the source (Fig. 7.5(a)), $i_b = -I_L$ and $v_b = Z_c I_L$, where I_L is the load current in the line at the moment of disconnection.

The voltage distribution will be the sum of the previous voltage V_L and the backward-traveling wave:

$$v(x, t) = V_L + Z_c I_L\left(t + \frac{x - \ell}{v}\right), \tag{7.40}$$

as shown in Fig. 7.5(b). Note that disconnecting the load results in a voltage increase, which in power TL can be significant.

Using the above studied technique, situations that are more complicated can be solved as in the following example.

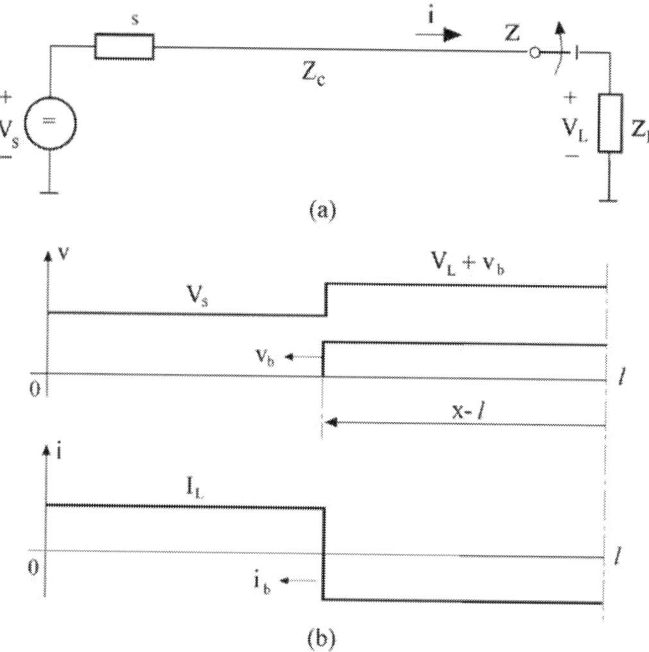

Figure 7.5 Traveling waves by load disconnection: circuit diagram (a); voltage and current distribution (b).

Example 7.2

Determine the voltage and current waves due to the connection of the resistive load $R_L = 300\,\Omega$ at the arbitrary point of the TL shown in Fig. 7.6(a). The characteristic impedance of TL is $400\,\Omega$ and it is charged with initial voltage $V_0 = 20$ kV and initial current $I_0 = 50$ A.

Solution

Since both directions of wave propagation (to the left and right from the connection point) are symmetrical, both current waves will be equal to each other, i.e., $i_f = i_b$ and $v_f = v_b$. Applying KCL and Ohm's law,

$$i_L = -(i_f + i_b) = -2i_f = -2i_b, \quad v_L = R_L i_L = V_0 + v_f = V_0 + v_b. \qquad (7.41)$$

Solving equation 7.41 with the relation $v_f = Z_c i_f$, we obtain

$$i_b = i_f = -\frac{V_0}{2R_L + Z_c} = -\frac{20 \cdot 10^3}{600 + 400} = -20\text{ A}$$

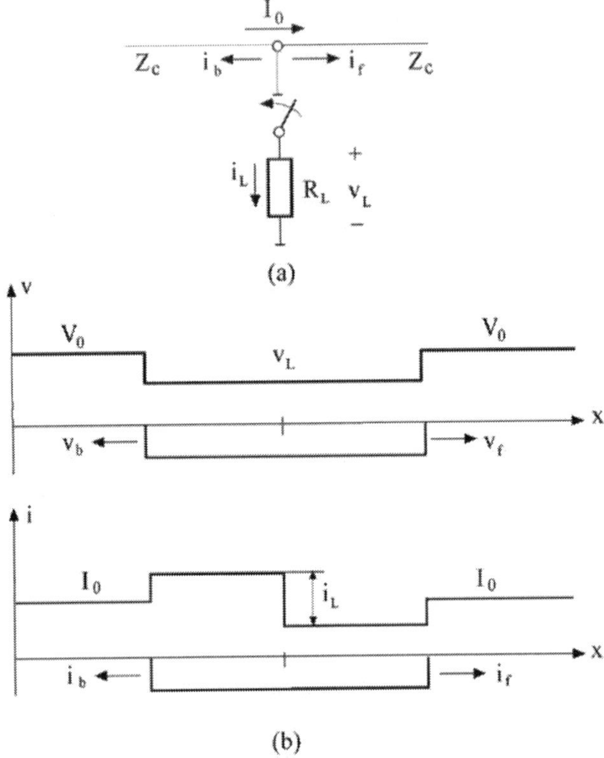

Figure 7.6 Traveling waves by load connection at an arbitrary point on the line: circuit diagram (a); voltage and current distribution (b).

and

$$v_b = v_f = Z_c i_f = -\frac{V_0 Z_c}{2R_L + Z_c} = -20\frac{400}{1000} = -8 \text{ kV}.$$

The voltage and current distribution along the TL is shown in Fig. 7.6(b).

7.4.3 A common method of determining traveling waves by any kind of connection

Consider an active network connecting to the junction of two lines, as shown in Fig. 7.7(a). The forward-traveling wave will appear on the right line and the backward-traveling wave on the left line. Both current waves can be determined from the equivalent circuit in which the active network is presented by its *Thévenin equivalent* and the two lines by their characteristic impedances, as shown in Fig. 7.7(b). Note that the voltage source of the Thévenin equivalent is simply the voltage across the switch at the zero initial condition of the lines.

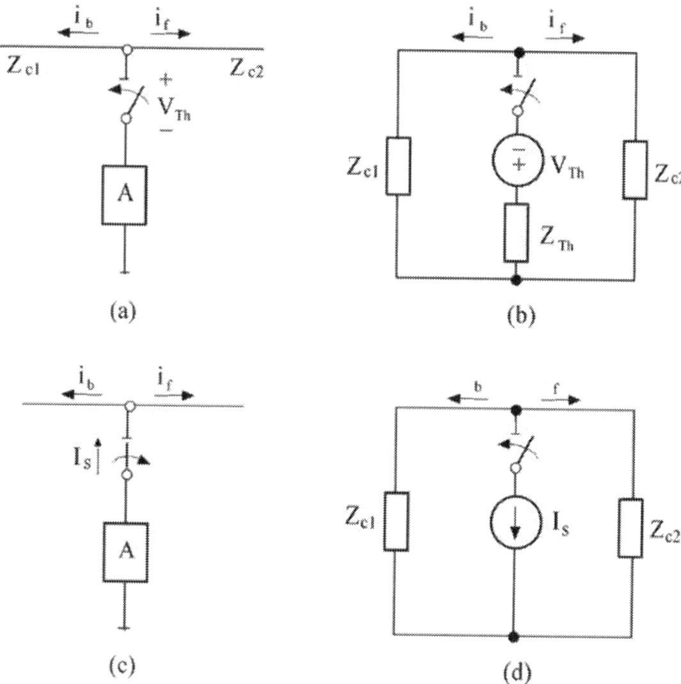

Figure 7.7 Active network connection and disconnection: circuit diagrams (a) and (c); the equivalent circuits for connection and disconnection (b) and (d).

If the lines were initially charged, the final voltage and current distribution will simply be the superposition of the initial values and the traveling waves.

If the switch is opening, i.e., the network is disconnecting, it has to be represented in the equivalent circuit by an ideal current source as shown in Fig. 7.7(c) and (d). Note that the value of the current source is equal to the current which flowed through the switch just before it opened.

As an example, consider the line connecting to the voltage source with inductive-resistive impedance, as shown in Fig. 7.8(a). Therefore, the equivalent circuit will simply be the series connection of the source and the characteristic impedance of the line as shown in Fig. 7.8b. The transient response of this circuit gives the forward-traveling current wave at the sending-end as

$$i_f = \frac{V_0}{R_0 + Z_c}\left(1 - e^{-\frac{t}{\tau}}\right),$$

where $\tau = L_0/(R_0 + Z_c)$ is the time constant of the circuit. The current distribution along the line will be

$$i_f(x, t) = \frac{V_0}{R_0 + Z_c}\left(1 - e^{-\frac{t - x/v}{\tau}}\right),$$

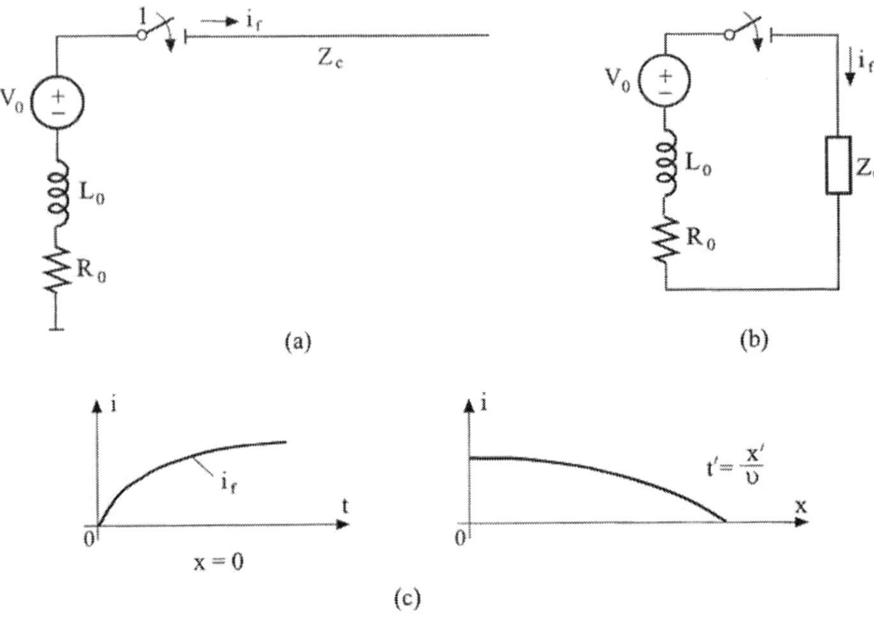

Figure 7.8 Voltage source with input impedance connecting to the TL: circuit diagram (a); the equivalent circuit (b); the current curve versus time, and current distribution along the line (c).

and is shown in Fig. 7.8(c), which corresponds to the moment of the arrival of the wave at point x'. The voltage wave is proportional to the current wave $v_f = Z_c i_f$.

7.5 WAVE REFLECTIONS IN TRANSMISSION LINES

Consider, at first, a step-function forward-traveling wave or **incident wave**. The moment that this wave reaches the receiving-end of the line (point 2) the **reflecting wave** will appear. In the general case of a line terminated in impedance Z_T, the boundary condition is simply Ohm's law[(*)]

$$v_T = Z_T i_T. \tag{7.42}$$

Let the front of the incident wave in expression 7.27 be V_0 ($v_f = V_0$), then we obtain

$$Z_T i_T + Z_c i_T = 2V_0, \tag{7.43}$$

[(*)] In cases where the line terminations consist of inductances or/and capacitances, expression (7.42) and subsequent ones are solved by means of the Laplace-transform method (see Chapter 3).

or

$$i_{\mathrm{T}} = \frac{2V_0}{Z_{\mathrm{T}} + Z_{\mathrm{c}}}. \tag{7.44}$$

Expression 7.44 shows that the current at the receiving-end of the line can be determined from the *equivalent lumped-impedance circuit* in which the line and its termination are represented by their impedances connected in series, while the circuit is activated by the double value of the incident wave, as shown in Fig. 7.9. The voltage at the receiving-end (equation 7.42) can be expressed as

$$v_{\mathrm{T}} = \frac{2Z_{\mathrm{T}}}{Z_{\mathrm{T}} + Z_{\mathrm{c}}} V_0 = \rho_{\mathrm{ref}} V_0, \tag{7.45}$$

where ρ_{ref} is the **refraction coefficient** or **transmission factor**

$$\rho_{\tau} = \frac{2Z_{\mathrm{T}}}{Z_{\mathrm{T}} + Z_{\mathrm{c}}}. \tag{7.46}$$

The reflecting or backward-traveling wave are easily obtained as

$$v_{\mathrm{b}} = v_{\mathrm{T}} - v_{\mathrm{f}} = \frac{2Z_{\mathrm{T}}}{Z_{\mathrm{T}} + Z_{\mathrm{c}}} V_0 - V_0 = \frac{Z_{\mathrm{T}} - Z_{\mathrm{c}}}{Z_{\mathrm{T}} + Z_{\mathrm{c}}} V_0 = \rho_{\mathrm{r}} V_0, \tag{7.47}$$

where ρ_{r} is the **receiving-end reflection coefficient**

$$\rho_{\mathrm{r}} = \frac{Z_{\mathrm{T}} - Z_{\mathrm{c}}}{Z_{\mathrm{T}} + Z_{\mathrm{c}}}. \tag{7.48}$$

When the reflecting wave arrives at the sending-end, it reflects again. The **sending-end reflection coefficient** will then be similar, i.e.,

$$\rho_{\mathrm{s}} = \frac{Z_{\mathrm{s}} - Z_{\mathrm{c}}}{Z_{\mathrm{s}} + Z_{\mathrm{c}}}, \tag{7.49}$$

where Z_{s} is the sending-end termination impedance or the generator input impedance.

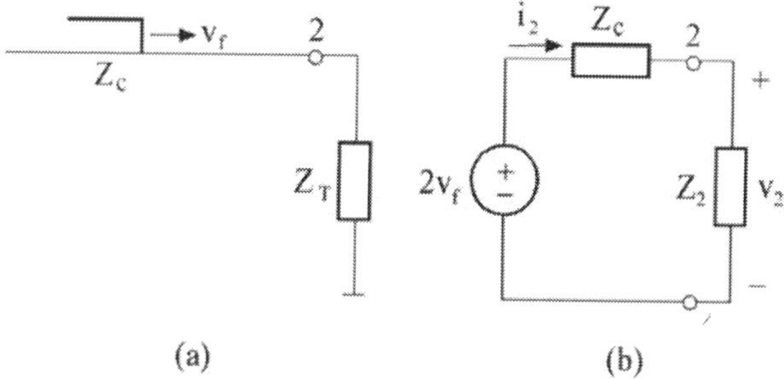

Figure 7.9 The incident wave arriving at the line termination: line diagram (a); equivalent circuit (b).

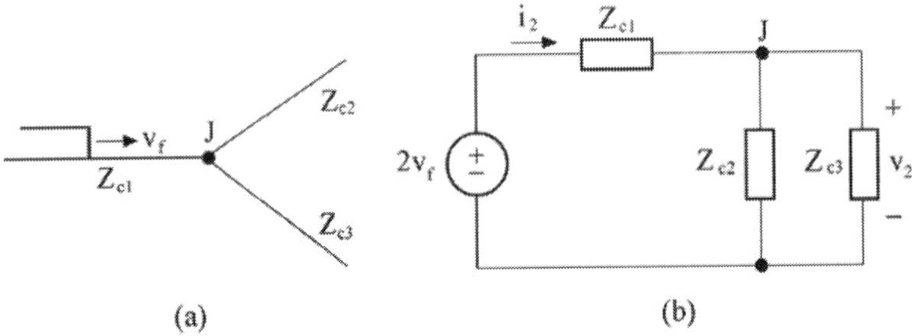

Figure 7.10 The junction of two outgoing lines: line diagram (a); equivalent circuit (b).

The concept of an equivalent circuit can be used for any kind of discontinuity. For example, in Fig. 7.10(a) the junction of three transmission lines is shown. The equivalent circuit is shown in Fig. 7.10(b) in which the two outgoing lines are represented by their characteristic impedances in parallel. The second example of an inductance connected between two transmission lines is shown in Figs. 7.11(a) and (b). Here the elements, which formed the junction of discontinuity, are represented in the equivalent circuit by their impedances in series.

In general, the equivalent circuit of any junction of discontinuity consists of lumped impedances, which represent the elements connected to the junction, and of the characteristic impedances of the lines. The circuit is driven by a voltage source of a double value of the incident wave voltage function.

Let us examine several particular kinds of TL terminations.

7.5.1 Line terminated in resistance

In this case the reflecting wave has the same shape as the incident wave, i.e. the shape of a step-function. (Note that the characteristic impedance of a loss-less line is also pure resistance.) The reflection wave is determined by the reflection

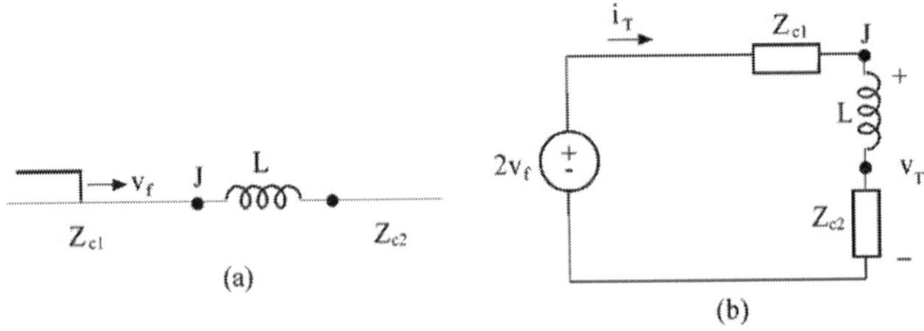

Figure 7.11 The connection of lumped-impedance in between two lines: line diagram (a); equivalent circuit (b).

coefficient (equation 7.47)

$$v_b = \rho_r V_f = \frac{R_T - Z_c}{R_T + Z_c} v_f. \tag{7.50}$$

The reflection coefficient ρ_r can be positive or negative, depending on the relative values of R_T and Z_c; it varies between ± 1 including zero, i.e., when $R_T = Z_c$ **(natural termination)**.

The current reflected wave (equation 7.18) is

$$i_b = -(v_b/Z_c).$$

Then, the voltage v_T and current i_T at the receiving-end are simply the sum of both the incident and reflected waves (equation 7.17):

$$v_T = v_f + v_b, \quad i_T = i_f + i_b. \tag{7.51}$$

Figure 7.12 shows the analysis of traveling waves when the line is terminated in a resistance that is larger than the line characteristic impedance (i.e., $R_T > Z_c$). Thus, v_b is positive and i_b is negative. Therefore, the traveling wave arrival at the line termination results in increased voltage and reduced current, as shown in Fig. 7.12(b). The opposite case when $R_T < Z_c$ is shown in Fig. 7.13. Here the

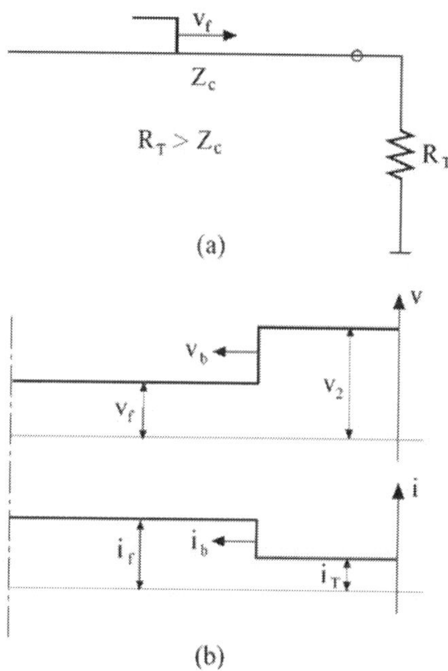

Figure 7.12 Traveling waves after arrival at termination in which $R_T > Z_c$: circuit diagram (a); voltage and current distributions (b).

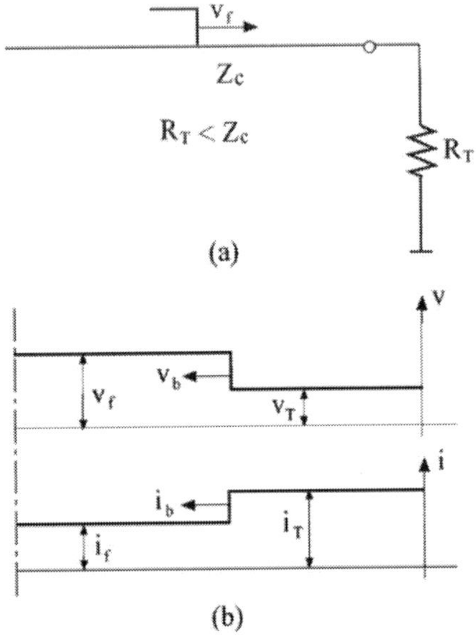

Figure 7.13 Traveling waves after arrival at termination in which $R_T < Z_c$: circuit diagram (a); voltage and current distributions (b).

traveling wave arrival at the line termination results in reduced voltage and increased currents as shown in Fig. 7.13(b).

Example 7.3

A line has a characteristic impedance of 400 Ω and a terminating resistance of 600 Ω. Assuming that the incident voltage wave is 100 kV, determine the following: (a) The reflection coefficient of the voltage wave; (b) The reflection coefficient of the current wave; (c) The backward-traveling voltage and current waves; (d) The voltage across and current through the resistor.

Solution

(a) $$\rho_{rv} = \frac{R_T - Z_c}{R_T + Z_c} = \frac{600 - 400}{600 + 400} = 0.2.$$

(b) $$\rho_{ri} = \frac{i_b}{i_f} = -\frac{v_b}{Z_c} \bigg/ \frac{v_f}{Z_c} = -\frac{v_b}{v_f} = -\rho_{rv} = -0.2.$$

(c) $$v_b = \rho_{rv} v_f = 0.2 \cdot 100 = 20\ \text{kV}$$

$$i_b = -\frac{v_b}{Z_c} = -\frac{20 \cdot 10^3}{400} = -50\ \text{A}.$$

(d) $v_T = v_f + v_b = 100 + 20 = 120\ \mathrm{kV}$

$$i_T = \frac{v_T}{R_T} = \frac{120 \cdot 10^3}{600} = 200\ \mathrm{A}.$$

7.5.2 Open- and short-circuit line termination

The boundary condition for the current in an open-circuit termination is $i_T = 0$. Therefore,

$$i_b = -i_f. \tag{7.52}$$

Using equation 7.18 yields

$$v_b = -Z_c i_b = Z_c i_f = v_f. \tag{7.53}$$

The same results, of course, can be obtained with a reflection coefficient. Since the *open-circuit termination* is an extreme termination in impedance $Z_T \to \infty$, the reflection coefficient is unity and $v_b = \rho_r v_f = v_f$. The total voltage at the open-end is $v_T = v_f + v_b = 2v_f$. Therefore, the voltage at the receiving-end is twice the forward voltage wave and this doubled value propagates on the line, as shown in Fig. 7.14(a).

The boundary condition for the voltage at the short-circuit termination is $v_T = 0$ and therefore

$$v_b = -v_f. \tag{7.54}$$

The *short-circuit termination* can be treated as the dual of the open-circuit termination. Therefore, the previous results for voltage traveling waves are now related to the current traveling waves and vice versa, concluding that the current at the short-circuited end of the line is twice the forward current wave as shown in Fig. 7.14(b).

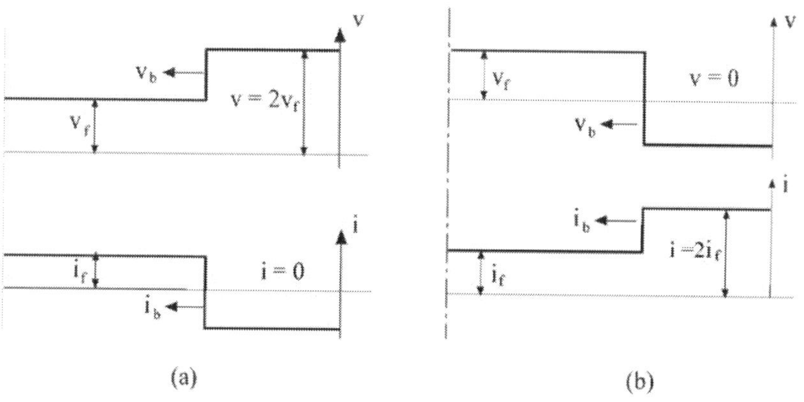

Figure 7.14 Traveling waves pattern for: open-circuit line termination (a); short-circuit line termination (b).

7.5.3 Junction of two lines

Applying the concept of an equivalent circuit to the junction of two lines (see Fig. 7.15(b)), we can conclude that this case is similar to the line terminated in resistance. Assume that $Z_{c1} > Z_{c2}$ where Z_{c1} and Z_{c2} are the characteristic impedances of the first and second lines, respectively. For example, it might represent the junction between an overhead line and an underground cable. If a voltage surge of a step function form approaches such a junction along the overhead line, the voltage at the junction decreases relative to the value of the increment wave. The voltage surge along the cable will be in accordance with the refraction coefficient (i.e. transmission factor):

$$v_{f2} = \rho_{ref} v_{f1} = \frac{2Z_{c2}}{Z_{c2} + Z_{c1}} v_{f1}. \tag{7.55}$$

The reflection, i.e. backward-traveling wave, in accordance with the reflection coefficient, is

$$v_{b1} = \rho_r v_{f1} = \frac{Z_{c2} - Z_{c1}}{Z_{c2} + Z_{c1}} v_{f1}. \tag{7.56}$$

Fig. 7.15(c) shows the waves occurring at the junction. It can be seen that the wave, which is refracted or transmitted to the cable, is equal to the sum of the forward and backward waves.

This property of cables to reduce the voltage surge is used in practice. When an overhead line is terminated by a transformer, the incident of a voltage surge on a transformer winding results in a very high voltage gradient at the winding turns nearest to the line conductor, and may lead to the breakdown of the

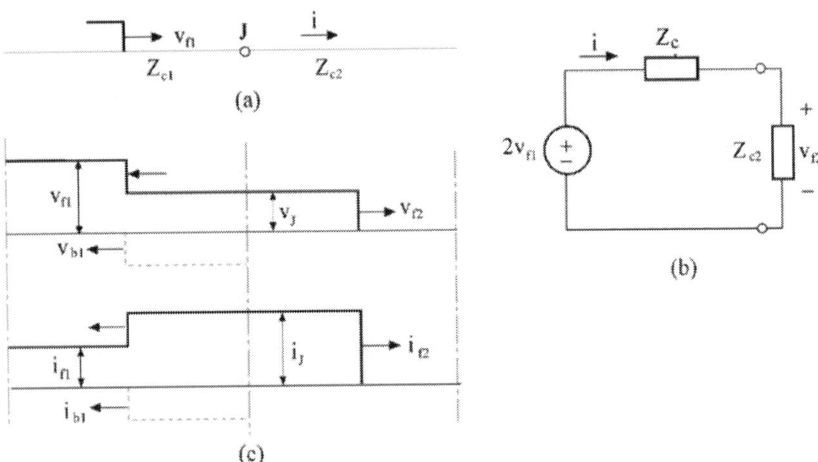

Figure 7.15 Traveling voltage and current waves at junction of two lines: circuit diagram (a); the equivalent circuit (b); the voltage and current distributions along two lines (c).

insulation. By putting in a short cable between the overhead line and the transformer, the magnitude of the voltage surge can be reduced before it reaches the transformer.

Example 7.4

The characteristic impedances of an overhead line and underground cable connected in series (Fig. 7.15(a)) are 400 Ω and 50 Ω respectively. The incident surge voltage of 800 kV rms is traveling on the overhead line toward the junction. Determine: (a) the surge voltage transmitted into the cable; (b) the surge current transmitted into the cable; (c) the surge voltage reflected back along the overhead line; (d) the power in the forward wave arriving at the junction and the transmitted wave power.

Solution

(a) $v_{\mathrm{f2}} = \rho_{\mathrm{ref}} v_{\mathrm{f1}} = \dfrac{2Z_{\mathrm{c2}}}{Z_{\mathrm{c2}} + Z_{\mathrm{c1}}} v_{\mathrm{f1}} = \dfrac{2 \cdot 50}{50 + 400} 800 = 178\ \mathrm{kV(rms)}.$

(b) $i_{\mathrm{f2}} = \dfrac{v_{\mathrm{f2}}}{Z_{\mathrm{c2}}} = \dfrac{178}{50} = 3.56\ \mathrm{kA(rms)}.$

(c) $v_{\mathrm{b1}} = v_{\mathrm{f2}} - v_{\mathrm{f1}} = 178 - 800 = -622\ \mathrm{kV(rms)}.$

(d) $P_{\mathrm{f1}} = \dfrac{v_{\mathrm{f1}}^2}{Z_{\mathrm{c1}}} = \dfrac{800^2}{400} = 1600\ \mathrm{MW}$

$$P_{\mathrm{f2}} = \frac{v_{\mathrm{f2}}^2}{Z_{\mathrm{c2}}} = \frac{178^2}{50} = 634\ \mathrm{MW}.$$

7.5.4 Capacitance connected at the junction of two lines

Figure 7.16(a) shows two lines connected in tandem and the *capacitance connected in parallel* to both lines at the junction J. Such a connection may represent two lines (incoming and outgoing) terminated in a transformer, since the behavior of a transformer at the first instance of wave arrival is as a capacitance. The equivalent circuit of the junction is shown in Fig. 7.16(b). Using the Laplace transform method, we obtain the expression of the voltage across the capacitance as

$$V_{\mathrm{C}}(s) = \frac{2Z_{\mathrm{par}}}{Z_{\mathrm{c1}} + Z_{\mathrm{par}}} V_{\mathrm{f1}}(s), \tag{7.57}$$

where

$$Z_{\mathrm{par}} = \frac{Z_{\mathrm{c2}}(1/sC)}{Z_{\mathrm{c2}} + 1/sC}$$

is the impedance of the parallel connection of C and Z_{c2}. Substituting Z_{par} into

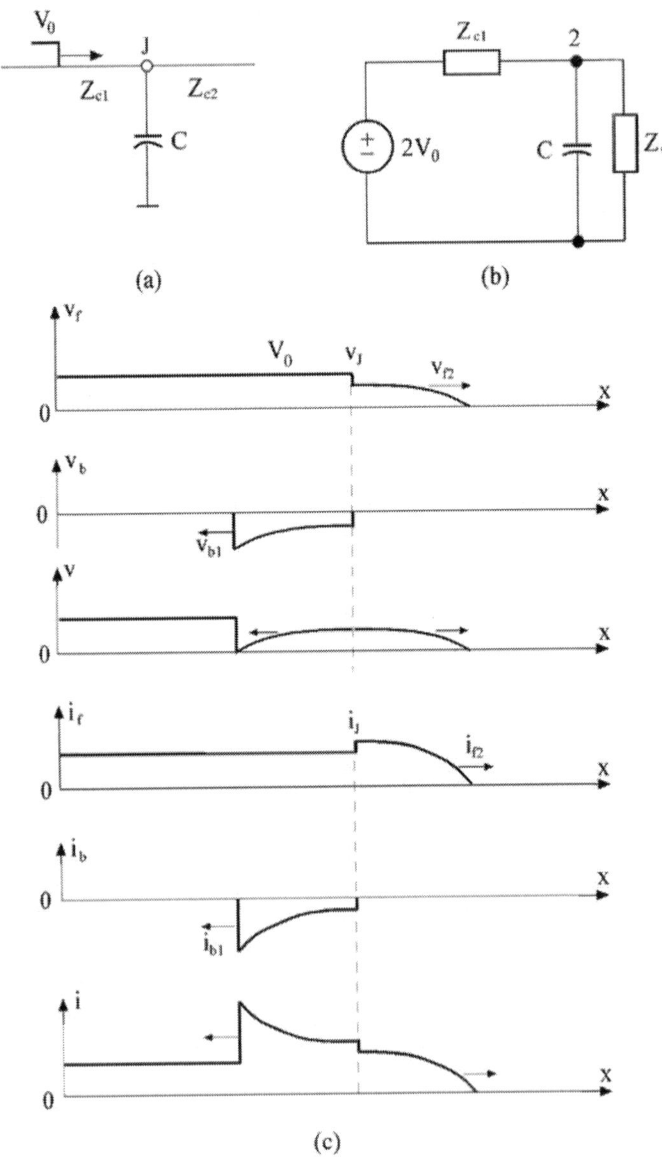

Figure 7.16 Traveling waves on two lines terminated by capacitance: circuit diagram (a); equivalent circuit (b); voltage and current waves on both lines (c).

equation 7.57 yields

$$V_C(s) = \frac{2Z_{c2}}{Z_{c1} + Z_2 + sCZ_{c1}Z_{c2}} V_{f1}(s), \tag{7.58}$$

or

$$V_C(s) = \rho_{ref} V_{f1}(s) \frac{1/T}{s + 1/T} \tag{7.59}$$

where $\rho_{ref} = 2Z_{c2}/(Z_{c1} + Z_{c2})$ is the transmission factor at infinite time ($t \to \infty$) after the capacitance is charged and $T = Z_{eq}C$ (where $Z_{eq} = Z_{c1}Z_{c2}/(Z_{c1} + Z_{c2})$ is the time constant of the circuit.

Considering the step-function incident wave and substituting its Laplace transform V_0/s into equation 7.59 gives

$$V_C(s) = \rho_{ref} V_0 \frac{1/T}{s(s + 1/T)}. \tag{7.60}$$

Taking the inverse Laplace transform, the time function of the capacitance voltage becomes

$$v_C(t) = \rho_{ref} V_0 (1 - e^{-(t/T)}). \tag{7.61}$$

Therefore the voltage distribution in the outgoing line will be

$$v_{f2}(x, t) = \rho_{ref} V_0 (1 - e^{-(t - x/v)/T}). \tag{7.62}$$

The reflected wave in the first line can be obtained as

$$v_{b1} = v_C - v_{f1} = (\rho_{ref} - 1)V_0 - \rho_{ref} V_0 e^{-(t/T)}.$$

Therefore the backward-traveling wave along the line is

$$v_{b1}(x, t) = \rho_r V_0 - \rho_{ref} V_0 e^{-(t - x/v)/T}. \tag{7.63}$$

The current traveling waves in both directions are the same shape as the voltage traveling waves and their value is related to them in the characteristic impedance of the corresponding line. The voltage and current distribution in both lines are shown in Fig. 7.16(c).

In most practical analyses, the *shape of the voltage/current waves caused by lightning is considered as an impulse* as shown in Fig. 7.17(a). In such an impulse, the voltage/current rises quickly to a maximum value and then decays slowly to zero. The crest time in which the voltage reaches its maximum value is about 1.2 μs.

When the first moment of the wave incident is of interest, the shape of the wave is simplified, having a constant value and ramped front as shown in Fig. 7.17(b). When the relatively long-term ($t \gg t_{cr}$) response is of interest, the voltage impulse of wave tail shape is considered, Fig. 7.17(c).

Now assume that the ramped front function incident wave reaches the junction of two lines with a *capacitance connected to the junction.* The ramped front

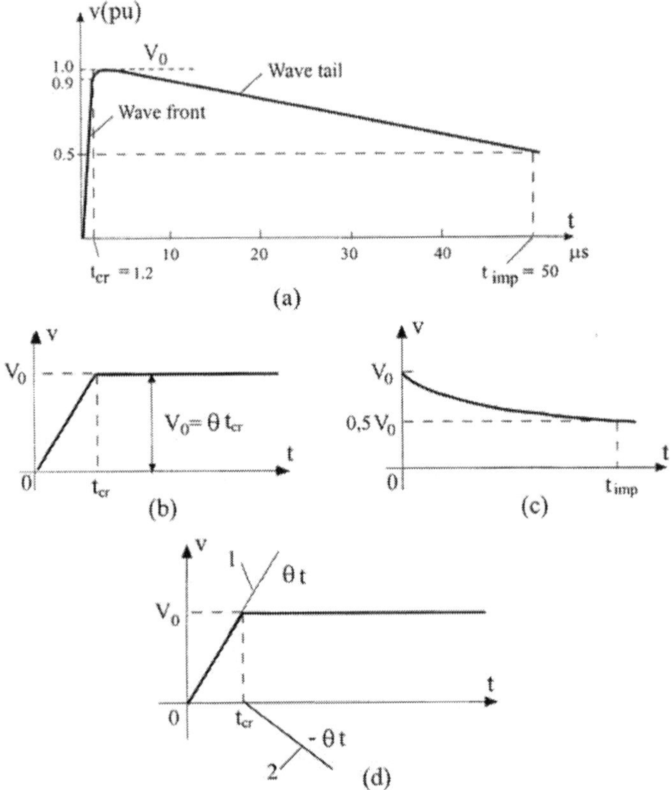

Figure 7.17 Impulse voltage waveform: standard impulse (a); ramped front function approximation (b); tail-shaped approximation (c); the superposition of two ramp functions (d).

function can be superposed from two ramp functions of slope $\theta = V_0/t_{cr}$: one positive and one negative, while the letter is shifted by the crest time, t_{cr}:

$$v_f(t) = \theta t - \theta(t - t_{cr}), \tag{7.64}$$

as shown in Fig. 7.17(d). Substituting the Laplace transform of the ramp function $v_r = \theta t \leftrightarrow \theta/s^2$ into equation 7.59 gives

$$V_{Cr}(s) = \rho_\tau \theta \frac{1/T}{s^2(s + 1/T)}. \tag{7.65}$$

Taking the inverse Laplace transform, the capacitance voltage in the time domain for time less than t_{cr} becomes

$$v_{Cr}(t) = \rho_\tau \theta [t - T(1 - e^{-(t/T)})], \quad t \le t_{cr}. \tag{7.66}$$

In accordance with equation 7.64, the capacitance voltage for the time larger then t_{cr} will be

$$v_{Cr}(t) = \rho_\tau V_0 \left[1 + \frac{T}{t_{cr}} (e^{-(t/T)} - e^{-(t-t_{cr})/T} \right], \quad t \geq t_{cr}. \tag{7.66a}$$

The capacitance voltage change for both instances of time is shown in Fig. 7.18(a) and (b). First, notice that for $t \to \infty$ the capacitance voltage is $\rho_\tau V_0$ which is the same as in the previous case of the step function wave response (see equation 7.61). Secondly, as can also be seen from Fig. 7.18(b), the slope of the resulting voltage has changed relative to the slope of the incident voltage wave. In order to estimate this change, let us determine the equivalent slope as

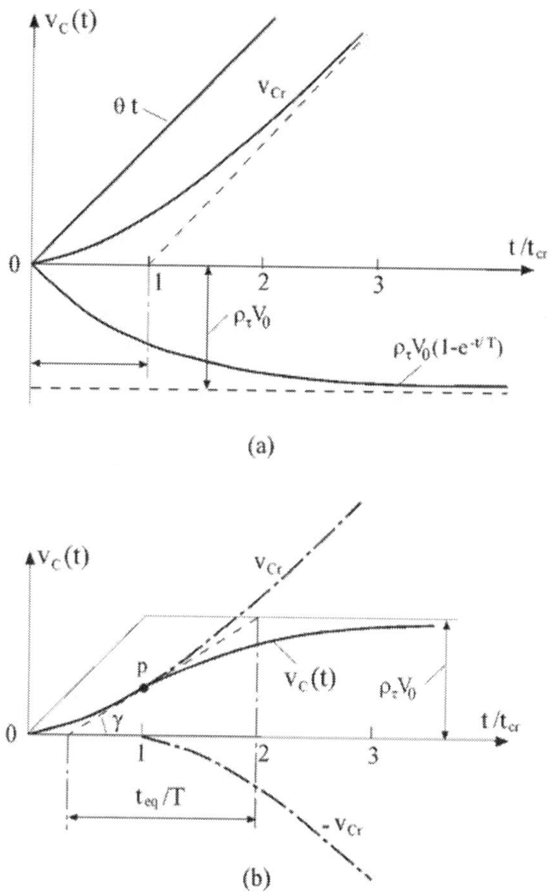

Figure 7.18 The voltage change across the capacitance as a response to: infinite ramp function incident wave (a); ramped front incident wave (b).

a $\tan\gamma$, where the angle γ is of a tangent drawn through a point p $(t = t_{cr})$:

$$\tan\gamma = \left[\frac{dv_C}{dt}\right]_{t=t_{cr}} = \rho_\tau V_0 \frac{1}{t_{cr}}(1 - e^{-t_{cr}/T}).$$

Therefore, the equivalent crest time of the capacitance voltage can be expressed as

$$t_{eq} = \frac{\rho_\tau V_0}{\tan\gamma} = \frac{t_{cr}}{1 - e^{-t_{cr}/T}}, \tag{7.67}$$

i.e., the bigger T is in relation to t_{cr}, the greater the change in the slope. (For $T \leq 1/3t_{cr}$ the change in slope is not significant.)

When the relatively long-term $(t \gg t_{cr})$ response is of interest, the voltage impulse of the shape shown in Fig. 7.17(c) is considered. In simple terms this is the decreasing exponential

$$v_f(t) = V_0 e^{-t/T_0}, \tag{7.68}$$

and T_0 is estimated in accordance with $T_0 = t_{imp}/0.7$ where t_{imp} is the time in which the maximum value of the impulse in Fig. 7.17(a) decreases by half.

Substituting the Laplace transform of an exponential into equation 7.59 yields

$$V_{C,exp} = \rho_\tau V_0 \frac{1}{T(s + 1/T_0)(s + 1/T)}, \tag{7.69}$$

and with the inverse Laplace transform

$$v_{C,exp}(t) = \rho_\tau V_0 \frac{T_0}{T_0 - T}(e^{-t/T_0} - e^{-t/T}). \tag{7.70}$$

The two exponential terms of equation 7.70 (see broken lines), the resulting voltage (1) and the voltage wave impulse (2), are shown in Fig. 7.19. Equating the derivative of equation 7.70 to zero yields the time in which the capacitance voltage reaches its maximum,

$$t_{(max)} = \frac{T_0 T}{T_0 - T} \ln\frac{T}{T_0} = \alpha \ln\frac{T}{T_0}, \tag{7.71}$$

and the scaled value of the maximum voltage is

$$\frac{V_{2max}}{\rho_\tau V_0} = \left(\frac{T}{T_0}\right)^\alpha, \tag{7.72}$$

where

$$\alpha = \frac{T/T_0}{1 - T/T_0},$$

i.e., the maximum voltage is dependent on the ratio T/T_0. If the ratio is $T/T_0 > 0.5$, then the capacitance results in reducing the maximum voltage by more than half.

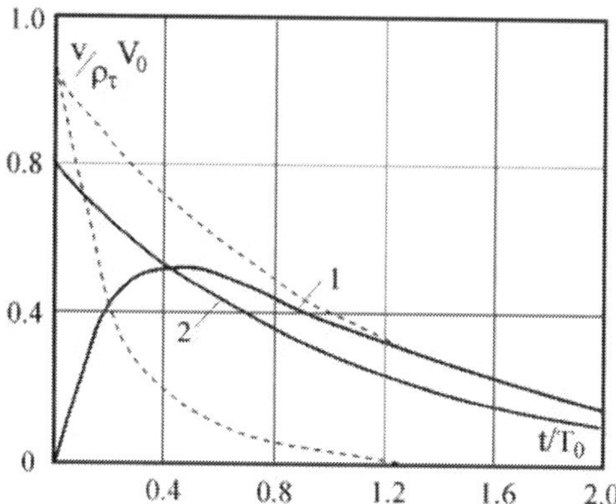

Figure 7.19 The voltage change across the capacitance as a response to the exponential function incident wave: curve 1, resulting voltage; curve 2, voltage incident wave. The broken lines are two exponential terms.

7.6 SUCCESSIVE REFLECTIONS OF WAVES

Consider the TL, which is terminated in the generator input impedance Z_s by the sending-end and in load impedance Z_T by the receiving-end (see the footnote in section 7.5 on p. 480). Neither of them is equal to the characteristic impedance Z_c, so in theory an infinite succession of reflected waves results.

The first forward-traveling (f.t.) wave (a step-function voltage source is assumed) will be (7.33)

$$v_{f1}(0, t) = \rho_\tau V_s = \frac{Z_c}{Z_s + Z_c} V_0 u(t), \tag{7.73}$$

where $u(t)$ is a unit function. The first backward-traveling (b.t.) wave appears after the first f.t. wave reaches the receiving end,

$$v_{b1}(\ell, t) = \rho_r v_{f1} = \rho_r \rho_\tau V_0 u(t - t_r), \tag{7.74}$$

where $t_r = \ell / v$ is the delay time in which the f.t. wave reaches the receiving-end of the line.

The second f.t. wave appears after the first b.t. wave reaches the sending-end of the line and it can be found in accordance with the sending-end reflection coefficient ρ_s (equation 7.49):

$$v_{f2}(0, t) = \rho_s \rho_r \rho_\tau V_0 u(t - 2t_r). \tag{7.75a}$$

In a similar way, the second b.t. wave becomes

$$v_{b2}(\ell, t) = \rho_s \rho_r^2 \rho_\tau V_0 u(t - 3t_r), \tag{7.75b}$$

or for kth incident $t > kt_r$

$$v_{f,k}(x,t) = (\rho_s\rho_r)^{k-1}\rho_\tau V_0 u\left(2(k-1)t_r - \frac{x}{v}\right)$$
$$v_{b,k}(x,t) = (\rho_s\rho_r)^{k-1}\rho_r\rho_\tau V_0 u\left(t - 2(k-1)t_r - \frac{x}{v}\right). \tag{7.76}$$

The current waves are simply related to the voltage waves by a characteristic impedance

$$i_{f,k} = \frac{v_{f,k}}{Z_c}, \quad i_{b,k} = -\frac{v_{b,k}}{Z_c}.$$

Thus, the complete response consists of an infinite series of voltage and current step-function waves which are added successively as the wave front travels from the source to its terminated end and back. Each of the forward- and backward-traveling wave series can be treated as infinitely decreasing geometric progressions having the ratio $\rho_s\rho_r$ (which is less than one) and the first terms $\rho_\tau V_0$ and $\rho_r\rho_\tau V_0$, respectively. Hence, the final value of the line voltage at $t \to \infty$ can be expressed as the sums of these two progressions,

$$v(x,t) = \frac{\rho_\tau V_0}{1-\rho_s\rho_r} + \frac{\rho_r\rho_\tau V_0}{1-\rho_s\rho_r} = V_0\frac{Z_T}{Z_s + Z_T}, \tag{7.77}$$

i.e., the steady-state voltage at the receiving-end of the line (note that the source is simply a d.c. quantity and the line is loss-less).

7.6.1 Lattice diagram

The voltage at a given point and time can be determined graphically with the help of the lattice diagram, suggested by Bewley(*). It gives a visual track representation of a traveling voltage or current wave as it reflects back and forth from the ends of the line, as shown in Fig. 7.20.

In the lattice diagram, the distance between the sending- and receiving-ends is represented by the horizontal line and time is represented by two vertical lines (t_r is the time for a wave to travel the line length). The diagonal zigzag line represents the wave as it travels back and forth between the ends or points of discontinuities: the line sloping to the right gives a forward-traveling wave in the increasing x direction, whereas the line sloping to the left gives the backward-traveling wave in the decreasing x direction. The slopes of the zigzag lines give the times corresponding to the distances traveled.

The value of each wave has been written above the corresponding line; each reflection is determined by multiplying the incident wave by the appropriate reflection coefficient ρ_r or ρ_τ. Of course, the same lattice diagram can also

(*) Bewley, L.V. (1951) *Traveling Waves on Transmission Systems.* Wiley, New York.

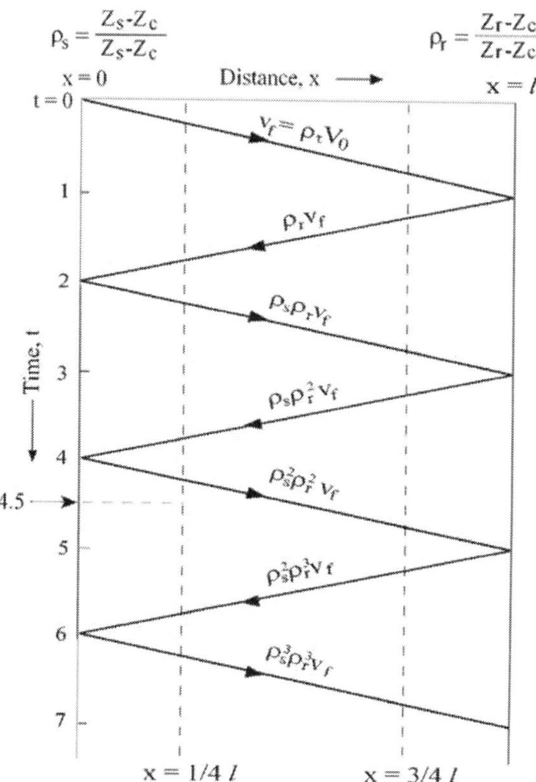

Figure 7.20 Lattice diagram for transmission line.

represent the current traveling waves. Nevertheless, the fact that the reflection coefficient for the current is always the negative of the reflection coefficient for the voltage should be taken into account. The voltage and current at a given point with the coordinates of time and distance along the line may then be determined by drawing a vertical line through this point and adding all the terms that are directly above that point corresponding to the intersections of the sloping lines with the given vertical line. For example, the voltage at $t = 4.5t_r$ and $x = (1/4)\ell$ is

$$v(0.25\ell, 4.5t_r) = \rho_\tau V_0(1 + \rho_r + \rho_s\rho_r + \rho_s\rho_r^2 + \rho_s^2\rho_r^2).$$

Example 7.5

Consider an underground cable line 1.6 km long with a characteristic impedance of 50 Ω and a wave propagation velocity of $1.6 \cdot 10^8$ m/s. The line is connected to the d.c. ideal $Z_s = 0$ voltage source $V_0 = 1000$ V and terminated in a 200 Ω resistor. (a) Determine the reflection coefficients at the sending- and receiving-ends; (b) Draw the appropriate lattice diagram for voltage and current; (c) Determine the value of voltage and current at $t = 5.5t_r$ and $x = (1/4)\ell$; (d) Plot the voltage and current versus time at line point $x = (1/2)\ell$.

Solution

(a) The reflection coefficients are

$$\rho_s = \frac{Z_s - Z_c}{Z_s + Z_c} = \frac{0 - 50}{0 + 50} = -1, \quad \rho_r = \frac{Z_T - Z_c}{Z_T + Z_c} = \frac{200 - 50}{200 + 50} = 0.6.$$

(b) The traveling time is

$$t_r = \frac{1.6 \cdot 10^3}{1.6 \cdot 10^8} = 10\ \mu s.$$

The lattice diagram is shown in Fig. 7.21(a). The values of the voltage/current traveling waves are written above the arrows.

(c) From the lattice diagram the voltage at the point A is

$$v\left(\frac{1}{4}\ell, 5.5t_r\right) = 1000 + 600 - 600 - 360 + 360 = 1000\ \text{V},$$

and the current is

$$i\left(\frac{1}{4}\ell, 5.5t_r\right) = \frac{1}{50}(1000 - 600 - 600 + 360 + 360) = 10.4\ \text{A}.$$

(d) The plot of the voltage and current at the line point $x = 1/2\ell$ versus time is shown in Fig. 7.21(b) and (c).

7.6.2 Bergeron diagram

Another convenient way of graphically determining voltages and currents at the two ends of the line as a result of an incident wave reflection is by a diagram attributed to Bergeron(*). This method is based on a graphical solution of a system of two linear equations and can be explained by a single example of the equivalent circuit of Fig. 7.2(b) shown again in Fig. 7.22(a), where the source is designated by V_0 and its input impedance by R_s. The following two equations express the two voltage-current loci (see Fig. 7.22(b)), one for the source (generator) and the other for the load (which here is the characteristic impedance Z_c):

$$v = V_0 - R_s i \tag{7.78a}$$

(*) Bergeron, L. (1961) *Water Hammer in Hydraulics and Wave Surges in Electricity*. The American Society of Mechanical Engineers, New York.

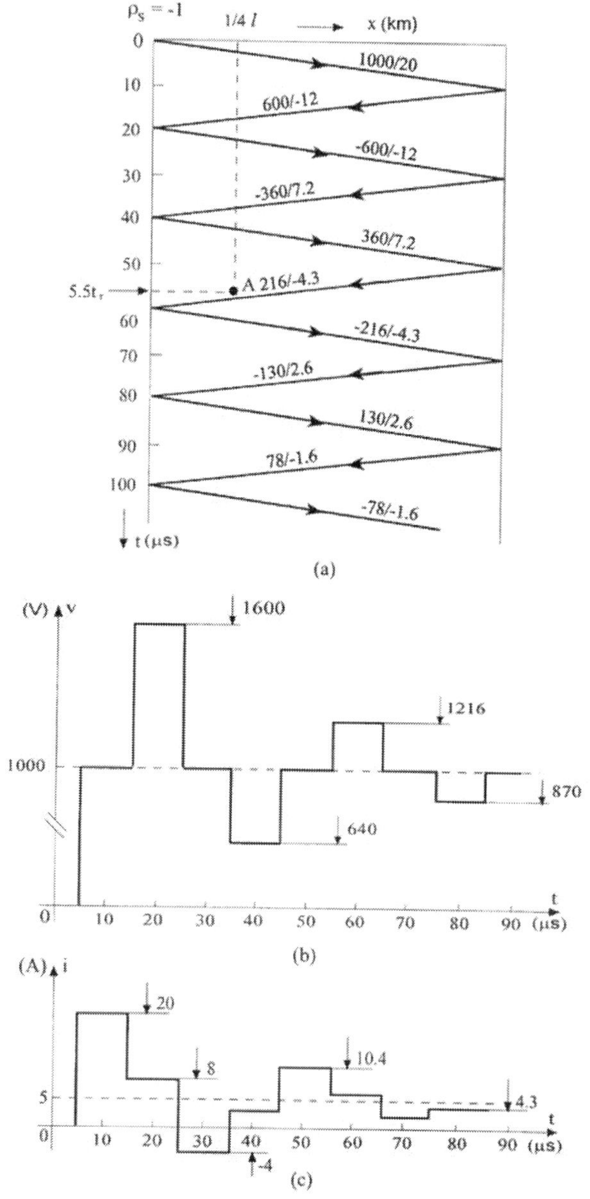

Figure 7.21 Lattice diagram (a) and voltage (b) and current (c) plots of Example 7.5.

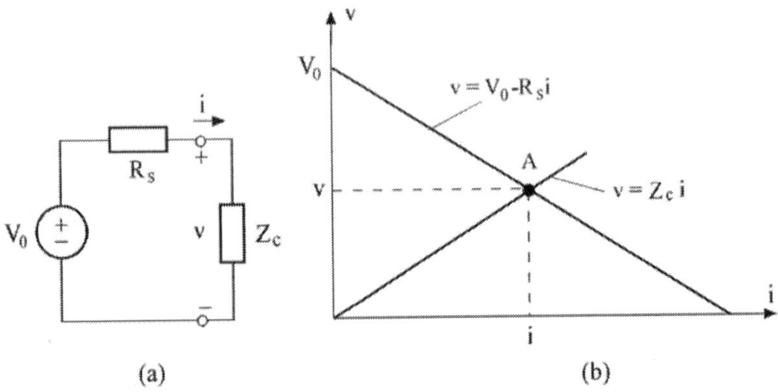

Figure 7.22 A graphical solution of series source-load circuit: the circuit diagram (a); the voltage-current loci (b).

and

$$v = Z_c i. \tag{7.78b}$$

Point A of their intersection gives the graphical solution of these two equations as shown in Fig. 7.22(b). Note that the slopes of these two lines are given by $-R_s$ and Z_c. It should also be noted that this graphical method is especially appropriate for nonlinear elements, like surge arresters, for example.

The complete solution for a line terminated in the voltage source (V_0, R_s) at the sending-end and in the resistor (R_T) at the receiving-end is given in Fig. 7.23(a). Point A_1 corresponds to the voltage and current at the sending-end after the source switching $t = 0_+$ (just like in Fig. 22(b)). Note that this voltage gives the first forward-traveling wave. The voltage and current at the receiving-end can be determined in accordance with the equivalent circuit similar to the one shown in Fig. 7.9(b) since the line is initially quiescent. The line drawing through point A_1 and sloping in accordance to $-Z_c$ represents the voltage-current locus of the source whose value is $2v_{f1}$ and whose input impedance is Z_c. (Note that in order to draw this locus it is not necessary to start with the point which lies on the ordinate axis, i.e. of $2v_{f1}$, and $i_{f1} = 0$, but it can be drawn through any point which belongs to this locus, for example, point A_1.) The intersection of this locus with the locus of the load resistor R_T, point B_1, yields the resultant voltage v_T and current i_T at the receiving-end immediately after the arrival of $2v_{f1}$. The first backward-traveling wave can then be obtained as

$$v_{b1} = v_{B1} - v_{f1}, \quad i_{b1} = i_{B1} - i_{f1}. \tag{7.79}$$

The equivalent circuit for calculating the next value of voltage and current at the sending-end is shown in Fig. 7.23(b). In accordance with the above, the locus, which determines the voltage and current at the sending-end, is parallel

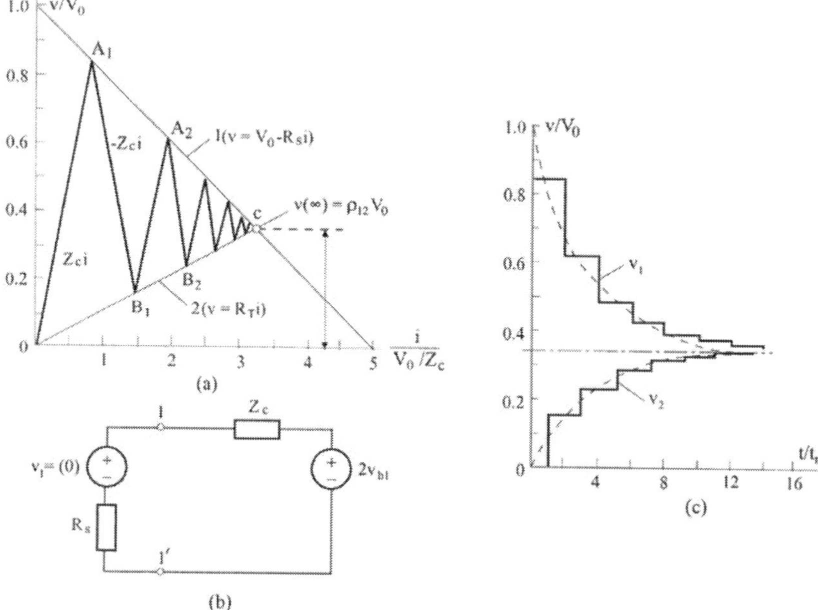

Figure 7.23 Bergeron diagram for TL with terminations $R_s = 0.2$, $R_T = 0.1Z_c$: Bergeron diagram (a); equivalent circuit for the sending-end (b); the plot of the sending- and receiving-end voltages versus time (c).

to the locus $v = Z_c i$ and passes through point B_1. (Note again that in order to draw this locus it is not necessary to start with the point on the ordinate axis, which is $2v_{b1}$, $i_{b1} = 0$.) The intersection of this locus with the sending-end voltage source locus, point A_2, gives the resultant voltage and current at $x = 0$ at the time of the arrival of v_{b1} and i_{b1}. The next forward-traveling wave will be

$$v_{f2} = v_{A2} - v_{b1}, \quad i_{f2} = i_{A2} - i_{b1}. \tag{7.80}$$

This process may be continued for any desired number of intervals. The intersection of two loci, point C, which represent both ends, yields the limiting values of line voltage and current as $t \to \infty$ (which is also in accordance with equation 7.77). The plot of both the sending- and receiving-end voltages versus time obtained with the help of the Bergeron diagram is shown in Fig. 7.23(c).

7.6.3 Nonlinear resistive terminations

The Bergeron diagram is the most suitable for reflecting wave determination when the transmission line is terminated in a nonlinear element. One example of such elements can be a surge arrester, which consists of an air gap and a nonlinear resistor. Fig. 7.24(a) shows the equivalent circuit of the line terminated in a surge arrester (SA), and Fig. 7.2(b) shows how the discharge voltage curve

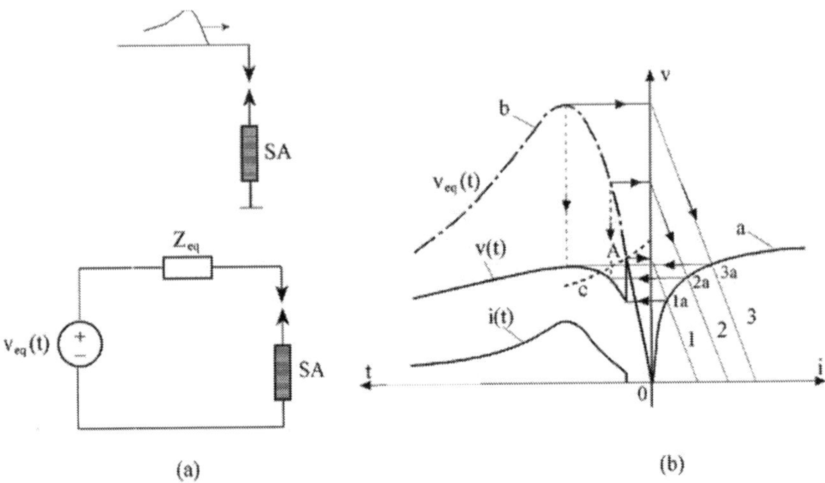

Figure 7.24 Transmission line terminated in a nonlinear resistor (surge arrester): equivalent circuit (a); arrester characteristic construction (b).

of the arrester can be built using the graphical solution. Since the line has been assumed to be initially quiescent, the voltage-current (v-i) locus of a nonlinear resistor (curve a in Fig. 7.24(b)) is drawn from the origin. In the left-half plane the equivalent surge voltage versus time (curve b) and volt-second characteristic of an arrester's air gap (curve c) are plotted. The intersection of these two curves, point A, determines the initial voltage, which activates the surge arrester (see the equivalent circuit (a) in Fig. 7.24)). The intersection of a sloped (in relation to Z_{eq}) line "1" and the v-i locus (curve a), point 1a, gives the voltage drop across the arrester, i.e., across the nonlinear resistor. The next voltage drops across the nonlinear resistor in accordance with the surge voltage change can be obtained in the same way (see points 2a, 3a and so on). Transferring these points to the left-half plane in accordance with the appropriate time results in the discharge voltage characteristic of an arrester: curve $v(t)$. Note that, because of nonlinearity, the voltage across the arrester hardly changes, being much lower than the surge voltage and thereby protecting the high voltage equipment.

7.7 LAPLACE TRANSFORM ANALYSIS OF TRANSIENTS IN TRANSMISSION LINES

As is known, any circuit equation, written in phasor notation, can be converted into a Laplace transform equation by simply replacing $j\omega$ with s. To use this procedure, let us first consider a transmission line activated at its sending-end by a sinusoidal voltage source. Applying the current and voltage phasors:

$\tilde{I} = I_{\mathrm{m}} e^{j\psi_i}$ and $\tilde{V} = e^{j\psi_v}$, which become functions of x only (and since the multiplier $e^{j\omega t}$ is crossed throughout the equations), partial derivative equations 7.1 convert into ordinary differential equations:(*)

$$-\frac{dV}{dx} = (R + j\omega L)I = ZI \tag{7.81a}$$

$$-\frac{dI}{dx} = (G + j\omega C)V = YV, \tag{7.81b}$$

where $Z = R + j\omega L$ and $Y = G + j\omega C$ are the impedance and admittance per unit length, respectively. Differentiating equations 7.81 with respect to x gives

$$-\frac{d^2V}{dx^2} = Z\frac{dI}{dx}, \quad -\frac{d^2I}{dx^2} = Y\frac{dV}{dx},$$

and substituting the values of dI/dx and dV/dx according to equations 7.81, we obtain

$$\frac{d^2V}{dx^2} = ZYV \tag{7.82a}$$

$$\frac{d^2I}{dx^2} = ZYI. \tag{7.82b}$$

Two ordinary second-order differential equations 7.82, which define the current/voltage phasors change along the line, are similar (from a mathematical point of view). Therefore, it is sufficient to solve one of them, for example equation 7.82a for the voltage and realize the current from equation 7.81a.

The solution of the ordinary second-order differential equation 7.82a for the voltage is of the form

$$V(x) = A_1 e^{-\gamma x} + A_2 e^{\gamma x}, \tag{7.83a}$$

and for the current I, from equation 7.81a with equation 7.83, is

$$I = \frac{\gamma}{Z}(A_1 e^{-\gamma x} - A_2 e^{\gamma x}),$$

or

$$I = \frac{1}{Z_{\mathrm{c}}}(A_1 e^{-\gamma x} - A_2 e^{\gamma x}). \tag{7.83b}$$

(*)For the simplification of formula, writing the superscript "∼", for denoting phasors, is omitted throughout this chapter.

The exponent power coefficient γ is found as a root of the characteristic equation $s^2 = ZY$,

$$\pm\gamma = \pm\sqrt{ZY} = \pm\sqrt{(R + j\omega L)(G + j\omega C)}, \tag{7.84a}$$

which is a complex quantity, called the *propagation constant*, and

$$Z_c = \frac{Z}{\gamma} = \sqrt{\frac{Z}{Y}} = \sqrt{\frac{R + j\omega L}{G + j\omega C}} = |Z_c|e^{j\theta} \tag{7.84b}$$

is the *characteristic impedance* of the line with a magnitude $|Z_c|$ and argument (angle) θ:

$$|Z_c| = \sqrt[4]{\frac{R^2 + \omega^2 L^2}{G^2 + \omega^2 C^2}}, \quad \theta = \frac{1}{2}\tan^{-1}\frac{\omega(GL - RC)}{RG + \omega^2 LC}.$$

(Here, the resistivity (R) and conductivity (G) of a TL are taken into account. By neglecting R and G, argument θ turns into zero, and $|Z_c|$ turns into the previously obtained quantity, given by expression 7.13, i.e., $Z_c = \sqrt{L/C}$.)

In equations 7.83, A_1 and A_2 are the arbitrary constants, which have to be selected to conform to the boundary conditions, and they are complex quantities:

$$A_1 = |A_1|e^{j\psi_1}, \quad A_2 = |A_2|e^{j\psi_2}.$$

To solve equations 7.83 constants A_1 and A_2 shall be found from the known boundary conditions. Let V_1 and I_1 be the voltage and current of the sending-end ($x = 0$) of the line. According to (7.83) for $x = 0$ we have $V_1 = A_1 + A_2$ and $I_1 Z_c = A_1 - A_2$. Therefore,

$$A_1 = \frac{1}{2}(V_1 + Z_c I_1), \quad A_2 = \frac{1}{2}(V_1 - Z_c I_1). \tag{7.85}$$

Substituting equations 7.85 into 7.83, we obtain voltage V and current I in any point of the line at the distance of x from its sending-end

$$\begin{aligned} V(x) &= \frac{1}{2}(V_1 + Z_c I_1)e^{-\gamma x} + \frac{1}{2}(V_1 - Z_c I_1)e^{\gamma x} \\ I(x) &= \frac{1}{2}\left(\frac{V_1}{Z_c} + I_1\right)e^{-\gamma x} - \frac{1}{2}\left(\frac{V_1}{Z_c} - I_1\right)e^{\gamma x}. \end{aligned} \tag{7.86a}$$

Equations 7.86a are known as the equations of the transmission line in exponential form. Combining similar terms in equation 7.86a,

$$\begin{aligned} V(x) &= \frac{e^{\gamma x} + e^{-\gamma x}}{2} V_1 - \frac{e^{\gamma x} - e^{-\gamma x}}{2} Z_c I_1 \\ I(x) &= -\frac{e^{\gamma x} - e^{-\gamma x}}{2}\frac{1}{Z_c} V_1 + \frac{e^{\gamma x} + e^{-\gamma x}}{2} I_1, \end{aligned}$$

and using the hyperbolic functions, these equations can be written in *hyperbolic form*

$$V(x) = (\cosh \gamma x)V_1 - (Z_c \sinh \gamma x)I_1$$
$$I(x) = \left(-\frac{1}{Z_c}\sinh \gamma x\right)V_1 + (\cosh \gamma x)I_1. \tag{7.86b}$$

Now consider the case where the voltage V_2 and the current I_2 of the receiving-end of the line are known. Let x' be the distance between the receiving-end of the line and the observed point. Since $x = \ell - x'$ (ℓ is the length of the line), equations 7.83 will be

$$V = A_1 e^{-\gamma\ell} e^{\gamma x'} + A_2 e^{\gamma\ell} e^{-\gamma x'}, \quad I = \frac{1}{Z_c}(A_1 e^{-\gamma\ell} e^{\gamma x'} - A_2 e^{\gamma\ell} e^{-\gamma x'}).$$

Let $A_3 = A_1 e^{-\gamma\ell}$ and $A_4 = A_2 e^{\gamma\ell}$ be the new boundary constants. Omitting the prime-sign in x', but taking into consideration that the variable x is reckoned from the receiving-end, we obtain

$$V = A_3 e^{\gamma x} + A_4 e^{-\gamma x}, \quad I = \frac{1}{Z_c}(A_3 e^{\gamma x} - A_4 e^{-\gamma x}). \tag{7.87}$$

Substituting $x = 0$ allows for determining the boundary constants A_3 and A_4

$$A_3 = \frac{1}{2}(V_2 + Z_c I_2), \quad A_4 = \frac{1}{2}(V_2 - Z_c I_2).$$

Equations 7.87 with the above arbitrary constants give the equations of a transmission line when the receiving-end boundary conditions are known

$$V(x) = \frac{1}{2}(V_2 + Z_c I_2)e^{\gamma x} + \frac{1}{2}(V_2 - Z_c I_2)e^{-\gamma x}$$
$$I(x) = \frac{1}{2}\left(\frac{V_2}{Z_c} + I_2\right)e^{\gamma x} - \frac{1}{2}\left(\frac{V_2}{Z_c} - I_2\right)e^{-\gamma x}. \tag{7.88a}$$

Combining the similar terms in equations 7.88a, the transmission line equations can be obtained in hyperbolic form

$$V(x) = (\cosh \gamma x)V_2 + (Z_c \sinh \gamma x)I_2$$
$$I(x) = \left(\frac{1}{Z_c}\sinh \gamma x\right)V_2 + (\cosh \gamma x)I_2. \tag{7.88b}$$

For the sending-end of the line, i.e. when $x = \ell$, equations 7.88 become

$$V_1 = (\cosh \gamma\ell)V_2 + (Z_c \sinh \gamma\ell)I_2$$
$$I_1 = \left(\frac{1}{Z_c}\sinh \gamma\ell\right)V_2 + (\cosh \gamma\ell)I_2. \tag{7.89}$$

Equations 7.89 express the voltage and current phasors of the sending-end of the line in terms of the voltage and current phasors of the receiving-end.

7.7.1 Loss-less *LC* line

By replacing $j\omega$ with s the above-obtained phasor equations become the Laplace-transform equations. Thus, equations 7.89 become

$$V(x,s) = V_2 \cosh \gamma\ell + Z_c I_2 \sinh \gamma\ell$$
$$I(x,s) = \frac{V_2}{Z_c} \sinh \gamma\ell + I_2 \cosh \gamma\ell. \tag{7.90}$$

where γ is in accordance with equation 7.84 and after replacing $j\omega$ with s becomes

$$\gamma = \sqrt{(R + sL)(G + sC)}. \tag{7.91a}$$

For a loss-less line ($R = 0$ and $G = 0$) it turns to

$$\gamma = s\sqrt{LC} = s/v, \tag{7.91b}$$

and the phasor equations 7.90 become the Laplace equations:

$$V(x,s) = V_2 \cosh st_r + Z_c I_2 \sinh st_r$$
$$I(x,s) = \frac{V_2}{Z_c} \sinh st_r + I_2 \cosh st_r, \tag{7.92}$$

where $t_r = \ell/v$ is the wave traveling time along the line. In order to find the solution of (7.92), the boundary equations for voltage vs. current (or vice versa) have to be taken into account.

7.7.2 Line terminated in capacitance

Consider the loss-less line, terminated in the capacitance and connected to the ideal source d.c. as shown in Fig. 7.25(a). For this line the boundary equations are:

$$V(0,s) = V_0/s \quad \text{for the sending-end} \tag{7.93a}$$

$$I(\ell,s) = sCV(\ell,s) \quad \text{for the receiving-end.} \tag{7.93b}$$

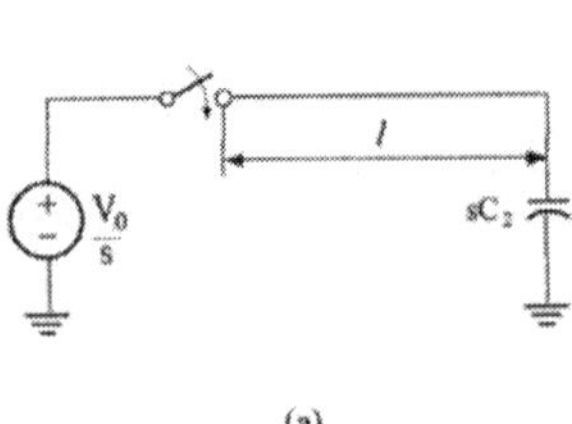

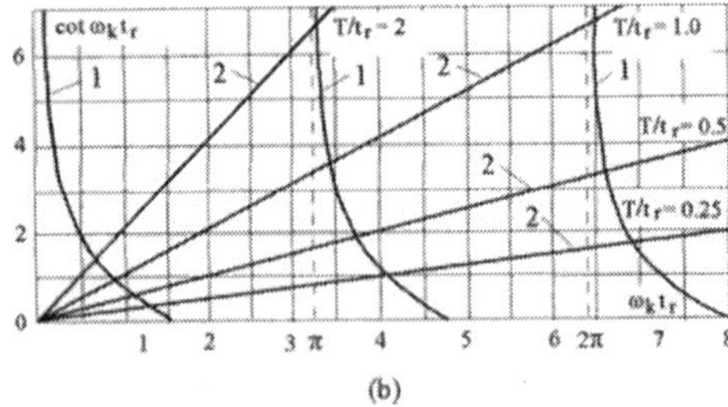

Figure 7.25 The circuit diagram of the TL terminated in capacitance (a) and the graphical solution of the characteristic equation (b) for a different ratio T/t_r.

Substituting these boundary conditions in the first equation 7.92 yields

$$V(\ell, s) = \frac{V_0}{s(\cosh st_r + sT \sinh st_r)} = \frac{V_0}{sF(s)} \tag{7.94}$$

where $T = C_2 Z_c$ is a time constant.

In order to obtain the receiving-end voltage in the time domain, we obtain the partial fraction expansions in the following form:

$$v(\ell, t) = \frac{V_0}{F(0)} + \sum_{-\infty}^{\infty} \frac{V_0}{s_k F'(s_k)} e^{s_k t}, \tag{7.95}$$

where s_k are roots (or the network poles) of the characteristic equation

$$\cosh st_r + sT \sinh st_r = 0. \tag{7.96}$$

Since the loss-less line only consists of inductances and capacitances, the roots of equation 7.96 are pure imaginary values ($s_k = \pm j\omega_k$). Therefore equation 7.96 changes into a trigonometrical equation

$$\cot \omega_k t_r = \frac{T}{t_r} \omega_k t_r. \tag{7.97}$$

The graphical solution of equation 7.97 is shown in Fig. 7.25(b). The intersection points of two plots, versus the variable $\omega_k t_r$, which represent two sides of equation 7.97, give the characteristic equation roots (the radiant frequency ω_k results in the quotient obtained by the division of the abscissa of the intersection points by t_r).

For any conjugate pair $\pm j\omega_k$ the sum of equation 7.95 consists of the kth cosine term

$$A_k e^{j\omega t} + \overset{*}{A}_k e^{-j\omega t} = 2A_k \cos \omega_k t, \tag{7.98}$$

where A_k and $\overset{*}{A}_k$ are also a conjugate pair. In order to determine A_k, the derivative of $F(s)$ has to be found

$$F'(s)\Big|_{s=\pm j\omega} = \pm jt_r \left[\left(1 + \frac{T}{t_r}\right) \sin \omega_k t_r + \omega_k t_r \frac{T}{t_r} \cos \omega_k t_r \right]. \tag{7.99}$$

Then in accordance with equation 7.95 and using equation 7.99, we obtain

$$\begin{aligned} -A_k &= \frac{V_0}{\omega_k t_r \left[\left(1 + \frac{T}{t_r}\right) \sin \omega_k t_r + \omega_k t_r \frac{T}{t_r} \cos \omega_k t_r \right]} \\ &= \frac{V_0}{\frac{\omega_k t_r}{\sin \omega_k t_r} + \cos \omega_k t_r}. \end{aligned} \tag{7.100}$$

The first term of equation 7.95 is

$$\frac{V_0}{F(0)} = \left[\frac{V_0}{\cosh st_r + sT \sinh st_r} \right]_{s=0} = V_0. \tag{7.101}$$

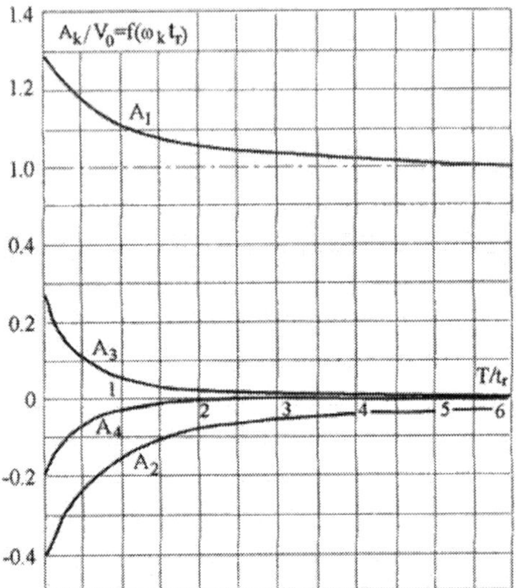

Figure 7.26 The dependence of magnitudes A_1–A_4 by the ratio T/t_r.

Therefore the complete receiving-end voltage response is

$$v(\ell, t) = V_0\left(1 - \sum_1^{\infty} 2A_k \cos \omega_k t\right). \tag{7.102}$$

Figure 7.26 shows the dependency of the magnitude A_k by the ratio T/t_r. It can be concluded that for large ratios of T/t_r (i.e. the capacitance is relatively big and/or the line is short) the high harmonic magnitudes are negligibly small and the first magnitude A_1 approaches unity. This means that the line terminating in a big capacitance behaves as a lumped LC circuit. Two plots of $v(\ell, t)$ versus t/t_r for $T/t_r = 0.5$ and $T/t_r = 2$, are shown in Fig. 7.27. Note that for the ratio $T/t_r = 2$ the voltage response curve is very close to the sinusoidal function. The points of non-continuity of the first curve (in Fig. 7.27(a)) represent the arrival of the incident waves.

It should be emphasized that solution of equation 7.92 immediately gives the complete voltage compared to other techniques in which the voltage or current are determined by means of the sum of the traveling waves.

7.7.3 A solution as a sum of delayed waves

As a further example of using Laplace transform techniques, let us again consider the TL equations, for the phasor representation, given in exponential form (see equations 7.88a). These equations can be rewritten in the following form:

$$V(x) = \frac{V_2 + Z_c I_2}{2}\left(e^{\gamma x} + \rho_2(s) e^{-\gamma x}\right) \tag{7.103a}$$

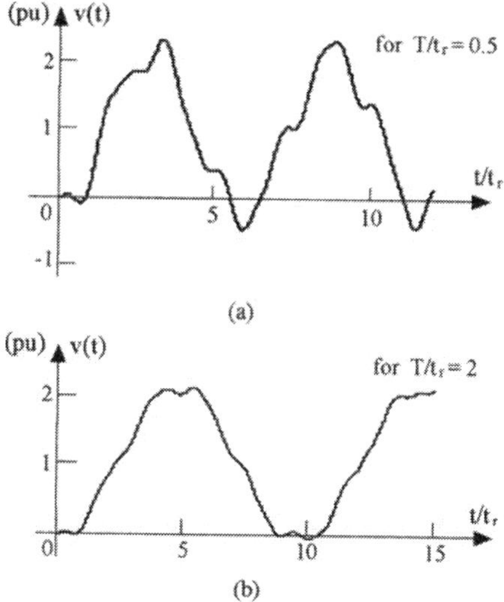

Figure 7.27 The receiving-end voltage curve versus the scaled time t/t_r: for the ratio $T/t_r = 0.5$ (a); for the ratio $T/t_r = 2$ (b).

$$I(x) = \frac{V_2 + Z_c I_2}{2Z_c}(e^{\gamma x} - \rho_2(s)e^{-\gamma x}), \tag{7.103b}$$

where $\rho_2 = (Z_2 - Z_c)/(Z_2 + Z_c)$ is the receiving-end reflection coefficient for the phasor quantities and $Z_2 = V_2/I_2$. Considering the voltage and current phasors as Laplace transform quantities (i.e., by replacing $j\omega$ by s) the above equations become the Laplace transform equations.

To simplify the following analysis we will assume, as before, that the TL is loss-less, therefore $\gamma = s\sqrt{LC} = s/v$ and equations 7.103 become

$$V(x, s) = \frac{V_2 + Z_c I_2}{2}(e^{sx/v} + \rho_2(s)e^{-sx/v}) \tag{7.104a}$$

$$I(x, s) = \frac{V_2 + Z_c I_2}{2Z_c}(e^{sx/v} - \rho_2(s)e^{-sx/v}). \tag{7.104b}$$

Now consider the boundary conditions

$$V_1(s) = \frac{V_0}{s} \quad (Z_1 = 0), \quad V_2(s) = Z_2(s)I_2(s). \tag{7.105}$$

Then equation 7.104a yields

$$V_1(s) = \frac{V_2(s) + Z_c V_2(s)/Z_2(s)}{2}(e^{s\ell/v} + \rho_2(s)e^{-s\ell/v}),$$

which allows the voltage transfer function to be determined:

$$H_v(s) = \frac{V_2(s)}{V_1(s)} = \frac{2}{1 + Z_c/Z_2} \frac{1}{e^{s\ell/v} + \rho_2(s)e^{-s\ell/v}}. \tag{7.106}$$

For the given boundary conditions, equation 7.105, and after substituting $I_2 = V_2/Z_2$ and $V_2 = V_1(s)H_v(s)$ (from equation 7.106) into equation 7.104a, this equation can be expressed as

$$V(x, s) = \frac{V_0}{s} \frac{e^{s(x-\ell)/v} + \rho_2(s)e^{-s(x+\ell)/v}}{1 + \rho_2 e^{-2s\ell/v}}. \tag{7.107}$$

Assuming again that x is reckoned from the sending-end, i.e., $x' = \ell - x$, but omitting the prime-sign in x', after interchanging, we obtain

$$V(x, s) = \frac{V_0}{s} \frac{e^{-sx/v} + \rho_2(s)e^{-s(2\ell-x)/v}}{1 + \rho_2 e^{-2s\ell/v}}, \tag{7.108a}$$

and similarly for the current

$$I(x, s) = \frac{V_0}{sZ_c} \frac{e^{-sx/v} + \rho_2(s)e^{-s(2\ell-x)/v}}{1 + \rho_2 e^{-2s\ell/v}}, \tag{7.108b}$$

where x is reckoned from the sending-end. To simplify the solution and better understand these techniques, consider an open-circuited receiving-end $Z_2 \to \infty$, i.e., $\rho_2(s) = 1$. Therefore, equation 7.108a yields

$$V(x, s) = \frac{V_0}{s} \frac{e^{-sx/v} + e^{-s(2\ell-x)/v}}{1 + e^{-2s\ell/v}}. \tag{7.109}$$

In order to find the inverse Laplace transform we note that the expression $1/(1 + e^{-2s\ell/v})$ can be treated as the sum of the infinitely decreasing geometric progression of the ratio $q = -e^{-2s\ell/v}$, i.e.,

$$\frac{1}{1 + e^{-2s\ell/v}} = 1 - e^{-2s\ell/v} + e^{-4s\ell/v} - e^{-6s\ell/v} + \cdots.$$

Then equation 7.109 becomes

$$V(s, x) = \frac{V_0}{s}(e^{-sx/v} + e^{-s(2\ell-x)/v} - e^{-s(2\ell+x)/v} - e^{-s(4\ell-x)/v} + e^{-s(4\ell+x)/v} + \cdots). \tag{7.110}$$

In accordance with the time-shift theorem, the time domain voltage is simply the infinite sum of the delayed step-functions of V_0

$$v(t, x) = V_0 \left[u\left(t - \frac{x}{v}\right) + u\left(t - \frac{2\ell - x}{v}\right) - u\left(t - \frac{2\ell + x}{v}\right) - u\left(t - \frac{4\ell - x}{v}\right) + \cdots \right], \tag{7.111a}$$

and similarly the time domain current is

$$i(t,x) = \frac{V_0}{Z_c}\left[u\left(t-\frac{x}{v}\right) - u\left(t-\frac{2\ell - x}{v}\right) - u\left(t-\frac{2\ell + x}{v}\right) + u\left(t-\frac{4\ell - x}{v}\right) + \cdots\right]. \tag{7.111b}$$

In equation 7.111 the terms $+u[t-(k2\ell - x)/v]$, $(k=0, 1, 2, \ldots)$ represent the unit step-functions delayed by the time $(k2\ell + x)/v$ in which the kth forward-traveling wave arrives the point x and the terms $-u[t-(k2\ell - x)/v]$ represent the unit step-functions delayed by the time $(k2\ell - x)/v$ in which the kth backward-traveling wave arrives at point x. Figure 7.28 shows the voltage and current changing in time at a half line distance.

As a next example, consider the line terminated in the capacitance. The Laplace transform reflection coefficient in this case is

$$\rho_2(s) = \frac{1/sC - Z_c}{1/sC + Z_c} = -\frac{s-\alpha}{s+\alpha}, \tag{7.112}$$

where $\alpha = 1/T = 1/CZ_c$. In accordance with equation 7.112 equation 7.108 yields

$$V(s,x) = \frac{V_0/s}{1-[(s-\alpha)/(s+\alpha)]e^{-2s\ell/v}}\left(e^{-sx/v} - \frac{s-\alpha}{s+\alpha}e^{-s(2\ell - x)/v}\right). \tag{7.113}$$

Again we will treat the expression

$$\frac{1}{1-[(s-\alpha)/(s+\alpha)]/e^{-2s\ell/v}}$$

as the sum of the infinitely decreasing geometric progression having the ratio $[(s-\alpha)/(s+\alpha)]e^{-2s\ell/v}$ (note that for $\mathrm{Re}[s] > 0$ the magnitude $|(s-\alpha)/(s+\alpha)| < 1$)

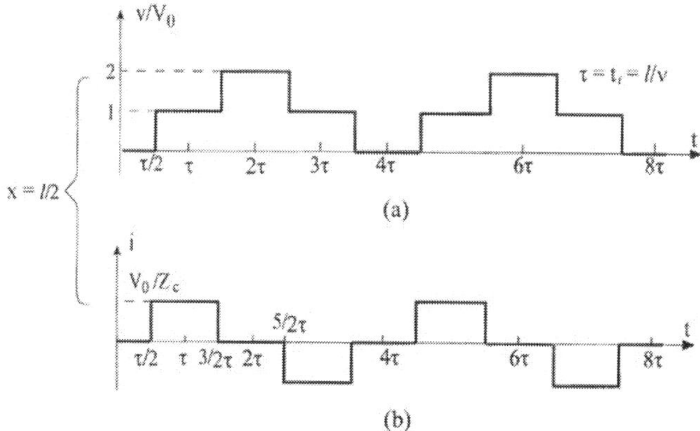

Figure 7.28 Voltage (a) and current (b) plots versus time at the line point $x=\ell/2$.

and the first member of unity. Therefore,

$$\frac{1}{1-[(s-\alpha)/(s+\alpha)]e^{-2s\ell/v}}=1+\frac{s-\alpha}{s+\alpha}e^{-2s\ell/v}+\left(\frac{s-\alpha}{s+\alpha}\right)^2 e^{-4s\ell/v}+\cdots$$

and

$$V(s,x)=\frac{V_0}{s}\left[e^{-sx/v}-\frac{s-\alpha}{s+\alpha}e^{-s(2\ell-x)/v}+\frac{s-\alpha}{s+\alpha}e^{-s(2\ell+x)/v}-\left(\frac{s-\alpha}{s+\alpha}\right)^2 e^{-s(4\ell-x)/v}\right.$$
$$\left.+\left(\frac{s-\alpha}{s+\alpha}\right)^2 e^{-s(4\ell+x)/v}-\left(\frac{s-\alpha}{s+\alpha}\right)^3 e^{-s(6\ell-x)/v}+\cdots\right]. \qquad (7.114)$$

The Laplace transform of the receiving-end voltage, i.e., voltage across the capacitance, becomes

$$V(s,\ell)=\frac{V_0}{s}\left[e^{-s\ell/v}-\frac{s-\alpha}{s+\alpha}e^{-s\ell/v}+\frac{s-\alpha}{s+\alpha}e^{-3s\ell/v}-\left(\frac{s-\alpha}{s+\alpha}\right)^2 e^{-3s\ell/v}\right.$$
$$\left.+\left(\frac{s-\alpha}{s+\alpha}\right)^2 e^{-5s\ell/v}-\left(\frac{s-\alpha}{s+\alpha}\right)^3 e^{-5s\ell/v}+\cdots\right]. \qquad (7.115)$$

Note that the inverse Laplace transforms of the terms are

$$\frac{1}{s}\frac{s-\alpha}{s+\alpha}\leftrightarrow -1+2e^{-\alpha t}$$

$$\frac{1}{s}\left(\frac{s-\alpha}{s+\alpha}\right)^2\leftrightarrow 1-4\alpha t e^{-\alpha t}$$

$$\frac{1}{s}\left(\frac{s-\alpha}{s+\alpha}\right)^3\leftrightarrow -1+2(1-2\alpha t+2\alpha^2 t^2)e^{-\alpha t}.$$

$$\vdots$$

and therefore with the time-shift theorem we obtain

$$v(t,\ell)=V_0\Big(u(t-t_r)-(-1+2e^{-\alpha(t-t_r)})u(t-t_r)+(-1+2e^{-\alpha(t-3t_r)})u(t-3t_r)$$
$$-[1-4\alpha(t-3t_r)e^{-\alpha(t-3t_r)}]u(t-3t_r)$$
$$+\{-1+2[1-2\alpha(t-5t_r)+2\alpha^2(t-t_r)^2]e^{-\alpha(t-5t_r)}\}u(t-5t_r)-\cdots\Big), \qquad (7.116)$$

where $t=\ell/v$ is the wave traveling time along the line. The capacitance voltage plots versus t in accordance with equation 7.116, for the ratios $T/t_r=0.5$ and $T/t_r=0.2$ are shown in Fig. 7.29. (Note the similarity of the curves in Fig. 7.29 and Fig. 7.27.) However, since only a few first terms in equation 7.116 have

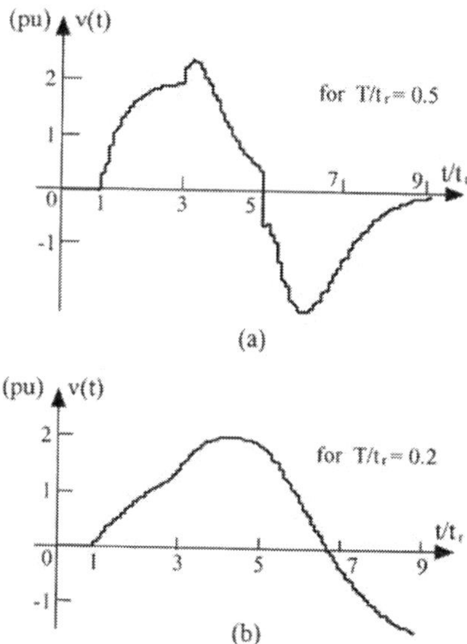

Figure 7.29 Voltage across the capacitance as versus t for the ratios $T/t_r = 0.5$ (a) and $T/t_r = 0.2$ (b).

been taken into consideration, the curves in Fig. 7.29 give just a rough approximation of the voltages after only a couple of reflections.

7.8 LINE WITH ONLY *LG* OR *CR* PARAMETERS

In some practical applications of electrical engineering techniques we consider networks in which the only significant parameters are L and G or C and R. An example of the former is ground rods used for grounding line towers and other power station (and substation) equipment. Under conditions of lightning impulse stress, the rods have to be treated as a network with distributed parameters as shown in Fig. 7.30(a). An example of the latter is an underground cable whose insulation is very good ($G = 0$) and inductance is negligible ($L = 0$) as shown in Fig. 7.30(b). The propagation constant and characteristic impedance in these cases are

$$\gamma(s) = \sqrt{LGs} \quad \text{or} \quad \gamma(s) = \sqrt{CRs} \tag{7.117a}$$

$$Z_c = \sqrt{\frac{sL}{G}} \quad \text{or} \quad Z_c = \sqrt{\frac{R}{sC}}. \tag{7.117b}$$

The differential equations 7.1 in such cases are simplified:

for the ground rod to

$$-\frac{\partial v}{\partial x} = L\frac{\partial i}{\partial t}, \quad -\frac{\partial i}{\partial x} = Gv \tag{7.118}$$

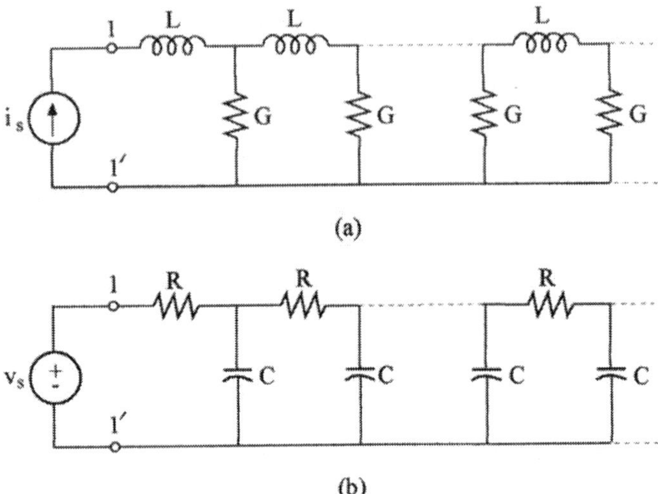

Figure 7.30 The equivalent circuits of: ground rod (a); underground cable (b).

and for the underground cable to

$$-\frac{\partial i}{\partial x} = C\frac{\partial v}{\partial t}, \quad -\frac{\partial v}{\partial x} = Ri. \tag{7.119}$$

Let us consider the case of an underground cable in more detail.

7.8.1 Underground cable

A transmission line, which behaves in accordance with differential equations 7.119, is an underground cable in which L and G can be neglected. Therefore its propagation constant and characteristic impedance in Laplace transform equations (see 7.84) are

$$\gamma = \sqrt{CRs} = \frac{\sqrt{s}}{a} \tag{7.120a}$$

$$Z_c = \sqrt{\frac{R}{C}}\frac{1}{\sqrt{s}}, \tag{7.120b}$$

where $a = 1/\sqrt{CR}$.

Applying the step voltage function $V_0u(t)$ at the sending-end of an infinite cable, and knowing that $V_1 = Z_c I_1$, equation 7.186a yields

$$V(x, s) = \frac{V_0}{s} e^{-\frac{x}{a}\sqrt{s}}. \tag{7.121}$$

Now the inverse Laplace transform gives

$$v(x, t) = V_0 \left(1 - \operatorname{erfc} \frac{x}{2a\sqrt{t}}\right) \tag{7.122}$$

where $\operatorname{erfc}(u) = \int_0^u e^{-\tau^2} d\tau$ is the error function. The Laplace transform of the current is

$$I(x, s) = \frac{V(x, s)}{Z_c} = V_0 \sqrt{\frac{C}{R}} \frac{1}{\sqrt{s}} e^{-\frac{x}{a}\sqrt{s}} \tag{7.123}$$

which gives the current in the time domain

$$i(x, t) = V_0 \sqrt{\frac{C}{R}} \frac{1}{\sqrt{\pi t}} e^{-\frac{x^2}{4a^2 t}}. \tag{7.124}$$

Example 7.6

An underground cable ("very long") has distributed parameters $R = 1\ \Omega$/km and $C = 0.1\ \mu$F/km. Assuming that at time $t = 0$ a step voltage source $v_s = 500u(t)$ V connects to the cable, find the voltage and current distribution along the cable line at $t_1 = 10\ \mu$s and at $t_1 = 50\ \mu$s.

Solution

The parameters of the cable are $a = 1/\sqrt{CR} = 3.16 \cdot 10^3$ and $\sqrt{C/R} = 3.16 \cdot 10^{-4}$. In accordance with equation 7.122 the voltage distribution along the line for $t_1 = 10\ \mu$s is

$$v(x) = 500 \left(1 - \operatorname{erfc} \frac{x}{20}\right) \text{V}.$$

In accordance with equation 7.124 the current distribution along the line is

$$i(x) = 500 \cdot 3.16 \cdot 10^{-4} \frac{1}{\sqrt{\pi 10 \cdot 10^{-3}}} e^{-(x^2/400)} = 28.21 e^{-(x^2/400)} \text{ A}.$$

Similarly we may calculate the voltage and current curves versus x for the second moment of time t_2. The resulting curves for both moments of time are shown in Fig. 7.31(a) and (b).

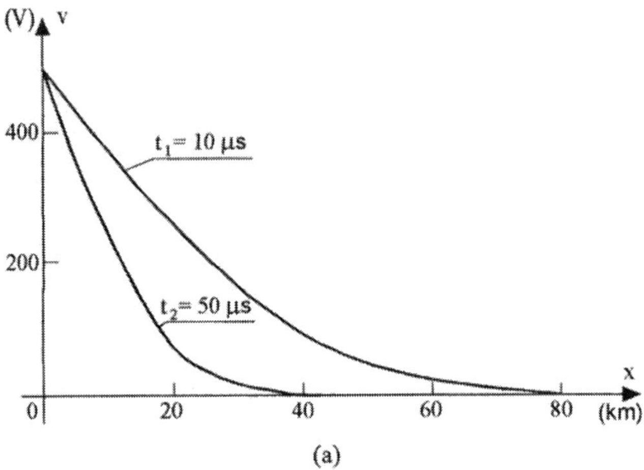

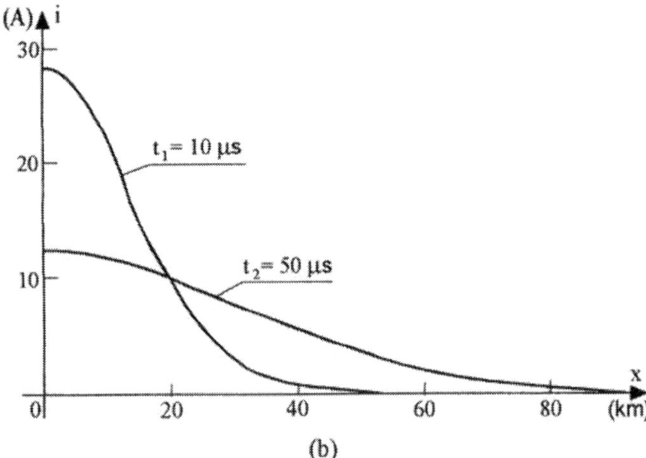

Figure 7.31 The voltage (a) and current (b) distribution along the underground cable for two different moments of time.

In conclusion consider an underground cable of length ℓ having a short-circuited receiving-end and connecting to a voltage source of step function $v_1 = V_0 u(t)$. In accordance with equations 7.88b and 7.99 and assuming $Z_2 = 0$ the voltage Laplace transform at any point on the line is expressed in terms of V_1 (but x is reckoned from the receiving-end):

$$V(x, s) = V_1(s)\frac{\sinh \gamma x}{\sinh \gamma \ell}. \tag{7.125}$$

With the Laplace transform of the sending-end voltage source $V_1(s) = V_0/s$

equation 7.125 becomes

$$V_1(x, s) = V_0 \frac{1}{s} \frac{\sinh(x\sqrt{s}/a)}{\sinh(\ell\sqrt{s}/a)} = \frac{F_1(s)}{F_2(s)}. \tag{7.126}$$

Using the partial fraction expansion formula, we obtain

$$v(x, t) = V_0 \left[\frac{F_1(0)}{F_2(0)} + \sum_{k=1}^{\infty} \frac{\sinh \gamma x e^{s_k t}}{s_k F_2'(s)} \right], \tag{7.127}$$

where s_k are the roots of the characteristic equation $\sinh(\ell/a)\sqrt{s} = 0$.
Therefore,

$$\gamma_k \ell = \frac{\ell}{a} \sqrt{s_k} = jk\pi, \tag{7.128}$$

or

$$s_k = -\frac{k^2\pi^2 a^2}{\ell^2}. \tag{7.129}$$

Evaluating the terms of equation 7.127 yields

$$\frac{F_1(0)}{F_2(0)} = \lim_{s\to 0} \frac{\sinh \frac{x}{a}\sqrt{s}}{\sinh \frac{\ell}{a}\sqrt{s}} \to \left[\frac{\frac{x}{2a\sqrt{s}} \cosh \frac{x}{a}\sqrt{s}}{\frac{\ell}{2a\sqrt{s}} \cosh \frac{\ell}{a}\sqrt{s}} \right]_{s=0} = \frac{x}{\ell} \tag{7.130}$$

$$F_1(s_k) = \sinh \frac{x}{\ell} jk\pi = j \sin \frac{k\pi x}{\ell}$$

$$F_2'(s_k) = \frac{d}{ds} \sinh \frac{\ell}{a} \sqrt{s} \bigg|_{s=s_k} = \frac{\ell}{2a\sqrt{s}} \cosh \frac{\ell}{a} \sqrt{s} \bigg|_{s=s_k} = \frac{\ell^2 \cos k\pi}{2a^2 jk\pi}. \tag{7.131}$$

Substituting equations 7.131, 7.130 and 7.129 into equation 7.127 yields

$$v(x, t) = V_0 \left[\frac{x}{\ell} + \frac{2}{\pi} \sum_{k=1}^{\infty} \frac{(-1)^k}{k} \sin \frac{k\pi x}{\ell} e^{-(k\pi a/\ell)^2 t} \right]. \tag{7.132}$$

Note that equation 7.132 gives the voltage at the receiving-end $x = 0$ equal to 0 at any time and the voltage at the sending-end $(x = \ell)$ equal to V_0. The time constants of the exponentials in equation 7.132 are proportional to the ratio $\ell^2/a^2 = \ell C \ell R = C_\ell R_\ell$, i.e., they are equal to the product of the complete capacitance and complete resistance of the cable.

Now, in accordance with the second equation of 7.119, we can obtain the current

$$i(x, t) = -\frac{1}{R} \frac{\partial v}{\partial x} = -\frac{V_0}{R\ell} \left[1 + 2 \sum_{k=1}^{\infty} (-1)^k \cos \frac{k\pi x}{\ell} e^{-(k\pi a/\ell)^2 t} \right]. \tag{7.133}$$

The voltage and current distributions along the cable are shown in Figs. 7.32(a) and (b). Other problems of the transient behavior of cables having different terminations and different sending-end conditions can be solved in a similar way.

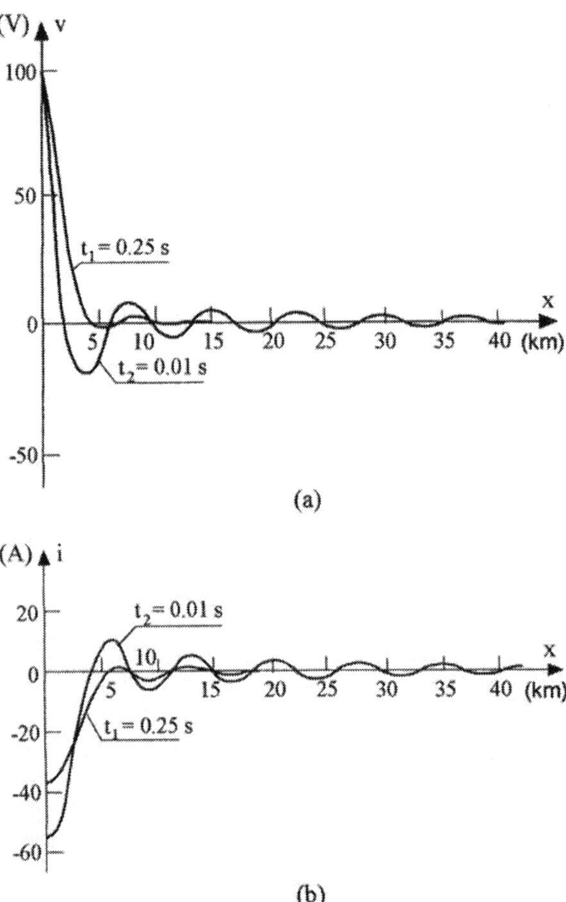

Figure 7.32 The voltage (a) and current (b) distribution along the short-circuited underground cable.

Chapter #8

STATIC AND DYNAMIC STABILITY OF POWER SYSTEMS

8.1 INTRODUCTION

Today's power systems are interconnected networks of transmission lines linking generators and loads into large integrated systems, some of which span entire countries and even continents. The main requirement for the reliable operation of such systems is to keep the synchronous generators running in parallel and with an adequate capacity to meet the load demand. When synchronous machines are electrically tied in parallel they must operate at the same frequency, i.e. they must all operate at the same speed (measured in electrical radians per second), which is called **being in synchronism.** If at any time a generator increases the speed and the rotor advances beyond a certain critical angle, counted between the rotor axis (usually the *d*-axis) and the system voltage phasor (the power angle δ), the magnetic coupling between the rotor and the stator fails. In such a situation the rotor rotates relatively to the field of the stator currents rather than being tied to this field, and pole slipping occurs, i.e., the generator **loses its synchronism** (falls out of step) with the rest of the system. Each time the generator speed changes, stability problems arise. The disturbance of the stability of the synchronous generators operating in parallel is one of the most arduous faults of power systems and may result in outage of entire regions.

8.2 DEFINITION OF STABILITY

Synchronous machines do not easily fall out of step under normal conditions. If a machine tends to speed up or slow down, synchronizing forces (see further on) tend to keep it in step. However, certain conditions may arise, in which the synchronizing forces for one or more generators may not be adequate and small impacts on the system may cause these generators to lose synchronism. On the other hand, if following an imbalance between the supply and demand created by a change in the load, in the generation or in the network conditions, all interconnected synchronous machines remain in synchronism adjusting themselves to a new state of operation, then the system is stable and the generators continue to operate at the same speed.

The perturbation could be of a major disturbance such as the loss of a generator, a fault or the loss of a line, or it could be of small, random load changes occurring under normal operation. The transients following system perturbations are oscillatory in nature, but if these oscillations are damped toward a new quiescent operating condition, we say that the system is stable. Thus, we may state that: **If the oscillatory response of a power system during the transient period is damped and the system settles in a finite time to a new steady-state condition, the system is stable.** Otherwise, the system is considered unstable.

The stability problems may be divided into two kinds: steady-state, or **static stability** and transient, or **dynamic stability**. The former is concerned with the effect of gradual infinitesimal power changes and is defined as the ability of a synchronous generator to reestablish its given state of operation after such changes. The latter, transient stability, deals with the effect of large, sudden disturbances such as line faults, the sudden switching of lines, the sudden application or removal of large loads, etc. The ability of the power system to retain synchronism when subject to such disturbances is considered as dynamic stability. Thus, the main criterion for stability in both regimes is that synchronous machines maintain synchronism at the end of the period of small as well as large disturbances.

8.3 STEADY-STATE STABILITY

Power systems form groups of synchronous generators (power station) interconnected by transmission lines. Experience in operating and theoretical study reveal that such transmission lines with synchronous machinery at both ends show that there are definite limits beyond which the operation becomes unstable, resulting in the loss of synchronism between the sending- and receiving-ends. This problem is termed the stability of the tie line, even though in reality it reflects the stability of two groups of machines. In order to understand this problem we shall introduce a transmission line power-transfer characteristic.

8.3.1 Power-transfer characteristic

Consider a group of synchronous generators, which is connected through a transmission line to a large system as shown in Fig. 8.1.

Here, the group of generators (power station) is represented by a single equivalent synchronous generator, operating with the phase EMF (along the quadrature axis) E_{ph}, and the system is represented by the **infinite bus**, whose voltage is kept constant regardless of any changes in the system behavior and is taken as the reference. The total reactance of the equivalent circuit is

$$X = X_d + X_{T1} + X_\ell + X_{T2}.$$

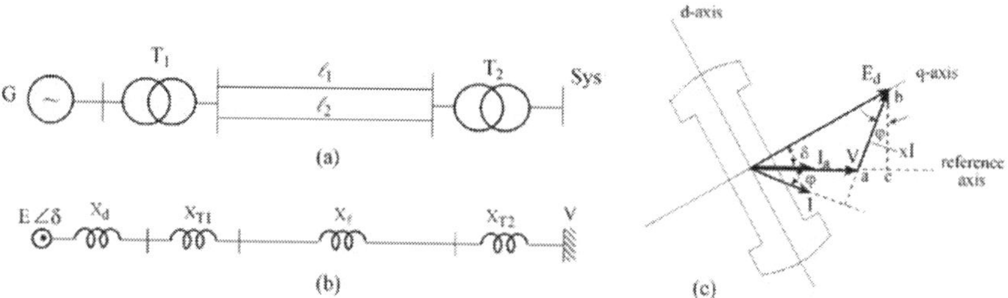

Figure 8.1 A group of generators connected to the system through a transmission line: one-line diagram (a), equivalent circuit (b) and the phasor diagram (c).

In accordance with the phasor diagram in Fig. 8.1(c) we have

$$\overline{bc} = E_d \sin \delta = \overline{ab} \cos \varphi = XI \cos \varphi,$$

where δ is the angle between the induced voltage of the generator E and the reference voltage V.

Thus,

$$I_a = I \cos \varphi = \frac{E_{ph}}{X} \sin \delta,$$

and the generator power transmitted by a transmission line is

$$P_e = 3V_{ph} I_a = 3\frac{V_{ph}E_{ph}}{X} \sin \delta = \frac{VE}{X} \sin \delta, \qquad (8.1)$$

where E and V are the line voltages.

The physical significance of angle δ is understandable from the phasor diagram, in Fig. 8.1(c), where the rotor position, in relation to the phasors, is also shown by a thin line. Subsequently, we realize that this angle is not only the *electrical angle* between E and V, but it is also the *mechanical angle* between the rotor q-axis and the reference axes. At no-load operation the rotor q-axis and the reference axis coincide. With increasing shaft or input power, the rotor advances (in the direction of the rotation) by angle δ, which is therefore called the power angle. The relationship of the power, developed by the generator, versus δ is given in equation 8.1 and plotted in Fig. 8.2. This plot, assuming E and V are constant, is a pure sinusoid, having the amplitude of $P_{\max} = EV/X$, and is called the **power-transfer curve**, or **power-angle curve**. It should be noted that for motor action the rotor is retarded relative to the reference axis and δ becomes negative.

For the given constant values of the generator EMF E and the receiving-end voltage V, the load of the transmission line can be gradually increased until a condition is reached corresponding to point A in Fig. 8.2. At this point the power transmitted is maximum and corresponds to angle $\delta = 90°$ (since the resistances are neglected) and represents the **static limit of stability** (i.e. for a gradually applied load) and any attempt to impose any additional load on the

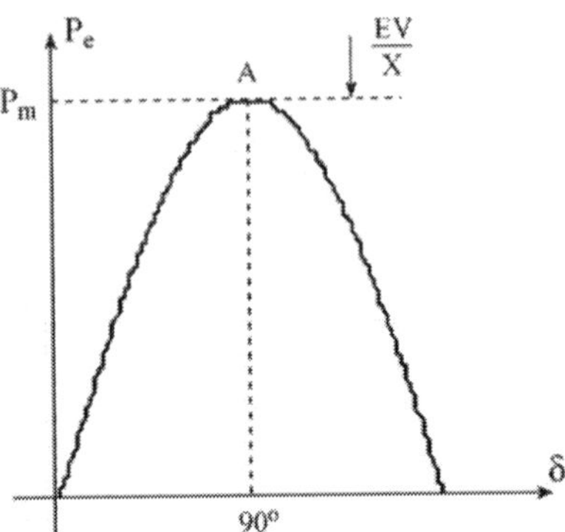

Figure 8.2 The power-angle curve.

line will result in the loss of the synchronism between the generator and the system. However, since today's generators are equipped with an AVR system, the terminal voltage of such generators is kept constant during the period of load changes. Hence, when increasing the load, the generator EMF will also increase and the operating power-angle curve will change, resulting in a higher limit of stability. This limit of stability may exceed the initial one by 30–40%.

Until now, system resistances have been neglected. However, sometimes transmission line resistances are relatively significant and should be considered. In this case, with the total impedance $Z = R + jX$, the line current is

$$\tilde{I} = \frac{\tilde{E} - \tilde{V}}{\sqrt{3}Z},$$

and the transmitted power will be

$$S = \sqrt{3}\overset{*}{E}\tilde{I} = \sqrt{3}\overset{*}{E}\frac{\tilde{E} - \tilde{V}}{\sqrt{3}Z}.$$

Substituting the polar forms of the quantities $\tilde{E} = E\angle\delta$, $\overset{*}{E} = E\angle -\delta$, $Z = z\angle\varphi$ in the above expressions after simplification, yields

$$S = \frac{E^2}{z}(\cos\varphi - j\sin\varphi) - \frac{EV}{z}[\cos(\delta + \varphi) - j\sin(\delta + \varphi)].$$

The real part of this expression gives the active power

$$P = \frac{E^2}{z}\cos\varphi - \frac{EV}{z}\cos(\delta + \varphi). \qquad (8.2a)$$

For an easier comparison of this expression to the previous one (equation 8.1) we assign an additional angle $d = 90° - \varphi$ to obtain

$$P = \frac{EV}{z}\sin(\delta - \alpha) + \frac{E^2}{z}\sin\alpha. \qquad (8.2b)$$

This power-angle curve is shown in Fig. 8.3. The maximum power transferred, or static limit if stability, in this case is

$$P_{\max} = \frac{EV}{z} + \frac{E^2}{z} \sin \alpha,$$

which is higher than in the case where the resistances are neglected. The critical angle here is also larger than in the previous case, i.e., $\delta_{cr} = (90^\circ + \alpha) > 90^\circ$.

The power-angle characteristic of the turbine, which governs the generator, is a straight line, as shown in Fig. 8.4, since the power developed by the turbine does not depend on angle δ, which is a pure electrical parameter. At a steady-state operation the mechanical power P_m is equal to the electrical one and, as can be seen from Fig. 8.4, for the given mechanical power P_m, there are two points of equilibrium, a and b, on the intersection of the turbine and generator characteristics. This means that two steady-state regimes are possible at each

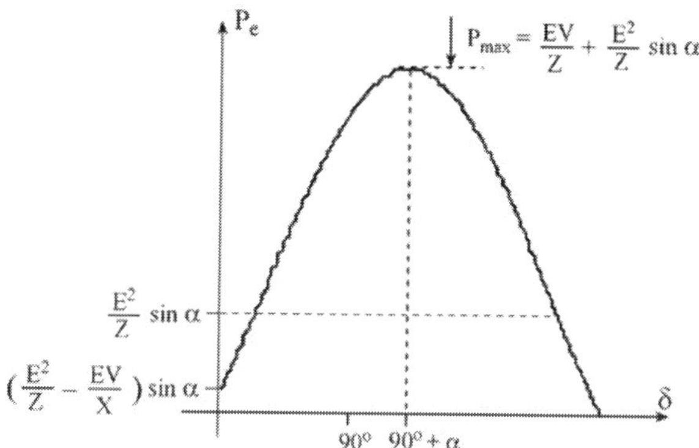

Figure 8.3 Power-angle curve for a system in which the resistances are considered.

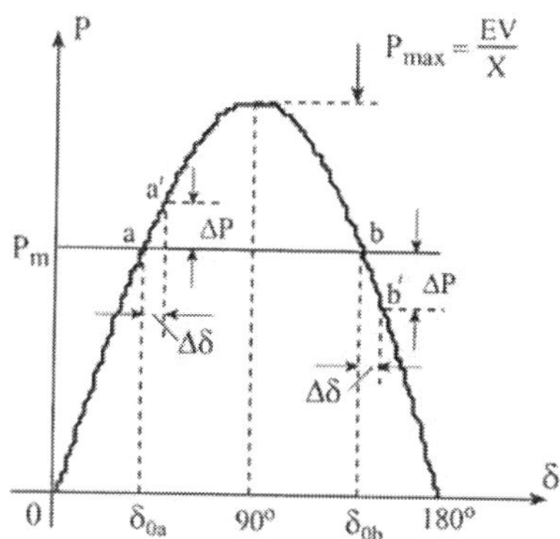

Figure 8.4 Steady-state stability of a synchronous generator.

of these points. However, the stable operation of the generator is possible only at point *a*. Indeed, assuming that angle δ is randomly increased by $\Delta\delta$, then the generator has to transmit power $P_e = P_m + \Delta P$ while the turbine power remains the same, P_m. The difference between P_m and P_e is an accelerated power P_a. Therefore, the acceleration power in this case becomes negative: $P_m - P_e = -\Delta P < 0$. This causes the generator to decelerate and return back to point *a*. With an analogous assumption for point *b* we realize that increasing δ by $\Delta\delta$, the generator power will be decreased and the acceleration power will, therefore, be positive. This causes the rotor of the generator to accelerate and δ will increase even more, actually up to 180°. The generator EMF E is at this position in the opposite phase with the system voltage, and this situation is equivalent to the short-circuit fault. The generator power drops to zero and it falls out of step.

It is self-evident that assuming a random decrease of δ at point *a*, we will arrive at the same conclusion, i.e., the generator returns back to point *a*. However, at point *b*, the generator, as previously, does not return to point *b*, but its rotor decelerates until it reaches point *a*. Hence, the stable operation of a generator in parallel with the system is possible only at point *a*, or in the increasing part of the power-angle curve, i.e., when $0 < \delta < 90°$. At angle $\delta = 90°$ $(\partial P/\partial\delta)_{90°} = 0$, i.e., the system is at the limit, and the operation at this point cannot be stable. The stability of the operation is often estimated by the assurance factor

$$k_{as} = \frac{P_{max} - P_m}{P_m}. \tag{8.3}$$

The steady-state **stability limit** is the maximum power that can be transmitted in a network between sources and loads when the system is subject to small disturbances. The stability is assured if the generator operates within the "safe area" of the power characteristic, which is in about a 20% margin lower than the steady-state stability limit. It should be noted that, as already has been mentioned, this limit may be extended by the use of an automatic voltage regulator (AVR). Besides the AVR, by analyzing the system stability, the effects of machine inertia and governor action should be taken into consideration. These functions greatly increase the complexity of the analysis (this is beyond the scope of this text). For more details the reader is referred to the book by Anderson and Fouad(*).

Usually the normal operating load angle for modern machines is in the order of 60 electrical degrees, and for the limiting value of 90°, this leaves 30° for the "safe area", to cover the large disturbances in the transmission line (see further on).

In a system with several generators/power stations and loads the common procedure is to reduce the network to the simplest form in which only the

(*)P. M. Anderson and A. A. Fouad (1980) *Power System Control and Stability*, Iowa State University Press.

relevant generators are connected to each other and which then allows the transfer reactances to be calculated. The values of the load, the power angles and the voltage are then calculated for the given conditions, and the steady-state stability limit and the assurance factor are determined for each machine. If those stability criteria are satisfactory, the loading is increased and the process repeated. If the voltages change appreciably, the $P-V$, $Q-V$ characteristics of the load should be used with the redistribution of the power.

Example 8.1

A synchronous generator having a local load, represented by constant impedance, is connected to an infinite bus through a transformer and a double circuit transmission line. The direct axis generator synchronous reactance is 1.2 pu, the load impedance is $Z_{load} = 2\angle 36°$ pu and the rest of the parameters are shown in Fig. 8.5. Check the steady-state stability of the given system, if the power transmitted to the systems is 0.5 pu and the generator terminal voltage is kept as 1.1 pu.

Solution

First we find the angle of the generator terminal voltage $V_T = V_2$. The power-angle equation is

$$P = \frac{V_T V}{X_{23}} \sin \delta_{23} \quad \text{or} \quad 0.5 = \frac{1.1 \cdot 1}{0.6} \sin \delta_{23},$$

where $X_{23} = 0.1 + 1.0/2 = 0.6$. Then

$$\sin \delta_{23} = 0.273 \quad \text{and} \quad \delta_{23} = 15.8°.$$

The current is found as

$$\tilde{I}_{23} = \frac{\tilde{V}_2 - \tilde{V}_3}{X_{23}} = \frac{1.1\angle 15.8° - 1}{0.6\angle 90°} = 0.5 - j0.1 \text{ pu}.$$

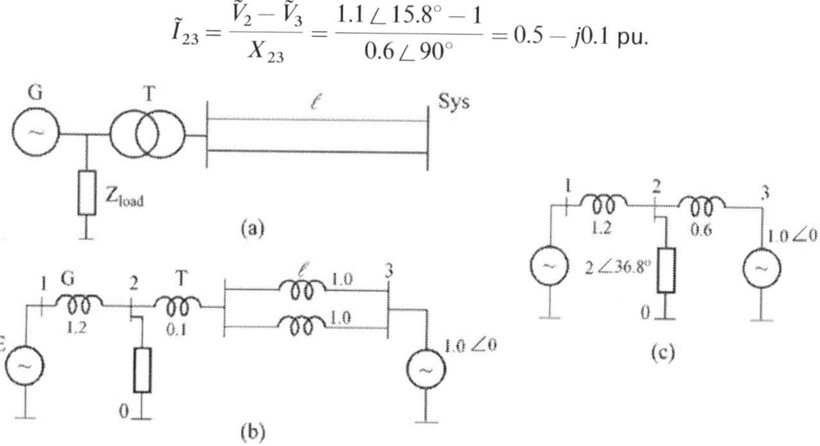

Figure 8.5 Network of Example 8.1: one-line diagram (a), equivalent circuit (b) and its simplification (c).

The load current is found as

$$\tilde{I}_{20} = \frac{1.1\angle 15.8^\circ}{2\angle 36.8^\circ} = 0.55\angle -21^\circ = 0.513 - j0.197 \text{ pu},$$

and the generator current is

$$\tilde{I}_G = \tilde{I}_{20} + \tilde{I}_{23} = 0.513 - j0.197 + 0.5 - j0.1 = 1.05\angle -16.13^\circ \text{ pu}.$$

Then, the internal generator voltage (EMF) is

$$E = 1.1\angle 15.8^\circ + (1.05\angle -16.3^\circ)(1.2\angle 90^\circ) = 1.41 + j1.51 = 2.07\angle 47.12^\circ \text{ pu}.$$

Hence, the angle between E and V is 47.12°. Since this angle is less than 90° with a safe area of about 40°, the system is stable. The active power produced by the generator is

$$P_G = EI\cos\varphi = 2.07\cdot 1.05\cos[47.12^\circ - (-16.3^\circ)] \cong 1.0 \text{ pu}.$$

8.3.2 Swing equation and criterion of stability

Assume that upon some change in the system operation the balance between the driving or mechanical input of the turbine P_m and the electrical power of the generator P_e is disturbed, so that $P_e < P_m$. Then, the additional kinetic energy will be stored by the rotated rotor, namely

$$J\frac{\omega_m^2 - \omega_{m0}^2}{2} = \int_0^t (P_m - P_e)dt = \int_0^t P_a dt, \tag{8.4}$$

where J is the moment of inertia (in kg/m^2) of all the rotating masses attached to the shaft, $\omega_m = d\delta_m/dt$ is an angular velocity of the shaft/rotor (in mechanical rad/sec) and P_a is an acceleration power. (In our future study we shall distinguish between the *electrical angle* δ_e, or just δ and the *mechanical angle* δ_m, i.e., $\delta \equiv \delta_e = (p/2)\delta_m$, where p is the number of poles, or $\delta_e = p\delta_m$, where p is the pole pairs, as it is adopted in some technical books.)

By differentiation, equation 8.4, we have

$$J\omega_m\frac{d\omega_m}{dt} = P_a. \tag{8.5a}$$

The change in the angular velocity about its initial or rated value is $\omega_m = \omega_{m0} + (d\delta_m/dt)$, then $(d\omega_m/dt) = (d^2\delta_m/dt^2)$ and

$$J\omega_m\frac{d^2\delta_m}{dt^2} = P_a. \tag{8.5b}$$

This equation, which governs the motion of the rotor of a synchronous machine, represents the power-angle δ change versus time, expressing the accelerating power applied to the shaft, and is called a **swing equation**. Usually it is written in a slightly different form, namely

$$M\frac{d^2\delta_m}{dt^2}=P_m-P_e=P_a, \tag{8.6a}$$

where

$$M=J\omega_m \tag{8.6b}$$

is an *angular momentum*, or *moment of inertia* (in joule·s/rad).

For a generator connected to an infinite bus, with operation at P_0 and δ_0, and small changes in δ and in P (so that linearity may be assumed), we can write

$$M\frac{d^2\Delta\delta}{dt^2}=-\Delta P=-\left(\frac{\partial P}{\partial\delta}\right)_0\Delta\delta \quad \text{or} \quad M\frac{d^2\Delta\delta}{dt^2}+\left(\frac{\partial P}{\partial\delta}\right)_0\Delta\delta=0. \tag{8.7}$$

The expression

$$\left.\frac{\partial P}{\partial\delta}\right|_{\delta=\delta_0}=P_m\cos\delta_0$$

is defined as the synchronizing power coefficient or just **synchronizing power** and is designated P_s, i.e., $P_s=(\partial P/\partial\delta)_0$. The characteristic equation of differential equation 8.7 is then

$$Ms^2+P_s=0, \tag{8.8}$$

which has two roots

$$s_{1,2}=\pm\sqrt{-\frac{P_s}{M}}. \tag{8.9}$$

If $(\partial P/\partial\delta)_0$ is positive, then both roots are imaginary numbers. In this case the solution of equation 8.7 is oscillatory undamped (since the resistances are neglected). Practically the oscillations decay and the stability is held (point *a* in Fig. 8.4). However, if $(\partial P/\partial\delta)_0$ is negative, both roots are real and one of them is positive, causing an unlimited increase in δ. In this case the stability is lost (point *b* in Fig. 8.4). If damping is present (i.e. the resistances are taken into consideration), equation 8.7 becomes

$$M\frac{d^2\Delta\delta}{dt^2}+K_d\frac{d\Delta\delta}{dt}+\left(\frac{\partial P}{\partial\delta}\right)_0\Delta\delta=0, \tag{8.10}$$

which results in a characteristic equation

$$Ms^2+K_ds+P_s=0. \tag{8.11}$$

Again, if $(\partial P/\partial\delta)_0$ is positive, the solution is a damped sinusoid and the operation is stable; in the opposite case, if $(\partial P/\partial\delta)_0$ is negative, the stability is lost, i.e., the sign of the sinchronizing power provides the *criterion of stability*. The derivative

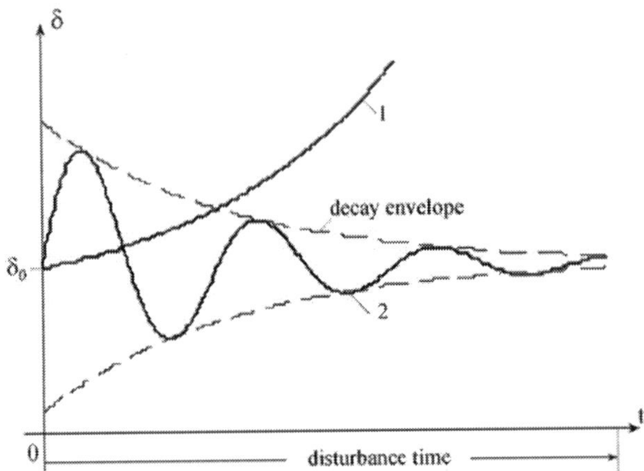

Figure 8.6 Two kinds of responses to a disturbance in the synchronous generator: 1) unstable, 2) stable.

$\partial P/\partial \delta$ is positive on the increasing branch of the power-angle curve and negative on the decreasing one, which confirms our previous conclusion that the operation at point *a* is stable, but at point *b* it is not. The two kinds of rotor motions, according to two kinds of solutions, are shown in Fig. 8.6.

For practical purposes (as is convenient in analyzing power systems) we shall normalize the swing equation. After dividing it by the rated power of the synchronous machine S_r, we have

$$\frac{M}{S_r}\frac{d^2\delta_m}{dt^2} = \frac{P_a}{S_r} = P_{an}, \tag{8.12}$$

where P_{an} is a normalized accelerating power in p.u. We shall next introduce the **inertia constant** H, which is one of the very important parameters of synchronous machines. It is defined as a quotient of the *kinetic energy* W_k, stored in the rotating rotor at rated angular velocity, and the rated power S_r:

$$H = \frac{W_k}{S_r} = \frac{1}{2}\frac{J\omega_{m,r}^2}{S_r} = \frac{1}{2}\frac{M}{S_r}\frac{\omega_{m,r}^2}{\omega_m}. \tag{8.13}$$

By using equation 8.13, equation 8.12 becomes

$$\frac{2H}{\omega_{m,r}}\frac{\omega_m}{\omega_{m,r}}\frac{d^2\delta_m}{dt^2} = P_{an}. \tag{8.14}$$

Since the change in angular velocity during the transient is relatively small, $(d\delta_m/dt) \ll \omega_{m,r}$ we may conclude that $\omega_m \cong \omega_{m,r}$(*) and equation 8.14 simplifies

(*)The angular frequency/velocity cannot change by a significant value before stability is lost. Thus for 60 Hz, $\omega_r = 377$ rad/s, and a 1% change in ω_m, i.e., 3.77 rad/s, will change the angle δ by 3.77 rad. Certainly, this would lead to a loss of synchronism.

to the swing equation in the form:

$$\frac{2H}{\omega_r}\frac{d^2\delta}{dt^2} = P_{an}. \tag{8.15}$$

Note that in equation 8.15 both angle δ and the angular velocity ω_r can be measured in electrical rad/s as well as in mechanical rad/s. We will treat this equation as written for electrical angle δ and electrical angular frequency.

The inertia constant H is somewhat similar to a per-unit quantity even though it is not a pure number. Since the quantities in the ratio, which express H (equation 8.13) do not have the same units, namely the kinetic energy is measured in MJ and the rated power in MVA, the unit of H is seconds. The value of H is usually in the range of 1–5 s (the smaller numbers are for small generators). The quantities of H given for a single generator may be modified for use in studies of a system with many generators by converting from the rating power S_r to the system base power (as in our previous study):

$$H_{sys} = H\frac{S_r}{S_b}. \tag{8.16}$$

The physical meaning of the inertia constant is that its value in seconds gives the time needed to accelerate the synchronous machine from zero speed to its rated value when the rated input power is applied.

As an example of using the swing equation, let us calculate the natural oscillations of a synchronous machine being subject to a small disturbance about the equilibrium point, like point a in Fig. 8.3. Assume that a small change in speed is given to the machine, i.e., $\omega = \omega_0 + \Delta\omega_0 u(t)$, where $\Delta\omega_0$ is the small change in speed and $u(t)$ is a unit step function. As a result of the change in speed, there will be a change in angle δ, i.e., $\delta = \delta_0 + \delta_\Delta$ and, in accordance with the power-angle curve, the electric power will be $P_e = P_{e0} + P_{e\Delta}$, while the mechanical power P_m remains constant and equal to P_{e0}. Then, the accelerating power $P_a = P_m - P_{e0} - P_{e\Delta} = -P_{e\Delta}$ and the *swing equation for the small changes becomes*

$$\frac{2H}{\omega_r}\frac{d^2\delta_\Delta}{dt^2} = -P_{e\Delta} = -P_s\delta_\Delta, \tag{8.17a}$$

or

$$\frac{2H}{\omega_r}\frac{d^2\delta_\Delta}{dt^2} + P_s\delta_\Delta = 0, \tag{8.17b}$$

where P_s is the synchronizing power, and as has been shown is

$$P_s = \left.\frac{\partial P}{\partial \delta}\right|_{\delta_0} = P_m \cos\delta_0.$$

The swing equation here is a second-order differential equation (when the damping is neglected) and the characteristic equation of which is

$$\frac{2H}{\omega_r}s^2 + P_s = 0. \tag{8.18}$$

The two roots of this equation are

$$s_{1,2} = \pm\sqrt{-P_s\omega_r/2H} = \pm j\omega_n, \tag{8.19}$$

where $\omega_n = \sqrt{P_s\omega_r/2H}$ is the *natural frequency* of the synchronous machine oscillations. Since the roots are imaginary numbers, the solution is pure sinusoid, i.e. an undamped oscillation:

$$\delta_\Delta(t) = A\sin(\omega_n t + \alpha).$$

To find the two unknown constants of integration, A and α, we shall determine two initial conditions which, obviously, are

$$\delta_\Delta(0) = 0, \quad \left.\frac{d\delta_\Delta}{dt}\right|_{t=0} = \omega_\Delta(0). \tag{8.20}$$

Thus,

$$A\sin(\omega_n + \alpha)|_{t=0} = 0 \quad \text{or} \quad \alpha = 0$$

and

$$\omega_n A\cos\omega_n t|_{t=0} = \omega_\Delta \quad \text{or} \quad A = \frac{\omega_\Delta}{\omega_n}.$$

Finally, we have

$$\delta_\Delta(t) = \frac{\omega_\Delta}{\omega_n}\sin\omega_n t. \tag{8.21}$$

Since the damping conditions are always present, these oscillations will decay (as shown in Fig. 8.6, curve 2) and the synchronous machine will return to point *a* of operation. The stability is held.

Example 8.2

A synchronous generator of reactance 1.25 pu is connected to an infinite bus bar system of $V = 1$ pu through a line and transformers of a total reactance of 0.5 pu. The generator's inertia constant is $H = 5$ s and EMF is 2.5 pu, and it operates at a load angle of 47°. Find the expression of the oscillations set up when the generator is subject to a sudden change of $+0.5\%$ of its speed.

Solution

The transmitted power of the generator is

$$P_e = \frac{EV}{X}\sin\delta = \frac{2.5\cdot 1}{1.25 + 0.5}\sin 47^\circ = 1.04 \text{ pu},$$

and the synchronized power is

$$P_s = \frac{2.5 \cdot 1}{1.25 + 0.5} \cos 47^\circ = 0.974 \text{ pu.}$$

Therefore, the angular frequency of oscillation is

$$\omega_n = \sqrt{0.974 \cdot (\pi \cdot 60)/5} = 6.06 \text{ rad/s,}$$

or

$$f_n = 6.06/2\pi = 0.96 \text{ Hz.}$$

The amplitude of oscillation will be

$$A = \frac{\omega_\Delta}{\omega_n} = \frac{0.005 \cdot 377}{6.06} = 0.31 \text{ rad.}$$

Thus

$$\delta_\Delta(t) = 0.31 \sin 6.06t \text{ rad.}$$

8.4 TRANSIENT STABILITY

Transient stability is concerned with the effect of large disturbances, which are usually due to faults, the most severe of which is a three-phase short-circuit (since by this kind of short-circuit the three-phase voltage may drop down to zero) and the most frequent is the single-line-to-earth fault. Some other kinds of such disturbances are line switching, sudden load changes, etc.

These kinds of disturbances are of a critical nature since they entail the sudden change of electrical output while the mechanical input from the turbine does not have time to change, during the relatively short period of fault, and remains practically constant. As a result, the rotor of the machine endeavors to gain speed and to store the excess energy. If the fault persists long enough the rotor angle will increase continuously and synchronism will be lost. Hence the time of operation of the protection devices and circuit breakers is of great importance.

The stability of the system may also be achieved using **autoreclosing circuit breakers.** The circuit breakers open when the fault is detected and automatically reclose after a prescribed period (usually less than 1 s). Due to the transitory nature of most faults (especially in the case of the single-line-to-earth fault) the circuit breaker often successfully recloses and the stability is held. However, if the fault persists, sometimes an autoreclosing is repeated. The length of the autoreclosing operation must be considered when the transient stability limits are calculated.

The transient stability of a power system is a function of the type and location of the disturbance to which the system is subjected. For instance, if two sections of a system are connected by a pair of lines, one of which is switched out, the power-angle characteristic is changed, having a lower power peak. The balance between the mechanical and electrical powers is disturbed, which causes transient stability problems. A more severe test of system stability is a short-circuit fault on one line followed by its being switched out.

One of the purposes of the analysis of the system transient stability is to determine a stability limit, usually in terms of a critical fault clearance and/or autoreclosing time t_{cr}. If, however, t_{cr} is given, being the minimal available time (as the fastest relay protection and circuit breakers have been anyway used) the system stability test in this case is an estimation of the maximum load which the system can carry without losing transient stability. In our next study using the following examples we shall illustrate how to check the transient stability of a system by solving the swing equation.

Therewith, some further assumptions and simplifications will be made. The cylindrical rotor machine is assumed and, therefore, the direct and quadrature axis reactances are assumed equal. Direct-axis transient reactance X'_d and transient EMF E'_d will be used for the machine representation. The input power for the first seconds following a disturbance remains constant. (This assumption is often made, as previously mentioned, considering that the mechanical system of governors, steam or hydraulic valves, and the like, is relatively sluggish with respect to the fast-changing electrical quantities. However, with improvements in mechanical equipment, such as fast valving, electronic regulators etc., the assumption of a constant input will not be valid and its appropriate change should be taken into consideration.)

Example 8.3

Assume that, at the sending-end of one of the transmission lines in the system shown in Fig. 8.7, a three-phase fault occurs. Develop and solve the swing equation of the system, if the fault reactance is 0.07 pu. The inertia constant of the generator is $H = 0.5$ s and the frequency $f = 60$ Hz. Initially the generator delivers a 0.8 pu power with a transient EMF of 1.22 pu.

Solution

The power-angle characteristic prior to the fault is

$$P_e = \frac{1.22 \cdot 1}{0.6} \sin\delta = 2.03 \sin\delta,$$

where $E'_d = 1.22$ and $X_{tot} = 0.6$ (see Fig. 8.7(b)).

Performing $Y \to \Delta$ transformation in the given circuit we obtain

$$Y_{12} = \frac{Y_1 Y_2}{Y_1 + Y_2 + Y_3} = \frac{(1/0.3)(1/0.3)}{1/0.3 + 1/0.3 + 1/0.07} = 0.528.$$

The electrical power output of the sending-end at the fault is

$$P_e = 1.22 \cdot 1 \cdot 0.528 \sin\delta = 0.644 \sin\delta.$$

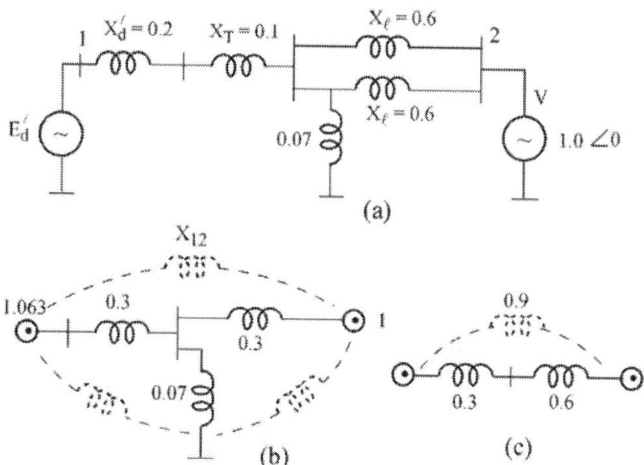

Figure 8.7 An equivalent circuit of the faulted network for Example 8.3 (a) and its simplifications (b) and(c).

Therefore, with equation 8.15 the swing equation is

$$\frac{d^2\delta}{dt^2} = \frac{\omega_r}{2H}(P_m - P_e),$$

or

$$\frac{d^2\delta}{dt^2} = \frac{377}{10}(0.8 - 0.644 \sin \delta) = 37.7(0.8 - 0.644 \sin \delta).$$

The solution of this nonlinear equation is obtained by the MATCAD program, which for the initial conditions of $\delta_0 = \sin^{-1}(0.8/2.03) = 23.2^\circ$ (0.405 rad) and $\omega(0) = \omega_0 - \omega_r = 0$ is shown in Fig. 8.8. As can be seen, the power angle increases indefinitely and the system is unstable.

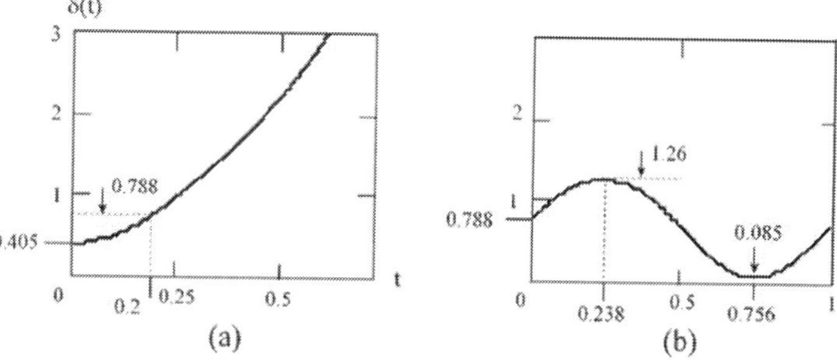

Figure 8.8 Angle-time curve for the faulted network of Example 8.3: at the first moment of the fault (a) and after clearing the fault (b).

Assume now that the fault is cleared in 0.2 s by opening the faulted line.

The equivalent circuit is now shown in Fig. 8.7(c), in which the total reactance is

$$X_{12} = 0.3 + 0.6 = 0.9\,\text{pu},$$

and the transient power will be

$$P_e = \frac{1.22}{0.9} \sin\delta = 1.36 \sin\delta.$$

Thus, the swing equation is

$$\frac{d^2\delta}{dt^2} = 37.7(0.8 - 1.36 \sin\delta).$$

The initial values of δ and ω are calculated with the previous solution for time 0.2 s, which gives $\delta_{0.2} = 0.788$ rad and $\omega_{0.2} = 3.56$, and the time solution is now shown in Fig. 8.9. (Note that for the new solution $\delta_{0.2} \equiv \delta_0$ and $\omega_{0.2} \equiv \omega_0$.) As can be seen the time change of the power angle is oscillatory and the first peak of about $\delta_{max} = 1.26$ rad is reached at $t = 0.238$ s after which δ is decreased until

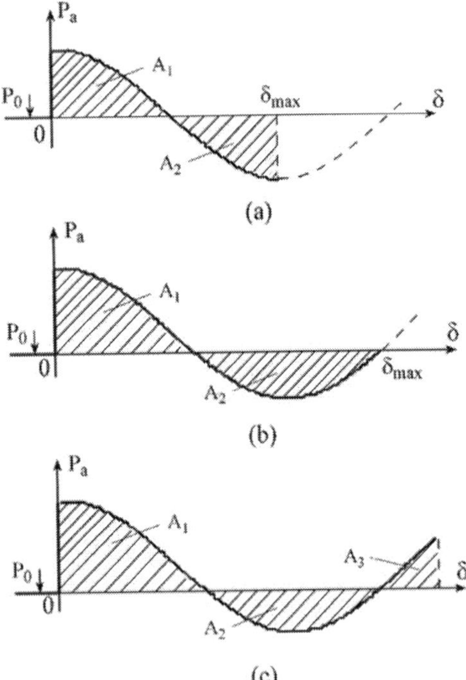

Figure 8.9 Equal-area criteria for a stable system (a), critical case (b) and for an unstable system (c).

it reaches a minimum value of about $\delta_{min} = 0.085\,\mathrm{rad}$ at $t = 0.756\,\mathrm{s}$ and the oscillations of the rotor angle continue until they decay due to the damping effect. For the system under study and for the given fault the synchronism is not lost and the machine is stable.

8.4.1 Equal-area criterion

Consider once again the swing equation derived previously in the form

$$\frac{d^2\delta}{dt^2} = \frac{\omega_r}{2H} P_a. \tag{8.22}$$

Multiplying both sides by $2(d\delta/dt)$, we have

$$\left(2\frac{d\delta}{dt}\right)\frac{d^2\delta}{dt^2} = \frac{\omega_r}{2H} P_a \left(2\frac{d\delta}{dt}\right),$$

or

$$d\left[\left(\frac{d\delta}{dt}\right)^2\right] = \frac{\omega_r}{H} P_a d\delta.$$

Integrating both sides gives

$$\left(\frac{d\delta}{dt}\right)^2 = \frac{\omega_r}{H}\int_{\delta_0}^{\delta} P_a d\delta. \tag{8.23}$$

This equation determines the quadrature of the relative speed of the machine (with respect to a reference frame moving at a constant speed, for instance, like the infinite bus) as proportional to the integral of P_a versus δ. For a rotor that is accelerating, the condition of stability is that this speed becomes zero, or negative, causing the motor to slow down. In other words, the increasing of angle δ is restricted and after reaching some maximal value, δ_{max}, the angle decreases. Thus, we may conclude that δ_{max} exists and it is given by the condition

$$\int_{\delta_0}^{\delta} P_a d\delta \leq 0. \tag{8.24}$$

In the opposite case $d\delta/dt$ does not become zero, the rotor will continue to move and synchronism is lost (the angle increases unlimitedly).

The integral of $P_a d\delta$ in equation 8.24 represents an area on the $P - \delta$ diagram. Hence, the criterion for stability is that the area between the $P - \delta$ curve and the line of the power input P_m (or P_0) must be zero. The difference between the $P - \delta$ curve and the input power, i.e., the accelerating power, might be also represented as a curve, $P_a(\delta)$, as shown in Fig. 8.9. Then the area under this curve must be zero, which again means that the positive and negative areas are equal. This is known as the **equal-area criterion.** Physically, this criterion means that the rotor must be able to return to the system all the energy gained from the turbine during the acceleration period. This is shown in Fig. 8.9. In figure (a) the positive area A_1 is equal to the negative area A_2 at the angle δ_{max}, at which the accelerating power is negative and the rotor slows down. Therefore, the system is stable and δ_{max} is the maximum rotor angle reached during the swing. In figure (b) the positive and negative areas are equal at the point where P_a reverses its sign, which means that it is a critical case: the oscillations will continue. However, due to the damping effect they will decay. The system is stable and angle δ_{max} is again the maximum rotor angle reached during the first swing.

If the accelerating power reverses it sign before the two areas A_1 and A_2 are equal, as in figure (c), angle δ continues to increase and synchronism is lost. The equal-area criterion is usually applied to the power-angle curve, where the electrical and mechanical powers are plotted as a function of δ. Note that the accelerating power curve could have discontinuities due to the switching of the network, faults occurring and the like.

A simple example of the equal-area criterion may be introduced by an examination of the system stability if one of the two parallel lines, which connect the generator to an infinite bus bar, is switched out (disconnected). The two power-angle curves pertaining to a normal (curve 1) and one line (curve 2) operation of the system are shown in Fig. 8.10.

The shaded area A_1 is proportional to the kinetic energy stored in the rotor, when the input power P_0 is larger than the electrical power delivered by the generator in accordance with curve 2, and in this case the rotor accelerates, Fig. 8.10(a). The shaded area A_2 represents the amount of energy, which the rotor returns to the system. Since these two areas are equal the rotor initially comes to rest at angle δ_{max} (point c) whereupon its speed is again synchronous. Having returned all of its extra kinetic energy back to the electrical circuit, the rotor continues to decelerate ($P_e > P_m$) falling through point b and back towards point a. Such oscillations will continue until completely damped at the new angle δ_1 ($\delta_1 > \delta_0$, point b). However, if the initial operating power P'_0 and angle δ'_0 are increased to such values that the area between δ'_0 and δ'_1 (A_1) is just

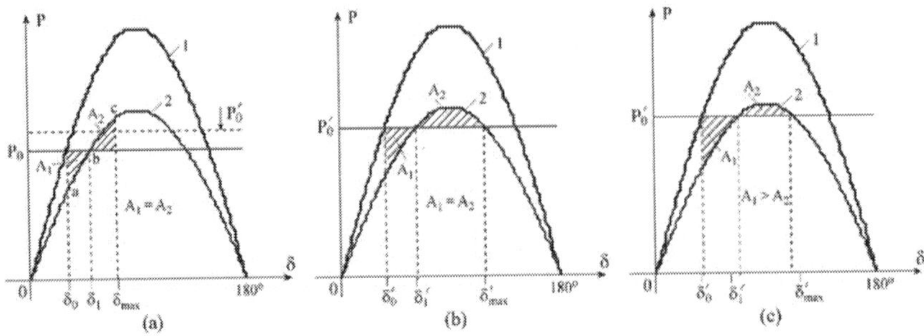

Figure 8.10 Power-angle curves for two lines in parallel (curve 1) and one line (curve 2) and equal area criterion: stable operation (a), critical operation (b) and nonstable operation (c).

equal to the available area between δ_1' and $\delta_{\max}'$, where $\delta_{\max} = 180° - \delta_1'$, it will be the critical operation, Fig. 8.10(b). This would be the condition for maximum input ($P_{0,\max}$) power. If the input power is larger than $P_{0,\max}$, then the accelerating energy (A_1) will be bigger than the available decelerating energy ($A_1 > A_{2,avail}$). The excess kinetic energy will cause δ to continue increasing beyond $\delta_{\max}'$ and the energy would again be absorbed by the rotor (since P_e is now decreasing with an increase in δ, i.e., the slope is negative) and stability will be lost, Fig. 8.10(c). The coefficient

$$k = \frac{A_{2,avail} - A_1}{A_1}$$

is sometimes defined as the **transient stability security factor**. Notice that (as can be seen from Fig. 8.10) it is permissible for the rotor to oscillate past the point where $\delta = 90°$, as long as the equal-area criterion is met.

As another example of using the equal-area criterion, let us consider the fault on one of two parallel lines as in Example 8.3. The power-angle curves pertaining to a fault on one of two parallel lines are shown in Fig. 8.11. The fault is cleared in a time corresponding to δ_1, and the shaded area δ_0 to δ_1, between the P_0-line and power-angle curve for the fault, A_1, indicates the energy stored. The rotor swings until it reaches δ_2 so that $A_1 = A_2$, where A_2 is the shaded area δ_1 to δ_2 between the P_0-line and the power-angle curve for one line after the faulted line has been switched out. Since δ_2 is less than the critical condition angle $\delta_{2\max}$ the system is stable. Critical conditions are reached when

$$\delta_2 = 180° - \sin^{-1}(P_0/P_2). \tag{8.25}$$

The time corresponding to the critical clearing angle is called the **critical clearing time** for a particular (normally full-load) value of power input. This time is of great importance for system protection and to switchgear designers,

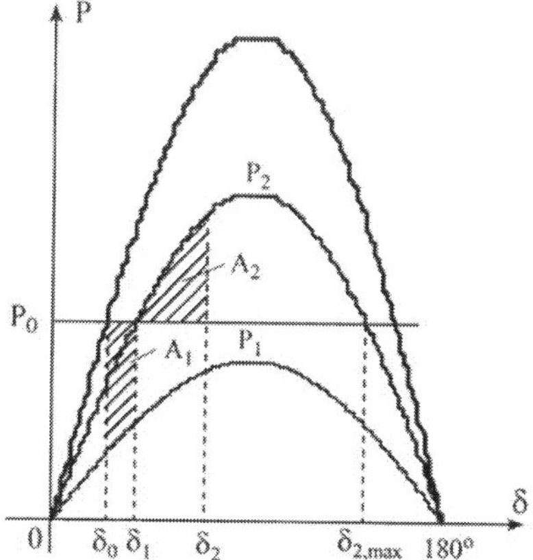

Figure 8.11 Equal-area criterion for the fault of one of two lines in parallel.

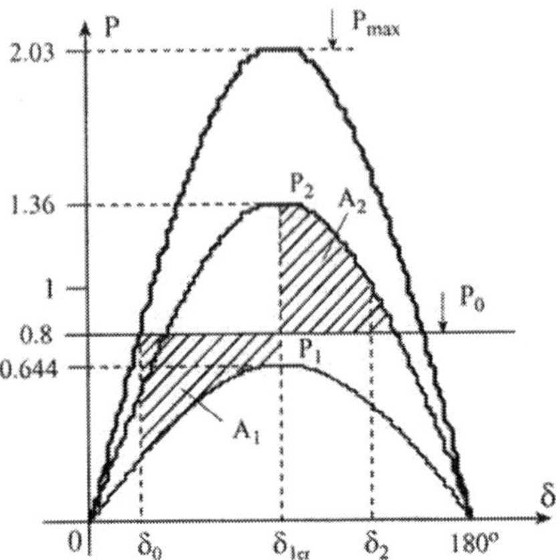

Figure 8.12 Application of the equal area criterion to a critically cleared system.

as it is the maximum time allowable for the equipment to operate without losing stability. The critical clearing angle for a fault on one of two parallel lines, for instance, may be determined as follows. With the equal-area criterion, as shown in Fig. 8.12, we have

$$\int_{\delta_0}^{\delta_{1cr}} (P_0 - P_1 \sin \delta)d\delta = -\int_{\delta_{1cr}}^{\delta_2} (P_0 - P_2 \sin \delta)d\delta,$$

and after integration

$$P_0(\delta_{1cr} - \delta_0) + P_1(\cos \delta_{1cr} - \cos \delta_0) = -P_0(\delta_2 - \delta_{1cr}) - P_2(\cos \delta_2 - \cos \delta_{1cr}),$$

from which the critical clearing angle is

$$\cos \delta_{1cr} = \frac{P_0(\delta_0 - \delta_2) + P_1 \cos \delta_0 - P_2 \cos \delta_2}{P_1 - P_2}. \tag{8.26}$$

where, with equation 8.25, $\delta_2 = 180° - \sin^{-1}(P_0/P_2)$. Knowing a critical angle and swing frequency, the critical clearing time can be readily obtained.

Example 8.4

Apply the equal-area criterion to the system of Example 8.3.

Solution

We may calculate the critical clearing angle as follows. For this system we have $\delta_0 = 23.2°$, $P_0 = 0.8$ pu, $P_1 = 0.644$ pu, $P_2 = 1.36$ pu and $\delta_2 = 180° - \sin^{-1}(0.8/1.36) = 144.0°$.

Calculation using equation 8.26 gives (note that electrical degrees must be expressed in radians)

$$\cos\delta_{1cr} = \frac{0.8\cdot(0.404 - 2.51) + 0.644\cos 23.2^\circ - 1.36\cos 144.0^\circ}{0.644 - 1.36} = -0.007,$$

or

$$\delta_{1cr} = \cos^{-1}(-0.007) = 90.4^\circ.$$

This situation is illustrated in Fig. 8.12.

8.5 REDUCTION TO A SIMPLE SYSTEM

When a number of generators are connected to the same bus bar, they can be represented by a single equivalent machine. E_{eq} and X_{eq} may be found as explained in section 6.2.2. The inertia constant H of the equivalent machine can then be evaluated by equating the stored energy of the equivalent machine to the total of the individual machines, which yields

$$H_{eq} = H_1\frac{S_1}{S_b} + H_2\frac{S_2}{S_b} + \cdots + H_n\frac{S_n}{S_b}, \qquad (8.27)$$

where $S_1 \cdots S_n$ are MVA powers of the generators and S_b is the base power. So, consider, for example, a power station, which consists of three generators of 60 MVA, 100 MVA and 300 MVA, having an H of 5 s, 6 s and 8 s respectively. Making the base power equal to 100 MVA, the inertia constant of an equivalent machine will be

$$H_{eq} = 5\frac{60}{100} + 6\frac{100}{100} + 8\frac{300}{100} = 33\ \text{s}.$$

Consider two machines, having M_1 and M_2,which are connected through transformers' and lines' impedances/reactances. The equations of motion for small changes are

$$\begin{aligned} M_1\frac{d^2\Delta\delta_1}{dt^2} + \left(\frac{\partial P_1}{\partial\delta_{12}}\right)_0\Delta\delta_{12} = 0 \\ M_2\frac{d^2\Delta\delta_2}{dt^2} + \left(\frac{\partial P_2}{\partial\delta_{12}}\right)_0\Delta\delta_{12} = 0, \end{aligned} \qquad (8.28)$$

where $\Delta\delta_{12} = \Delta\delta_1 - \Delta\delta_2$.

By subtracting these two equations we may obtain a single equation of the relative motion

$$\frac{d^2\Delta\delta_{12}}{dt} + \left[\frac{(\partial P_1/\partial\delta_{12})_0}{M_1} - \frac{(\partial P_2/\partial\delta_{12})_0}{M_2}\right]\Delta\delta_{12} = 0, \qquad (8.29)$$

for which the characteristic equation has two roots

$$s_{1,2} = \pm j\sqrt{\frac{(\partial P_1/\partial\delta_{12})_0}{M_1} - \frac{(\partial P_2/\partial\delta_{12})_0}{M_2}}. \tag{8.30}$$

As was previously shown, if the quantity under the square root is positive, the stability of both generators is assured.

Consider now, two synchronous generators (or group of generators) connected by a reactance. In this case they may be reduced to one equivalent machine connected through the reactance to an infinite bus bar system. The transient equations of motion for the generators are

$$\frac{d^2\delta_1}{dt^2} = \frac{\Delta P_1}{M_1} \quad \text{and} \quad \frac{d^2\delta_2}{dt^2} = \frac{\Delta P_2}{M_2}. \tag{8.31}$$

Then the equation of motion for the two-machine system is

$$\frac{d^2\delta}{dt^2} = \frac{\Delta P_1}{M_1} - \frac{\Delta P_2}{M_2} = \left(\frac{1}{M_1} - \frac{1}{M_2}\right)(P_0 - P_{e,\max}\sin\delta), \tag{8.32}$$

where $\delta = \delta_1 - \delta_2$ is the relative angle between the machines and $d\delta/dt$ is the relative velocity of the two groups with respect to each other. Also note that $\Delta P_1 = -\Delta P_2 = P_0 - P_{e,\max}\sin\delta$, where P_0 is the input power and $P_{e,\max}$ is the maximum transmittable power. For a single generator of M_{eq} and the same input power connected to the infinite bus bar system we have

$$M_{eq}\frac{d^2\delta}{dt^2} = P_0 - P_{e,\max}\sin\delta. \tag{8.33}$$

Therefore, we may conclude that

$$M_{eq} = \frac{M_1 M_2}{M_1 + M_2}, \tag{8.34}$$

and that this equivalent generator has the same mechanical input as the actual machines and that the load angle δ in equation 8.33 is the angle between the rotors of the two machines.

The most useful method of network reduction is by nodal elimination, in which the network is finally represented by only the transfer reactances between the reduced nodes, as any shunt impedances at these nodes do not influence the power transferred.

The electrical network for the transient stability analysis will then obtain n generator buses, to which the voltages (i.e. the internal generator transient EMF's) behind their transient reactances are applied. The values of EMF's are determined, as in the one-machine system, from the pre-transient conditions.

Loads are represented by passive admittances i.e., $G_L = P_L/V_r^2$ and $B_L = Q_L/V_r^2$, which are connected at the load nodes. Note that such a representation is very simplified. Since a network fault usually causes a reduction in the voltages near the fault location, this will result in a decrease in the load power proportional to V^2. In the real system, however, the decrease in power is likely to be less than this, but to occur in a more complicated manner. Since the load usually contains a large proportion of non-static elements, such as induction motors, the nature of the load characteristics is such that beyond the critical point the motors will run down to a standstill and stall. For a more precise analysis of system stability, therefore, it is important to consider the actual load characteristic (see the next section), which makes such an analysis much more complicated.

Passive impedances are connected between various nodes of the network (representing transformers, lines, etc.), and the reference nodes of the active elements and loads are connected in a common reference bus.

Now, let us say that in the given network there are n generator, or active element, nodes and r (remaining) nodes with passive elements. Then the network admittance matrix $\mathbf{Y}$ may be partitioned as

$$\mathbf{Y} = \left[\begin{array}{c:c} \mathbf{Y}_{nn} & \mathbf{Y}_{nr} \\ \hdashline \mathbf{Y}_{rn} & \mathbf{Y}_{rr} \end{array}\right]. \tag{8.35}$$

It can be shown that the matrix for the reduced network, which has only the active nodes, is

$$\mathbf{Y}_{red} = \mathbf{Y}_{nn} - \mathbf{Y}_{nr}\mathbf{Y}_{rr}^{-1}\mathbf{Y}_{rn}, \tag{8.36}$$

which is of dimension $(n \times n)$, where n is the number of generators.

The maximum powers transferable between the relevant generators, before and during a fault, can now be calculated from this reduced configuration of a network. In accordance with equation 8.2a (note that here $E \equiv E_i$, $V \equiv E_j$ and $(1/z)\cos\varphi = G$) we may write for the power of the i-ts generator

$$P_{ei} = E_i^2 G_{ii} + \sum_{\substack{j=1 \\ j \neq i}} E_i E_j Y_{ij} \cos(\delta_{ij} - \varphi_{ij}), \quad i = 1, 2, \ldots, n, \tag{8.37}$$

where $\mathbf{Y}_{ii} = Y_{ii} \angle -\varphi_{ii} = G_{ii} - jB_{ii}$ is the admittance for node i and $\mathbf{Y}_{ij} = Y_{ij} \angle -\varphi_{ij}$ is the negative of the transfer admittance between nodes i and j. The equations of motion are then given as

$$\begin{aligned} \frac{2H_i}{\omega_r}\frac{d\omega_i}{dt} + D_i\omega_i &= P_{mi} - \left[E_i^2 G_{ii} + \sum_{\substack{j=1 \\ j \neq i}} E_i E_j Y_{ij} \cos(\delta_{ij} - \varphi_{ij}) \right] \\ \frac{d\delta_i}{dt} &= \omega_i - \omega_r \qquad\qquad i = 1, 2, \ldots, n. \end{aligned} \tag{8.38}$$

The damping coefficient term $D\omega$, which represents the turbine damping, generator electrical damping and the damping effect of electrical loads, is frequently added in the swing equation. (Values of the damping coefficient used in the stability studies are in the range of 1–3 pu.)

8.6 STABILITY OF LOADS AND VOLTAGE COLLAPSE

If the load is purely static, i.e. represented by an impedance, the system will operate stably even at low voltages. However, in reality the load contains non-static elements such as induction motors. The nature of the load Q-V characteristics, when they include a large proportion of induction motors, is non-linear. Both characteristics P-V and Q-V for such a load are shown in Fig. 8.13. While the active power characteristic is almost always a straight line, the *reactive power Q-V characteristic* is a curve having a minimum or critical point and two branches with positive and negative slopes.

Beyond the critical point c even a very small decrease in voltage causes an increase in the reactive power Q, which in turn results in a decrease in the voltage (since the voltage drop depends on the power $\Delta V_{rct} = QX/V$) and so forth. This process of voltage collapse is mathematically defined as $dQ/dV \to \infty$, i.e., on the left branch of the Q-V curve. The physical explanation of the voltage collapse may be found in the behavior of induction motors. Beyond some critical voltage the motors will run down to a standstill or stall. In this situation induction motors consume pure reactive power, which at low voltage causes very large currents similar to short-circuit currents. Finally, this results in very low voltage.

In the power system the problem arises due to the relatively high impedance of the connection between the load and the feeding bus, which can be considered as an infinite bus bar. This happens when one line of two or more forming the load connection is suddenly lost. Consider a simple network, shown in Fig. 8.14, where the non-static load is supplied through a reactance from a constant voltage source E. In this circuit

$$E = \sqrt{\left(V + \frac{QX}{V}\right)^2 + \left(\frac{PX}{V}\right)^2}, \tag{8.39}$$

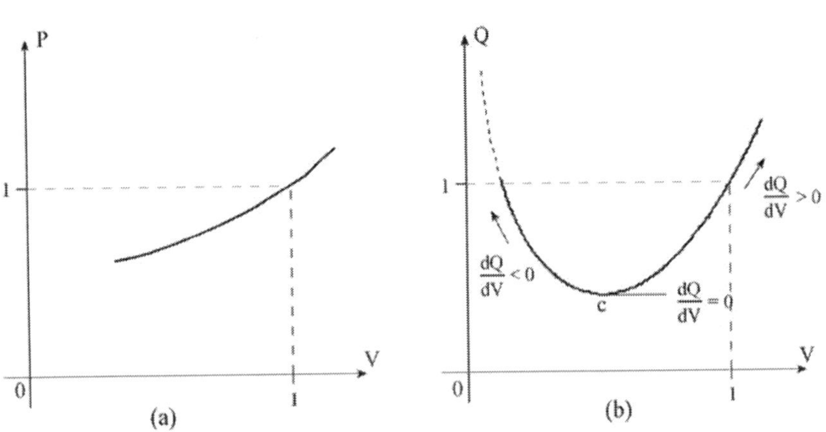

Figure 8.13 Non-static power characteristics: active power-voltage characteristic (a) and reactive power-voltage characteristic (b).

Figure 8.14 A network with a load dependent on voltage.

or if

$$\frac{PX}{V} \ll \frac{V^2 + QX}{V}$$

(which usually takes place)

$$E = V + \frac{QX}{V}. \tag{8.40}$$

From the system viewpoint, it is worthwhile to develop a voltage stability criterion with a dependency on E versus V, i.e., by using the *E-V curve*. This curve (equation 8.40) is plotted in Fig. 8.15, where the two components V and QX/V are also shown. Performing the differentiation of equation 8.40 with respect to V yields

$$\frac{dE}{dV} = 1 + \left(\frac{dQ}{dV}XV - QX\right)\frac{1}{V^2} = 1 + \left(\frac{dQ}{dV} - \frac{Q}{V}\right)\frac{X}{V}. \tag{8.41}$$

Here, when $dQ/dV \to -\infty$, the term in the parentheses and, therefore, the entire expression (8.41) approaches a negative infinity. Thus, we may conclude that if $dE/dV \to -\infty$, then the system is unstable and the *voltage collapse takes place.* When dQ/dV is positive, the term in the parentheses is also positive ($dQ/dV > Q/V$) and therefore $dE/dV > 0$, i.e. the system is stable, *which means*

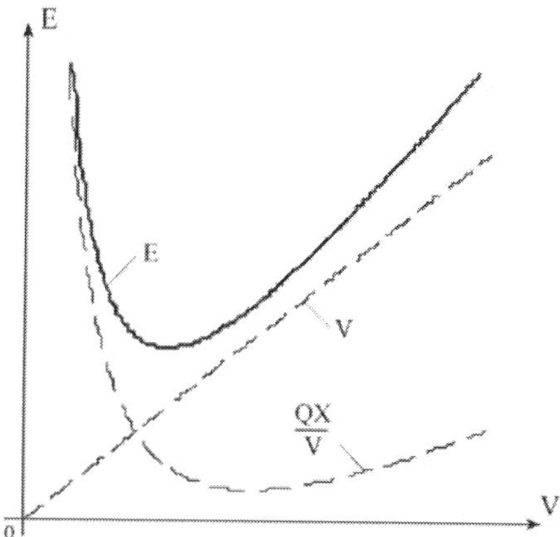

Figure 8.15 *E-V* curve and its two components.

that the sign of the derivative dE/dV provides the criterion of load stability. At the critical point on the E-V curve, i.e., where $dE/dV=0$, we have

$$\frac{dQ}{dV}=\frac{Q}{V}-\frac{V}{X}, \tag{8.42}$$

or, since $QX/V=E-V$ we have

$$\frac{Q}{V}=\frac{E}{X}-\frac{V}{X}$$

and

$$\frac{dQ}{dV}=\frac{E}{X}-\frac{2V}{X}. \tag{8.43}$$

This means that at $dE/dV=0$, the sign of dQ/dV is defined by the sign of $(E-2V)$ and if it is negative, the system is unstable.

Example 8.5

Examine the voltage stability of a non-static load supplied from a 275 kV infinite bus bar through a line of reactance 50 Ω per phase. The load consists of a constant active power of 200 MW and 200 MVAr rating reactive power, which is related to the voltage by the equation (in pu) $Q=5(V-0.7)^2+0.8$.

Solution

With $S_b=200$ MVA and $V_b=275$ kV the pu value of the line reactance is

$$X=\frac{50\cdot 200}{275^2}=0.132 \text{ pu}.$$

The load voltage can then be found from the equation

$$E=V+\frac{QX}{V}.$$

Thus,

$$E=V+\frac{0.132}{V}[5(V-0.7)^2+0.8].$$

Since $E=1$ after simplification we have

$$V^2-1.159V+0.258=0.$$

Thus, the roots are

$$V_{1,2}=0.858;\ 0.301.$$

Taking the upper value (which is suitable to a physical reality) we obtain

$$Q=5(0.858-0.7)^2+0.8=0.925,$$

and

$$\frac{dQ}{dV} = 10(V - 0.7) = 1.58.$$

Then,

$$\frac{dE}{dV} = 1 + \left(1.58 - \frac{0.925}{0.858}\right)\frac{0.132}{0.858} = 1.077.$$

Since the result is positive, the system is stable. (Note that $PX/V = 1 \cdot 0.132/0.858 = 0.15$, which is much less than $(V^2 + QX)/V = (0.858^2 + 0.924 \cdot 0.132)/0.858 \cong 1$.)

The reader can now convince himself that, if one of the two parallel lines is lost and the reactance changed to 0.264, dE/dV becomes negative and the system is unstable, i.e., the voltage collapses.

APPENDIX I

SOLVING EXAMPLE 5.6 USING THE MATHCAD PROGRAM[(*)]

Definition of the array first element subscript

$$\mathrm{ORIGIN} \equiv 1$$

Data:

$$C1 := 1 \quad C2 := 2 \quad L4 := 1 \quad G3 := 1 \quad R5 := 1$$

$$R6 := \frac{2}{7} \quad R7 := \frac{1}{3} \quad a := \frac{1}{(1 + R5 \cdot G3)}$$

Matrix **A**

$$\begin{bmatrix} \dfrac{-(1 + a \cdot R6 \cdot G3)}{R6 \cdot C1} & a \cdot \dfrac{G3}{C1} & \dfrac{a}{C1} \\ a \cdot \dfrac{G3}{C2} & \dfrac{-(1 + a \cdot R7 \cdot G3)}{R7 \cdot C2} & \dfrac{(-1 + a)}{C2} \\ \dfrac{a}{L4} & \dfrac{(1 - a)}{L4} & -a \cdot \dfrac{R5}{L4} \end{bmatrix}$$

$$= \begin{pmatrix} -4 & 0.5 & -0.5 \\ 0.25 & -1.75 & -0.25 \\ 0.5 & 0.5 & -0.5 \end{pmatrix}$$

The characteristic equation

$$\left| \begin{pmatrix} \lambda + 4 & -0.5 & 0.5 \\ -0.25 & \lambda + 1.75 & 0.25 \\ -0.5 & -0.5 & \lambda + 0.5 \end{pmatrix} \right| \rightarrow \lambda^3 + 6.25\lambda^2 + 10.125\lambda + 4.500$$

Finding the roots

$$\mathrm{coef} := \begin{pmatrix} 4.5 \\ 10.13 \\ 6.25 \\ 1 \end{pmatrix} \quad \mathrm{polyroots}\,(\mathrm{coef}) = \begin{pmatrix} -3.998 \\ -1.504 \\ -0.748 \end{pmatrix}$$

(*)See page 307.

Unity, initial, **b** and **w** matrixes

$$U:=\begin{pmatrix}1 & 0 & 0\\ 0 & 1 & 0\\ 0 & 0 & 1\end{pmatrix}\quad X0:=\begin{pmatrix}0.5\\ 1.5\\ 1\end{pmatrix}\quad b:=\begin{pmatrix}3.5 & 0\\ 0 & 1.5\\ 0 & 0\end{pmatrix}\quad w:=\begin{pmatrix}1\\ 1\end{pmatrix}$$

The matrix for finding β coefficients

$$d:=\begin{pmatrix}1 & -0.75 & 0.5625\\ 1 & -1.5 & 2.25\\ 1 & -4 & 16\end{pmatrix}$$

$$d^{-1}\cdot\begin{pmatrix}\exp(-0.75t)\\ \exp(-1.5t)\\ \exp(-4t)\end{pmatrix}\text{ float, 4}$$

$$\rightarrow\begin{pmatrix}2.462\exp(-0.75t)-1.600\exp(-1.5t)+0.1385\exp(-4t)\\ 2.256\exp(-0.75t)-2.533\exp(-1.5t)+0.2769\cdot\exp(-4t)\\ 0.4103\exp(-0.75t)-0.5333\exp(-1.5t)+0.1231\exp(-4t)\end{pmatrix}$$

The β-coefficients

$$\beta:=\begin{pmatrix}2.462\exp(-0.75t)-1.600\exp(-1.5t)+0.1385\exp(-4t)\\ 2.256\exp(-0.75t)-2.534\exp(-1.5t)+0.2770\cdot\exp(-4t)\\ 0.4103\exp(-0.75t)-0.5333\exp(-1.5t)+0.1231\exp(-4t)\end{pmatrix}$$

Calculating exp(At)

$$A:=\begin{pmatrix}-4 & 0.5 & -0.5\\ 0.25 & -1.75 & -0.25\\ 0.5 & 0.5 & -0.5\end{pmatrix}\quad A^2=\begin{pmatrix}15.875 & -3.125 & 2.125\\ -1.563 & 3.063 & 0.438\\ -2.125 & -0.875 & -0.125\end{pmatrix}$$

$$A^{-1}=\begin{pmatrix}-0.222 & 0 & 0.222\\ 0 & -0.5 & 0.25\\ -0.222 & -0.5 & -1.528\end{pmatrix}$$

$$\exp(At)=(U\cdot\beta_1+A\cdot\beta_2+A^2\cdot\beta_3)$$

Equation to calculate X_{nat}:

$$(U\cdot\beta_1+A\cdot\beta_2+A^2\cdot\beta_3)\cdot X0$$

which results in

$$X_{nat} := \begin{pmatrix} -0.5101\exp(-0.75t) + 0.7695\exp(-1.5t) + 0.2466\exp(-4t) \\ -0.7.67\exp(-0.75t) + 2.303\exp(-1.5t) - 3.079\times10^{-2}\exp(-4t) \\ 2.565\exp(-0.75t) - 1.534\exp(-1.5t) - 3.074\times10^{-2}\exp(-4t) \end{pmatrix}$$

Equation to calculate X_{part}:

$$A^{-1}\cdot[(U\cdot\beta_1 + A\cdot\beta_2 + A^2\cdot\beta_3 - U)\cdot b\cdot w]$$

which results in

$$X_{part} := \begin{pmatrix} 0.5426\exp(-0.75t) - 0.5596\exp(-1.5t) - 0.7699\exp(-4t) + 0.7777 \\ 0.7500 + 0.8196\exp(-0.75t) - 1.670\exp(-1.5t) + 9.632\times10^{-2}\exp(-4t) \\ -2.738\exp(-0.75t) + 1.112\exp(-1.5t) + 9.59\times10^{-2}\exp(-4t) + 1.528 \end{pmatrix}$$

The total response $X(t) = X_{nat} + X_{part}$:

$$X(t) := \begin{pmatrix} 3.25\times10^{-2}\exp(-0.75t) + 0.2099\exp(-1.5t) - 0.5233\exp(-4t) + 0.7777 \\ 5.29\times10^{-2}\exp(-0.75t) + 0.633\exp(-1.5t) + 6.553\times10^{-2}\exp(-4t) + 0.7500 \\ -0.173\exp(-0.75t) - 0.422\exp(-1.5t) + 6.516\times10^{-2}\exp(-4t) + 1.528 \end{pmatrix}$$

APPENDIX II

THE CALCULATION OF THE p.u.VALUES FOR A GIVEN NETWORK(*)

The 1-line diagram of the network of example 6.2 is shown in Fig. AII-1. The parameters of the network elements are as follows.

Generators: G_1- $S_n = 470$ MVA, $V_n = 15.75$ kV, $E'_d = 1.25$, $X'_d = 0.3$,
$G_2 = G_3$- $S_n = 2 \times 118$ MVA, $V_n = 13.8$ kV, $E'_d = 1.33$, $X'_d = 0.38$.

System- $E_s = 1$, $X_s = 83\ \Omega$.

Transformers: T1- $S_n = 250$ MVA, $V_n = 242/13.8$ kV, $X_{s.c.} = 11\%$,
T2- $S_n = 120$ MVA, $V_n = 220/11.0$ kV, $X_{s.c.} = 12\%$.

Autotransformers:
AT1- $S_n = 480$ MVA, $V_n = 242/121/15.75$ kV, $X_{s.c.} = 13.5/12.5/18.8\%$,
AT2- $S_n = 360$ MVA, $V_n = 525/242/13.8$ kV, $X_{s.c.} = 8.4/28.4/19.0\%$.

Transmission lines: $\ell_1 = 120$ km, $\ell_2 = 95$ km, $\ell_3 = 80$ km, $x_0 = 0.4\ \Omega$/km.

Load Ld- $S_n = 250$ MVA, $X_n = 1.2$.

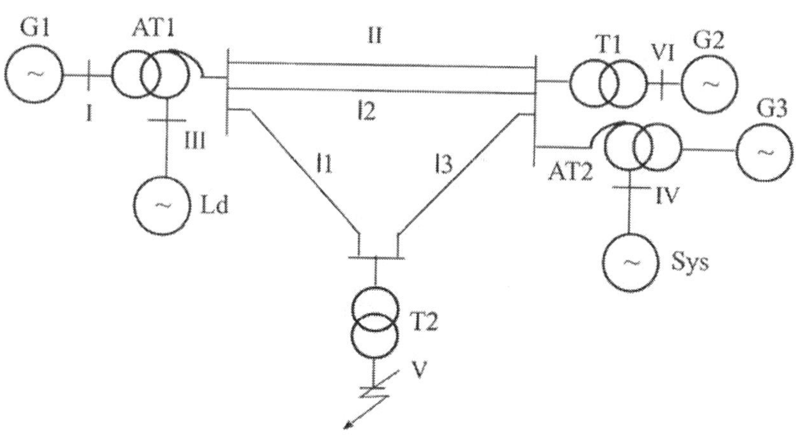

Fig. AII-1.

(*)See page 331.

The base values are chosen as:

$$S_b = 1000 \text{ MVA}, \quad V_{bI} = 15.75 \text{ kV (main level)}, \quad V_{bII} = 15.75\,\frac{242}{15.75} = 242 \text{ kV},$$

$$V_{bIII} = 121 \text{ kV}, \quad V_{bIV} = 242\,\frac{525}{242} = 242 \text{ kV}, \quad V_{bV} = 242\,\frac{11}{220} = 12.1 \text{ kV}.$$

The p.u.values of the network elements are (see Fig. 6.8a):

G_1- $\quad X_1 = 0.3\,\frac{1000}{470} = 0.64.$

G_2, G_3- $\quad X_2 = X_3 = 0.38\,\frac{1000}{2\cdot 118} = 1.61.$

Sys- $\quad X_4 = 83\,\frac{1000}{525^2} = 0.3.$

AT1- $\quad X_5 = \frac{1}{2}\cdot\frac{13.5 + 12.5 - 18.8}{100}\cdot\frac{1000}{480} = 0.075.$

$$X_6 = \frac{1}{2}\cdot\frac{13.5 - 12.5 + 18.8}{100}\cdot\frac{1000}{480} = 0.206.$$

$$X_7 = \frac{1}{2}\cdot\frac{-13.5 + 12.5 + 18.8}{100}\cdot\frac{1000}{480} = 0.185.$$

AT2- $\quad X_8 = \frac{1}{2}\cdot\frac{8.4 + 28.4 - 19.0}{100}\cdot\frac{1000}{360} = 0.247.$

$$X_9 = \frac{1}{2}\cdot\frac{8.4 - 28.4 - 19.0}{100}\cdot\frac{1000}{360} \cong 0.$$

$$X_{10} = \frac{1}{2}\cdot\frac{-8.4 + 28.4 + 19.0}{100}\cdot\frac{1000}{360} = 0.542.$$

T1- $\quad X_{11} = \frac{11}{100}\cdot\frac{1000}{250} = 0.44.$

T2- $\quad X_{12} = \frac{12}{100}\cdot\frac{1000}{120}\left(\frac{220}{242}\right)^2 = 0.83.$

$\ell 1$- $\quad X_{13} = 0.4\cdot 120\,\frac{1000}{242^2} = 0.82.$

$\ell 2$- $\quad X_{14} = \frac{1}{2}\,0.4\cdot 95\,\frac{1000}{242^2} = 0.32.$

$\ell 3$- $\quad X_{15} = 0.4\cdot 80\,\frac{1000}{242^2} = 0.54.$

Ld- $\quad X_{16} = 1.2\,\frac{1000}{250}\left(\frac{115}{121}\right)^2 = 4.34.$

APPENDIX III

AN EXAMPLE OF A SHORT-CIRCUIT FAULT CALCULATION IN A POWER NETWORK[(*)]

Find the short-circuit current at the fault point F using the linearization approach for two cases: a) the AVRs are not activated; b) the AVRs are activated.

The one-line diagram of the network is shown in Fig. AIII-1. The parameters of the network elements are as follows.

Turbo-generators: G_1 and G_2-	$S_n = 15$ MVA, $V_n = 6.3$ kV, $SCR = 0.68$, $I_f = 2.1$, $I_{f\,\max} = 4$.
Hydro-generator: G_3-	$S = 60$ MVA, $V_n = 10.5$ kV, $SCR = 1$, $I_f = 1.75$, $I_{f\,\max} = 3.1$.
Transformers: T1-	$S_n = 10$ MVA, $V_n = 6/37$ kV, $X_{s.c.} = 7.5\%$,
T2-	$S_n = 40.5$ MVA, $V_n = 121/37.5/10.5$ kV, $X_{s.c.1} = 11\%$, $X_{s.c.2} = 6\%$, $X_{s.c.3} = 0$.

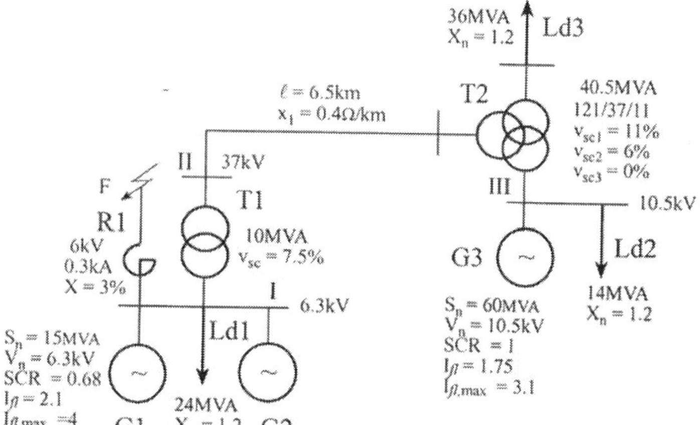

Fig. AIII-1.

(*)See page 370.

Transmission line- $\ell = 6.5\text{km},\ x_0 = 0.4\ \Omega/\text{km}.$

Loads: Ld1- $S_n = 24\ \text{MVA},\ X_n = 1.2,$
Ld2- $S_n = 14\ \text{MVA},\ X_n = 1.2,$
Ld3- $S_n = 36\ \text{MVA},\ X_n = 1.2.$

Reactor R1- $I_n = 0.3\ \text{kA},\ V_n = 6.0\ \text{kV},\ X_{x.c.} = 4\%.$

The base values are chosen as:

$$S_b = 100\ \text{MVA},\quad V_b = V_{level},\quad I_{bI} = \frac{100}{\sqrt{3}\cdot 6.3} = 9.22\ \text{kA}.$$

a) The p.u. values of the network elements, if the AVRs are not activated, are (see Fig. AIII-2(a)):

G_1, G_2- $\quad E_1 = 0.2 + 0.8\cdot 2.1 = 1.88,\quad X_1 = \dfrac{1.88}{0.68\cdot 2.1}\,\dfrac{100}{30} = 4.4;$

G_3- $\quad E_2 = 0.2 + 0.8\cdot 1.75 = 1.6,\quad X_2 = \dfrac{1.6}{1\cdot 1.75}\,\dfrac{100}{60} = 1.52;$

T1- $\quad X_3 = \dfrac{7.5}{100}\,\dfrac{100}{10} = 0.75;$

T2- $\quad X_4 = \dfrac{11}{100}\cdot\dfrac{100}{40.5} = 0.27,\quad X_5 = \dfrac{6}{100}\cdot\dfrac{100}{40.5} = 0.15,\quad X_6 = 0;$

ℓ- $\quad X_7 = 0.4\cdot 6.5\,\dfrac{1000}{37^2} = 0.19;$

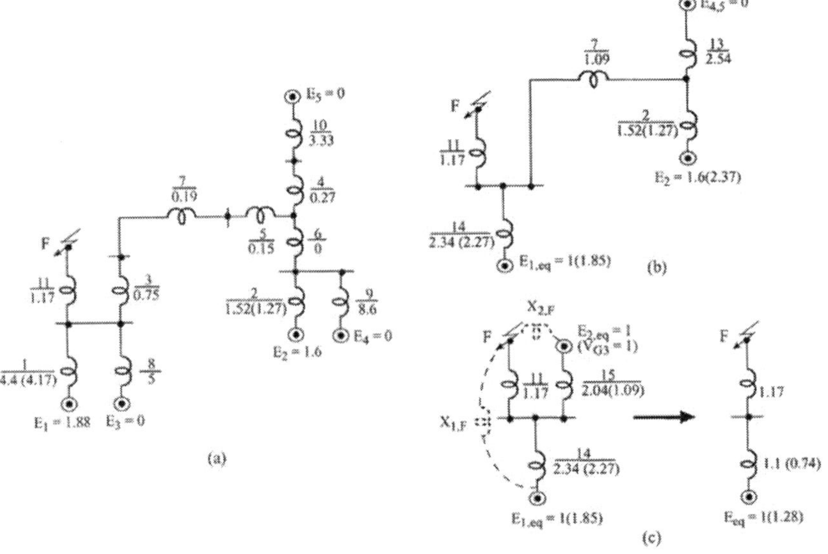

Fig. AIII-2.

Lds- $X_8 = 1.2\frac{100}{24} = 5, \quad X_9 = 1.2\frac{100}{14} = 8.6, \quad X_{10} = 1.2\frac{100}{36} = 3.33;$

R1- $X_{11} = \frac{4}{100} \cdot \frac{9.2}{0.3} \cdot \frac{6}{6.3} = 1.17 \quad \text{and} \quad E_3 = E_4 = E_5 = 0.$

The equivalent circuit of the network with the p.u. values of the network elements is shown in Fig. AIII.2(a). By simplifying this circuit we obtain:

$$X_{12} = 0.75 + 0.19 + 0.15 = 1.09, \quad X_{13} = (3.33 + 0.27)//8.6 = \frac{3.6 \cdot 8.6}{3.6 + 8.6} = 2.54$$

and $E_{4,5} = 0$;

$$X_{14} = 5//4.4 = 2.34 \quad \text{and} \quad E_{1,eq} = \frac{1.88 \cdot 5}{5 + 4.4} \cong 1 \quad \text{(Fig. AIII.2(b)).}$$

In the next step we have

$$X_{15} = X_{12} + X_2//X_{13} = 1.09 + \frac{1.52 \cdot 2.54}{1.52 + 2.54} = 2.04,$$

$$E_{2,eq} = \frac{1.6/1.52}{1/1.52 + 1/2.54} \cong 1 \quad \text{(Fig. AIII.2(c)),}$$

and finally

$$X_{eq} = 2.04//2.34 + 1.17 = 2.27 \quad \text{and} \quad E_{eq} = E_{1,eq}//E_{2,eq} = 1.$$

Thus, the short-circuit current is $I_{s.c.} = (1/2.27) = 0.441$, or in amperes $I_{s.c.} = 0.441 \cdot 9.2 = 4.06$ kA.

The terminal voltage of G1 will be $V_1 = 1.17 \cdot 0.441 = 0.516$ and the currents of each of the generators G1 and G2 will be

$$I_{G1} = \frac{E_1 - V_1}{X_1} = \frac{1.88 - 0.516}{4.4} = 0.31$$

and

$$I_{G2} = \frac{E_2 - V_2}{X_{15}} = \frac{1.6 - 0.775}{2.04} = 0.542,$$

where the voltage of V_2 can be found as

$$V_2 = V_1 + X_{12}I_{21} = 0.516 + 1.09\frac{1 - 0.516}{2.04} - 0.775.$$

To find the partial currents of each of the generators, which make up the fault current we shall first transfer the Y-connection of the circuit, Fig. AIII.2(c), into the Δ-connection (see the dash lines). Thus,

$$X_{1,F} = 1.17 + 2.34 + \frac{1.17 \cdot 2.34}{2.04} = 4.85 \quad \text{and} \quad I_{G1,F} = \frac{1}{4.85} = 0.206,$$

$$X_{1,F} = 1.17 + 2.04 + \frac{1.17 \cdot 2.04}{2.34} = 4.23 \quad \text{and} \quad I_{G2,F} = \frac{1}{4.23} = 0.236.$$

We may solve the above problem using the straightforward method. For this purpose we have to write and solve the matrix equation:

$$[Im] = [Zm]^{-1}[Em],$$

where the matrixes are: mesh-current, mesh-impedance and mesh-EMF matrixes.

The circuit in Fig. AIII.2(b), which has been obtained just after the trivial stages of simplification, is redrawn in a slightly different way in Fig. AIII.3(a). The four meshes of this circuit are chosen in such a way (note that the branches 0n–1n and 1n–2n form a tree of the circuit graph) that the mesh currents depict the short-circuit current ($I_3 \equiv I_{sc}$)and the first and the second generator currents ($I_1 \equiv I_{G1}$ and $I_2 \equiv I_{G2}$). The following solution is performed by the MATHCAD program.

ORIGIN: = 1

The mesh matrixes are

$$\text{Zm} := \begin{pmatrix} 9.4 & 5 & -5 & 5 \\ 5 & 7.61 & -5 & 6.09 \\ -5 & -5 & 6.17 & -5 \\ 5 & 6.09 & -5 & 8.63 \end{pmatrix} \quad \text{Em} := \begin{pmatrix} 1.88 \\ 1.6 \\ 0 \\ 0 \end{pmatrix},$$

and the solution of the above matrix equation is

$$\text{Im} := \text{Zm}^{-1} \cdot \text{E}_\text{m}.$$

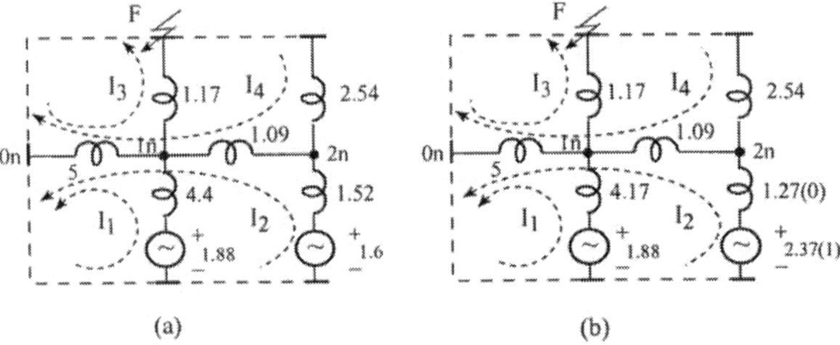

Fig. AIII-3.

Thus, the mesh currents are

$$\mathrm{Im} = \begin{pmatrix} 0.31 \\ 0.542 \\ 0.443 \\ -0.305 \end{pmatrix},$$

where the short-circuit current is $I_{s.c.} = 0.443$, and the generators currents are $I_{G1} = 0.31$ and $I_{G1} = 0.542$, i.e., as previously calculated.

To find the generators' partial currents of the short-circuit current, we shall find the determinant of the impedance matrix and its two cofactors:

$$\left|\begin{pmatrix} 9.4 & 5 & -5 & 5 \\ 5 & 7.61 & -5 & 6.09 \\ -5 & -5 & 6.17 & -5 \\ 5 & 6.09 & -5 & 8.63 \end{pmatrix}\right| = 377.932 \quad \left|\begin{pmatrix} 5 & -5 & 5 \\ 7.61 & -5 & 6.09 \\ 6.09 & -5 & 8.63 \end{pmatrix}\right| = 41.431$$

$$-\left|\begin{pmatrix} 9.4 & -5 & 5 \\ 5 & -5 & 6.09 \\ 5 & -5 & 8.63 \end{pmatrix}\right| = 55.88.$$

Now with equation 6.51 we have

$$I_{G1,F} = B_{F,1}E_1 = 0.109 \cdot 1.88 = 0.206, \quad \text{where } B_{F,1} = \frac{\Delta_{31}}{\Delta} = \frac{41.43}{377.9} = 0.109$$

and

$$I_{G2,F} = B_{F,2}E_2 = 0.148 \cdot 1.6 = 0.237, \quad \text{where } B_{F,2} = \frac{\Delta_{32}}{\Delta} = \frac{55.9}{377.9} = 0.148.$$

b) The p.u. values of the network elements, if the AVRs are activated, are (see the numbers in the parenthesis in Fig. AIII-2(a), (b) and (c)):

$$E_1 = 0.2 + 0.8 \cdot 4 = 3.4, \quad X_1 = \frac{4.17}{0.68 \cdot 4}\frac{100}{30} = 4.17 \quad \text{and} \quad X_{cr,1} = \frac{4.17}{3.4 - 1} = 1.74;$$

$$E_2 = 0.2 + 0.7 \cdot 3.1 = 2.37, \quad X_2 = \frac{2.37}{1 \cdot 3.1}\frac{100}{60} = 1.27 \quad \text{and} \quad X_{cr,2} = \frac{1.27}{2.73 - 1} = 0.93.$$

Thus, the critical currents are:

$$I_{cr,1} = \frac{1}{1.74} = 0.58 \quad \text{and} \quad I_{cr,2} = \frac{1}{0.93} = 1.08.$$

All other parameters are as in the previous calculation. By circuit simplification we assumed that generators G_1 and G_2 are operated in the maximal field regime (since they are relatively "close" to the fault point) and generator G_3 is operated in the nominal voltage regime (since it is relatively "far" from the fault point). In accordance with the circuit in Fig. AIII.2(c) we have

$$E_{eq} = E_{1,eq}//V_{G3} = \frac{1.85 \cdot 1.09 + 1 \cdot 2.27}{2.27 + 1.09} = 1.28, \quad \text{where } E_{1,eq} = \frac{3.4 \cdot 5}{4.17 + 5} = 1.85,$$

and

$$X_{eq} = X_{14}//X_{12} + X_{11} = \frac{2.27 \cdot 1.09}{2.27 + 1.09} + 1.17 = 1.91.$$

Thus, the short-circuit current is $I_{s.c.} = (1.28/1.91) = 0.67$, or in amperes $I_{s.c.} = 0.67 \cdot 9.2 = 6.2$ kA.

The terminal voltage of G1 will be $V_{G1} = 0.67 \cdot 1.17 = 0.78$ and the currents of each of the generators G1 and G2 will be

$$I_{G1} = \frac{E_1 - V_1}{X_1} = \frac{3.4 - 0.78}{4.17} = 0.628$$

and

$$I_{G2} = I_{L2} + I_{G2,F} = \frac{1}{2.54} + \frac{1 - 0.78}{1.09} = 0.596.$$

Since $I_{G1} > I_{cr1}$ and $I_{G2} \le I_{cr2}$, the assumption about their regimes was correct.

With the straightforward method, applied to the circuit in Fig. AIII.3(b), we can find the generators' currents and therefore will know in which of the two regimes the generators are operated. Thus, by applying the MATHCAD program, we have:

The mesh-impedance and mesh-voltage matrixes are

$$\text{Zm1} := \begin{pmatrix} 9.17 & 5 & -5 & 5 \\ 5 & 7.36 & -5 & 6.09 \\ -5 & -5 & 6.17 & -5 \\ 5 & 6.09 & -5 & 8.63 \end{pmatrix} \quad \text{Em1} := \begin{pmatrix} 3.4 \\ 2.37 \\ 0 \\ 0 \end{pmatrix},$$

and the solution is

$$\text{Im1} := \text{Zm1}^{-1} \cdot \text{Em1},$$

which gives

$$\text{Im1} = \begin{pmatrix} 0.599 \\ 0.856 \\ 0.77 \\ -0.505 \end{pmatrix}.$$

Since the current of the first generator, $I_{G1} = 0.599$, is larger than the critical one, it is operated in the maximal field regime. However, since the current of the second generator, $I_{G2} = 0.856$, is smaller than the critical one, it is operated in the nominal (rated) voltage regime and its parameters should be changed to $V_{G2} = 1$ and $X_2 = 0$. (See the numbers in the parentheses in Fig. AIII.3(b).) Now the new mesh matrixes are

$$\text{Zm2} := \begin{pmatrix} 9.17 & 5 & -5 & 5 \\ 5 & 6.09 & -5 & 6.09 \\ -5 & -5 & 6.17 & -5 \\ 5 & 6.09 & -5 & 8.63 \end{pmatrix} \quad \text{Em2} := \begin{pmatrix} 3.4 \\ 1 \\ 0 \\ 0 \end{pmatrix},$$

and the solution is

$$\text{Im2} = \begin{pmatrix} 0.627 \\ 0.592 \\ 0.67 \\ -0.394 \end{pmatrix}.$$

Note that the short-circuit current, 0.67, and the currents of both generators, 0.627 and 0.592, are as previously calculated.

INDEX

17420523R00313

Made in the USA
Lexington, KY
14 September 2012